112 Springer Series in Solid-State Sciences

Edited by Manuel Cardona

Springer Series in Solid-State Sciences

Editors: M. Cardona P. Fulde K. von Klitzing H.-J. Queisser

Managing Editor: H. K.V. Lotsch Volumes 1–89 are listed at the end of the book

90 **Earlier and Recent Aspects of Superconductivity** Editors: J. G. Bednorz and K. A. Müller

91 **Electronic Properties of Conjugated Polymers III** Basic Models and Applications
Editors: H. Kuzmany, M. Mehring, and S. Roth

92 **Physics and Engineering Applications of Magnetism** Editors: Y. Ishikawa and N. Miura

93 **Quasicrystals** Editors: T. Fujiwara and T. Ogawa

94 **Electronic Conduction in Oxides** By N. Tsuda, K. Nasu, A. Yanase, and K. Siratori

95 **Electronic Materials** A New Era in Materials Science
Editors: J. R. Chelikowsky and A. Franciosi

96 **Electron Liquids** By A. Isihara

97 **Localization and Confinement of Electrons in Semiconductors**
Editors: F. Kuchar, H. Heinrich, and G. Bauer

98 **Magnetism and the Electronic Structure of Crystals**
By V. A. Gubanov, A. I. Liechtenstein, and A.V. Postnikov

99 **Electronic Properties of High-T_c Superconductors and Related Compounds**
Editors: H. Kuzmany, M. Mehring, and J. Fink

100 **Electron Correlations in Molecules and Solids** 2nd. Ed. By P. Fulde

101 **High Magnetic Fields in Semiconductor Physics III**
Quantum Hall Effect, Transport and Optics By G. Landwehr

102 **Conjugated Conducting Polymers** Editor: H. Kiess

103 **Molecular Dynamics Simulations** Editor: F. Yonezawa

104 **Products of Random Matrices in Statistical Physics**
By A. Crisanti, G. Paladin, and A. Vulpiani

105 **Self-Trapped Excitons** By K. S. Song and R. T. Williams

106 **Physics of High-Temperature Superconductors** Editors: S. Maekawa and M. Sato

107 **Electronic Properties of Polymers** Orientation and Dimensionality of Conjugated Systems
Editors: H. Kuzmany, M. Mehring, and S. Roth

108 **Site Symmetry in Crystals** Theory and Applications
By R. A. Evarestov and V. P. Smirnov

109 **Transport Phenomena in Mesoscopic Systems** Editors: H. Fukuyama and T. Ando

110 **Symmetry and Optical Phenomena in Superlattices and Other Heterostructures**
By E. L. Ivchenko and G. E. Pikus

111 **Low-Dimensional Electronic Systems** New Concepts
Editors: G. Bauer, F. Kuchar, and H. Heinrich

112 **Phonon Scattering in Condensed Matter VII** Editors: M.Meissner and R. O. Pohl

113 **Electronic Properties of High-T_c Superconductors**
Editors: K. Kuzmany, M. Mehring, and J. Fink

M. Meissner · R. O. Pohl (Eds.)

Phonon Scattering in Condensed Matter VII

Proceedings of the Seventh International Conference, Cornell University, Ithaca, New York, August 3–7, 1992

With 305 Figures

Springer-Verlag
Berlin Heidelberg New York
London Paris Tokyo
Hong Kong Barcelona
Budapest

Dr. Michael Meissner
Hahn-Meitner Institute, Glienicker Strasse 100
W-1000 Berlin 39, Germany

Professor Robert O. Pohl
Department of Physics, Cornell University
Ithaca, NY 14853-2501, USA

Local Committee

M. Meissner and R. O. Pohl (Chair)

International Advisory Committee

A. C. Anderson USA
W. E. Bron USA
L. J. Challis UK
W. Eisenmenger Germany
J. P. Harrison Canada
S. Hunklinger Germany
J. Joffrin France
A. A. Kaplyanskii Russia
H. Kinder Germany
K. Lassmann Germany
H. J. Maris USA
L. P. Mezhov-Deglin Russia
T. Nakayama Japan
V. Narayanamurti USA
A. K. Raychaudhuri India
K. F. Renk Germany
R.-J. Wei P. R. China
O. Weis Germany
H. W. de Wijn The Netherlands
J. P. Wolfe USA
A. F. G. Wyatt UK

ISBN-13:978-3-642-84890-2 e-ISBN-13:978-3-642-84888-9
DOI: 10.1007/978-3-642-84888-9

Library of Congress Cataloging-in-Publication Data. Phonon scattering in condensed matter VII : proceedings of the seventh international conference, Cornell University, Ithaca, New York, August 3-7, 1992 / M. Meissner, R.O. Pohl, eds. p. cm. – (Springer series in solid-state sciences ; 112) Includes bibliographical references and index.
ISBN-13:978-3-642-84890-2 1. Condensed matter–
Congresses. 2. Phonons–Scattering–Congresses. I. Meissner, M. (Michael), 1944- . II. Pohl, Robert O. III. Series. QC173.4.C65P453 1993 530.4'16–dc20 92-44418

Typesetting: Camera-ready by authors/editors
54/3140 - 5 4 3 2 1 0 - Printed on acid-free paper

Preface

This volume contains the proceedings of the Seventh International Conference on Phonon Scattering in Condensed Matter held August 3–7, 1992, at Cornell University in Ithaca, NY, USA. The preceding conferences were held at:

St. Maxime and Paris (France) 1972,
Nottingham (UK) 1975,
Providence (USA) 1979,
Stuttgart (Germany) 1983,
Urbana (USA) 1986, and
Heidelberg (Germany) 1989.

The Heidelberg conference was held jointly with the Third International Conference on Phonon Physics. The next conference, to be held in August, 1995, in Sapporo, Japan, and hosted by Professor T. Nakayama and his colleagues, will also be such a joint conference.

This conference was attended by 227 scientists from 27 countries, and covered all aspects of phonon scattering in condensed matter, ranging from the more traditional topics of thermal conductivity, Kapitza resistance, and ballistic phonon propagation to the recently added topics, such as electron–phonon interaction in high-T_C superconductors, the use of phonons in particle detection, and phonons in confined geometries. The 207 papers arranged in 11 chapters in this volume are a cross section of the present activities in the quite obviously vibrant field of phonons and their interactions.

85 papers were presented orally, in lectures ranging from 20 to 30 minutes duration, some in two parallel sessions. 130 papers were presented in poster sessions during three evenings. The papers for each poster session were selected and arranged according to specific and rather narrow topics, in order to achieve maximum interaction between the presenters and the audience, and each session was chaired by a rapporteur whose task was to provide the stimulus for the exchange of ideas.

On three afternoons, workshops were held on topics that appeared to be of particular current interest. Contributors were encouraged to emphasize open questions and disagreeing viewpoints. The titles of these workshops, and their chairmen, were:

"Phonon Scattering in High-T_c Materials" (V. Narayanamurti and K.F. Renk),

"Thermal Conduction in Glasses Above ~ 10 K" (D.G. Cahill and Ping Sheng),

"Phonon Thermalization and Particle Detectors" (M. Meissner and B. Sadoulet).

The conference was opened with a lecture by J.A. Krumhansl entitled "Phonons in Retrospect – 1952 Onward"; a modified version of this talk is contained in these proceedings. The concluding remarks were made by W. Eisenmenger. He emphasized the growth that the fledgling topic of phonon scattering has experienced in the two decades since being the subject of the first conference under this title. While continuing to be a topic of fundamental interest, it has now also found numerous practical applications. A conference covering all of these aspects continues to be of great importance.

A session on phonon scattering in diamond provided the background to honor Dr. Robert Berman from Oxford University for his seminal work on phonon scattering through measurements of the thermal conductivity in solids, which he began in the late forties. Dr. Berman's lecture was introduced by R.L. Sproull, who in turn had started the work on thermal conductivity at Cornell in 1952, together with his student G.A. Slack. Sproull's introduction is reproduced here:

"It is a distinct pleasure and privilege to introduce Dr. Robert Berman.

The reason this conference in 1992 is at Cornell is that Robert Berman listened to Sir Francis Simon at Oxford in 1948. Let me explain.

Dr. Berman in his 1976 book reports that F.E. Simon, then the high priest of the Clarendon Laboratory's distinguished low-temperature group, suggested that thermal conductivity measurements could be used to investigate lattice defects in non-metals. By 1950 Dr. Berman was already publishing the results of measurements on neutron-irradiated quartz. That and other seminal and suggestive work by Dr. Berman was reported in an invited paper by Dr. W.W. Tyler at the American Physical Society meeting in Columbus in the spring of 1952.

Now I and others of my generation had been weaned on the attitude that phonons *interfered* with really interesting processes, such as electronic conduction in metals or the shapes of optical absorption lines in insulators. Dr. Berman's work and Dr. Tyler's paper, along with the nearly simultaneous measurements of the isotope effect in superconductors, convinced me that *phonons could be friends*.

At Cornell in 1952 we *thought* we had some capability for introducing known defects into synthetic non-metallic crystals, although the capability now looks primitive by today's standards. The natural approach to follow up Dr. Berman's ideas was thus to study the thermal conductivity of systems in which the defects were quantitatively specified. Dr. Glen Slack's thesis launched this program auspiciously in 1954, but only when Dr. Robert Pohl arrived in 1958 did it attain the richness and content which brings this conference to Cornell. Dr. Pohl and his students developed a real capability for introducing known defects and conducted far-reaching and imaginative experiments on thermal conductivity and other phonon processes.

Meanwhile Dr. Berman's central position in phonon processes in non-metals was maintained and enhanced. His experimental program flourished, and his publications also exhibited an important theoretical content, harvesting and accommodating the theoretical contributions of Debye, Peierls, Pomeranchuk, Ziman, Krumhansl, Callaway, and others. His 1953 *Advances in Physics* review effectively established the field under its title "Thermal Conductivity of Dielectric Solids at Low Temperatures", reinforced by his 1976 book *Thermal Conduction in Solids*. His classic 1965 papers on the isotope effect in LiF and solid He, along with his earlier experiments on diamond, are again in the news with his 1991 comment on the thermal conductivity of isotopically enriched diamonds.

On behalf of the organizers of this conference, it is my privilege to present to Dr. Berman this plaque: "The Seventh International Conference on Phonon Scattering in Condensed Matter honors Robert Berman for his pioneering contributions to the physics of phonon scattering. Presented on August 5, 1992, Cornell University, Ithaca, New York".

In order that this will not be a "dry promotion", the organizers are also presenting this Steuben bowl with their congratulations.

Dr. Berman's title is "It's Quite Normal not to (um)Klapp",

Robert L. Sproull,
President Emeritus, University of Rochester.

This conference, and in particular the truly international composition of its participants would not have been possible without the generous contributions from its sponsors:

International Union of Pure and Applied Physics
National Science Foundation
U.S. Army Research Office
U.S. Office of Naval Research
Vice President for Research, Cornell University
Materials Science Center, Cornell University
Physics Department, Cornell University
Laboratory of Atomic & Solid State Physics, Cornell University.

Their help is gratefully acknowledged, as is a donation made by Professor Emeritus J.A. Krumhansl.

Several corporations have sponsored conference fellowships to help young scientists who had made significant contributions to the field of phonon scattering to attend this conference. The conference fellows and their sponsors are:

J. Classen	(Univ. Heidelberg)	Corning Inc.
L.I. Deich	(Kirensky Inst., Krasnoyarsk)	
M. Gienger	(Univ. Stuttgart)	Eastman-Kodak Co.
G.N. Gol'tsman	(Moscow State Pedagogical Univ.)	Xerox Corp.
T.J. Gramila	(AT&T Bell Labs)	Exxon Corp.
U. Happek	(Univ. Regensburg, Cornell Univ.)	Canon Inc.
K. McEnaney	(Univ. Pierre et Marie Curie, Paris)	Canon Inc.
K. Nandakumar	(Cochin Univ.)	Canon Inc.
S. Sahling	(T.U. Dresden)	Hahn-Meitner Institut
A.S. Semenov	(Kapitza Inst., Moscow)	IBM Corp.
A. Smontara	(Univ. Zagreb)	Exxon Corp.

R.J. Stoner	(Brown Univ., Providence)	3M Corp.
S.K. Watson	(U.C. San Diego)	GE Co.
A. Wixforth	(Univ. München)	IBM Corp.

We combine our expression of appreciation for the gift by the sponsors with our best wishes for successful scientific careers for the fellows!

Finally, we wish to acknowledge the help of the International Advisory Committee, in particular H.J. Maris and V. Narayanamurti for the help in planning the program of this conference, and also the local committee, in particular Liane Strantz as Conference Secretary, and our wives, Ute Meissner and Karin Pohl, in seeing the Conference to a successful completion.

Berlin, Ithaca
October 1992

M. Meissner
R.O. Pohl

Contents

Part II Ballistic Propagation and Imaging

Part III **Phonon Spectroscopy**

Part V High-Temperature Superconductors

Part VI Disordered Systems

Part VII Two-Dimensional Electron Systems

Part XI Lattice Dynamics

Part I

Heat Transport in Dielectric Crystals

On the General Physics of Heat Conduction in Ordered and Disordered Solids

J.A. Krumhansl

Laboratory of Atomic & Solid State Physics,
Cornell University, Emeritus*

Abstract

Heat conduction is temperature times entropy transport. In good crystalline material both the entropy content and transport is contained in the same entity i.e. phonon. In disordered systems there may be localized excitations as well as collective excitations; in this case heat can be transported both by hopping (diffusion) of entropy between localized states or by the collective excitations, which may be sound waves, spin waves or the like. In this latter case, the heat current can be thought of as a mixture of "ballistic" (i.e. collective) and "'stochastic" (i.e. hopping) transport. Some history and general phenomenological features are discussed.

Introduction

In the 1950's and 1960's at Cornell both experimental and theoretical research on thermal conduction at low temperatures was carried out.[1,2] During the first period of that program highly quantitative work was done to refine and perfect crystals, then characterize added defects using thermal conduction as a tool. Some of this work continues. In the mid 1960's, following the discovery that (for example) the small ion Li^+ was often "off-center" if mixed in other alkalide halides with large ions there followed a number of studies on the tunneling dynamics of these and molecular ion substitutions.[3–5] A theoretical program on tunneling states was also carried out at Cornell during that period.[6,7] The situation of a distribution of tunneling states and its implications regarding thermal conductivity and other properties was considered by Sussman.[8] Proceeding along similar lines Halperin and Anderson[9] explained the linear heat capacity in amorphous solids also by postulating a uniform distribution of tunneling states, (of unspecified origin), and provided

* Present address. 515 Station Road, Amherst, MA 01002,USA

Springer Series in Solid-State Sciences, Volume 112
Phonon Scattering in Condensed Matter VII Eds.: M. Meissner and R.O. Pohl

a heuristic model of thermal conductivity with resonant scattering by phonon coupling to the tunneling excitations. Though we attempted a more formal and quantitative many-body theory of thermal transport[7] at the time, we had little success. However, the phenomenological physical concepts enlightened in hindsight by many further studies since that time, appear still to have relevance, so we present them here briefly and phenomenologically.

Heat Transport, Generally

In entirely general, basic terms a heat current $J_Q = TJ_s$ where T = temperature J_Q = heat current and J_s = entropy transfer current. This expression can be applied in great generality; schematically in Fig. 1 we show two (sub)systems at different temperatures, which can transfer excitations to one another, as determined by the distribution of energy

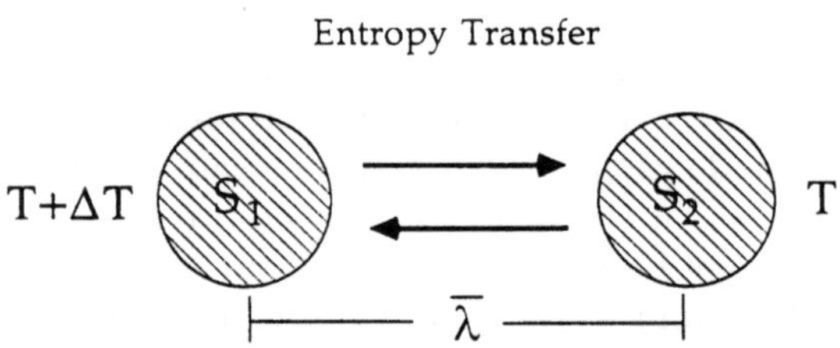

Fig. 1 Schematic of entropy transfer between two subsystems at different temperatures.

among excitation levels (therefore subsystem entropy). [This model is essentially that used in introducing irreversible thermodynamics; c.f., I. Prigogine, "Introduction to Irreversible Thermodynamics," Interscience, New York, 1967].

The transfer of excitations may take place by a variety of mechanisms, depending on the physical situation under discussion: by photon transport for "radiative" conduction; by phonon transport for "ballistic" transport in crystals; by phonon (or photon) assisted transfer between localized subsystems (states) [in effect, virtual phonon (photon) transfer] for "stochastic" transport. Generally, exact calculations of these processes, except for perfect crystals (vacuum), are not possible. That does not vitiate the correctness of the physics, however. Realistic parameters can be assigned from experiment.

Defining the transfer rates out of each system into the other to be $R_1(T + \Delta T)$ and $R_2(T)$ respectively, entropy densities $S_1(T + \Delta T)$ and $S_2(T)$ we define entropy currents

$$J_s^{(12)}(T + \Delta T) = R_1(T + \Delta T)S_1(T + \Delta T) \quad ,$$

$$J_s^{(21)}(T) = R_2(T)S_2(T) \quad . \qquad (1)$$

Then

$$Q = T[J_s^{(12)}(T + \Delta T) - J_s^{(21)}(T)] \quad . \tag{2}$$

Phonon Transport in Crystals

While detailed theoretical models of many sorts can be carried through to various degrees, the essence of the result, in general accord with experiments, is that one can imagine entropy exchange between regions separated by a nominal phonon men free path $\bar{\lambda}$. Of course if one breaks up S into $\sum_i S_i$ over all modes "i,", $\lambda \equiv \{\lambda_i\}$, then the details may be refined but the principles are unchanged.

Noting also that ΔT across a distance $\bar{\lambda}$ is $\Delta T \simeq \bar{\lambda}\frac{dT}{dX}$, we can reduce

$$[TS(T + \Delta T) - TS(T)] = T\frac{dS}{dT}\Delta T \simeq C_v(T)\bar{\lambda}\frac{dT}{dX} \tag{3}$$

and assuming for an isotropic crystal $R = R_1 = R_2 \simeq \frac{1}{3}\bar{V}_g$, the average group velocity, we get directly

$$Q = \frac{1}{3}C_v(T)\bar{V}_g\bar{\lambda}(\frac{dT}{dX}) \tag{4}$$

$$= \kappa(\frac{dT}{dX}) \quad .$$

All this is familiar; however, it may be useful to emphasize a few points, which need *not* hold in general:
(1) The entropy content in excitations other than lattice modes is negligible.
(2) The transfer mechanism, phonon propagation with group velocity $\bar{V}_g$ is the only entropy transfer method, i.e. "ballistic."
(3) Defect scattering is only weak, in the sense that these conditions (1) and (2) are very good approximations to reality.

Glasses, Disordered and Infected Solids

When we tried to apply this general approach back in the 60's to off-center tunneling systems, crystals with localized modes and so forth, and to do so in the very formal many-body tradition of the times we were unsuccessful. Since that time two things have taken place: theoretically, heuristic but sound physical models are back in style, and, experimentally, we now know of many kinds of highly localized excitations which can live in (i.e. "infect") *either* ordered or disordered systems.

For purposes of discussion let me coin the phrase "*localon*". This may refer to an off-center tunneling configuration, to the rotational states of a substitutional molecular ion, to traditional harmonic resonant modes or localized states, to a localized fracton in a demonstrably fractal structure, to an atomic rearrangement caused by "hole-burning," to a localized lattice instability which is a precursor to a structural transformation, and others probably. An interesting model for localons which can produce either 2-level or anharmonic local states has been discussed by Buchenau at this meeting. Experimental data has shown that excitations in these states may have very long lifetimes (e.g. years for "hole-burned" rearrangements) compared to a Debye period. Thus "localons" can be considered as any of many separate localized subsystems, generally weakly coupled to the phonon system. The latter may act as a temperature bath, however, and a thermal equilibrium distribution of excitations within the "localon" becomes meaningful. The localon has its own entropy and specific heat. For example: $C_{loc} = C_v(\text{oscillator}), C_{loc} = C_v(\text{rotor}), C_{loc} = C_v(\text{2-level})$, etc. It is easy to get a linear T dependence of specific heat with a suitably assumed distribution of localons of any type; to invert the reasoning and infer one specific mechanism among the many possible from that property *only* is unjustified, and certainly not unique.

The matter of thermal conductivity is more subtle, however. The image we have for the localon heat transport is quite different from the traditional phonon mechanism. The entropy transport rate will not generally be determined by the phonon travel time between localons; that can be very fast, indeed possibly not relevant at all, as will be explained. The model for transport is now as follows:

(i) Excitations at one localon site are transferred (i.e. hop) to another site by some virtual mechanism (e.g. local atomic rearrangement, virtual phonon, infrared radiation, etc). The mean life for hopping is τ_H. Assuming that the hopping distance is roughly the spacing S_l between localons, the *effective* localon transport velocity is

$$\bar{V}_l \simeq (S_l/\tau_H) \quad . \tag{5}$$

This velocity $\bar{V}_l$ is not related in any direct way to phonon velocities; it is expected to be generally lower.

(ii) Now we can easily get the "stochastic" localon thermal conductivity κ_{loc}, assuming $\lambda \simeq S_l$:

$$\begin{aligned} \kappa_{loc} &\simeq \sum_l C_{loc}(S_l^2/\tau_H) \\ &= \sum_l C_{loc}\, \bar{V}_l S_l \end{aligned} \tag{6}$$

the sum indicating addition over a distribution of localons.

The Total Thermal Conductivity of a Disordered Crystal

It should be clear, but we explicate for emphasis, that it is easily possible to have localons within real materials (e.g. glasses with flipping between directed bond configurations, off-center ions) which are disjoint, topologically and energetically, from the collective phonon excitations. The localons are in effect a separate phase as far as excitations are concerned. The total thermal conductivity is then that of a two phase mixture

$$\kappa = \kappa_{phon} + \kappa_{loc} \quad . \tag{7}$$

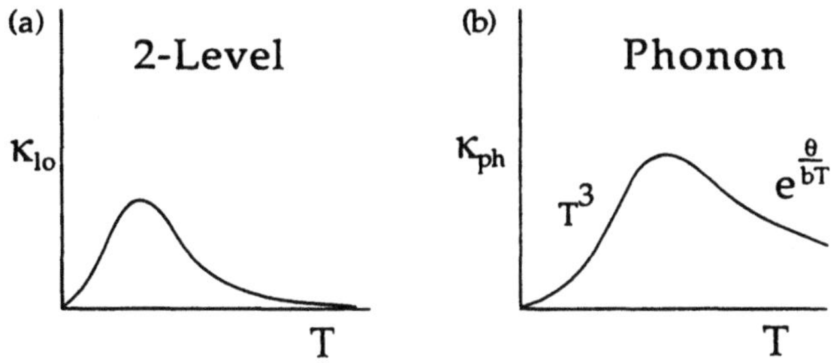

Fig. 2 Prototypical thermal conductivity:

(a) "Stochastic" in 2-level systems;
(b) "Ballistic" conventional phonon.

One can easily imagine that, with "loving" parameters a combined conductivity of many forms can be obtained.
The former would be prompt on the scale of sound wave (phonon) transfer, the latter might be much slower.

As an example, consider a set of two-level localons, with similar splitting energy $\Delta\epsilon_o$. The heat capacity is

$$C_v = kx\frac{e^x}{(1+e^x)^2} \quad , \qquad x = (\Delta\epsilon_o/kT) \quad . \tag{8}$$

Assuming a stochastic velocity V_l and spacing S_l between localon sites we would get a localon conductivity as shown in Fig. 2(a). The phonon thermal conductivity would look pretty much like the usual Fig. 2(b).

Summary

Going back to the fundamental idea that heat flow is T times entropy current, based on the concepts of irreversible thermodynamics, it is possible to address the question of thermal conductivity in solids for both nice and variously infected (disordered and worse) materials. It is essential to realize then that heat can be simultaneously transported by two independent mechanisms; "ballistically" by collective, extended modes (e.g. phonons), and "stochastically" by hopping of the entropy in localized excitations ("localons"). Steady state heat flow can not easily disentangle their contributions. However, time dependent experiments may be able to distinguish the two in terms of time scales, or, alternatively, finite temperature molecular dynamics simulation studies of vibrational modes into which distinct excitations, beyond simple mass or spring defects, such as two-level systems are connected at various sites. Perhaps inelastic neutron scattering studies for small Δq (momentum) and $\Delta\epsilon$ (energy) transfer could show the diffusion of localon excitations.

The interpretation of ultrasonic experiments also needs to be readdressed if localon excitations are prevalent.

Acknowledgments

I am most pleased to acknowledge many discussions over the years on these topics with Bobby Pohl, as well as those earlier discussions with my students Sam Bowen and Manuel Gomez. Thanks particularly to Bobby for inviting me to participate in this conference, thereby stimulating this writeup of old and new thoughts.

References

1. V. Narayanamurti and R.O. Pohl, Rev. Mod. Phys. 42, 201 (1970).
2. P. Carruthers, Rev. Mod. Physics. 33, 92 (1961).
3. N.E. Byer and H.S. Sack, Phys. Rev. Lett. 17, 72 (1966).
4. R.O. Pohl in Localized Excitation in Solids, R. Wallis, ed., Plenum Press, 1968, p. 434.
5. M. Gomez, S.P. Bowen, and J.A. Krumhansl, Phys. Rev. 153, 1009 (1967).
6. M. Gomez-Rodriguez, Ph.D. Thesis, Cornell University, June 1968, unpublished.
7. S.P. Bowen, Ph.D. Thesis, Cornell University, Sept. 1967, unpublished.
8. J.A. Sussmann, Phys. Kondens. Materie 2, 146 (1964).
9. P.W. Anderson, B.I. Halperin, and C.M. Varma, Phil. Mag. 25, 1 (1972).

It's Quite Normal not to (um)Klapp

R. Berman

Clarendon Laboratory, Parks Road, Oxford OX1 3PU, UK

1. Introduction

Interest in the problems of predicting and analysing phonon thermal conductivity of crystals has been rekindled by the measurement of the thermal conductivity of isotopically enriched diamonds. If the special role of three-phonon Normal (N-) processes is not taken into account in interpreting or predicting the experimental results they seem very surprising. It is therefore relevant to give a brief review of the theory and of experiments to verify the effect of N-processes.

2. Recent Publications

The authors of an article [1] prior to any measurements on enriched diamonds predicted that a pure diamond with no ^{13}C would have a conductivity at room temperature only 5% greater than one with the natural isotopic constitution (1.1% ^{13}C), while in an article [2] after the measurements it was suggested that the 50% increase observed was due either to the removal of defects present in 'natural' diamond or to isotopes producing phonon scattering five times greater than would be calculated by accepted scattering theory (as given by Lord Rayleigh [3] or by more modern treatments [4]). In neither of these articles were N-processes treated in the manner suggested in several theories which have appeared since Klemens [5] took into account their role in limiting the effective relaxation time of low-frequency phonons. In a crystal with no resistive scattering other than that produced by point defects the low-frequency phonons, in the absence of N-processes, would 'run away' with the heat, since the rapid rise in relaxation time with decreasing frequency would outweigh the decrease in their number. Since N-processes cause the thermal energy to be interchanged among phonons of different frequencies, none of it can be carried continuously by

Springer Series in Solid-State Sciences, Volume 112
Phonon Scattering in Condensed Matter VII Eds.: M. Meissner and R.O. Pohl

low-frequency phonons and thus escape the scattering. On the other hand, N-processes do not directly cause thermal resistance since they tend to restore a disturbed phonon distribution to one which still corresponds to the heat flow taking place and has the form of a displaced Planck distribution.

3. Outline of Theories

Klemens [5] was the first to point out the principle of the way in which N-processes could be taken into account, but the most successful theory in the analysis of phonon thermal conductivity is that due to Callaway [6]. He distinguished between intrinsically resistive scattering processes, which tend to restore a disturbed phonon distribution to the equilibrium Planck form corresponding to zero heat flow, and N-processes which restore a disturbed distribution to a form corresponding to the heat flow occurring but is unaffected further by N-processes. This is analogous to the case of a gas flowing in a smooth tube, where the collisions among the gas molecules do not produce flow resistance but just ensure that the velocity distribution has the form of a displaced Maxwell distribution. However, in a rough tube these collisions transmit to the bulk of the gas 'knowledge' of a layer with zero drift velocity at the walls, and thus limit the drift velocity of gas within the tube.

The conductivity derived from the Callaway theory can be expressed as the sum of two terms. The first looks like the conductivity which would result from lumping the relaxation rate for N-processes with that for truly resistive processes, while the second can be regarded as a correction for thereby overestimating the resistive effect of N-processes. An interesting consequence of the expression obtained is that if N-processes are the dominant source of scattering, then the conductivity is given entirely by the 'correction' term and thermal resistivities are additive. The same result is obtained for this limiting case by the theories of both Ziman [7] and of Guyer and Krumhansl [8].

The Callaway expression and its limiting form for dominant N-processes have been tested in the interpretation of measurements of the conductivity of crystals with different isotopic compositions. The conductivity-temperature curve for a single specimen can be represented by a number of combinations of relaxation rates, which may, or may not, correspond with the truth. However, if a series of crystals is measured in which only one type of defect is varied, only a correct analysis is likely to lead to a good fit to the experimental curves by varying just the relaxation rate ascribed to the defects, in accordance with their concentration. For the 'classical' crystal LiF the scattering rate which had to be ascribed to the presence of isotopes [9] agreed with the theoretical value, and the N-process relaxation rate was in conformity with the results of de Klerk and Klemens on ultrasonic attenuation [10]. For the quantum crystal helium, the isotope scattering for low densities appeared to be greater than the 'classical' value [11], which is understandable, and the N-process scattering rate was in reasonable agreement with values found by other means (second sound and Poiseuille flow of phonons). The enhanced isotope scattering was confirmed in experiments by Lawson and Fairbank [12] who introduced such small quantities of ^{3}He into ^{4}He crystals that it could be assumed that N-processes *were* dominant, so that resistances are additive and the relaxation rate for isotopes can be calculated easily.

4. Application to Diamond

In the absence of measurements over a wide temperature range, the easiest way to compare theory with the measurements of isotopically enriched diamonds is to assume that only N-processes and isotopes need be considered. Nepsha et al. [13] have assumed that for all ^{13}C concentrations up to the natural 1.1%, N-processes are indeed dominant, so that the limiting form of the Callaway (and other theories) expression can be used. They find that this expression gives results for the differences between the resistivities in quite good agreement with experiment [14]. Using a variational treatment due to Ziman [7], Sheard and Ziman (see Berman et al. [15]) have calculated by how much thermal resistivity due to isotopes is less than the value given by the limiting expression when N-processes are *not* dominant. The departure from additivity which then occurs accounts for deviations of Nepsha et al.'s calculated resistivity differences from those measured, while Berman [16] has discussed the uncertainties involved in trying to refine the analysis to the case of N-processes not dominant, and showed that in the limit of nearly zero ^{13}C concentration the extrapolated isotope resistance is close to the N-process-dominant value.

5. Conclusions

We may conclude that the various post-1951 theories are essentially correct and can account for the effect of isotopes on thermal conductivity if N-processes are taken into account. It is only in cases where there is a very great amount of resistive scattering that they can be ignored with impunity in any attempt to interpret experimental results.

References

[1] G.A. Slack, R.A. Tanzilli, R.O. Pohl, J.W. VanderSande: J. Phys. Chem. Solids **48**, 641 (1987).

[2] G. Onn, A. Witek, Y.Z. Qiu, T.R. Anthony, W.F. Banholzer: Phys. Rev. Lett. **68**, 2806 (1992).

[3] Lord Rayleigh, *Theory of Sound* (2nd) edn. (1896) Vol. ii. Dover Publications, New York (1945).
[4] P.G. Klemens: Proc. Phys. Soc. A**68**, 1113 (1955).
[5] P.G. Klemens: Proc. R. Soc. London A**208**, 108 (1951).
[6] J. Callaway: Phys. Rev. **113**, 1046 (1959).
[7] J.M. Ziman: Canad. J. Phys. **34**, 1256 (1956).
[8] R.A. Guyer, J.A. Krumhansl: Phys. Rev. **148**, 766 (1966).
[9] R. Berman, J.C.F. Brock: Proc. R. Soc. London A**289**, 46 (1965).
[10] J. de Klerk, P.G. Klemens: Phys. Rev. **147**, 585 (1966).
[11] R. Berman, C.L. Bounds, S.J. Rogers: Proc. R. Soc. London A**289**, 66 (1965).
[12] D.T. Lawson, H.A. Fairbank: J. Low Temp. Phys. **11**, 363 (1973).
[13] V.I. Nepsha, V.R. Grinberg, Yu. A. Klyuev, A.M. Naletov, G.B. Bokiĭ: Doklady Akad. Nauk SSSR **317**, 96 (1991) [Sov. Phys. Dokl. **36**, 228 (1991)].
[14] T.R. Anthony, W.F. Banholzer, J.F. Fleischer, L. Wei, P.K. Kuo, R.L. Thomas, R.W. Pryor: Phys. Rev. B**42**, 1104 (1990).
[15] R. Berman, P.T. Nettley, F.W. Sheard, A.N. Spencer, R.W.H. Stevenson, J.M. Ziman: Proc. R. Soc. London A**253**, 403 (1959).
[16] R. Berman: Phys. Rev. B**45**, 5726 (1992).

Umklapp Processes at Low Temperatures

*P.G. Klemens and Yong-Jin Han**

Department of Physics and Institute of Materials Science,
University of Connecticut, Storrs, CT 06269, USA
*Present address: Department of Physics, Soonchunhyang University,
Choongchungnam 336-600, Korea

The intrinsic thermal resistivity at low temperatures is normally thought of as arising from three-phonon U-processes of the type $\underline{q} + \underline{q}' \to \underline{q}''$, where the phonon $\underline{q}$ is of low thermal frequency, and $\underline{q}'$, $\underline{q}''$ lie near the zone boundary, so that

$$\underline{q} + \underline{q}' = \underline{q}'' + \underline{b} \qquad (1)$$

$$f + f' = f'' \qquad (2)$$

Thus f', f'' have frequencies close to f_z, and the intrinsic resistivity has the form

$$W_U = \phi(T)\, \exp(-\theta_z/T) \qquad (3)$$

where $k\theta_z = hf_z$. The magnitude and form of $\phi(T)$ is not predicted clearly (e.g. Berman /1/), but the exponent θ_z should be given by the lowest zone boundary frequency.

It has been pointed out /2/ that these processes are not allowed by the requirements (1) and (2) for low frequency modes f, since the group velocity at the zone boundary lies well below the phase velocity of the thermal mode. The dissipation of excess phonon momentum takes place in two steps: first a normal process, transferring momentum to modes of some intermediate frequency f_i, followed by an allowed U-process with $f=f_i$, $f'=f_1$, $f''=f_2$. Here f_1 and f_2 are the lowest and next lowest frequencies near the zone boundary, and belong to different polarization branches. Thus U-processes are allowed only if f exceeds f_2-f_1. The value of θ_z in (3) should be identified with f_i. Alternatively, the second step is scattering by residual point defects, weak at thermal frequencies, but possibly significant at intermediate frequencies. It so happens that none of the crystals which have been measured, except solid helium, were sufficiently free of point defects for the second step to be due to U-processes alone.

The resistance arises by the transfer of momentum from thermal modes to a reservoir of modes $f > f_R$ by N-processes. The reservoir modes have a small heat capacity $C(f_R) \sim C\exp(-\theta_R/T)$, where C is

Springer Series in Solid-State Sciences, Volume 112
Phonon Scattering in Condensed Matter VII Eds.: M. Meissner and R.O. Pohl

the total heat capacity; the N-process rate into those modes is similarly reduced. The second step has a relaxation time $\tau'(f_R)$, where τ' is the combined U-process and point defect relaxation time. Also, from a momentum balance argument,

$$\tau(f) = [C/C(f_R)] (\tau_N(f) + \tau'(f_R)) \quad (4)$$

where $1/\tau_N(f)$, the relaxation rate of the thermal mode, is proportional to fCT^4 /3/. The reservoir frequency f_R is given by the condition $\tau_N(f_R)= \tau'(f_R)$, and is thus below f_i if point defect scattering is strong.

The U-process relaxation rate is obtained for modes lying in principal directions b by integrating over a surface q' which is simply taken to be parallel to the zone boundary, using matrix elements given in /3/. Actual dispersion relations are used to determine f' and f". The only adjustable parameter, both in τ_N and τ_U, is the Grüneisen constant.

Point defect scattering for each sample is obtained from the conductivity curve at temperatures well below the maximum, where the departures from a T^3 dependence can be ascribed to point defect scattering /4/.

Thermal conductivities were calculated numerically for LiF, NaF, diamond, Ge and Si. It was possible to obtaon a good fit over a wide temperatue range, as long as the thermal spectral range is well seperated from the reservoir range above f_R. In the case of diamond this is about 250K. The case of diamond is intersting because the thermal conductivity is still sensitive to point defects at that temperature, even though the point defect strength seems larger than that due to isotopes.

While other authors have also obtained good fits, our fit uses known phonon dispersion curves, rather than an empirical exponent θ_z in the intrinsic resistivity component of equation (3).

Details of the theory and calculations can be found in a dissertation /5/, and will be published elsewhere.

References

/1/ R. Berman, Thermal Conduction in Solids, Oxford, University Press, 1976.

/2/ P.G. Klemens, in "Thermal Conductivity 19", p.453, Plenum Press, New York, 1988.

/3/ P.G. Klemens, in "Solid State Physics", vol.7, p.1, Academic Press, New York, 1958.

/4/ P.G. Klemens, Proc. Roy. Soc (London) A208, 109 (1951).

/5/ Y.-J. Han, Theory of Phonon Thermal Conductivity at Low Temperatures. Dissertation, University of Connecticut, 1990.

Grueneisen Gammas and Phonon Scattering

P.G. Klemens

Department of Physics, University of Connecticut,
Storrs, CT 06269-3046, USA

The anharmonic properties are described by Grueneisen parameters γ. One must distinguish between different γ's, which are related to one another.

For a dilatation Δ the frequency of a normal mode shifts by $\delta\omega = -\gamma\omega\Delta$, resulting in an outward pressure

$$P = \sum_i C_i \gamma_i \delta T$$

where C_i is the contribution of the mode to the specific heat C. The thermal expansivity thus becomes

$$d\Delta/dT = P/K_B = C/K_B \gamma_X$$

where K_B is the bulk modulus, and the expansion gamma is

$$\gamma_X = C^{-1} \sum_i C_i \gamma_i = \langle\gamma_i\rangle .$$

The gammas are weighted according to their specific heat contributions. Other strains also cause frequency shifts, but in cubic crystals they average to zero.

The perturbation of the phonon Hamiltonian due to a strain ε which is uniform or slowly varying is of the form

$$H' = 2\omega\delta\omega M a^* a$$

where a^*, a are time-dependent amplitudes or phonon creation or annihilation operators. Again $\delta\omega = -\gamma\omega\varepsilon$, where ε is is either static or the dynamic strain of a third phonon[1]. The γ for phonon scattering, γ_{PS} thus becomes

$$\gamma_{PS}^2 = C^{-1} \sum_i C_i (\gamma_i)^2$$

since scattering is proportional to $(H')^2$.

The ultrasonic attenuation in the Akhieser [2] limit arises because an ultrasonic strain ε changes the temperature of a group of modes by $\delta T_i = T\varepsilon\gamma_i$ while the average change is given by $\langle\gamma_i\rangle$. The dparture from the average is given by $\gamma_i - \langle\gamma\rangle$ and the anharmonic interactions transfer energy in or out of each mode. The rate of entropy production is

$$T^2 \, dS/dt = \varepsilon^2/\tau \sum_i C_i [\delta T_i - \langle\delta T\rangle]^2$$

Springer Series in Solid-State Sciences, Volume 112
Phonon Scattering in Condensed Matter VII Eds.: M. Meissner and R.O. Pohl

where τ is the thermal mode relaxation time. After averaging over a cycle and noting the phase differences, one finds that the ultrasonic attenuation becomes

$$\alpha = 2\ CT/K \langle (\gamma - \langle\gamma\rangle)^2 \rangle \omega^2 \tau$$

where K is the appropriate modulus. Writing

$$\alpha = 2\ CT/K\ \gamma_A^2 \omega^2 \tau$$

$$\gamma_A^2 = \langle (\gamma_i - \langle\gamma\rangle)^2 \rangle = \gamma_{PS}^2 - \gamma_X^2$$

The final form follows from the second moment theorem.

Thus the phonon scattering gamma becomes

$$\gamma_{PS}^2 = \gamma_X^2 + \gamma_A^2$$

and since the second term must be positive, the value of γ_{PS}, which enters in the phonon scattering matrix element, must always exceed in magnitude the value γ_X derived from the thermal expansion.

References

1 P.G. Klemens, in Solid State Physics, ed. by F. Seitz and D. Turnbull, vol. 7, p.1, Academic Press, New York, 1958

2 A.Akhieser, J. Phys. (U.S.S.R.) 1, 277 (1939)

Temperature-Dependence of the Velocity of Second Sound and the Determination of Phonon Lifetimes from Thermal Conductivity

R.J. Stoner and H.J. Maris

Department of Physics, Brown University, Providence, RI 02912, USA

The traditional method for the determination of lifetimes of thermal phonons in crystals is based on the analysis of the thermal conductivity κ in terms of the kinetic formula

$$\kappa(T) = \frac{1}{3}\, C(T)\, v_D^2\, \tau_R \tag{1}$$

where $C(T)$ is the specific heat, v_D is the Debye velocity, and τ_R is the average resistive (non–momentum conserving) phonon relaxation time. Examples of results obtained for τ_R in this way are shown in Fig 1 [1]. This approach has the advantage of simplicity, but in several instances the results obtained have been clearly inconsistent with other data. As examples we mention the following:

1) Estimates of the average phonon lifetime in silicon (see Fig. 1) give a room temperature value of around 8 psecs. However, direct measurement of the lifetime of the zone center LO phonon gives a lifetime of 13 psecs [2]. Since phonon lifetimes generally decrease with increasing phonon frequency it is hard to understand why the phonon of highest frequency in silicon should have a lifetime less than the average. The decay of the LO phonon by the 3–phonon process is not forbidden by any conservation law.

2) Numerical calculations of the Umklapp scattering rate [3] give lifetimes which, although subject to uncertainties associated with difficulties of estimating the interatomic potential, are considerably longer than the results obtained via Eq. 1.

3) Finally, we mention that even some more sophisticated analyses of the thermal conductivity, for example based on the Callaway model, lead to difficulties regarding the relative rates of N and U processes. For $T < \Theta_D$, τ_N should always be less than τ_U, and at high temperatures the two times should be roughly equal. The models do not always succeed in giving these results [4], thereby raising doubts about the approach.

To try to understand these problems we have reconsidered heat transport in the presence of a large rate for N–processes. First we note that the velocity v_2 of second sound is [5]

$$v_2^2 = \frac{\left[\int d^3k\ n_k(n_k+1) v_{kz} k_z \omega_k\right]^2}{\int d^3k\ n_k(n_k+1)\omega_k^2 \int d^3k\ n_k(n_k+1) k_z^2} \tag{2}$$

Springer Series in Solid-State Sciences, Volume 112
Phonon Scattering in Condensed Matter VII   Eds.: M. Meissner and R.O. Pohl

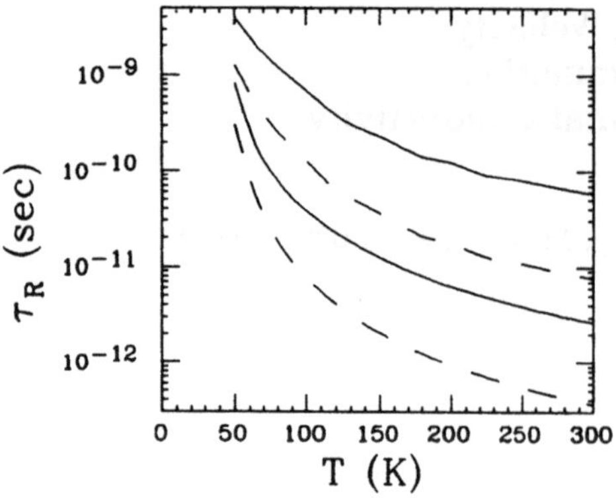

Fig.1. Estimates of τ_R from kinetic formula Eq. 1 (dashed) and from Eq. 3 (solid) for LiF (lower curves) and Si (upper curves).

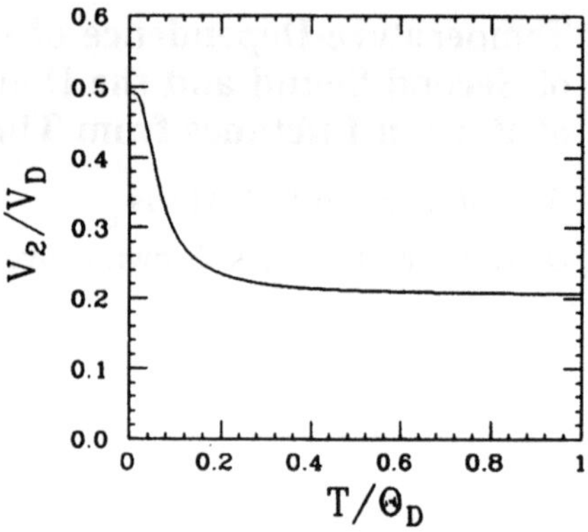

Fig.2. Velocity of second sound v_2 as a function of temperature for a model of an fcc crystal.

where n_k is the Bose distribution, and v_{kz} is the z-component of the phonon group velocity. We have evaluated these integrals for an fcc crystal with nearest neighbor central forces and the results are shown in Fig. 2. It is found that because of the dispersion in the phonon group velocity the velocity v_2 drops from its low temperature limiting value in a narrow temperature range between 0.05 and 0.2 Θ_D, i.e the decrease occurs at a surprisingly low temperature. A second result is that when allowance is made for phonon dispersion Eq. 1 is replaced by the more general result

$$\kappa(T) = C(T)\, v_2^2(T)\tau_R\,. \tag{3}$$

The results for τ_R calculated from this equation are included in Fig. 1. For T above about 0.1 Θ_D the effect is to increase the estimated values of τ_R by about a factor of 6 relative to those determined from Eq. 1. While it is not unexpected that group velocity dispersion should modify Eq. 1 and increase the estimated value of τ_R, it is interesting that the effect is so large. It appears that this factor may well resolve the inconsistencies referred to at the beginning of this paper. We will describe these calculations, their limitations, and the significance for particular crystals in more detail in a subsequent paper [6].

This work was supported in part by the Department of Energy through grant DE-FG02-86ER45267.

[1] Thermal conductivities and specific heats are taken from Y.S. Toulookian, Thermophysical Properties of Matter, (Plenum, New York, 1970). We have taken the highest measured thermal conductivity for each crystal.
[2] J. Menendez and M. Cardona, Phys. Rev. **29**, 2051 (1984).
[3] S. Tamura and H.J. Maris, to be published.
[4] See, for example, R. Berman and J.C.F. Brock, Proc. Roy. Soc. Lond. **289** ,46 (1965), especially p. 60.
[5] H.J. Maris, Phys. Rev. **B 24**, 1205 (1981).
[6] R.J. Stoner and H.J. Maris, to be published.

Simulation of Phonon State and Phonon Transport in Model Crystals

Y. Hiki and S. Ozawa

Applied Physics Group, Faculty of Engineering,
Ibaraki University, Nakanarusawa, Hitachi 316, Japan

The authors are developing their studies to simulate phonon transport in solids by utilizing a molecular dynamics (MD) computer simulation using one-dimensional mass-spring model crystals [1-3]. The present study is divided into two parts: the preparation of a crystal having an equilibrium phonon state or a definite temperature; and the observation of propagating pulsed phonons with a definite frequency in the crystal.

A homogeneous one-dimensional model crystal composed of equal masses and nonlinear springs is firstly considered. The number of masses is 550; and the lattice anharmonicity up to the fourth order is considered, where the ratio of the 2nd, 3rd, and 4th order elastic constants is 1 : -10 : 100. An input phonon pulse is applied to an end of the crystal. The pulse is composed of vibrations with frequency ν_{in} modulated by step functions with a triangular envelope. A Fourier analysis is made for the vibrational state of the lattice to obtain an energy spectrum. Examples are presented in Fig. 1, where the spectra in an early (a) and a later (b) computational steps are shown. The ratio ν_{in}/ν_D is 0.061, where ν_D is the Debye frequency of the crystal. The vibrational energy initially concentrated around the frequency ν_{in} and its overtones is finally homogenized. The inclusion of the lattice anharmonicity, and the use of input phonons not so high in frequencies, are needed for the phonon

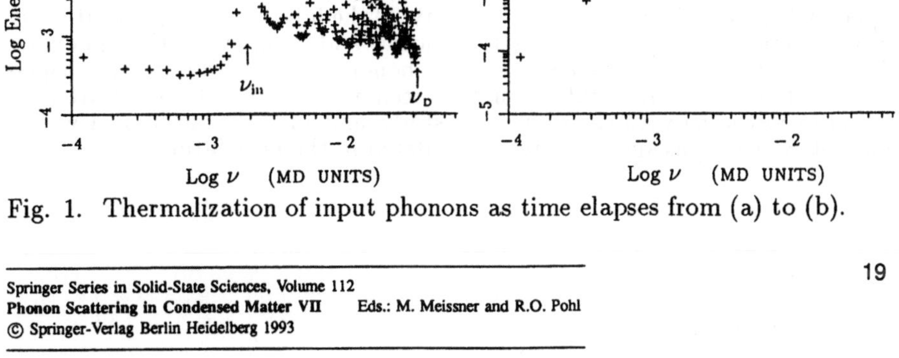

Fig. 1. Thermalization of input phonons as time elapses from (a) to (b).

Springer Series in Solid-State Sciences, Volume 112
Phonon Scattering in Condensed Matter VII Eds.: M. Meissner and R.O. Pohl

thermalization. By inputting various amounts of phonon energies, crystals of different temperatures can be prepared.

The propagation of pulsed phonons in imperfect crystals is now simulated, and here we adopt a crystal containing mass defects. The crystal is composed of atoms with masses m(body) + atoms with m(defect). The defect atoms are randomly distributed among the body atoms, and the total number of atoms is 550. Further, a row of 150 heat sink atoms with masses m(body) is attached to the right end of the crystal. Then, the triangular phonon pulse is applied to the left end of the crystal. The phonons propagate through the crystal, and are finally absorbed by the heat sink. At a fixed point in the crystal, the kinetic energy K at the position is recorded as a function of the computational time. Examples of the results are presented in Fig. 2. The conditions of the simulation are: $m(\text{defect})/m(\text{body}) = 3$; the defect concentration = 10 %; the input phonon frequencies are (a) $\nu_{\text{in}}/\nu_{\text{D}} = 0.061$, and (b) $\nu_{\text{in}}/\nu_{\text{D}} = 0.83$; the crystal has a definite temperature, say T_0; the energy around the 200th atom averaged over 10 neighboring atoms is computed. It can be seen that the energy increases and then gradually decreases with time, which is the typical behavior of the diffusion of heat in a medium having thermal losses to surroundings.

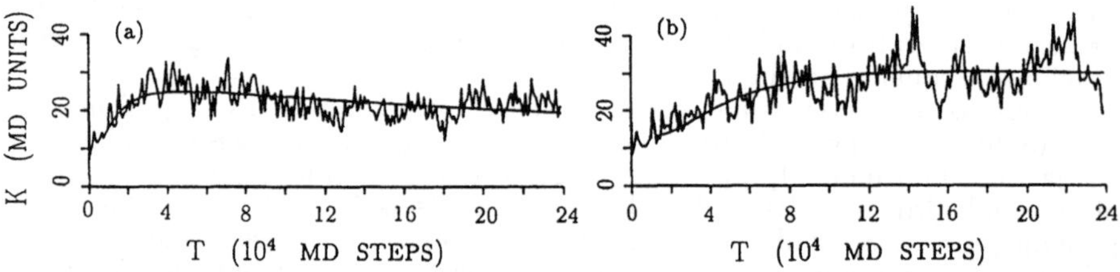

Fig. 2. Time variation of energy at a position in crystal for (a) lower and (b) higher frequency input phonons.

The above problem of thermal diffusion with losses in a solid has precisely been studied [2,4], and the temperature in a pulse-heated solid can be expressed as a function of position and time with the thermal diffusivity D of the solid as a parameter. The energy versus time relation obtained by the present simulation can be used to determine the thermal diffusivity in the model crystal. The solid lines in Fig. 2 are the fitted curves to determine the diffusivty D. The results were such that $D[\nu_{\text{in}}/\nu_{\text{D}} = 0.061]/D[\nu_{\text{in}}/\nu_{\text{D}} = 0.83] = 3.5$. Other cases were also adopted for the computation, and here the input phonon frequency was fixed as $\nu_{\text{in}}/\nu_{\text{D}} = 0.061$. Examples of the obtained results are: at a higher crystal temperature, $D[T_0]/D[5.3T_0] = 1.0$; for lighter mass defects, $D[m(\text{defect}) = 3m(\text{body})]/D[m(\text{defect}) = 1.5m(\text{body})] = 0.38$.

These results may qualitatively be explained by the relation $D \propto vl$, which comes from the gas kinetics equation $\kappa = (1/3)Cvl$ and the definition $D = \kappa/\rho C$, where κ is the thermal conductivity, C the heat capacity per unit mass, ρ the material density, v and l are the velocity and mean free path of phonons. The phonon velocity v depends on the phonon frequency due to the dispersion, and the mean free path l on the mechanism of the phonon scattering. We are proceeding with the simulation study, and an unified argument will be presented after consequences are accumulated in the near future.

References

[1] S. Ozawa and Y. Hiki, Jpn. J. Appl. Phys. **30**, Suppl. 30-1, 46 (1991).
[2] S. Ozawa and Y. Hiki, Jpn. J. Appl. Phys. **31**, Suppl. 31-1, 63 (1992).
[3] Y. Hiki and S. Ozawa, Thin Solid Films **210/211**, 634 (1992).
[4] Y. Kogure, T. Mugishima and Y. Hiki, J. Phys. Soc. Jpn. **55**, 3469 (1986).

Thermal Conductivity in Low-Dimensional Nonlinear Lattices

N. Nishiguchi[1], *N. Takahashi*[1;2], *Y. Kawada*[1;**], *and T. Sakuma*[1]

[1]Department of Engineering Science, Hokkaido University, Sapporo 060, Japan

*Present address: Hokkaido Tecsys Corp., Chuo-ku, Ohdori, W 14-7, Sapporo 060, Japan

**Present address: Advanced Software Development, Asia/Pacific Products, APTO, IBM Japan Ltd., Shimotsuruma, Yamato 242, Japan

Heat transport in an electrically insulating material is attributed to lattice vibrations and is described in terms of the phenomenological theory referred to as Fourier's law. A number of simulations using molecular dynamics have been performed to verify Fourier's law in low-dimensional lattices from first principles[1-4]. However, most attempts failed [1-4] to present the normal thermal conductivity without external disturbances since the resultant conductivity was dependent on system size.

We study the validity of Fourier's law for the low-dimensional nonlinear lattices using molecular dynamics techniques, paying special attention to the nonlinear temperature profiles[2,4] and to the non-diffusive energy flow[2,4] which have been overlooked so far to estimate the thermal conductivity.

One of the models to be studied here is a one-dimensional diatomic Toda lattice[5,7]. Another one is a quasi-periodic Toda lattice where the arrangement of masses obeys the Fibonacci sequence[6,7]. The last model is a two-dimensional monatomic Toda lattice[8]. We put buffer areas (integrable lattices) between the heat baths and the specimen lattice to reduce temperature discontinuity at interfaces between the heat baths and the specimen lattice.

The local temperature is defined by twice the averaged local kinetic energy. We investigate the heat transport above the critical temperature where the system can offer irreversibility. Figure1 exhibits the resultant temperature profiles of the one-dimensional diatomic and Fibonacci Toda lattices with N=250. We can easily see that the temperature profiles vary exponentially with respect to the distance. This is true independently of the lattice size and of the dimensionality. We can empirically approximate the temperature profiles with the exponential function of the form

$$T(x) = T_H (T_L / T_H)^{x / N_x}, \qquad (1)$$

where T_H and T_L are the temperatures at the specimen lattice ends and x measures the distance from the interface between the buffer area and the specimen lattice.

We suppose that the heat current J_x is composed of diffusive heat flow J_N and the ballistic or non-diffusive heat flow J_B as $J_x = J_N + J_B$. Only the normal heat current J_N is expected to obey Fourier's law $J_N = -\kappa\, dT/dx$. Here we assume that the ballistic energy flow depends on the temperature difference of the buffer areas and on the system size since the buffer areas act as the heat baths effectively.
From these considerations, we get

$$J_x = \kappa \log(T_H / T_L)\ T / N_x + J_B(T_{B,H} - T_{B,L}, N_x). \qquad (2)$$

Springer Series in Solid-State Sciences, Volume 112
Phonon Scattering in Condensed Matter VII Eds.: M. Meissner and R.O. Pohl

Figure 1

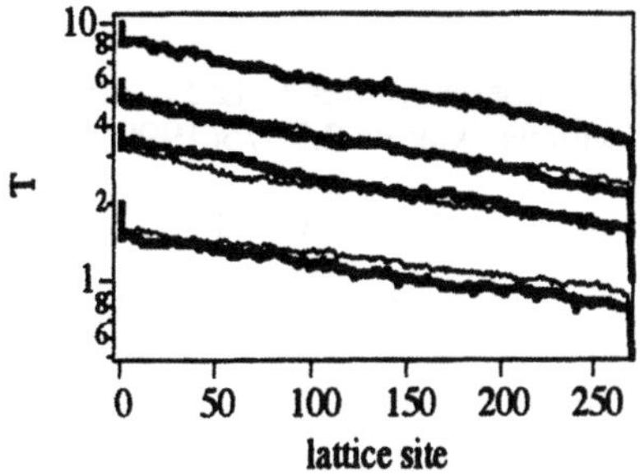

Figure 2

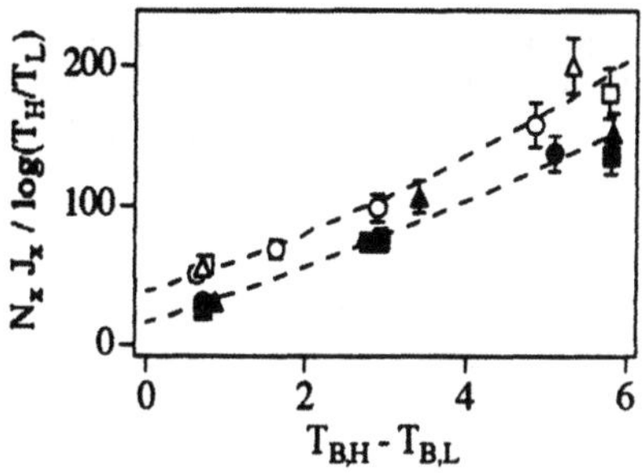

Figure 1 Temperature profiles for the one-dimensional diatomic(thin lines) and Fibonacci Toda lattice(thick lines).

Figure 2 The heat current per unit length $/\log(T_H/T_L)$ versus temperature difference of the buffer areas for the one-dimensional diatomic and Fibonacci Toda lattcies. Open(Solid) circles, squares, and triangles denote the data for the diatomic (Fibonacci) lattices with N=250, 350, and 450, respectively. The dashed lines are drawn with the least-squares method to exhibit the extrapolation of the data to zero temperature difference.

Equation (2) requires the thermal conductivity to be inversely proportional to local temperature to maintain the steady non-equilibrium state as follows:

$$\kappa = \xi / T \, , \qquad (3)$$

where ξ is a constant.

In order to estimate ξ from the measurements, we have to exclude the contribution of the ballistic heat current from the equation(2). We extrapolate ξ from the experimental data above the critical temperature, utilizing the difference of the temperature dependence of the heat currents J_N and J_B. Figure 2 shows the magnitude of $J_x N_x / \log(T_H/T_L)$ for the one-dimensional diatomic and Fobonacci Toda lattices. As the temperature difference decreases, the magnitude decreases drastically and seems to become a positive and finite value at zero temperature difference. In addition, it should be noted that the results are almost size-independent. These characteristic features are also identified in the two-dimensional monatomic Toda lattices. These facts state that the normal current exits and that the ξ is invariant for all the sizes. That is, we can conclude Fourier's law is satisfied in the one- and two-dimensional nonlinear lattices. The resultant coefficents ξ 's are $\xi_{1Diatomic}$=36.8 and $\xi_{1Fibonacci}$=15.9 for one-dimensional diatomic and Fibonacci lattices, respectively, and ξ_{2Toda}=78.0 for a two-dimensional moatomic Toda lattice.

Finally, we emphasize again that we should take account of the nonlinear temperature profile and the existence of the ballistic heat current to estimate the thermal conductivity.

[1] M.Toda, Phys. Scr. **20**,424(1979).
[2] M.Mareschal and A.Amellal, Phys. Rev. A**37**,2189(1988).
[3] F.Mokross and H. Büttner, J.Phys. C**16**,4539(1983).
[4] E. A. Jackson and A. D. Mistriotis, J. Phys. : Condens. Matter **1**, 1223 (1989).
[5] N. Nishiguchi and T. Sakuma, J. Phys. : Condens. Matter **2**,7575(1990); ibid **3**,10077(1992).
[6] N. Nishiguchi, N. Takahashi and T. Sakuma, J. Phys. : Condens. Matter **4**,1465(1992).
[7] N. Nishiguchi, N. Takahashi and T. Sakuma, Jpn J. Appl. Phys. Suppl. **31**, 179 (1992).
[8] N. Nishiguchi, Y. Kawada and T. Sakuma, (to be submitted to J. Phys. : Condens. Matter).

Phonons in Crystals, Quasicrystals and Topologically Disordered Lattices

*P. Erdös and J. Michalski**

Institute of Theoretical Physics, University of Lausanne, CH-1015 Lausanne, Switzerland
*Present address: Institute of Zoology and Animal Ecology, University of Lausanne, CH-1015 Lausanne, Switzerland

Using the molecular dynamics simulation we studied the phonon localization and the lattice thermal conductivity of systems with and without translational invariance. The localized vibrational modes, their spectral intensity, and the localization-induced anharmonicity has been investigated.

As a model we used interacting masses arranged on a two-dimensional Penrose tiling with and without defects [1]. We studied three versions of the model: periodic, aperiodic and quasiperiodic.

Thermal conductivities of the periodic and aperiodic models are presented in Fig. 1. We explain the difference in the conductivities by the hopping conductivity [2] in the aperiodic model [1,3]. The thermal conductivity of the quasiperiodic model is indistinguishable from that of the aperiodic model [3].

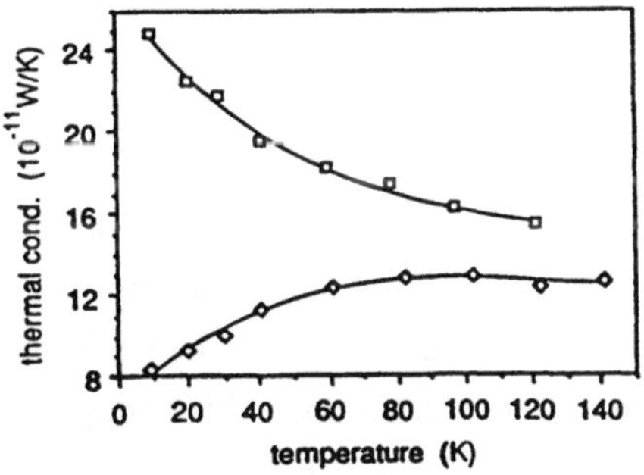

Fig. 1
Thermal conductivity as a function of temperature:
□ periodic model;
◇ aperiodic model.

The spectral intensities of the modes of the periodic model are all well-defined single peaks. The spectral intensities of the localized modes of the aperiodic model are different. Some of them have additional peaks (Fig. 2). This

Springer Series in Solid-State Sciences, Volume 112
Phonon Scattering in Condensed Matter VII Eds.: M. Meissner and R.O. Pohl

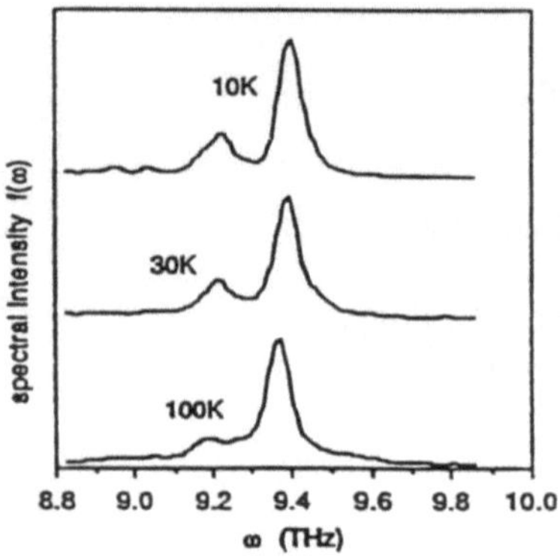

Fig. 2
Spectral intensity of the 7-th mode of the aperiodic model (AM mode 7 from Fig. 3).

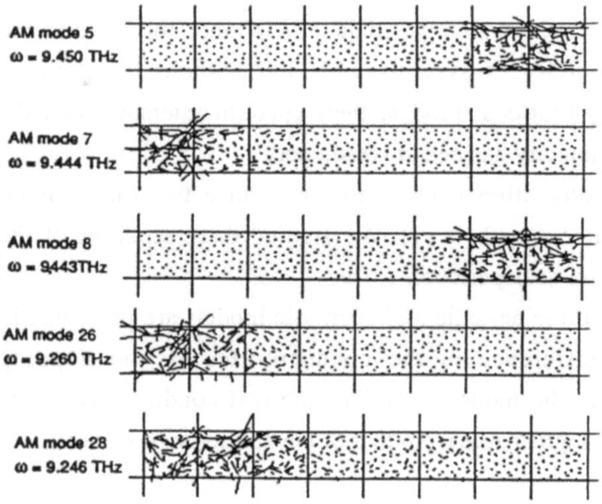

Fig. 3
Some high-frequency harmonic modes of the aperiodic model. (modes are numbered in order of decreasing frequency).

means that localized harmonic modes are not well-defined excitations of the aperiodic model, even at low temperature.

The additional peak at 9.25 THz in Fig. 2 is due to the strong interaction of the AM mode 7 with the AM modes 26 and 28, which are localized at the same place as the AM mode 7 (see Fig. 3). On the other hand, modes that have almost the same frequency as the AM mode 7 but which are localized at a different place (AM modes 5 and 8 in Fig. 3) do not interact with the AM mode 7.

Shortly: Modes that are localized at the same place interact strongly, even if their frequencies are not very close; modes that are localized at different places interact very weakly, even if their frequencies are very close.

1 J.Michalski and P.Erdös, Solid State Commun. **72**, 967 (1989).
2 A.Jagannathan, R.Orbach & O.Entin-Wohlman, Phys.Rev. B **39**, 13465 (1989)
3 J.Michalski, Phys.Rev. B **45**, 7054 (1992).

Scattering of Acoustic Phonons by Isotopic Impurities in a Cubic Medium: Singularities of the F-L Transforms of the Distribution Function

T. Paszkiewicz and M. Wilczyński

Institute of Theoretical Physics, Wroclaw University,
Pl. Maxa Borna 9, PL-52-404 Wroclaw, Poland

In our previous paper we studied the relaxation of an homogeneous initial state of a gas of long-wavelength acoustic phonons scattered by isotopic impurities embedded in a medium of cubic symmetry [1]. This problem is equivalent to the study of the spectrum of a suitable collision integral C[f] of the Boltzmann-Peierls equation (BPE). The linear operator $\mathbf{B}$ is defined as

$$\mathbf{B}f(\hat{K};\mathbf{r},t) = C[f(\hat{K};\mathbf{r},t)],$$

where $f(\hat{K};\mathbf{r},t)$ is the distribution function of phonons $\hat{K}$ (PDF, for short). $\hat{K}$ comprises both the direction of the wave vector $\hat{\mathbf{k}}$ and the polarization index $j(j=0,1,2)$). The PDF depends on the time t and the space variable $\mathbf{r}$. We obtained the spectral decomposition of $\mathbf{B}$ [1]

$$\mathbf{B} = 0\mathcal{P}_J + (-1/\tau_L)\mathcal{P}_L + (-1/\tau_M)\mathcal{P}_M + (-1/\tau)\mathcal{Q}.$$

The projection operators $\mathcal{P}_J$, $\mathcal{P}_L$, $\mathcal{P}_M$, $\mathcal{Q}$ are idempotent and project any function of $\hat{K}$ onto mutually orthogonal subspaces. The explicit closed analytic expressions for the inverse relaxation times τ_L, τ_M were obtained. They are proportional to τ^{-1} and depend on the numerical values of the elastic constants (cf. [1]). Our present numerical results correspond to the elastic constants of gallium arsenide (GaAs).

Here we shall consider spatially inhomogeneous initial states. One of the ways to tackle this problem is to use the Fourier-Laplace transform of the PDF. The BPE yields the expression for $f(\hat{K};\mathbf{q},z)$ which exhibits singularities in the plane of complex $\xi = z\tau$. These singularities are: three cuts along the interval $(-1-i\ \max|\mathbf{v}(\hat{K})\vec{\kappa}|, -1+i\ \max|\mathbf{v}(\hat{K})\vec{\kappa}|)$ and poles at

$$\xi_J(\vec{\kappa}), \xi_L(\vec{\kappa}), \xi_M(\vec{\kappa}) \qquad \text{and} \qquad \xi_Q = -1+i\ \mathbf{v}(\hat{K})\vec{\kappa},$$

where $\mathbf{v}(\hat{K})$ is the group velocity and $\vec{\kappa} = \mathbf{q}\tau$.

To establish the explicit dependence of the Fourier transform of the PDF on time one should calculate the inverse Laplace transform. With the help of the

Springer Series in Solid-State Sciences, Volume 112
Phonon Scattering in Condensed Matter VII Eds.: M. Meissner and R.O. Pohl

standard methods of complex analysis this reduces to calculations of suitable integrals around the above mentioned singularities. The pre-exponential factor depends on the initial conditions and does not grow faster in time than the suitable exponent. The contribution of cut corresponding to a given j is proportional to $\exp(-t/\tau)$. The contribution of the pole located at ξ_U $(U = J, L, M)$ is proportional to $\exp(\xi_U t/\tau)$.

For $q = |\mathbf{q}| \to 0$ the above poles move to points $\xi = 0, (-\tau/\tau_L), (-\tau/\tau_M), (-\tau/\tau)$ respectively. With growing $\kappa = |\vec{\kappa}|$ the poles at $\xi_J(\vec{\kappa})$, $\xi_L(\vec{\kappa})$, $\xi_M(\vec{\kappa})$ move in the plane of complex ξ towards the cut, merge with it for $\kappa = \kappa_m^{(U)}$ and vanish. However, their trajectories depend on $\hat{\kappa}$ - the direction of $\vec{\kappa}$. For $\hat{\kappa} =< 100 >$ all poles are real. We check that for other directions there exist $\kappa_c^{(U)}(\vec{\kappa})$ – the critical values of κ, such that the pole U become complex when $\kappa > \kappa_c$ (cf. Table 1). One should expect that due to oscillations the contribution for $\kappa > \kappa_c$ to the inverse Fourier transform is small. For small κ the pole $\xi_J \sim -\kappa^2 \sim -Dq^2$ (where D is the coefficient of diffusion), i.e. it is related to diffusion (the diffusive pole).

We also performed preliminary studies of the anisotropy of $Re\xi_U$. We chose the somewhat arbitrary $\kappa = 100 \cdot 10^{-8}$. All For growing κ the anisotropy emerges. For all poles $(U = J, L, M)$ the relaxation of Fourier harmonics is fastest for $\vec{\kappa}$ directed along the high symmetry directions (cf. Table 2).

Table 1

Values of $\kappa_c^{(U)}$ and $\kappa_m^{(U)}$ for poles J, L, M *for several $\hat{\kappa}$. The values of κ are multiplied by* 10^{-8}

Pole	*J*			*L*			*M*		
Direction	<100>	<110>	<111>	<100>	<110>	<111>	<100>	<110>	<111>
$\kappa_c^{(U)}$	-	-	381	-	-	146	-	157	-
$\xi_U(\kappa_c^{(U)})$	-	-	- 0.8	-	-	- 0.85	-	- 0.91	-
$\kappa_m^{(U)}$	1410	570	760	380	260	310	750	210	180

Table 2

Pole	*J*			*L*			*M*		
Direction	<100>	<110>	<111>	<100>	<110>	<111>	<100>	<110>	<111>
$-Re\ \xi_U$	0.036	0.036	0.037	0.678	0.683	0.687	0.692	0.706	0.707

Acknowledgements

This work is supported by a grant from the National Committee for Scientific Research of Poland under the contract No 397/2/91.

References

[1] T. Paszkiewicz, M. Wilczynski, Z. Phys. B-Condensed Matter, **88**, 5, 1992

Phonon–Phonon and Phonon–Defect Scattering in ^{4}He Crystals

L.P. Mezhov-Deglin and A.A. Levchenko
Institute of Solid State Physics, Russian Academy of Sciences,
142432-Chernogolovka, Moscow District, Russia

Abstract. The thermal conductivity measurements were made in pure 4He and 4He+3He liquid and solid mixtures under pressure up to 84 atm.The data cover the temperature range down to 0,4 K and include the results of investigations of recovery processes in plastically deformed crystals.The importance of different phonon relaxation mechanisms is estimated from the behavior of the phonon mean free path calculated from the thermal conductivity.The main results are:(1) the frequency of phonon-phonon scattering in volume is increasing nearly an order on solidifying the superfluid liquid; (2) in solid helium with rise the pressure from 25 to 100 atm the frequency of normal phonon-phonon scattering is decreasing an order at the same reduced temperature T/Q ;(3) the value of additional thermal resistivity in plastically deformed crystals is determined by a resonant phonon-dislocation scattering; (4) the activation energy of recovery processes in bent samples containing up to 0,5% of 3He atoms is close to the energy of vacancy creation.So the motion of freshly introduced dislocations is vacancy assisted,and concentration of vacancies in samples,grown at 31 or 26 atm. is less than $2*10^{-10}$% at temperatures T<0,45 K.(5) the activation energy of recovery processes falls more than twice on

Springer Series in Solid-State Sciences, Volume 112
Phonon Scattering in Condensed Matter VII Eds.: M. Meissner and R.O. Pohl

increasing the impurity content up to 5%,probably due to self-localization of 3He atoms.

Here we report liquid and solid 4He and 4He+3He mixtures thermal conductivity results obtained by using two different samples tubes ranged in diameter from 1,0 to 1,8 mm.The measurements include the observations of transition from Knudsen flow of the phonon gas to viscous Poiseuille flow of dense phonon gas in solid or phonon-roton gas in superfluid helium.Investigations of the recovery processes in plastically deformed (by bending the sample at T<0,45 K) crystals give us any information on mechanisms of phonon-dislocation scattering and on mechanisms of diffusion of freshly introduced dislocations.Moreover supposing that the main is vacancy assisted motion we can estimate the first time the concentration of thermal vacancies at T<0,45 K and state that concentration of zero-point vacancies in samples grown at 26 and 31 atm is less than $2*10^{-10}$ % [1,2].In additon we want to mention the results of qualitive measurements of thermal conductivity of liquid and solid 4He near the melting curve in a metal tube filled with 5 mkm superconducting wires which demonstated clearly the effects of super-cooling of liquid helium.

References

1.L.P.Mezhov-Deglin,in Proceedings of the Second Intern. Conference on PHONON PHYSICS,Budapest 1985(World Scientific,1985) p.855

A.A.Levchenko,L.P.Mezhov-Deglin, ibid p.859

2.A.A.Levchenko,L.P.Mezhov-Deglin,in Proceedings of the Fifth Intern. Symposium STRUCTURE AND PROPERTIES OF DISLOCATIONS,Moscow 1986 (Chernogolovka 1989) p.147

Thermal Conductivity of Diamond Gems and CVD Diamond Films: Temperature and Isotopic Composition Dependence

Y.Z. Qiu[1], *A. Witek*[1], *D.G. Onn*[1], *W.F. Banholzer*[2], *and T.R. Anthony*[2]

[1]Applied Thermal Physics Laboratory, Department of Physics and Astronomy, University of Delaware, Newark, DE 19716, USA

[2]Research and Development Center, General Electric Company, River Road, Schenectady, NY 12309, USA

In order to investigate the reported high thermal conductivity (k) of isotopically pure ^{12}C diamonds [1], we have measured k(T) of two synthetic gems of different isotopic compositions: (A) 1.1% ^{13}C and (B) 0.04% ^{13}C between 300 K and 550 K. The resultant k(T) data are shown in Fig. 1(a) (triangles) together with k(T) for natural Type IIa diamond from the literature [2]. Analysis of k(T) data was based on a comprehensive Klemens-Callaway model. The resultant phonon-impurity scattering contribution for the Type IIa data is within the uncertainty of Berman's previous analysis [2]. The attenuation of k with increasing ^{13}C concentration for our synthetic gem data (Fig. 1) can be attributed to the influence of the impurity term alone.

We also measured k(300) of several synthetic diamond gems of different isotopic concentrations (○ in FIg. 1b). The results are re-analyzed together with previously published data [2]. The impurity coefficient is five times larger than the Rayleigh scattering rate for the ^{13}C content. The Klemens-Callaway model, using the ^{13}C concentration, provides an upper limit for the data analyzed (Fig. 1b). We attribute the discrepancy to lattice vacancies or impurity clustering. The former are almost 150 times more effective phonon scatterers than ^{13}C. Hass [3] has suggested that normal processes explain the discrepancy. Our analysis suggests normal processes are not effective.

In addition we have measured k(T) for three CVD diamond films of different isotopic compositions [(1) 1.1%, (2) 0.04%, and (3) 49.6% ^{13}C] between 90 K and 550 K. Fig. 2 shows the data and the analysis curves, also based on the Klemens-Callaway model. The resulting scattering parameters are found to be physically reasonable based on the likely microstructure of the samples. The thermal conductivity for the 0.04% film falls below that of the 1.1% film due to a smaller grain size but the impurity term is similar for both. However the attenuation of k(T) in the 49.6% film is attributable to impurities rather than grain size. As in the gems, the impurity scattering may be due to single ^{13}C atoms, lattice vacancies or isotope clustering.

Analyses of our data are based on the re-analysis [2] of the k(T) data of Berman between 10 K and 300 K in combination with k(T) of Vandersande

Springer Series in Solid-State Sciences, Volume 112
Phonon Scattering in Condensed Matter VII Eds.: M. Meissner and R.O. Pohl

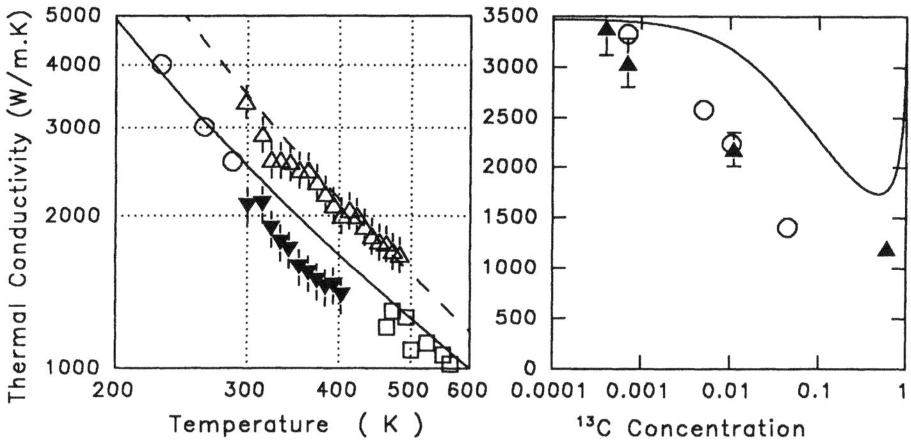

Figure 1. (a). The impurity effect in diamond gems showing data for 99.96 % ^{12}C gem (▵),the natural abundant gem (▾), part of Berman data (○), part of Vandersande data (□), and the fit curve to the two sets type IIa gem data (——) and the curve with point defect parameter set to zero (--). (b). A Klemens-Callaway fit (——) showing phonon-isotope point defect scattering versus ^{13}C concentration together with our data (▴) and data from literature (○), assuming that the other scattering processes are independent of isotope concentration in the solid.

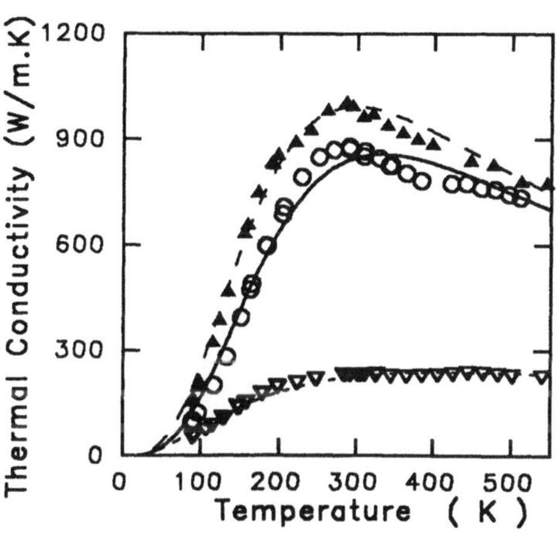

Figure 2. CVD diamond film data for natural abundance (▴), 99.96 % ^{12}C (○), and 50.4 % ^{12}C 49.6 % ^{13}C (▿) with corresponding Klemens-Callaway fit curves.

above 600 K for natural Type IIa diamond under the assumption that (1) the umklapp power (ξ = 3) is the same in all forms of diamond; (2) the normal process parameter is independent of the point defect scattering process; (3) the umklapp scattering parameters are the same in all forms of diamond.

References

(1) T. R. Anthony, W. F. Banholzer, J. F. Fleischer, L. H. Wei, P. K. Kou, R. L. Thomas, and R. W. Pryor, Phys. Rev. B **46**, 1104 (1990).
(2) A. Witek, Y. Z. Qiu, D. G. Onn, T. R. Anthony, and W. F. Banholzer, Phys. Rev. Lett., **68**, 18, 2806,(1992) and references therein.
(3) K. C. Hass, M. A. Tamor, T. R. Anthony, and W. F. Banholzer, Phys. Rev. B, **45**, 13, 7171(1992).

Thermal Conductivity of Diamond Between 170 and 1200 K and the Isotope Effect

J.W. Vandersande[1], *A. Zoltan*[1], *J.R. Olson*[2], *R.O. Pohl*[2], *T.R. Anthony*[3], *and W.F. Banholzer*[3]

[1]Jet Propulsion Laboratory, California Institute of Technology, Pasadena, CA 91109, USA
[2]Laboratory of Atomic and Solid State Physics, Cornell University, Ithaca, NY 14853-2501, USA
[3]General Electric Company, Research and Development Center, Schenectady, NY 12309, USA

1. Introduction

The thermal conductivity of single–crystal diamond has recently received increased attention resulting from an announcement by General Electric of a 50% enhancement in the thermal conductivity of synthetic diamond of high isotopic purity[1] at room temperature. This increase was unexpected by standard theory. Thermal conductivity measurements on natural diamond between 500 and 1200K, and on synthetic, nearly isotopically pure (0.07% ^{13}C) diamond between 170 and 320K are reported here. The large temperature range allows a more accurate determination of the isotope effect than previously possible[1,2].

2. Experimental Details

The natural diamond measured was supplied by DeBeers Corp., and was determined to be type IIa based on the UV absorption spectrum. The synthetic diamond measured was grown at General Electric using a technique described elsewhere[1], and is known to be high–quality single–crystal type IIa diamond[1].

Between 500K and 1200K, the thermal conductivity of the natural diamond was measured at the Jet Propulsion Laboratory (JPL) using the flash diffusivity method. Between 170 and 320K, the thermal conductivity of the synthetic diamond was measured at Cornell in an insertion–type cryostat using the 3ω technique.

3. Results

The thermal conductivity measured on natural and synthetic diamond is shown in Fig. 1. For comparison previously published data on natural diamond from Berman et al.[3], and synthetic diamond from Onn et al.[2] are included. The measurements on synthetic diamond agree very well with those of Onn et al. where they overlap, and seem to show the same temperature dependence. The values for the nearly isotopically pure samples lie ~35% above those of natural type IIa diamond, in agreement with previous measurements[1]. The data were analyzed using the Debye model of thermal conductivity in order

Springer Series in Solid-State Sciences, Volume 112
Phonon Scattering in Condensed Matter VII   Eds.: M. Meissner and R.O. Pohl

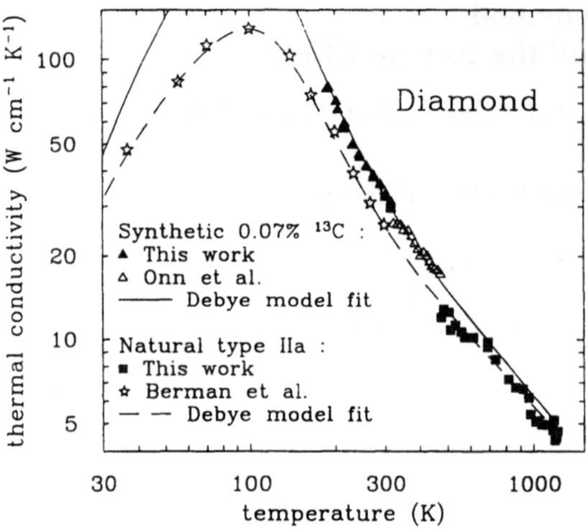

Figure 1. Thermal conductivity of natural type IIa diamond measured between 500 and 1200K, and synthetic, nearly isotopically pure diamond measured between 170 and 320K, together with measurements by Berman et al.[3] and Onn et al.[2]. The two curves are the Debye fits, varying only the point defect term.

to quantify the phonon scattering resulting from the isotopic impurities in natural diamond. The amount of point defect scattering found from the Debye integral calculations has been compared to that predicted by Turk and Klemens[4]. It is found that the ^{13}C impurities scatter 6.3× more strongly than predicted on the basis of mass and volume differences alone.

The research at Cornell was supported by the National Science Foundation, Grant No. DMR 91 15981. The research at the Jet Propulsion Laboratory was performed under contract with the National Aeronautics and Space Administration, and was partially funded by the Wright Research and Development Center.

[1] T.R. Anthony et al., Phys Rev. **42**, 1104 (1990).
[2] Onn et al., Phys. Rev. Lett. **68**, 2806 (1992).
[3] R. Berman et al. J. Phys. C:Solid State Phys. **8**, L430 (1975).
[4] Lawrence A.Turk et al., Phys. Rev. B **9**, 4422 (1974).

Phonon Scattering in Lightly-Neutron-Irradiated Diamond

D.T. Morelli and T.A. Perry

Physics Department, General Motors Research and Environmental Staff,
Warren, MI 48090-9055, USA

The role of defects in limiting the thermal conductivity of synthetic diamond films has not been fully explored. In order to shed light on phonon-defect interactions in this material, we have studied the thermal conductivity of single crystal diamond irradiated with fast neutrons. For low fluences neutrons produce lattice vacancies and small regions of disordered carbon [1], two important types of defects which occur in diamond films.

Figure 1 shows the thermal conductivity for the unirradiated type IIa single crystal diamond and for the specimen irradiated up to a total fluence of 4.5 x 10^{18} neutrons cm^{-2}. After the first irradiation the thermal conductivity is dramatically depressed throughout the entire temperature range and the maximum is displaced upward to 250 K; below 30 K, the temperature dependence of κ weakens significantly as the curve begins to approach that of the unirradiated material. After subsequent irradiations the

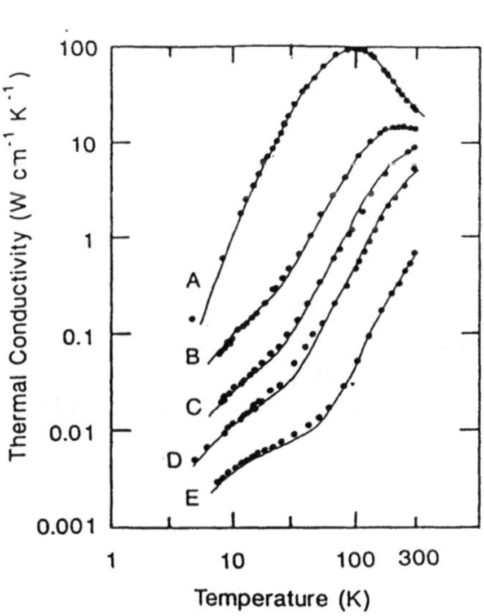

Fig. 1. Thermal conductivity of neutron irradiated diamond. A-unirradiated; B-1 unit; C-4 units; D-20 units; E-150 units (1 unit = 3 x 10^{16} n/cm^2). Solid lines are fits to the data discussed in the text.

Springer Series in Solid-State Sciences, Volume 112
Phonon Scattering in Condensed Matter VII Eds.: M. Meissner and R.O. Pohl

thermal conductivity is depressed even further, and the bending at low temperature is even more pronounced.

From a simple radiation damage model it is estimated that fast neutron irradiation of diamond produces on the order of 1000 displaced carbon atoms per incident neutron [2]. If these form into spherically-shaped disordered regions, the spatial extent of these would be about 20 A. Phonons will scatter from these extended defects according to [3]

$$\tau_e^{-1}(\omega) = Nv\pi a^2/4 \quad \text{for } qa > 1$$
$$= C\omega^4 \quad \text{for } qa < 1.$$

Here a is the spatial extent of the defect, N is the extended defect concentration, q is the phonon wavevector, and $C = N\pi a^6/4v^3$. Figure 1 shows a fit to the data achieved by combining the extended defect rate with boundary, Umklapp, and point defect rates [4] for the pristine sample. We find a ≃ 11-16 A for all irradiated samples, in good agreement with the estimated defect size. Detailed optical studies [1] indicate that irradiation introduces a host of defects, including isolated vacancies, nitrogen-vacancy complexes, and extended regions of disordered carbon and/or vacancy clusters.

A major impetus for carrying out the present study was to shed light on the role played by defects in the thermal conductivity of diamond films. Morelli et. al. [5] showed that the thermal conductivity of films grown by a hot filament (HF) process is lower than that of films grown by a microwave (M) process even though the two types of films had nearly the same crystallite size; The diminution of κ in the HF films was due mainly to a depression in the thermal conductivity in the range 15-60 K. It is clear from the above results that small regions of disordered carbon material or vacancy clusters in the diamond lattice are capable of producing precisely this effect.

We would like to thank Dr. Joseph Heremans and Professor Paul Klemens for stimulating discussions, and Mr. Philip Simpson of the University of Michigan and Dr. John Farmer of the University of Missouri-Columbia for the irradiations.

1. D.T. Morelli and T.A. Perry (to be published).
2. G.J. Dienes and G.H. Vineyard, Radiation Effects in Solids (Interscience, New York, 1957), p. 28.
3. J.W. Vandersande, Phys. Rev. B 15, 2355 (1977).
4. D.G. Onn, A. Witek, Y.Z. Qiu, T.R. Anthony, and W.F. Banholzer, Phys. Rev. Lett. 68, 2806 (1992).
5. D.T. Morelli, T.M. Hartnett, and C.J. Robinson, Appl. Phys. Lett. 59, 2112 (1991).

Extraction of a Local Thermal Conductivity in CVD Diamond Films

J.E. Graebner

AT&T Bell Laboratories, 600 Mountain Avenue,
Murray Hill, NJ 07974, USA

One might expect the columnar-grain microstructure of chemical-vapor-deposited (CVD) diamond films to have a strong effect on the thermal conductivity κ. It was recently demonstrated [1] that the local conductivity κ_{local} at room temperature of high-quality microwave-plasma-deposited diamond indeed exhibits a strong gradient with respect to position in the films. κ_{local} was deduced by studying a series of films of different thickness prepared under as identical conditions as possible, the only parameter differentiating the samples being the time of deposition. The local conductivity at height z above the substrate was obtained as the derivative of the measured conductance $Z\kappa_{obs}$: $\kappa_{local}(z) = d\,(Z\kappa_{obs})/dZ$, where the observed conductivity is necessarily averaged over the sample thickness Z by the nature of the measurement. The results, Fig. 1, show that κ_{local} increases by at least a factor of four going from the substrate surface to the top of a thick (~ 400 µm) sample. Here we present the results of such measurements over a wide temperature range, 6-400K, which allows us to extract the phonon scattering strengths as a function of position z within a sample.

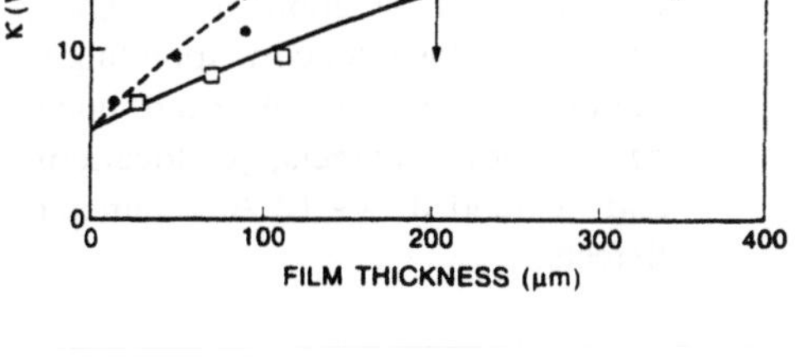

Fig. 1. Measured thermal conductivity κ_{obs} at 25°C for five samples (squares). The local conductivity κ_{local} is derived in two ways: by comparison of pairs of samples of different thickness (solid circles), and by appropriate differentiation (dashed curve) of a low-order polynomial (solid curve) least-squares-fit to the κ_{obs} data.

Springer Series in Solid-State Sciences, Volume 112
Phonon Scattering in Condensed Matter VII Eds.: M. Meissner and R.O. Pohl

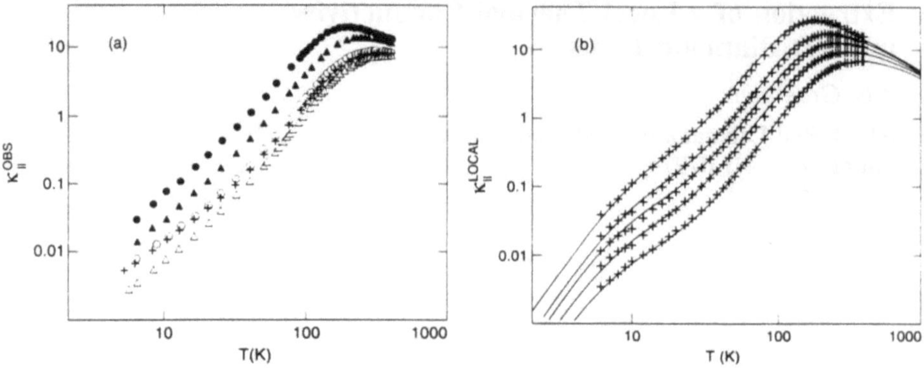

Fig. 2. (a) Measured thermal conductivity vs. temperature for five samples of CVD diamond with thicknesses of 28, 68, 111, 185, and 386 μm. A monotonically increasing conductivity as a function of film thickness is observed at all temperatures. (b) Local thermal conductivity κ_{local} vs. temperature at various values of z, the height above the substrate. The values (crosses) are obtained from the observed conductivity of Fig. 2a by the method described for Fig. 1. The solid lines are fits using the standard model of thermal conductivity.

The κ_{obs} data are shown in Fig. 2a and the calculated $\kappa_{local}(z,T)$ in Fig. 2b for selected values of z. The standard model for thermal conductivity using a Debye phonon spectrum and scattering from grain boundaries, point defects, dislocations, and extended defects [2] has been fit to each of the κ_{local} sets of data. The fitting parameters are shown in Fig. 3 as a function of height z. The physical significance of the absolute values of the scattering strengths is not yet determined, but the variation of each one with z is interpreted as a direct measure of the relative abundance of that scattering entity as a function of z. Figure 3 indicates the relatively greater crystalline perfection of the diamond material near the top of the thickest sample, and it is consistent with the gradient of the local conductivity observed in Fig. 1.

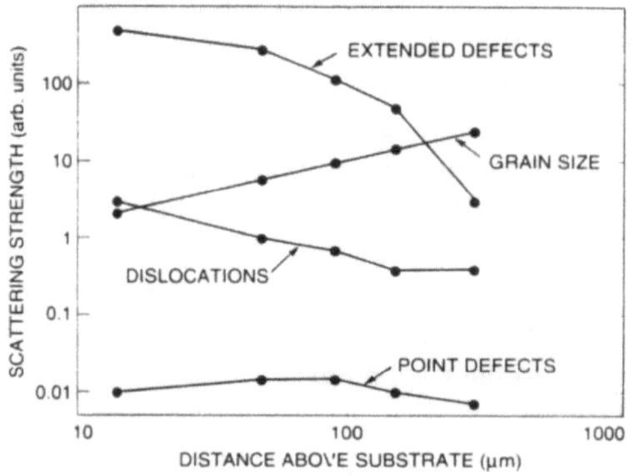

Fig. 3. Parameters of the fits in Fig. 2b as a function of height z above the substrate, corresponding to scattering strengths of grain boundaries, point defects, dislocations, and extended (~ 12 Å diameter) defects.

We are grateful to J. A. Herb and C. F. Gardinier of Crystallume, Inc., for careful preparation of the samples and S. Jin and G. W. Kammlott for analysis of the microstructure.

1. J. E. Graebner, S. Jin, G. W. Kammlott, J. A. Herb, and C. F. Gardinier, Appl. Phys. Lett. **60**, 1576 (1992).
2. J. E. Graebner and J. A. Herb, Diamond Films Technol. **1**, 155 (1992).

Specific Heat and Thermal Conductivity of Solid Fullerenes

J.R. Olson, K.A. Topp, and R.O. Pohl

Laboratory of Atomic and Solid State Physics, Cornell University, Ithaca, NY 14853-2501, USA

The first model for lattice vibrations of solids was proposed in 1907 by Einstein who applied the quantum concept to the mechanical motion of individual atoms in a crystal lattice which he assumed to be vibrating with random phases.[1] He subsequently found that this model was inadequate, its major shortcoming being that it led to a thermal conductivity which disagreed with the observation on crystals both in magnitude and temperature dependence.[2] This disagreement was removed by Debye[3] and by Born and von Karman[4] who demonstrated that in crystalline solids the atoms vibrate collectively as elastic waves.

In this note, we report on a crystal of a rather simple composition whose lattice vibrations are predominantly localized. Above ~4 K, both its specific heat and its thermal conductivity can be quantitatively described with Einstein's model,[1,2,5] thus showing clearly the limitations of the commonly used picture which is based on plane waves.

Starting material was microcrystalline commercial Fullerite powder containing 85% C_{60} and 15% C_{70} clusters (Buckyballs), which was compressed in a pellet press at 3000 Atm, to form pellets of mass density $\rho = 1.54$ g cm^{-3} (±10%), close to the theoretical density (ρ_{theor} = 1.676 g cm^{-3}) of the fcc lattice.

Between room temperature and ~80 K, the specific heat of these compacts is very similar to that of crystalline graphite (Fig. 1). This suggests that the modes excited in this temperature range are internal modes of the individual Fullerenes. Between 80 K and a few Kelvin, the specific heat greatly exceeds that of graphite, and is described with an Einstein specific heat, with the characteristic temperature Θ_E = 35 K and a number density of the modes equal to six times the number density of the Fullerenes. It is suggested that these modes result from vibrational and from torsional motions of the rigid Fullerenes.

This interpretation of the modes is confirmed by thermal conductivity measurements, see Fig. 2. Above 10 K, the conductivity is close to that predicted by Einstein's model of heat transport by randomly oscillating entities (the Fullerenes).

Evidence for collective excitations is found only below a few K. In this temperature range both specific heat and thermal conductivity resemble those of glasses; evidence is seen of elastic waves, with a Debye temperature Θ_D = 72 K, and of tunneling states, leading to a specific heat varying linearly with temperature below 1 K. A thermal conductivity varying as T^2, with magnitude close to that common to all glasses, is also observed. The origin for this glass-like behavior is probably structural disorder in the compacts.

The work was supported by the National Science Foundation, Grant No. DMR-91-15981, and the Cornell Materials Science Center.

Springer Series in Solid-State Sciences, Volume 112
Phonon Scattering in Condensed Matter VII Eds.: M. Meissner and R.O. Pohl

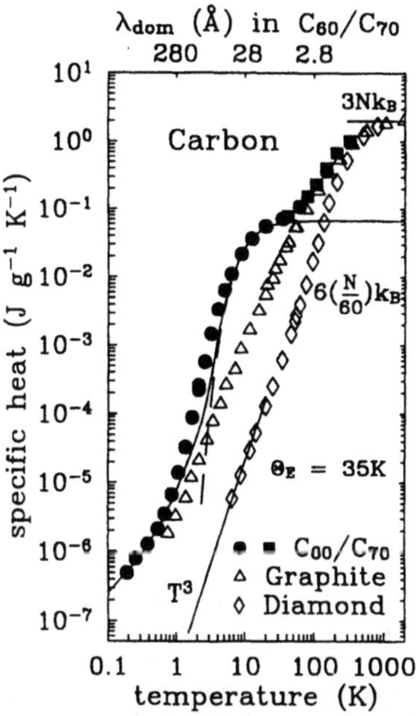

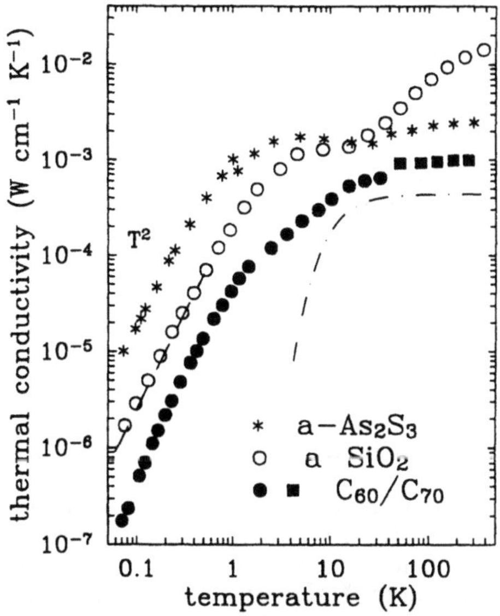

Fig. 1 (left). Specific heat of microcrystalline compacts of C_{60}/C_{70} clusters (Buckyballs) of composition 85%/15%, compared with that of diamond and graphite. Above ~40 K, the specific heat of the C_{60}/C_{70} compacts is close to that of graphite. The dashed line which merges above 10 K with a solid line is an Einstein specific heat with the characteristic temperature Θ_E = 35 K (ν_E = 7.2 × 10^{11}Hz). N is the number density (per unit mass) of carbon atoms, and $N/60$ is the approximate number density of the carbon clusters. $6(N/60)k_B$ is a modified Dulong-Petit specific heat assuming that the rigid clusters perform translational as well as torsional vibrations, with a total of six degrees of freedom. Below ~4 K, the solid line is of the form $c_1T + c_3T^3$. Assuming that the cubic term is caused by waves obeying the Debye law, a Debye temperature Θ_D = 72 K (based on the number density of the clusters) and a velocity v_D = 2.2 × 10^5 cm/sec can be determined. From v_D, the dominant phonon wavelength λ_{dom} in the C_{60}/C_{70} compact has been determined, shown on the upper horizontal scale.

Fig. 2 (right). Thermal conductivity of polycrystalline C_{60}/C_{70} compact compared with that of amorphous solids ; the dashed line shows a T^2 dependence.

References

1. A. Einstein, Ann. Phys. **22**, 180, (1907).
2. A. Einstein, Ann. Phys. **35**, 679, (1911).
3. P. Debye, Ann. Phys. **39**, 789, (1912).
4. M. Born and Th. V. Karman, Phys. Z. **13**, 297, (1912).
5. D.G. Cahill and R.O. Pohl, Solid State Comm. **70**, 927, (1989).

Phonon Scattering by Ultrafine Particulates in SiGe Alloys at High Temperatures

J.W. Vandersande[1], *J.-P. Fleurial*[1], *C.B. Vining*[1], *J. Beaty*[2], *J. Rolfe*[2], *and P. Klemens*[2]

[1]Jet Propulsion Laboratory, California Institute of Technology, Pasadena, CA 91109, USA
[2]Thermo Trex Corp., Waltham, MA 02254-9046, USA

1. Introduction and Theory

Silicon-Germanium (SiGe) alloys are used as thermoelectric materials in space power applications at high temperatures. The material conversion efficiency is unfortunately low so needs to be improved. The most promising approach to improving the figure of merit, Z, and hence the efficiency is to reduce the thermal conductivity of the SiGe alloy. A theoretical model has been developed that predicts that the thermal conductivity of these alloys can be reduced by adding appropriately sized particles (40-200Å in diameter) to act as scattering centers for those phonons that conduct most of the heat through the material [1, 2]. The mean-free-path of the high frequency phonons is reduced by point-defect (mass difference) scattering as a result of the use of a SiGe alloy as the basic material, whereas that of the low frequency phonons is reduced by a combination of grain boundary scattering and the interaction of phonons with electrons (or holes). This leaves unaffected a range of intermediate phonon frequencies to carry most of the heat. Figure 1 shows the calculated thermal conductivity as a function of inclusion concentration. A reduction of up to 40 percent is possible in the case of a 40Å diameter particulate that conducts no heat and has a concentration of about 7 volume percent.

To test this prediction, ultra-fine particles of BN and Si_3N_4 were made by eroding single crystal boron and silicon in a nitrogen atmosphere in an apparatus described elsewhere [3]. The particulates were introduced into ultra-fine p-type doped SiGe powder, uniformly dispersed, and hot pressed at 1525K into the alloy. Samples were heat treated at 1550K to grow the grains from submicron to greater than a micron in size.

2. Results and Discussion

The results of thermal conductivity measurements on a sample with Si_3N_4 particulates (TPF-766) and on one with BN particulates (TPF-790) are shown in Figure 2. Also shown is the thermal conductivity of a p-type SiGe sample made using standard fabrication techniques. The grain size of the standard material is several microns which is similar to the heat treated samples made from ultra-fine powders. The reduction in thermal conductivity (from 15 to 30 percent) of the two samples with the fine particulates can clearly be seen.

Springer Series in Solid-State Sciences, Volume 112
Phonon Scattering in Condensed Matter VII Eds.: M. Meissner and R.O. Pohl

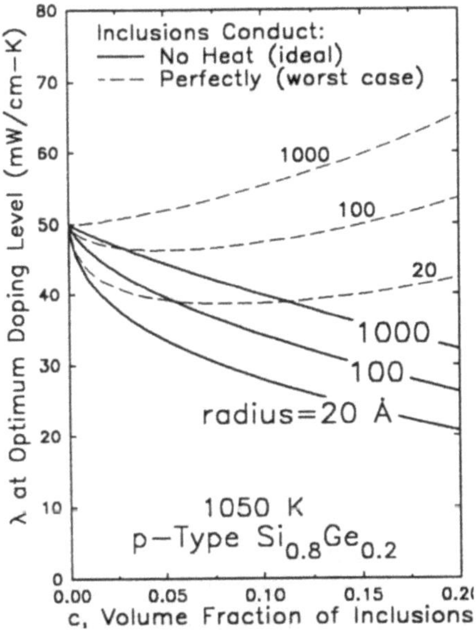

Figure 1: Thermal conductivity of SiGe alloy at 1050K as a function of inclusion concentration

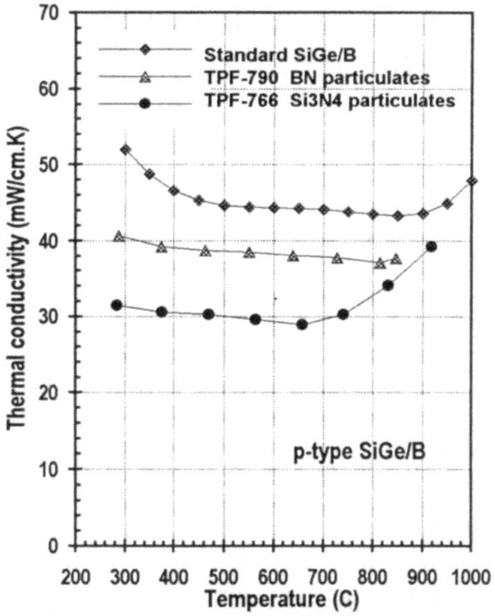

Figure 2: Thermal conductivity of samples with BN and Si_3N_4 particulates compared to a sample of standard p-type SiGe

The carrier concentrations of the samples varied by up to one order of magnitude but calculations showed that the effect of this on the total conductivity is at most a few percent. Sample TPF-790 was examined in the TEM. The grains were found to be around 2 μm in size and BN particulates of 50-200Å in diameter were found inside the grains. The concentration of BN particulates appeared to be at most a few percent which is probably the reason for only a 15 percent reduction in thermal conductivity.

These TEM observations and the measured conductivities are consistent with the theoretical predictions of increased scattering of intermediate frequency phonons, which carry most of the heat in SiGe alloys.

The work described in this paper was performed at ThermoTrex under a contract with the Jet Propulsion Laboratory and at JPL, California Institute of Technology under contract with the National Aeronautics and Space Administration.

3. References

1. Vining, C.B., Mat. Res. Soc. Symp. Proc. 234, pp. 95-104, 1991.
2. White, D.P. and Klemens, P.G., J. Appl. Phys. 71(9) 4258-4263 (1992).
3. Beaty, J.S., Rolfe, J.L. and Vandersande, J.W., Proc. 25th IECEC Conf., Publisher IEEE Publishing service, NY 902379 (1990) pp. 379-382.

Specular Phonon Scattering from Ferroelectric Domain Walls in KDP

M.A. Weilert, M.E. Msall, A.C. Anderson, and J.P. Wolfe

Physics Department and Materials Research Laboratory,
University of Illinois at Urbana-Champaign, Urbana, IL 61801, USA

Measurements of the low temperature (T<3K) thermal conductivity of KH_2PO_2 (KDP) indicate that scattering from domain walls has a large effect on phonon transport. KDP has a ferroelectric phase transition from tetragonal to orthorhombic structure at T_c=123K. Below T_c domains of opposite electric polarization and crystal orientation form unless the sample is cooled in an electric field. Thermal conductivity measured along the [100] (tetragonal) axis shows a 30% drop when domain walls are present, which is independent of sample size and temperature. This independence could be due to polarization dependent scattering at the domain boundaries. To investigate this possibility, we make use of ballistic phonon imaging techniques. [1]

The phonon images in fig. 1 probe directions near [001] and show a large change in the presence of domain walls. The most striking feature of the domained images is the large difference in the effect of the domain walls on different phonon modes. The FT and L structures are relatively unaffected by the domain walls, but there are large changes in the ST structures. Optical observations reveal that the intersections of the domain walls with the (001) surface are lines parallel to the horizontal images axis, y,

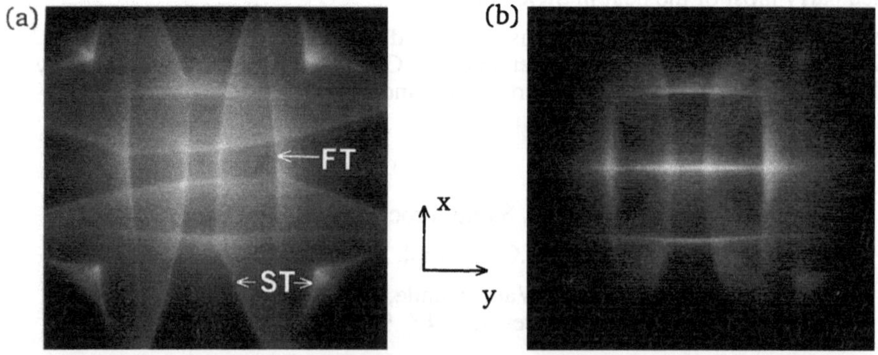

Fig. 1 Phonon images of (001) KDP (a) poled (b) with domain walls

Springer Series in Solid-State Sciences, Volume 112
Phonon Scattering in Condensed Matter VII Eds.: M. Meissner and R.O. Pohl

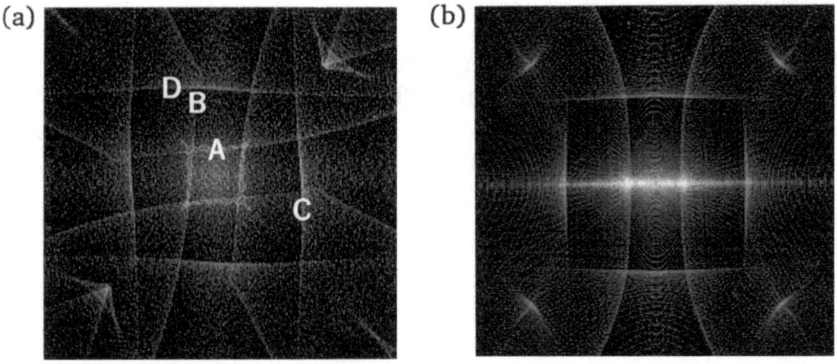

Fig. 2 Monte Carlo image simulations of KDP (a) poled (b) with domains

suggesting that the observed increased in ST phonon flux along y is primarily due to phonons which are reflected at domain walls.

Domain wall interactions are modelled using continuum acoustic reflection/refraction [4] at domain boundaries to determine the associated probabilities for phonon reflection and transmission. The elastic constants used in this calculation were obtained by a simple splitting of the known room temperature tetragonal elastic constants [3] into an orthorhombic set. The probability of reflection events varies with phonon mode and polarization. Table 1 gives a summary of the calculated reflection probability for phonons which arrive at a number of points indicated by letter on Fig. 2a. The table shows that ST phonons with large y-components of polarization can have large reflection probabilities, and will be strongly channelled. For phonons in the FT caustics, the reflection probability is very low, and they are not strongly channelled. Monte Carlo image simulations [2] in Fig. 2 accurately reproduce the features of the experimental images, including the 2-fold symmetry. In summary, we find that scattering by domain walls in KDP is highly polarization dependent and can be modelled by a simple acoustic mismatch model of the domain boundary. This work was supported in part by NSF grants DMR-89-19032 and DMR-89-20538.

Table 1 Phonon reflection probabilities for phonons from points in Fig. 2(a)

Point	Wavevector Components	Polarization Components	% Reflection
A	(.24,-.09,.97)	(-.21,.98,.04)	55.7
B	(.12,.24,.96)	(.99,-.04,-.07)	1.53
C	(-.03,-.83,.56)	(.46,-.38,-.93)	0.03
D	(.83,.03,.56)	(.37,.04,-.93)	0.05

References

1. G.A. Northrop and J.P. Wolfe, Phys. Rev. Lett. **43**, 1424 (1979).
2. G.A. Northrop, Comp. Phys. Comm. **28**, 103-107 (1982).
3. O. Delekta and A. Opilski, Archives of Acoustics **5**, 181-188 (1980).
4. O. Weis, Z. Physik B**34**, 55-63 (1979).

The Heat Pulse Transport in Single-Phase Ceramics

S.N. Ivanov, A.G. Kozorezov, E.N. Khazanov, and A.V. Taranov

Russian Academy of Sciences, Institute of Radioengineering and Electronics, Mokhovaya Str. 11, 103907 Moscow, GSP-3, Russia

The elastic and thermal properties of inhomogeneous materials are known to be quite different from that of homogeneous medium.

We discuss a simple qualitative model of the phonon transport in ceramics. The characteristic time for a given phonon to pass along the path R from one boundary of the grain to the other is either $t_b=R/v_s$, or $t_R \approx R^2/D_1$ dependent on whether the phonon moves ballistically or diffusively (v_s is the sound velocity, D_1 the diffusion coefficient in the grain material). At low temperatures both t_b and t_R are much smaller than phonon inelastic relaxation time τ_p. This suggests a low level of excitation which we assume to be the case. Let t_0 to be the phonon transition time from a given grain to the neighbouring one. If $t_0<\tau_p$ then for $t_0 \ll t \ll \tau_p$ phonon transport looks like hopping conductivity. This results in a diffusive motion of phonons with diffusion coefficient $D \sim R^2/t_0$. If the phonons move ballistically within the grain, then $t_0 \sim RS/v_s\Sigma f_\omega$. f_ω stands for the phonon transmission probability at frequency ω through the contact area and S/Σ is the ratio of grain surface to the total contact area per grain. The bolometer signal reaches it's maximum at $t_m \sim L^2/D \sim L^2 S/\Sigma v_s R f_\omega$, where L is the sample length.

More accurate analysis gives the diffusion coefficient in the form $D_{eff} \equiv D+V_0^2R_0^2/3D$, the expressions for D and the structural term $V_0^2R_0^2/3D$, which accounts for the fluctuations in grain sizes, being the statistically averaged combinations of $p(\upsilon \to \upsilon')$ - the transition probability for phonon to pass from the υ-th to the neighbouring υ'-th grain and $\mathbf{a}_{\upsilon\upsilon'}$ - the vector connecting these grains.

We now discuss the temperature dependence of t_m:

1. **Acoustic mismatch model.** The transition probability does not depend on the phonon frequency or nonequilibrium phonon temperature and t_m=const.

2. **Phonon diffraction.** This affects the phonon transport if the linear size of the contact area r_0 is less than the characteristic phonon wavelength $\overline{\lambda}$. Since $\overline{\lambda}$ gets smaller when as temperature increases, then $\partial t_m/\partial T < 0$.

3. **Defect or glassy contact.** If $r_0 \gg \overline{\lambda}$, then the dependence of transition probability upon $\overline{\lambda}$ might occur due to a large number of defects near the contact, the same is true if the contact area is amorphous. The incident phonon then has a considerable probability of being scattered back. If $\overline{\lambda}$ decreases (T increases) the phonon penetration rate gets smaller and $\partial t_m/\partial T > 0$.

Springer Series in Solid-State Sciences, Volume 112
Phonon Scattering in Condensed Matter VII Eds.: M. Meissner and R.O. Pohl

4.**Fractal contact.** If the contact has fractal structure and characteristic wavelengths $\bar{\lambda}$ correspond to the fracton part of the vibrational states, then phonon transmission through the contact might be due to hopping of localised excitations resulting in $\partial t_m/\partial T < 0$.

The main conclusion is therefore that for different ceramics at low temperatures if phonons move ballistically within grains one can observe any type of behaviour: t_m=const, $\partial t_m/\partial T < 0$ or $\partial t_m/\partial T > 0$.

We have studied experimentally the heat pulse transport in two different ceramics: Zirconium-Titanium-Lead-Lanthanum (PZTL) and Ruby 22HS. The samples were thin, flat, parallel slabs polished to optical precision with thickness 100-200 μ. Thin, ~1000Å, thermally sputtered Au film on one side of the slab was used as a heat pulse generator, while a superconducting In or Sn film at the sample's back was used as a bolometer. The dissipated power P_H in the Au film was less than 10^{-1} Wt/mm^2. The excess generator temperature with respect to heat bath was carefully controlled and did not exceed 0.2K in the whole temperature range of the experiment from 2.2 up to 3.8K.

Fig.1a shows the detected signal S(t) in PZTL. The curves are typical for the diffusion process. The measured dependence $t_m(T) \sim T^n$ with $n \geq 5$ which cannot be attributed to Rayleigh scattering of phonons at contacts is very close to what was observed under the same experimental conditions in glasses [1].Fig. 1b shows the detected signals in 22HS Ruby ceramics. In contrast to PZTL t_m in 22HS Ruby ceramics gets bigger if the temperature increases.

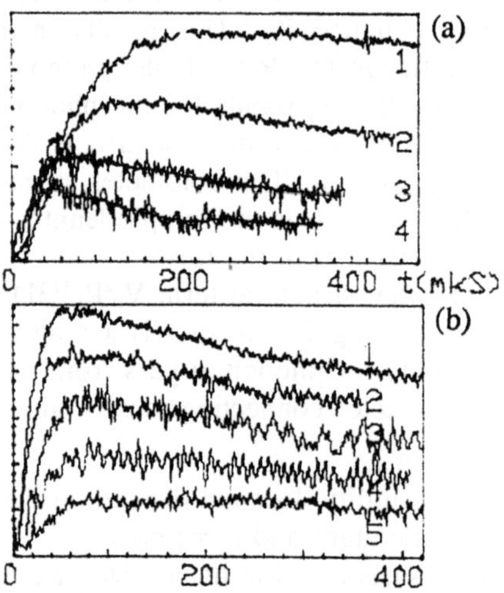

Fig.1:(a) The detected signals in PZTL. L=120μ. 1.T=3.82K; 2.T=3.4K; 3.T=3.01K; 4.T=2.2K. (b) The detected signal in Ruby ceramics. L=140μ. 1.T=3.8K; 2.T=3.65K; 3.T=3.37K; 4.T=3.13K, 5.T=2.44K.

References

1.S.N.Ivanov, E.N.Khazanov and A.V.Taranov, Proceedings. of the 14th USSR Conference on Acoustoelectronics and Physical Acousics in Solid State, Kishinev, **1**, 229, (1989).

Relationship Between the Thermal Conductivity and the Adiabatic Bulk Modulus of Epoxy Resins Loaded with Ceramic Particles

P. Bujard[1] *and A. Kuttler*[2]

[1]Ciba-Geigy SA, Research Center FO 5, CH-1723 Marly, Switzerland
[2]Ciba-Geigy SA, Mathematics and Operation Research Center, IS7.1, CH-4002 Basel, Switzerland

1 Introduction

The importance of heat diffusion manifests itself in connection with heat removal from plastic packaging of active semiconductors devices. The effective thermal conductivity $\lambda_{eff}(C)$ and the adiabatic bulk modulus $K_{eff}(C)$ of an epoxy resin ($\lambda_r = 0.186$ W/mK; $K_r = 6.01$ GPa) has been measured versus the volume loading ratio (hereafter V_r) C (0 to 0.7) of crystalline quartz ($\lambda_f = 14$ W/mK; $K_f = 57$ GPa) and fused quartz ($\lambda_f = 1.4$ W/mK; $K_f = 37$ GPa). Experimental data on boron nitride and aluminum nitride have already been published [1,2]. Effective medium approximations, percolation or lowermost and uppermost limits of Hashin-Shtrikman (H-S) [3] or Milton [4] are all classical models to predict λ_{eff} or K_{eff} of particulate filled plastics (hereafter PFP). Parameter fitting is generally required except for the H-S bounds.

2 Model, Experimental Results and Conclusions.

Experiments on electrically conductive PFP's have indicated the existence of a cluster (a set of particles in close contact) spreading throughout the PFP if $C \geq C_p$ (percolation threshold) [5]. Thus a PFP with $C \geq C_p$ can be visualized as a continuous phase A of thermal conductivity λ_A (cluster $V_r = C_{max}$) containing a discontinuous phase B of thermal conductivity λ_B (epoxy resin containing isolated filler particles $V_r = C_{H\text{-}S}$). C_{max} is the highest possible $V_r \cong$ (0.67 to 0.7), $C_{H\text{-}S} = C$ if $C \leq C_p$ and $C_{H\text{-}S} < C$ if $C > C_p$. The relative volumes C_A and C_B of phases A and B satisfy $(C_A + C_B) = 1$. The probability $P(C) = (C - C_p)^\beta / (C - C_{max})^\beta$ that a particle belongs to the cluster depends on the parameter β. β reflects the concentration dependence of P(C) for a particle to belong to the cluster. β may vary from 0.4 to 1 [5]. C_A, C_B and $C_{H\text{-}S}$ are deduced analytically from C, C_{max}, and β. λ_A is calculated analytically from a model of a chain of spheres in close contact (filler with λ_f) buried in a matrix of epoxy resin [6,7]. λ_A is given firstly by the H-S lower bound if $\lambda_f < 2$W/mK and secondly by $\lambda_A = (4\ln(\lambda_f/\lambda_r) - 7.1) \cdot \lambda_r$ if $\lambda_f > 5$ W/mK [7]. λ_B is given by the H-S lower limit of λ_f, λ_r and $C_{H\text{-}S}$.

$\lambda_{eff}(C)$ is given by the H-S uppermost limit of λ_A, $\lambda_B \leq \lambda_A$ and C_B. The same procedure is used for K_{eff}. The present model yields λ_{eff} and K_{eff} versus C within ±5% of the experimental data [7]. $\lambda_{eff}(\lambda r; C=\text{const.})$ can also be calculated within ±5% and experimentally $d(\lambda_{eff})/d(\lambda r) = \exp(2.77 \cdot C)$ for $\lambda_r \subseteq (0.15, 0.25)$ W/mK. Figure 1 shows K_{eff} (left scale) and C (right scale) versus λ_{eff}.

Springer Series in Solid-State Sciences, Volume 112
Phonon Scattering in Condensed Matter VII Eds.: M. Meissner and R.O. Pohl

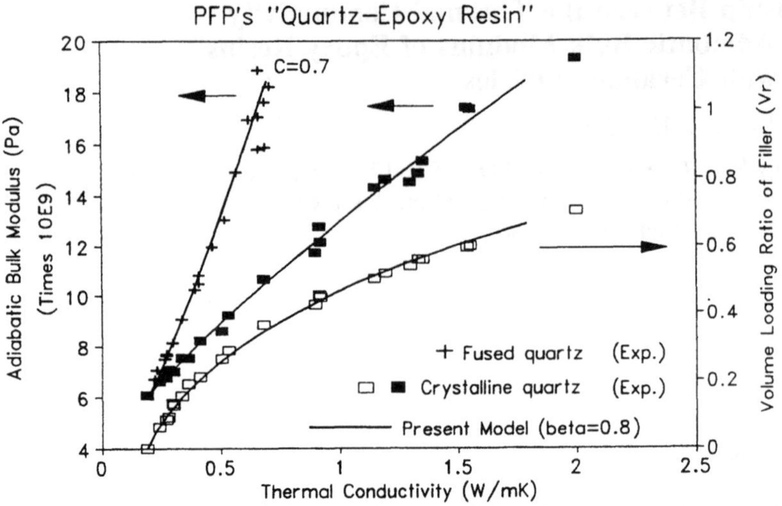

Figure 1: Adiabatic K_{eff} versus λ_{eff} of PFP's containing crystalline and fused quartz. K_{eff} is deduced from sample density and ultrasonic sound velocity (10MHz). To a first approximation K_{eff} appears directly proportional to λ_{eff}.

The extreme limits of the H-S uppermost and lowermost bounds are useful only if λ_A is close to λ_B [3], which is the case for the cluster (phase A) and its inclusions of phase B [7]. A value of Cp in the range 0.05 to 0.15 has a negligible influence on the calculated $\lambda_{eff}(C)$. A value of β in the range 0.4 to 1 alters λ_{eff} or K_{eff} within ±5%. The fact that λ_A is equal to λ_{eff} of a chain of spheres in close contacts (V_r=0.67) is the major result of the present investigation. Numerical analysis shows that most of the heat flow between two spheres in contact occurs within a solid angle of 15° as measured from the sphere centers [7]. It explains why the choice of the grade of quartz powder has no influence on the experimental λ_{eff}. Effectively, the number and distribution of contact points on a particle surface should not vary much for different particle shapes provided they are rounded and globular. The present model has the advantage to take into account that λ_r is not negligible with respect to λ_f. The temperature dependence of λ_{eff} can also be accounted for. Investigations on AlN and Al_2O_3 as fillers have led exactly to the same results.

3 References

1. P.Bujard,Proceedings I-Therm 1988, Los-Angeles, May 11-13, pps 41-49.
2. P.Bujard and JP.Ansermet, Proceedings SEMI-THERM 1989, San-Diego, February 7-9, pps. 126-130.
3. Z.Hashin, J.Appl.Mechanics, Vol.50, 1983, pps. 481-505.
4. GW.Milton, Phys.Rev.Letters, Vol.46, No.8, 1981, pps. 542-545.
5. D.Stauffer,"Introduction to Percolation Theory", Taylor and Francis, 1985.
6. GK.Batchelor and RW.O'Brien, Proceed.Royal.Soc.Lond.A355, 1977, pps. 313-333.
7. P.Bujard, D.Randazzo and A.Kuttler, to be published

Contribution of Phasons in the Thermal Conductivity of Charge-Density-Wave Materials

A. Smontara[1], *K. Biljaković*[1], *and S.N. Artemenko*[2]

[1]Institute of Physics of the University, P.O. Box 304, 41001 Zagreb, Croatia
[2]Russian Academy of Sciences, Institute of Radioengineering and Electronics, 103907 Moscow, Russia

We report on thermal conductivity λ of the inorganic quasi-one-dimensional charge-density-wave conductors (CDW) $K_{0.3}MoO_3$ and $(NbSe_4)_{10}I_3$ between 80K and 320K. The technique we used is the standard four contact method, relative to the constantan foil [1]. Both free carriers and phonons contribute to the total thermal conductivity λ. The free carriers contribution to the thermal conductivity λ_{fc} in CDW semiconductors for which the effective gap is temperature dependent is given by [2]

$$\lambda_{fc}=\frac{\pi^2}{3}\left(\frac{k_B}{e}\right)^2\cdot\sigma T+\frac{b}{(b+1)^2}\left(\frac{E_g}{k_BT}+4\right)\left(-T\frac{d(E_g/k_BT)}{d(T)}+4\right)\left(\frac{k_B}{e}\right)^2\cdot\sigma T$$

where e is the electronic charge, σ the electrical conductivity, E_g the temperature dependent effective Peierls gap [1], and b the electron-hole mobility ratio. The remained lattice thermal conductivity of $K_{0.3}MoO_3$ and $(NbSe_4)_{10}I_3$ is shown in Fig. 1. In both of cases there is a well defined minimum in the temperature region of the Peierls transition and a weak anomaly at T_c (Peierls transition temperature). The "usual" phonon contribution is known to obey the exponential law due to the Umklapp processes, $\lambda_u = A\exp(\theta_D/\alpha T)$. Such a variation (given by dashed line in Fig. 1) is followed for $K_{0.3}MoO_3$ and $(NbSe_4)_{10}I_3$ from 80K up to the well defined minimum at 20-30K below T_c (Fig.1a). Beyond this phonon contribution an extra thermal conductivity $\Delta\lambda_p$ is found below T_c (inset in Fig. 1a). The conductivity in exceees of the phonon and free carriers contribution may be naturally associated with excitations of the CDW systems, mainly phasons due to their high sound velocity. Their spectrum is strongly affected by the long-range Coulomb interactions [3]. Indeed, the

Springer Series in Solid-State Sciences, Volume 112
Phonon Scattering in Condensed Matter VII Eds.: M. Meissner and R.O. Pohl

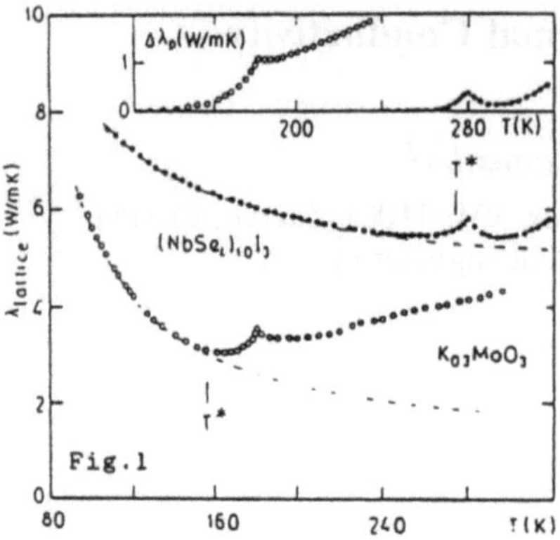

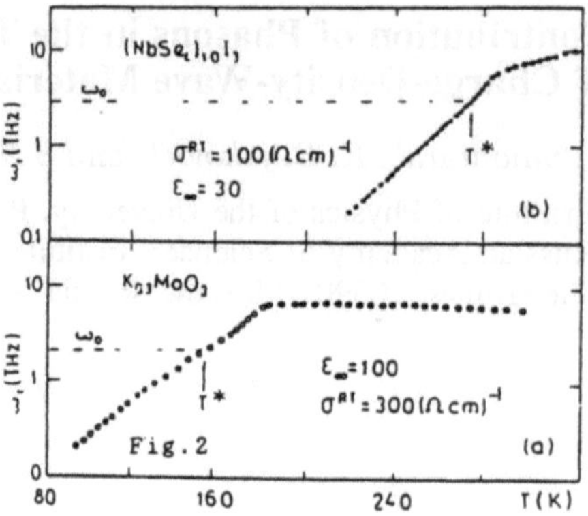

Fig.1 The lattice thermal conductivity vs T for $K_{0.3}MoO_3$ (o) and $(NbSe_4)_{10}I_3$ (•). The dashed line is the estimated regular contribution due to Umklapp processes. The inset shows the extra thermal conductivity attributed to the phase excitations.

Fig.2 The frequency of dielectric screening ω_r vs T for $K_{0.3}MoO_3$ (a) and $(NbSe_4)_{10}I_3$ (b) gives the temperature region for the phason thermal conductivity.

effect of screening by the quasiparticles enables the phason to contribute to the thermal conductivity at the temperature $T>T^*$ at which the frequency of dielectric screening $\omega_r=4\Pi\sigma/\varepsilon_\infty$ (σ is the electrical conductivity and ε_∞ the dielectric constant) is larger than the phason frequency in the unscreened regime ω_o (at T=0) as shown on Fig. 1b by arrows. The region of the thermal conductivity minimum covers the temperature range around T^* in which the condition $\omega_o \approx \omega_r$ is reached for both systems giving an convincing argument on the existence of the phason contribution to thermal condictivity.

[1] A. Smontara, K. Biljaković, L. Fórro and F. Lévy, Physica 143B, 264 (1986); A. Smontara, *Ph. D. thesis*, University of Zagreb, 1991, (unpublished)

[2] A. Smontara, K. Biljaković, and S. N. Artemenko, to be published.

[3] S. N. Artemenko and A. F. Volkov , in *Charge Density Waves in Solids* , ed. by L. P. Gor'kov and G. Grüner, (Elsevier Science Publishers B.V., 1989), p. 365.

Additional Scattering from a Low-Lying Mode in the Phonon Thermal Conductivity of the Quasi One-Dimensional Compound $(TaSe_4)_2I$

J.C. Lasjaunias[1], *A. Smontara*[2], *P. Monceau*[1], *and F. Lévy*[3]

[1]Centre de Recherches sur les Très Basses Températures, Laboratoire associé à l'Université Joseph Fourier, C.N.R.S., B.P. 166, F-38042 Grenoble Cédex 9, France

[2]Institute of Physics of the University, P.O. Box 304, 41001 Zagreb, Croatia

[3]Institut de Physique Appliquée, Ecole Polytechnique Fédérale de Lausanne, CH-1015 Lausanne, Switzerland

We report on the thermal conductivity κ of the quasi-1D compound $(TaSe_4)_2I$ which is characterized by the presence of a charge density wave (CDW) below the Peierls transition at $T_P \simeq 260$ K, and we compare it to a compound of similar structure, but without a CDW, $(NbSe_4)_3I$ [1]. The technique we used is the usual steady heat flow technique with one heater and two thermometers, the flow propagating along the chains (c-axis). Thermal conductivity of a single crystal with perfect crystallographic surfaces is reported in Fig. 1. It exhibits an unusual behaviour for an insulating single crystal. At low temperature, $\kappa(T)$ follows a classical T^3 law ($\simeq 18\,T^3$ $W.m^{-1}.K^{-1}$) which is in good numerical agreement with sound velocity obtained from the phonon specific heat or by acoustic or neutron measurements, and the mean width of the sample ($\simeq$ 0.7 mm). Above 1 K, κ is usually dominated by scattering by impurities and structural defects (dislocations, point defects), but in the case of $(TaSe_4)_2I$ a strong additional phonon scattering occurs which induces a sudden change of κ at 1.2 K, which decreases towards a broad minimum around 3 K. This property is peculiar to this compound among other quasi-1D materials : $(NbSe_4)_3I$ [1], $K_{0.3}MoO_3$ [2]. For example, in the Nb compound, we measured a $\kappa(T)$ which progressively deviates above 1 K from a low-T $T^{2.7}$ limit regime. We ascribe this anomaly to a strong phonon scattering due to low-lying transverse acoustic (TA) modes characterized by a very weak dispersion over a large part of the Brillouin zone and which lie at an unusual low frequency $\nu \simeq 0.15$ THz (see inset of Fig. 2) [3,4].

Usual kinetic formula gives :

$$\kappa = \frac{1}{3} C_{ph}\,\overline{v}^2\,\tau \qquad \text{with} \quad \tau^{-1} = \tau_B^{-1} + \tau_{TA}^{-1}$$

with B referring to boundary scattering, and TA to the second mechanism. We establish that τ_{TA} is proportional to phonon occupation number of this TA branch, that means a propagative contribution for the initial part of the branch ($q \leq \eta q_{BZ}$) and a purely scattering contribution for the flat part (for $q \geq \eta q_{BZ}$, $\eta \simeq 0.3$ from neutron data, q_{BZ} the average zone-boundary wave vector). In addition, we take into account that the frequency of this branch remains within two extreme values ν_o = 0.15 THz to $\nu_o+\delta \simeq 0.19$ THz, when q explores the basal plane between [100] and [110] [4]. Resulting fit, where parameters are ν_o, δ, η and N the total

Springer Series in Solid-State Sciences, Volume 112
Phonon Scattering in Condensed Matter VII Eds.: M. Meissner and R.O. Pohl

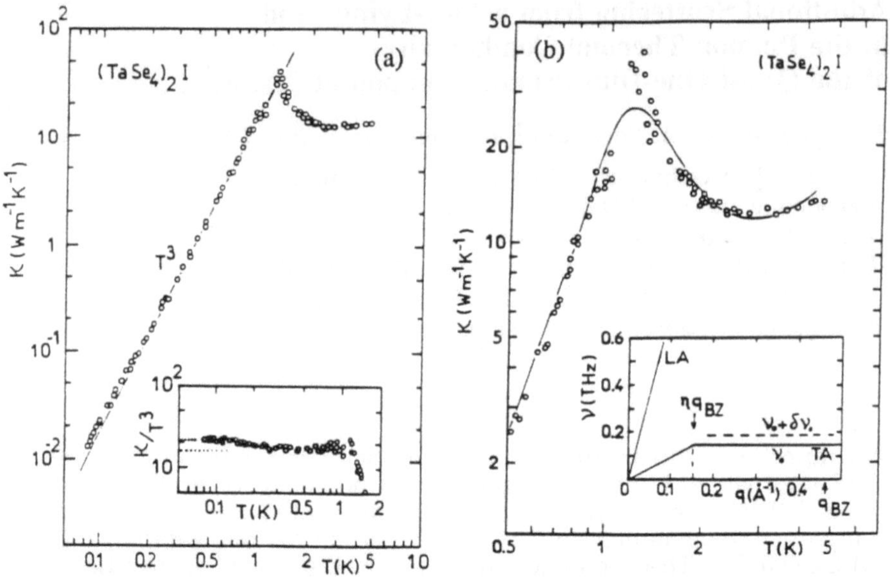

Fig. 1 : a) Thermal conductivity, κ, of quasi-1D compound $(TaSe_4)_2I$. Below 1.2 K κ varies closely to a cubic law, as indicated in the inset. Above 1.2 K a crossover between two different scattering regimes occurs.

b) One possible fit of experimental data above T = 0.5 K, ascribed to a strong scattering of phonons by the low-lying transverse acoustic modes as the origin of the deep minimum of κ around 3 K. The inset shows a schematic representation of dispersion curves of TA modes from ref. [3,4] with polarization along the c^*-axis.

number of modes in the TA branch, is given in Fig. 1b. A good fit is only possible for values of parameters ν_0, δ and η which are very close to data of neutron scattering ; only N is a free parameter. However, this model is unable to correctly describe the sharp maximum at T = 1.2 K and which is possibly due to some other mechanisms.

Finally, the unusual dispersionless low-lying TA branch is a specific property of the quasi-1D $(TaSe_4)_2I$: it reflects a strong decoupling of the chains which vibrate independently one from another at this characteristic low frequency ν_0. In a more general sense, the almost non-propagative excitations of this branch may be reminiscent of localized low-frequency modes due to defects, at the origin of a largely depressed thermal conductivity in crystalline materials in this T-range.

[1] A. Smontara, J.C. Lasjaunias, P. Monceau and F. Lévy, to be published.

[2] K.J. Dahlhauser, A.C. Anderson and G. Mozurkewich, Phys. Rev. B **34**, 4432 (1986).

[3] H. Fujishita, S.M. Shapiro, M. Soto and S. Hoshino, J. Phys. C : Solid State Phys. **19**, 3049 (1986).

[4] J.E. Lorenzo, R. Currat, P. Monceau *et al.*, to be published.

Photothermal Study of Anisotropic Media

B. Perrin[1], *X. Quelin*[1], *B. Mangeot*[2], *G. Louis*[1], *and P. Peretti*[2]

[1]Département de Recherches Physiques, CNRS UA71,
Université Pierre et Marie Curie, B.P. 136, T22-E4,
4 Place Jussieu, F-75252 Paris Cédex 05, France
[2]Laboratoire de Physique des Phases Condensées,
Université René Descartes, F-75270 Paris Cédex 06, France

Photothermal techniques are based on the illumination of the sample under study by a modulated light beam (pump beam). In absorbing media, light is converted into heat and induces thermal waves within the sample. This heat flow produces a refractive index gradient in the adjacent air medium, close to the sample surface. In the probe beam deflection technique [1], a second laser beam is directed through the heated region and is deflected by the *mirage effect*. The analysis of this deflection gives the temperature distribution in the sample and information about its thermal properties. The 3d photothermal deflection has been extensively reported in the literature for isotropic media [2] but only a few results are related to anisotropic ones [3].

A theoretical and experimental photothermal analysis of an orthorhombic medium is reported in this paper. The thermal conductivity tensor was measured in a polydiacetylene monocrystal [4] (thickness $d = 50\mu m$) and the case of a very thin film ($5000\mathring{A}$) of this material was also investigated.

The experimental setup (Fig. 1) is described in ref. [5] : the pump beam is an 8 mW argon laser beam tuned on 488 nm wavelength and modulated in the low frequency range $[0.1-10]Hz$. The probe beam deflection is detected with a position sensor and measured by a lock-in amplifier.

The 3d thermal conduction equation can be written as

$$\kappa_{//}\frac{\partial^2 T}{\partial x_1^2} + \kappa_{\perp}\frac{\partial^2 T}{\partial x_2^2} + \kappa_z\frac{\partial^2 T}{\partial z^2} + i\rho C\omega T = -\frac{\tau}{2\pi r^2}exp\left(-\frac{x_1^2+x_2^2}{2r^2}\right)exp(-\beta z),$$

in the reference frame $(//, \perp, z)$ for an orthorhombic system . $\kappa_{//}, \kappa_{\perp}$ and κ_z are the three components of the thermal conductivity tensor ; ρ, C, β are respectively the density, the specific heat and the optic absorption of the sample ; ω is the modulation frequency of the pump beam of radius r. The term on the RHS describes the heat source which keeps the cylindrical symmetry of the gaussian pump beam. In order to obtain the distribution temperature within the sample, we used a 2d Fourier transform in the $(//, \perp)$ plane taking account of the boundary conditions at the interfaces between the sample and the air. However, we are only interested by the deflection signal of the probe beam which is given by the integration of the temperature gradient along the whole optical path (y axis) on the sample surface ($z = 0$). The sample, being much larger than the thermal length, can be considered

Springer Series in Solid-State Sciences, Volume 112
Phonon Scattering in Condensed Matter VII Eds.: M. Meissner and R.O. Pohl

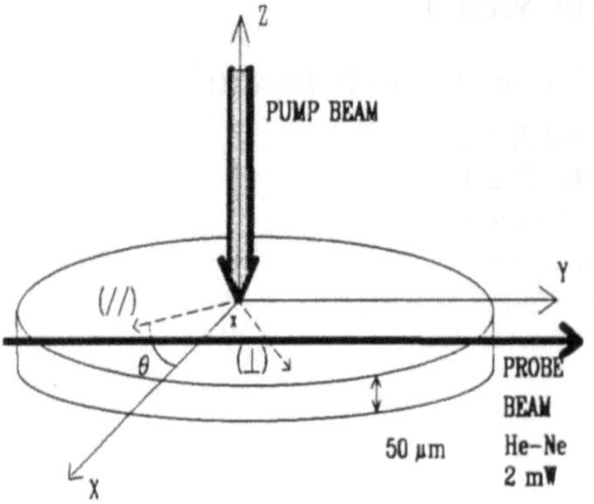

Fig.1 - Experimental configuration : // refers to the polymer chains x is the scanning direction

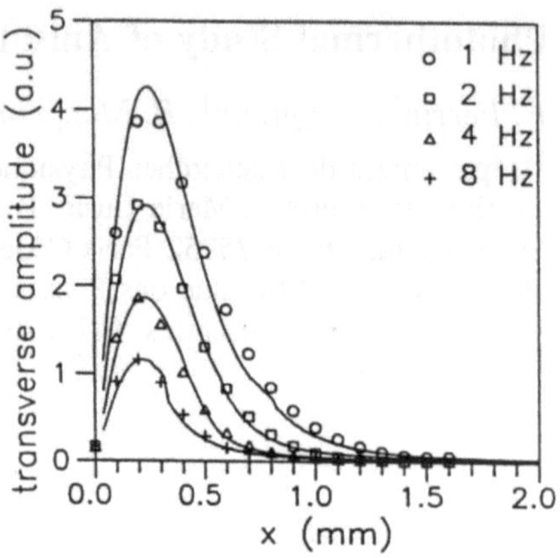

Fig.2 - Variation of Φ_t in the θ=90° direction ; fits (full lines) done with $\kappa_z = 0.05, \kappa_\perp = 0.5W/(m \times K)$

as infinite in the plane $(//, \perp)$: under such conditions we showed that the deflection angle is given by the same expression as in an isotropic medium [2] provided that an effective thermal conductivity $\kappa_\theta = cos^2\theta\kappa_{//} + sin^2\theta\kappa_\perp$, relative to the scanning direction x, and a rescaled sample thickness $\tilde{d} = \sqrt{\kappa_\theta/\kappa_z}d$, be defined. For our optically opaque medium $(\beta\tilde{d} \gg 1)$, the transverse component of the deflection beam is given by

$$\Phi_t(x) \sim \int dk k sin(kx) \frac{1 + exp(-2\sigma\tilde{d})}{1 - exp(-2\sigma\tilde{d})} \frac{1}{\sigma} \quad , \quad \sigma^2(k) = k^2 - 2il_\theta^{-2},$$

where $l^2 = 2\kappa/(\rho C\omega)$ defines the thermal length. In the asymptotic limits of a thermally thin $(\tilde{d} < l_z)$ and a thermally thick $(\tilde{d} > l_z)$ sample, the integrand, proportional to σ^{-2} and σ^{-1} respectively, is independent from κ_z and the value of κ_θ can be derived from the x dependence of Φ_t. Varying the frequency modulation in the range $[1-8]Hz$, the crossover between the two regimes is observed (Fig. 2) and an estimation of κ_z can be obtained.

The anisotropy was clearly seen in the $50\mu m$ thick sample $(k_{//} \simeq 2, k_\perp \simeq 0.5W/(m \times K))$ and to a lesser extent, in a thin film (5000Å) deposited on a pyrex substrate. With a good bonding, the heat would be carried mainly by the pyrex, therefore the anisotropic behavior shows a large thermal mismatch between the film and the substrate.

References

1. A.C. Boccara, D. Fournier, J. Badoz, Appl. Phys. Lett. **36**, 130 (1980)
2. A.Salazar, A.Sanchez-Lavega, J.Fernandez, J.Appl.Phys. **65**, 4150 (1989)
3. X.Quélin, C.Lapersonne-Meyer, J.Berrehar, G.Louis, P.Peretti, in Photo-acoustic and Phothermal Phenomena III, Springer Verlag (in press)
4. sample provided by C.Lapersonne and J.Berrehar, GPS, Paris 7 University
5. A.Hadj-Sahraoui, G.Louis, B.Mangeot, P.Peretti, J.Billard, Liq. Cryst. **5**, 579 (1989)

Recent Work with and Developments of the PTHS-(Pulse Transient Hot Strip) Technique

S.E. Gustafsson and E. Karawacki

Department of Physics, Chalmers University of Technology, S-41296 Gothenburg, Sweden

Abstract. The PTHS-method is a direct development of the Transient Hot Strip technique and specifically designed for thermal conductivity and thermal diffusivity measurements of non-conducting surface layers produced by PVD or similar techniques. In measurements performed so far an evaporated "micro-strip" of a Nobel metal has been used both as heat source and as temperature sensor.

Measurements have been performed on Vitreous Silica produced by PVD, Intrinsic Gallium Arsenide, Optical Glass and Liquid Crystal. For these measurements the probing depth has been estimated to vary from one to ten micrometers.

Springer Series in Solid-State Sciences, Volume 112
Phonon Scattering in Condensed Matter VII Eds.: M. Meissner and R.O. Pohl

Part II

Ballistic Propagation and Imaging

Phonon Imaging by Electron Beam Scanning

R.P. Huebener

Physikalisches Institut, Lehrstuhl Experimentalphysik II,
Universität Tübingen, W-7400 Tübingen, Fed. Rep. of Germany

1. Introduction

Following the pioneering experiments of von Gutfeld and Nethercot [1] more than 25 years ago, the propagation of heat pulses in solids at low temperatures has been intensively studied. In these experiments a region near the crystal surface is locally heated and the transport of this thermal energy across the crystal is measured with high time resolution. This energy transport is usually associated with phonons which propagate through the crystal ballistically and/or diffusively. Of course, the magnitude of the diffusive component depends upon the purity of the sample. Typical phonon frequencies are in the gigahertz to terahertz range. Following their passage through the crystal, the phonons are registered with a sensitive phonon detector.

The role of heat pulse experiments in phonon physics has expanded considerably, when scanning techniques have been applied for moving the phonon source two-dimensionally across the sample surface. Using a phonon detector of small area attached to the opposite sample surface, phonon transport through the crystal could now be studied as a function of its direction of propagation. In this way crystal properties could be imaged, and this technique is now referred to as "phonon imaging".

The field of phonon imaging has been started by Wolfe and coworkers who used laser beam scanning for moving the phonon source across the sample surface. Summaries of this approach and the important references can be found in a series of reviews [2-4]. Of course, for phonon imaging in principle any other scanning technique can be utilized which locally deposits energy at the crystal surface. Up to now, in addition to laser beam scanning, electron beam scanning has been used extensively. For both techniques adequate instrumentation is commercially available (the laser scanning microscope being of more recent origin than the scanning electron microscope). Of course, in both cases the scanning apparatus must be equipped with a low-temperature stage.

In this paper we discuss the role of electron beam scanning for phonon imaging. We only cover a few main points and refer to recent summaries for more details and additional references [5-7].

Springer Series in Solid-State Sciences, Volume 112
Phonon Scattering in Condensed Matter VII Eds.: M. Meissner and R.O. Pohl

2. Techniques

The principle of phonon imaging by electron beam scanning is shown in Fig. 1. The crystal to be investigated is mounted on the low-temperature stage of the scanning electron microscope (SEM) in such a way that its bottom side carrying the phonon detector is in direct contact with the liquid-He bath. The top side of the crystal is exposed to the vacuum of the beam chamber of the SEM and can be scanned directly with the electron beam. At the coordinate point of the beam focus the crystal is heated locally by means of the dissipated beam power (typically 0.01 - 1 mW). The local region heated by the beam acts as phonon source emitting phonons according to Planck's radiation law. For a typical effective source temperature $T^* = 10$ K the emitted phonons are concentrated around 500 GHz. The phonon detector attached to the opposite surface of the sample crystal usually consists of a superconducting thin-film bolometer or tunnel junction of small area. The operating temperature of the experiment can be finely adjusted for maximum detector sensitivity. By proper beam modulation signal detection with a lock-in amplifier becomes possible. However, only the time-integrated behavior can be found from this method. On the other hand, high time resolution, say, for discrimination of the different phonon modes, can be achieved using boxcar averaging. Computer control of the whole experiment and digital image processing has become an important feature of phonon imaging.

Of course, maximum angular resolution requires a small effective area of phonon source and detector. The source area is determined by the thermal sample properties. (The diameter of the electron beam can easily be made smaller than 1 μm and does not play a major role). By means of high-frequency beam modulation the modulated source area can be considerably reduced because of the thermal skin effect. [8] So far modulation frequencies up to about 20 MHz have been used, yielding an effective source diameter of typically 20 μm. Of course, the increase in modulation frequency for obtaining better resolution is limited by

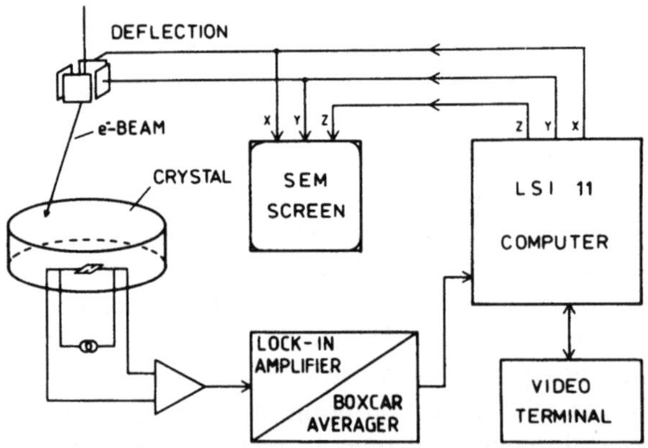

Fig. 1: Principle of phonon imaging by electron beam scanning

the fact that simultaneously the magnitude of the modulated signal is reduced. The superconducting thin-film phonon detectors are usually deposited directly on the bottom side of the sample crystal. Standard techniques for microfabrication presently yield an effective detector area as small as about 1 x 1 μm^2. Hence, the angular resolution of phonon imaging by means of electron beam scanning is limited by the effective phonon source area. Phonon imaging experiments performed with electrically insulating crystals benefit considerably from a thin metallic overlay film on the top sample surface, since in this way charging effects due to beam irradiation can be eliminated.

3. Phonon Focusing

By electron beam scanning the phonon focusing effect has been imaged for dielectric (α-quartz, sapphire) and semiconducting (Ge, Si, GaAs) single-crystals.[7] The phonon focusing effect refers to the anisotropy of the ballistic phonon energy flux through the crystal caused by the anisotropy of the elastic crystal properties. By comparing the measured phonon focusing patterns with theoretical calculations, extremely sensitive information on the elastic constants for the different crystal directions is obtained. From time-resolved measurements phonon focusing images of the individual phonon modes can be found. [9] Of course, the time-resolved experiments require a careful evaluation of the spatial extension of the hot phonon source region and of the corresponding separation of the phonon transit time into a diffusive and a ballistic part. [10]

In addition to the long-wavelength limit of the phonon focusing images, one can study dispersive effects expected at high phonon wave vectors. Superconducting tunnel junctions for phonon detection are well suitable for studying such dispersion since in this case the detector signal is caused by the breaking of Cooper pairs due to the phonons. This results in an energy threshold of the detector given by twice the energy gap of the superconducting electrodes of the tunnel junction. Dispersion of the phonon focusing pattern has been clearly observed in [100] GaAs. [11] From a detailed comparison of the experimental results with the calculations of Tamura et al. [12] the dominant frequency range contributing to the detector signal could be found.

The influence of phonon scattering on the phonon focusing effect represents a highly interesting subject. In Fig. 2 we show the phonon focusing pattern of two [001] oriented semi-insulating GaAs samples with the same thickness of 2.0 mm. [13] Both samples differ strongly in impurity concentration. The sample on the left is undoped and contains a low residual impurity concentration of 2-3 x 10^{15} cm^{-3} of C-atoms. The sample on the right is Cr-doped for compensating the higher residual impurity concentration of 2-3 x 10^{16} cm^{-3} of Si-atoms. One can clearly see how the distinct features of the phonon focusing pattern are washed out in the sample with less purity on the right. Of course, in highly impure crystals diffusive phonon propagation is increased at the expense of the ballistic one.

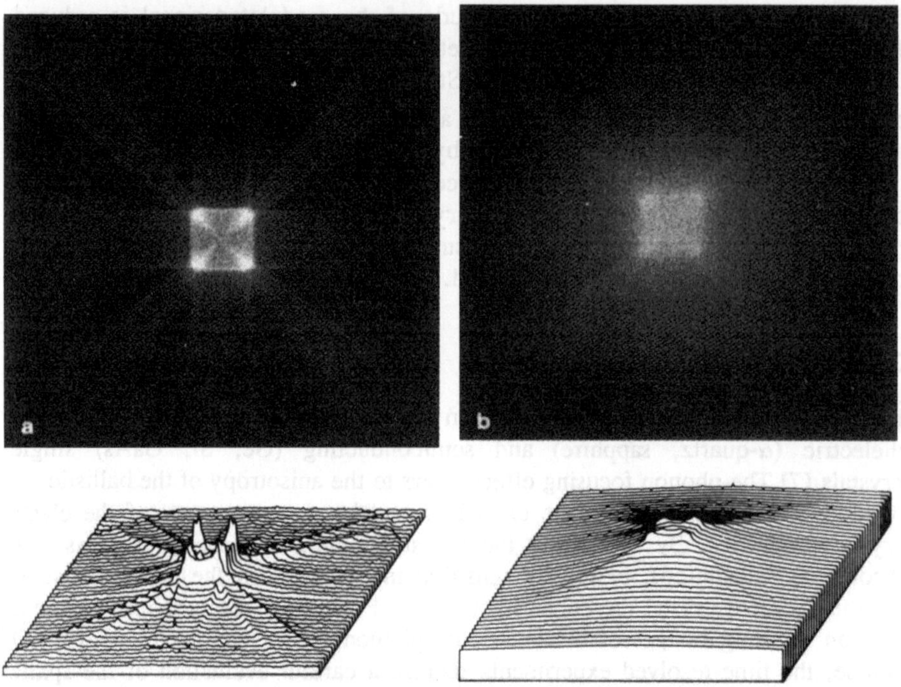

Fig. 2: Phonon focusing image of [001] oriented semi-insulating GaAs for a sample with low (left) and high (right) impurity concentration. Scanned area is 3.2 mm x 3.2 mm. Bottom part shows pseudo-3D plot of the images at the top. Further details are given in the text.

The phonon mean free path between two scattering events can be quantitatively determined using a Monte Carlo simulation for calculating the angular variation of the phonon intensities including the scattering processes. One only has to vary the phonon mean free path in these calculations and to compare the theoretical and experimental results. Further details on this approach can be found elsewhere. [13]

We see that a quantitative study of phonon focusing provides information on the crystal quality which can be particularly important for evaluating semiconductor materials.

4. Defect Structures and Acoustic Tomography

Phonon imaging can also be used for detecting defect structures and spatial inhomogeneities within the crystal. The defects are registered by means of the acoustic shadow they project on the location of the phonon detector. With the ballistic phonons being in the 100 GHz range, the phonon wave length is as short as only a few ten nm, and the acoustic shadow is usually defined by the principles of geometric optics. If two separate small-area phonon detectors are employed, stereoscopic imaging of the defect structure is possible. Of course, in the case of two or more detectors the general principles of tomography can be applied for the three-dimensional reconstruction of the defect structure in the crystal. The successful application of these ideas has been demonstrated using laser drilled holes with a diameter of 100 μm or smaller in single crystals of α-quartz and sapphire. [14-16] Of course, computer control and digital image processing are highly important for this acoustic tomography.

References

[1] R. J. von Gutfeld and A. H. Nethercot, Phys. Rev. Lett. 12, 641 (1964)

[2] J. P. Wolfe, Physics Today, December 1980, p. 1

[3] G. A. Northrop and J. P. Wolfe, in Proc. NATO Advanced Study Institute on Nonequilibrium Phonon Dynamics, W. E. Bron, ed., Les Arces, Plenum, New York, 1985

[4] J. P. Wolfe, Advances in Solid State Physics, Vieweg, Braunschweig, Vol. 29, p. 75 (1989)

[5] R. P. Huebener and W. Metzger, Scanning Electron Mciroscopy, 1985, II, p. 617

[6] R. P. Huebener, Advances in Electronics and Electron Physics, P. W. Hawkes, ed., Academic Press, San Diego, Vol. 70, p. 1 (1988)

[7] R. P. Huebener, E. Held, and W. Klein, Materials Science and Engineering B5, 157 (1990)

[8] J. R. Clem and R. P. Huebener, J. Appl. Phys. 51, 2764 (1980)

[9] R. Eichele, R. P. Huebener, and H. Seifert, Z. Phys. B - Cond. Matter 48, 89 (1982)

[10] W. Kieninger, W. Metzger, and R. P. Huebener, Appl. Phys. A41, 179 (1986)

[11] H. Kittel, E. Held, W. Klein, and R.P. Huebener, Z. Phys. B - Cond. Matter 77, 79 (1989)

[12] S. Tamura and T. Harada, Phys. Rev. B32, 5245 (1985)

[13] E. Held, W. Klein, and R. P. Huebener, Z. Phys. B - Cond. Matter 75, 17 (1989)

[14] W. Klein, E. Held, and R. P. Huebener, Z. Phys. B - Cond. Matter 69, 69 (1987)

[15] E. Held, W. Klein, and R. P. Huebener, Z. Phys. B - Cond. Matter 75, 223 (1989)

[16] E. Held, Thesis, University of Tübingen, 1989 (unpublished)

Imaging of Coherent Phonons

J. Wesner, K.U. Würz, K. Hillmann, and W. Grill

Physikalisches Institut der Johann Wolfgang Goethe-Universität,
Robert-Mayer-Str. 2–4, W-6000 Frankfurt, Fed. Rep. of Germany

Formerly used techniques of investigating acoustic anisotropy in crystals were based on thermally excited incoherent phonons in the THz-range at liquid helium temperatures [1]. As a new method of phonon-imaging in crystals we introduce a holographic detection scheme using monochromatic ultrasonic waves of frequencies up to 1 GHz. A *phase-sensitive* scanning acoustic microscope (PSAM) has been developed (details to be published elsewhere) with a lateral resolution in the range of 1 μm and a phase resolution better than a thousandth of a wavelength. Detecting *phase and amplitude* simultaneously, the PSAM shows the first acoustic holograms of anisotropic crystals and allows the investigation of acoustic anisotropy at room temperatures. Our method is partly similar to the experiments of Hauser et. al., who recently published first results on detecting *intensities* of coherent sound waves in anisotropic crystals at a frequency of 15 MHz [2].

As shown in fig. (1) a point source of rf ultrasonic waves is created by focusing a commercial ultrasonic lens on the surface of a single crystal. In the case of an isotropic material, spherical waves would propagate into the volume. In anisotropic materials the wave fronts can be strongly distorted. A second ultrasonic lens on the backside of the sample scans over the surface, recording the image of the consecutive wave fronts arriving at that plane (fig. (2)). Such an image resembles a hologram of a point source recorded through an anisotropic medium.

Although only longitudinal waves travel in the liquid coupling medium, transverse waves can be detected as well in the solid, since mode conversion at the interfaces

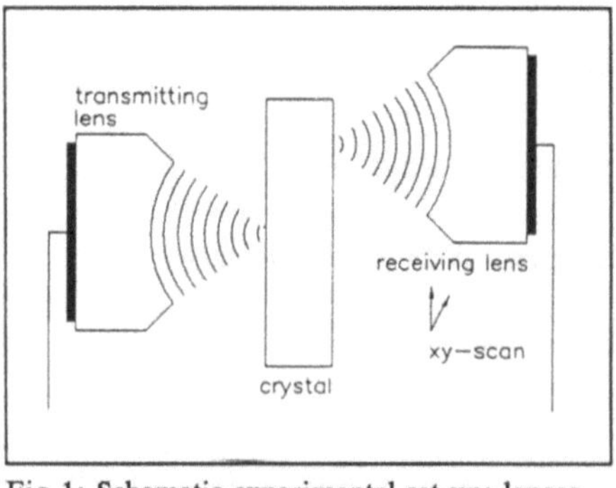

Fig. 1: Schematic experimental set-up: lenses and sample are immersed in water

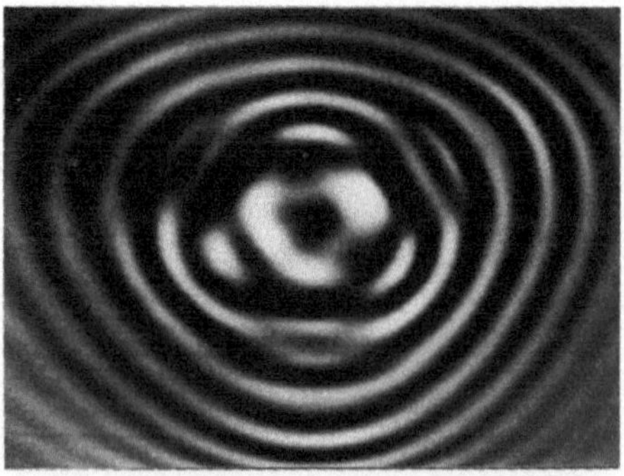

Fig. 2: Hologram of a $LiTaO_3$ single crystal at 392 MHz, detection area 300μm x 400μm, sample is 100μm thick

Springer Series in Solid-State Sciences, Volume 112
Phonon Scattering in Condensed Matter VII Eds.: M. Meissner and R.O. Pohl

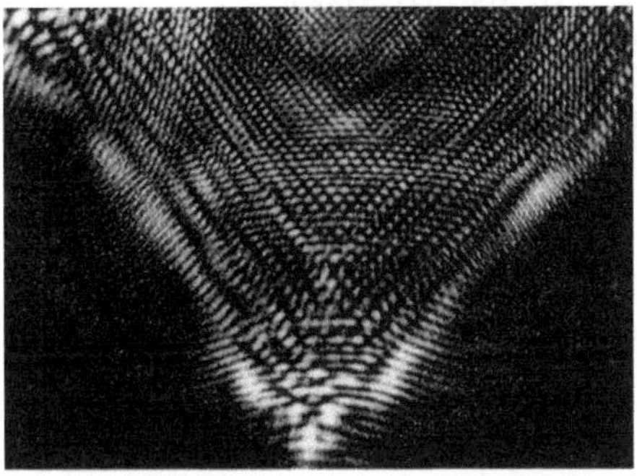

Fig. 3: GaAs (100)-plane, measured amplitude image, detection area 1.2 mm x 1.6 mm, sample is 4.8 mm thick

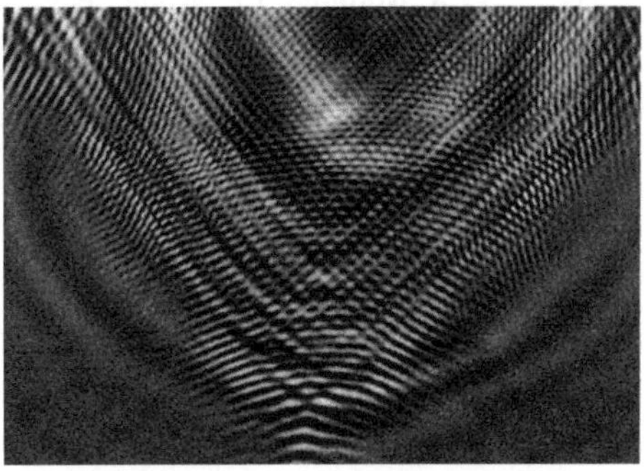

Fig. 4: GaAs, the corresponding measured hologram to fig. 3, amplitude and phase information still mixed

occurs. Quasi-longitudinal and quasi-transverse waves are separated by their different travel times even in rather thin crystals ($d = 100\ \mu m$). In the center of fig. (2) the slow transverse waves are selected by the fixed time gate, while the fast longitudinal waves have already arrived at the outer parts of the picture.

Phase and amplitude informations can be extracted separately from the holograms. Fig. (3) shows an example of the amplitude image of a GaAs single crystal at 392 MHz in the (100)-plane and its corresponding hologram in fig. (4). Only the lower half of the full symmetry is shown. As predicted by Maris [3] interference fringes occur, caused by the waves which travel in the same direction despite of different wave vectors $\underline{\mathbf{k}}$. The caustics caused by phonon focusing are not as sharp as in the short wavelength limit.

The measured phase images show the wave fronts travelling at the group velocity. By the phase-sensitive measurement of a quasi-continuous distribution of group velocities in a wide range of directions (depending on scan range and the number of data points taken) one gains a detailed knowledge of the acoustical properties of the crystal (see e.g. [4]). All group velocities can even be calculated absolutely if only one of them is measured absolutely (e.g. in a symmetry direction). This could be useful for a high-precision determination of the elastic constants of very small samples, in cases where large single crystals are not available.

Theoretical images (figs. (5) and (6)) were calculated using a modified Monte Carlo ray tracing method based on the principle of stationary phase [3]. In the vicinity of every calculated group velocity vector, only a neighbourhood of k-vectors is generated to contribute to the complex wave amplitude of the holograms.

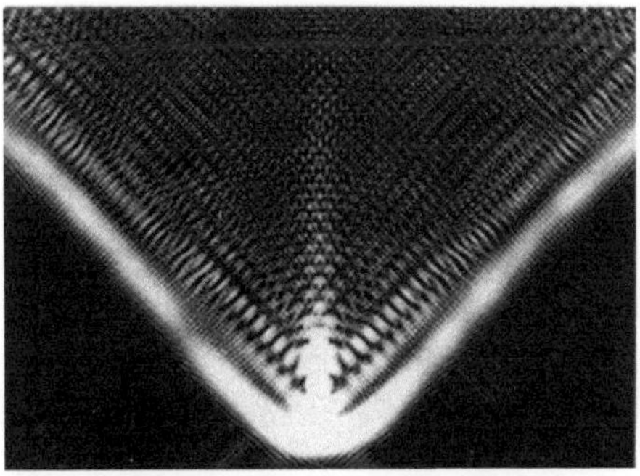

Fig. 5: GaAs, calculated amplitude image corresponding to fig. 3

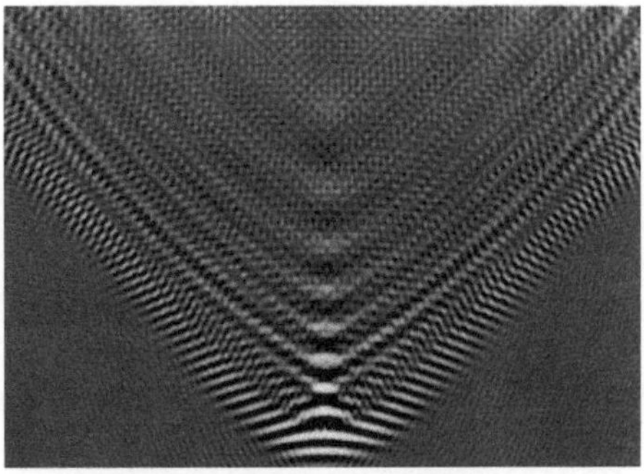

Fig. 6: GaAs, calculated hologram corresponding to fig. 4

[1] G.A. Northrop, J.P. Wolfe : Phys. Rev. B 22, 6196-6212 (1980)
[2] M.R. Hauser, R.L. Weaver, J.P. Wolfe : Phys. Rev. Lett. 68, 2604-2607 (1992)
[3] H.J. Maris : Phys. Rev. B 28, 7033-7037 (1983)
[4] B.A. Auld: "Acoustic Fields and Waves in Solids" (R.E. Krieger, Malabar 1990)

Specular Phonon Reflection in Anisotropic Media

R. Wichard and W. Dietsche

Max-Planck-Institut für Festkörperforschung,
Heisenbergstr. 1, W-7000 Stuttgart, Fed. Rep. of Germany

We report on our recent results about the phonon-reflection imaging experiments in Si. The setup was similar as the one used by Northrop and Wolfe [1] except that a superconducting Al tunnel junction was used as the detector. Heat pulses were generated by focusing a HeNe-laser beam onto an Al film which covered the remainder of the detector surface.

Images taken with (100)-oriented crystals showed similar features as the ones observed in transmission. But in addition there were sharp new lines between the pairs of the STA ridges which were not observed before [2]. They are due to phonons which were mode-converted from LA to STA ones during the reflection. The sharpness of the ridge was surprising at first, because the LA phonons do not show sharp structures in transmission images. Thus the reflected beam of the STA phonons must be convergent in exactly such a way to compensate for the spatial spread of the original LA ones. This behavior follows from the shapes of the respective slowness surfaces.

Numerical simulations were made and the sharp line was reproduced [2]. In other numerical simulations we increased the distance from the generator to the back surface and kept the distance to the detector fixed. Two caustic lines are now found by the simulation of the mode conversion. We have found that these two lines approach each other and merge if the two distances fom the reflecting surface are almost equal.

The (100) orientation is actually a very special because the reflecting surface is a mirror plane of the crystal. In this case, one of the three possible reflected phonon modes is identical to the incident one except for the sign reversal of the perpendicular component of momentum and group velocity. Therefore some phonon reflection structures are just mirror images of the familiar phonon transmission ones. In all other crystal cuts (except (110)) the concept of mirror

Springer Series in Solid-State Sciences, Volume 112
Phonon Scattering in Condensed Matter VII Eds.: M. Meissner and R.O. Pohl

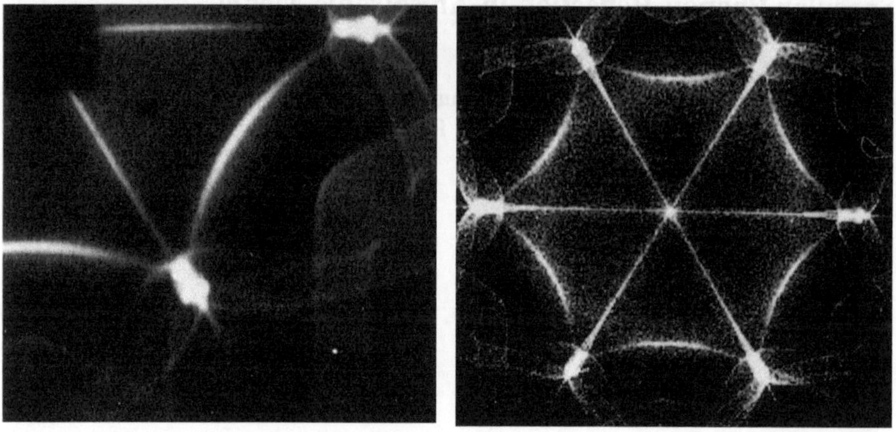

Fig.1 Phonon reflection image in a (111) Si sample. The experimental image (left) covers the lower right quadrant of the theoretical image(right).

reflection fails completely. Specular reflection requires mode conversion, and entirely different structures in the phonon flux result.

A particularly interesting case is the (111) orientation which was studied with a similar experimental setup. The detector is now in the center of the scanned area. The resulting reflection image is shown in Fig. 1. Most striking is the fact that the symmetry of the image is not threefold but rather sixfold. Clearly, this symmetry doubling must be a consequence of the specular reflection process. The reason is that two modes with $\pm \mathbf{q}_{\parallel}$ will, after specular reflection, hit the original surface at points which are inversion symmetric through the origin. The two group velocities of them are just interchanged after reflection leading to inversion symmetric trajectories. Numerical simulations prove this.

In conclusion we have shown that if the reflecting plane is a mirror plane of the crystal then reflection without mode conversion is possible and some of the original transmission features are reproduced like in an optical mirror. If the sample is cut along a non-mirror plane as the (111) then this is not the case. The specular reflection requires mode conversion, which leads to entirely new structures. Furthermore, the reflection images are always inversion symmetric.

[1] G.A. Northrop and J.P. Wolfe, Phys. Rev. Lett. **52**, 2156 (1984).

[2] R. Wichard and W. Dietsche, Phys. Rev. B 45, 9705 (1992).

Anisotropy Imaging of Materials

W. Sachse[1], *A.G. Every*[2], *and M.O. Thompson*[3]

[1]Department of Theortical and Applied Mechanics, Cornell University, Ithaca, NY 14853, USA

[2]Department of Materials Science and Engineering, Cornell University, Ithaca, NY 14853, USA

[3]University of the Witwatersrand, Johannesburg, South Africa

Abstract. We describe in this paper an imaging technique in which a pulsed laser, operating as an ultrasonic *point-source*, is scanned across the surface of a specimen and the generated elastic waves are detected by a *point-receiver*. The signals detected along a scan line, comprise a *scan image* which contains the spatial- and temporal-dependence of the elastic wave field in a specimen that can be processed to recover all the elastic constants of the material and to detect microstructural features, including damage. By monitoring the amplitude of a particular wave arrival over a number of scan lines, an *anisotropy image* is generated that is analogous to a ballistic–phonon image but differs from it in that the measurements are made at MHz frequencies, at ambient temperatures and in metals and non–metals alike. We show the results of measurements made on single crystal specimens of Si, zinc and in a number of polymeric composite materials. Anisotropy-induced focusing effects are observed in particular directions in each of these materials. An interpretation of these images has been obtained using *Monte Carlo* simulation.

Summary

Our goal is the development of non-destructive techniques suitable for the characterization of single crystal specimens and anisotropic composite materials. For this, we have implemented an ultrasonic *point-source/point-receiver* technique that uses a focused Nd:YAG laser pulse as the scanned thermoelastic source and a piezoelectric element (1.3 mm in diameter) as the detector. One can determine from the detected signals the signal arrival times and amplitudes of several wave modes propagating simultaneously along a number of directions in a specimen. We have developed signal processing algorithms for solving the inverse problem of recovering the elastic constants of a material from measurements of the group velocities along a number of non–principal directions in the material [1, 2]. A semi–inverse approach has been used to process the anisotropy images so that regions of inhomogeneity or damage in an anisotropic material can be detected and characterized.

Springer Series in Solid-State Sciences, Volume 112
Phonon Scattering in Condensed Matter VII Eds.: M. Meissner and R.O. Pohl

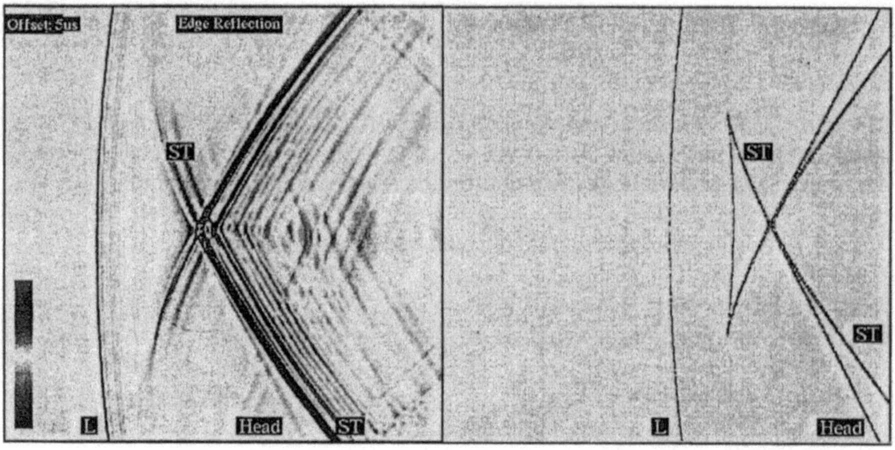

Fig. 1: (a) Measured scan image for a scan line lying in the plane of transverse isotropy in a single crystal specimen of zinc (Horizontal axis: *Time*: Range - 17.08 μs; Vertical Axis: *Source/receiver separation*: Center is epicenter). (b) Monte Carlo simulation of longitudinal, L, slow transverse, ST, and head wave arrivals.

Anisotropy images have been obtained in single crystal specimens of silicon whose orientation was $\langle 100 \rangle$, $\langle 110 \rangle$ and $\langle 111 \rangle$ [3]. In every case, the scan images exhibit a clearly defined structure corresponding to the arrivals of various wave modes propagating in the specimen. The anisotropy in the detected wave amplitudes is governed by the characteristics of the source and receiver as well as the focusing of acoustic energy flux into particular directions of the specimen. Using Monte Carlo simulations, features in the measured images can be understood [3]. Deviations between the measured images and the simulations provides a basis for detecting and identifying material anomalies. We have obtained scan images in a single crystal specimen of zinc and in several polymeric composite materials fabricated of a number of ply layers of varying lay–up configurations. Current work is focused on two–dimensional scans to generate images corresponding to the amplitudes of one or more wave arrivals.

Acknowledgements

This work has been supported by the Office of Naval Research (Physical Acoustics Program). The work of A. G. E. has been supported by the Materials Science Center at Cornell University and by the University of the Witwatersrand.

[1] W. Sachse, B. Castagnede, I. Grabec, K. Y. Kim and R. L. Weaver, *Ultrasonics*, **28**, 97-104 (1990).

[2] A. G. Every and W. Sachse, *Physical Review B*, **42**(13), 8196–8205 (1990).

[3] A. G. Every and W. Sachse, *Physical Review B*, **44**(13), 6689–6699 (1991).

Geometrical Dispersion of Ballistic Heat Pulses: Application to Paratellurite

A.G. Every

Department of Physics, University of the Witwatersrand,
P.O. Wits 2050, South Africa

Geometrical dispersion of ballistic phonon transit times arises partly from the spread in path lengths of individual phonon trajectories linking different points of the source and detector and partly from the spread in phonon group velocities within the trajectory solid angle due to acoustic anisotropy.

The time dependent phonon power P(t) transmitted through a slab–shape sample of thickness h consists of a sum of contributions over modes having energy flux vectors in the source–detector direction. Each takes the form of a convolution integral involving the fourth power of the time dependent source temperature T(t) and a transmission function

$$I(t)=\int I(\mathbf{R}'',\mathbf{R}')\ \delta(t-\tau(\mathbf{R}'',\mathbf{R}'))\ da'da'' \qquad (1)$$

where da' and da'' are areal elements of the source and detector respectively located at points $\mathbf{R}'=(X',Y')$ and $\mathbf{R}''=(X'',Y'')$, $\tau(\mathbf{R}'',\mathbf{R}')$ is the transit time and $I(\mathbf{R}'',\mathbf{R}')$ is an intensity factor, which incorporates the effects of phonon focusing. I(t) for each mode is non–zero only in an interval $\tau_{min} \lesssim t \lesssim \tau_{max}$ defined by the range of transit times for the given mode.

When the source and detector dimensions are small compared with h, τ is a linear function of $\mathbf{R}''$ and $\mathbf{R}'$. Using the well known identity $\mathbf{V}.\mathbf{s}=1$, where $\mathbf{s}$ is the slowness vector associated with the group velocity vector $\mathbf{V}$ pointing from $\mathbf{R}'$ to $\mathbf{R}''$, it is readily established that $\partial\tau/\partial X = s_x$ and $\partial\tau/\partial Y = s_y$ and that τ therefore is given by

$$\tau(\mathbf{R}'',\mathbf{R}') = \tau_0 + \mathbf{s}_{/\!/}.(\mathbf{R}''-\mathbf{R}'), \qquad (2)$$

where $\mathbf{s}_{/\!/}=(s_x,s_y)$ and τ_0 is a constant. The pulse width for each mode is thus

$$\Delta\tau = \tau_{max} - \tau_{min} = |\mathbf{s}_{/\!/}|.(\ell'' + \ell'), \qquad (3)$$

where ℓ'' and ℓ' are the lateral dimensions of the detector and source in the direction of $\mathbf{s}_{/\!/}$.

For circular source and detector $(\ell''+\ell')$ is the sum of the diameters, and $\Delta\tau$ can be identified with $|\mathbf{s}_{/\!/}|$ to within a constant factor. τ is stationary and $\Delta\tau=0$ for real space directions that $\mathbf{s}_{/\!/}=0$ is mapped onto, which do not in general correspond to V being stationary or to epicentral detection, as is the case for isotropic solids.

We report below calculations pertaining to the ballistic heat pulse experiments of Hurley et al. [1] on (110)–oriented TeO_2. Fig. 1 shows a Monte Carlo generated (001) section of the wave surface of TeO_2. Much of the complexity of this section can be ascribed to folding of this surface. Fig. 2 shows Monte Carlo simulated heat pulse profiles for directions 1 and 8 as

Springer Series in Solid-State Sciences, Volume 112
Phonon Scattering in Condensed Matter VII Eds.: M. Meissner and R.O. Pohl

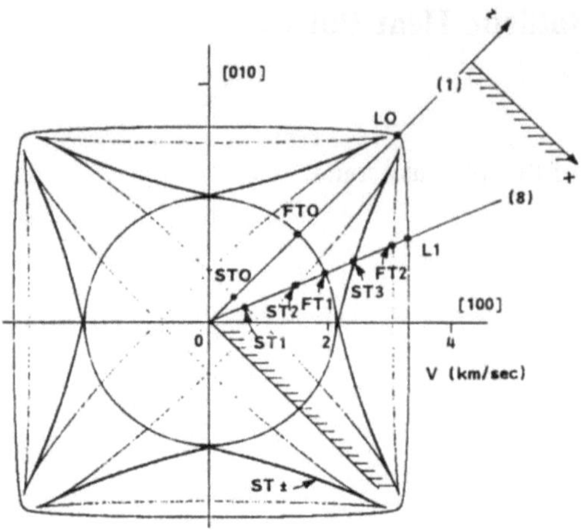

Fig. 1: Monte Carlo generated (001) section of the wave surface of paratellurite, TeO_2.

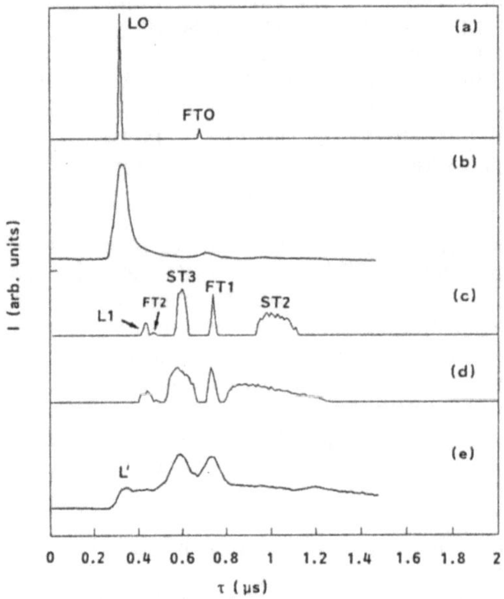

Fig. 2: Comparison between simulated heat pulse data and experimental results of [1] on (110)–oriented TeO_2.
(a) direction 1, ℓ=0.015mm,
(b) direction 1, experiment,
(c) direction 8, ℓ=0.06mm,
(d) direction 8, ℓ=0.15mm and
(e) direction 8, experiment.

defined in [1], assuming point–like source and various detector radii ℓ, compared with the experimental time traces of [1] for these directions. There is good agreement between theory and experiment with regard to the arrival times, widths and also the intensities of the various pulses. In particular, referring to Fig. 1, the lack of geometrical dispersion for direction 1 accounts for the narrowness of the L0 and FT0 pulses (ST0 lies off scale) and the pronounced geometrical dispersion of ST2 explains why this pulse is almost completely washed out in the experimental results (ST1 lies off scale).

Reference

1. D.C. Hurley, J.P. Wolfe and K.A. McCarthy, Phys. Rev. B33, 4189 (1986)

Internal Diffraction of Acoustic Waves in Silicon

M.R. Hauser[1], *R.L. Weaver*[2], *and J.P. Wolfe*[1]

[1]University of Illinois at Urbana-Champaign, Department of Physics, 1110 W. Green, Urbana, IL 61801, USA

[2]University of Illinois at Urbana-Champaign, Department of Theoretical and Applied Mechanics, 104 S. Wright, Urbana, IL 61801, USA

Phonon focusing has been extensively studied using phonons with frequencies of several hundred GHz [1]. We have recently reported a related technique for studying anisotropies of acoustic flux at sub-GHz frequencies [2]. In these experiments using coherent waves, ultrasound is generated and detected by focused immersion transducers in a water bath. The transducers act as a point-source/point-receiver pair, one of which is scanned to produce an ultrasound image. We present here the experimental results for Si along with a model which accurately reproduces the observed images.

In an elastically anisotropic material the group velocity and phase velocity of a wave of given **k** are in general not parallel. This gives rise to spatial focusing of the energy flux (phonon focusing), and may result in folded wave surfaces for the transverse modes. In directions where the wave surface is folded there will be multiple wave-front arrivals, and with narrow-band ultrasound these arrivals may overlap in time and interfere with one another. It is this interference effect that we have termed "internal diffraction." Figure 1a illustrates this effect with an experimental image of a 1-cm thick Si sample using tone-bursts of 15 MHz ultrasound. The intense horizontal and vertical lines of flux and the pattern of dots near the center are focused slow-transverse (ST) waves displaying interference between sheets of the ST wave surface. The bright ring is due to head waves, generated by surface skimming waves, which interfere with the ST wave.

We have modeled ultrasonic flux imaging (UFI) by decomposing the sound field created by the focused transducers into plane waves, and considering the propagation of each wave through the two fluid-solid interfaces and the intervening bulk of the sample. At each interface reflection and transmission coefficients are calculated. Only transmitted waves are considered -- the effects of reflections are neglected in the current model. The resulting plane wave amplitudes are summed at the detector position to obtain the transmitted amplitude. Figure 1b is a calculated image for a Si sample with the same geometry as that in Fig. 1a. It can be seen that the two images agree quite well. We postulate that the discrepancies are due to the neglected reflected waves. Figure 1c shows the experimental and calculated intensities

Springer Series in Solid-State Sciences, Volume 112
Phonon Scattering in Condensed Matter VII Eds.: M. Meissner and R.O. Pohl

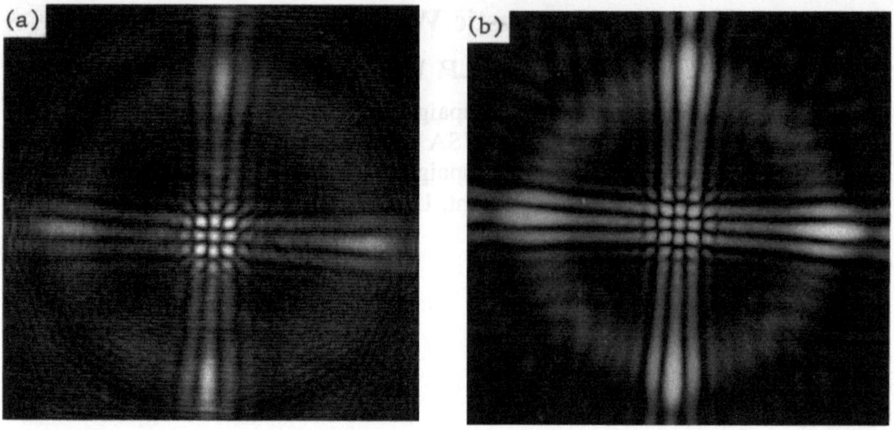

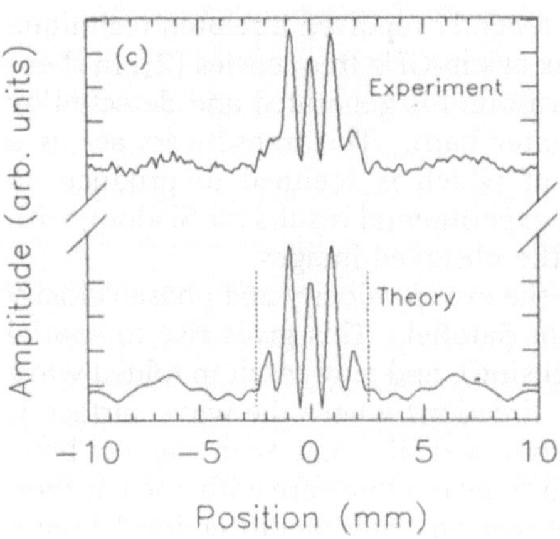

Figure 1. (a) Ultrasound image of a 1-cm thick (100)-oriented Si sample, covering a 2-cm scan range, taken with 500 ns tone-bursts of 15 MHz ultrasound. Horizontal and vertical axes are roughly <110>. (b) Calculated image for the same geometry as part a. (c) Comparison of the intensity in the images on a line along the [100] direction (roughly diagonal in the images.) The theoretical curve outside of the dashed line is the average of several adjacent lines in the image.

along a line passing through the central dot in the images and extending along the <100> direction.

This work is supported in part by the NSF the Materials Research Laboratory Grant No. DMR-89-20538 (JPW and MRH), and under the Solid Mechanics and Materials Program grant MSS-91-14360 (RLW); MRH is also partially supported by an IBM predoctoral fellowship.

References

[1] For a review, see G.A. Northrop and J.P. Wolfe in *Nonequilibrium Phonon Dynamics*, edited by W.E. Bron (Plenum, New York, 1985), Chap. 5.

[2] Matt R. Hauser, R.L. Weaver, J.P. Wolfe, Phys. Rev. Lett. **68**, 2604 (1992)

Measurements of Phonon Elastic Scattering Rates by Phonon Imaging and Monte-Carlo Simulation

S. Tamura[1], *J.A. Shields*[2], *M.T. Ramsbey*[2], *and J.P. Wolfe*[2]

[1]Department of Engineering Science, Hokkaido University, Sapporo 060, Japan

[2]Department of Physics and Materials Research Laboratory, University of Illinois at Urbana-Champaign, Urbana, IL 61801, USA

1. Introduction

The propagation of high-frequency phonons through crystals at low-temperature is characterized by both ballistic and scattering processes. The ballistic propagation is governed by the elastic anisotropy and is highly anisotropic, an effect known as phonon focusing[1]. The main scattering processes of phonons in pure dielectric crystals are anharmonic decay and isotope scattering. Historically, the most important method for characterizing the scattering of phonons in solids has been thermal conductivity which gives us information on scatterings averaged over phonon polarizations. With the advent of heat-pulse techniques, which have the ability to detect ballistic phonons of selected polarization and propagation direction, one might hope to make detailed measurements of the phonon scattering processes. For the sub-THz phonons detected predominantly in the current experiments, the isotope scattering usually dominates anharmonic decay. In the long-wavelength limit (usually valid up to about 1 THz) the individual scattering rate γ of a phonon $\lambda=(\boldsymbol{k},j)$ [$\boldsymbol{k}$ is the wave vector and j is the phonon polarization] to a phonon $\lambda'=(\boldsymbol{k}',j')$ and the total scattering rate Γ are given by[2]

$$\Gamma_\lambda = \sum_{j'} \int \frac{d\Omega_{\lambda'}}{4\pi}\gamma_{\lambda\lambda'} \;, \quad \gamma_{\lambda\lambda'} = \frac{V_0\, g\, \omega_\lambda{}^4 |\boldsymbol{e}_\lambda \cdot \boldsymbol{e}_{\lambda'}|^2}{4\pi v_{\lambda'}} \;, \tag{1}$$

where g is a factor measuring the magnitude of mass fluctuations caused by isotopic atoms, V_0 is the volume per atom, $\omega = 2\pi\nu$ is angular frequency, $\boldsymbol{e}$ is the polarization vector, and the integral is performed over the solid angle in the wave-vector space. The presence of the term $|\boldsymbol{e}\cdot\boldsymbol{e}'|^2$ in Eq.(1) means that this scattering is an anisotropic P-wave scattering. More specifically a phonon is scattered preferentially in the forward or backward direction for the intramode scattering process, whereas it is scattered mainly perpendicular to the initial propagation direction for the intermode process. As we shall see below, this kinematical constraint of the scattering combined with the phonon focusing effect leads to the channeling of phonons scattered a few times by isotopic atoms[3]. We also note that the scattering into a lower branch with slower phase velocity v dominates the scattering into an upper phonon branch. Thus, an each scattering process ($\lambda\rightarrow\lambda'$) is anisotropic and mode dependent. However, the total scattering rate obtained by summing $\gamma_{\lambda\lambda'}$ over λ', $\Gamma_\lambda = A_0\nu^4$ (A_0 is a constant.), is independent of phonon mode.

Springer Series in Solid-State Sciences, Volume 112
Phonon Scattering in Condensed Matter VII Eds.: M. Meissner and R.O. Pohl

2. Channeling effects

Because ballistic phonons display huge anisotropy in heat flux due to phonon focusing, we may expect that phonons scattered a few times still retain some amount of the initial directivity, which should be enhanced by the polarization selective scattering of phonons from mass defects. A striking example of this channeling of scattered phonons is predicted for FT phonons. The ridges of high focusing flux encompass (100) planes of the crystal, as illustrated in the diagram of Fig.1. Elasticity theory shows that the polarization of phonon propagating within the ridges is almost exactly perpendicular to the plane of the ridge. Thus, the $|\boldsymbol{e} \bullet \boldsymbol{e}'|^2$ selection rule implies that a scattered FT phonon will have a strong tendency to remain in the same ridge[3].

The phonon channeling effect is observed by physically blocking off a selected portion of the ballistic phonon image, leaving only the scattered phonon flux[4]. The sample geometry is depicted in Fig.2(a). A crystal of high-purity silicon is slotted with a "gentle" string-saw cut to minimize damage to the nearby region. A cavity-dumped argon-ion laser beam, which is focused to approximately a 10 μm spot with a pulse width of 12 ns, provides a point source of heat. The beam is raster-scanned across the face of the crystal opposite the detector while the crystal is immersed in superfluid helium at about 1.7 K. The detector is a 10 ×10 μm^2 PbBi superconducting tunnel junction which is sensitive to phonon frequencies above 700 GHz. The experimental phonon image for the slotted silicon crystal is shown in Fig.3(a). The striking result in the image is the pronounced continuation of the FT ridges behind the slot. This is precisely the predicted behavior for phonons scattered once or twice in the FT ridge.

The Monte Carlo simulation with the geometry close to Fig.3(a) is shown in Fig.3(b). Black dots represent phonons which reach the detector with t=0-1.5t_b (t_b is the ballistic time of flight of T phonons) and this theoretical image reproduces the channeling in the FT ridges quite well. A simple isotropic approximation for the scattering (corresponding to replacing $|\boldsymbol{e} \bullet \boldsymbol{e}'|^2$ by 1/3

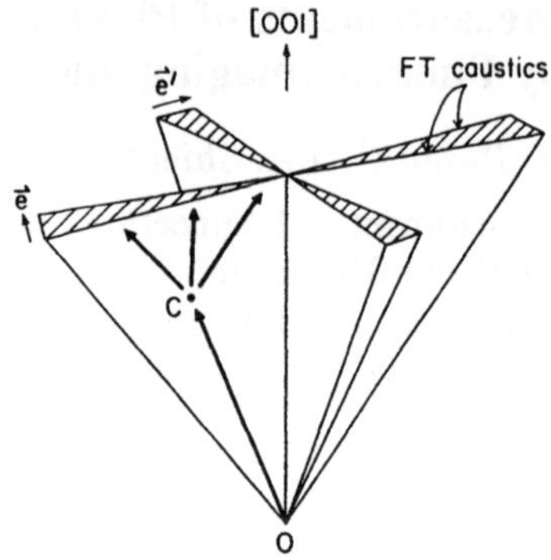

Fig.1 Schematic diagram of FT ridges.

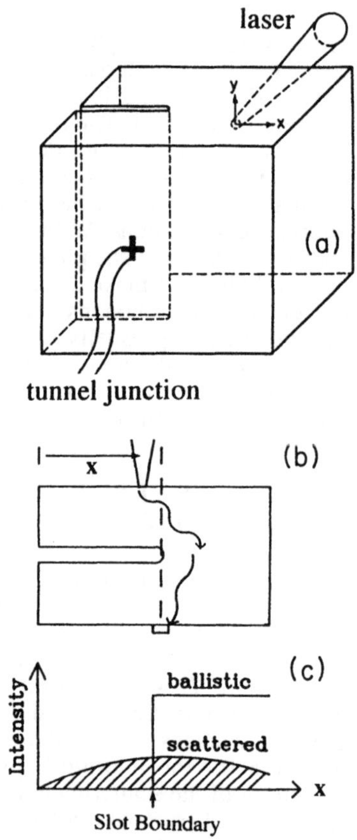

Fig.2 (a) Schematic of the "slotted" sample geometry. (b)Top view. (c) Schematic of the expected intensity distribution.

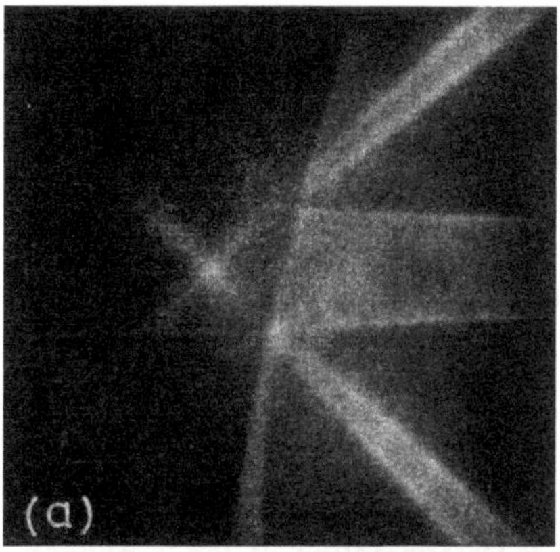

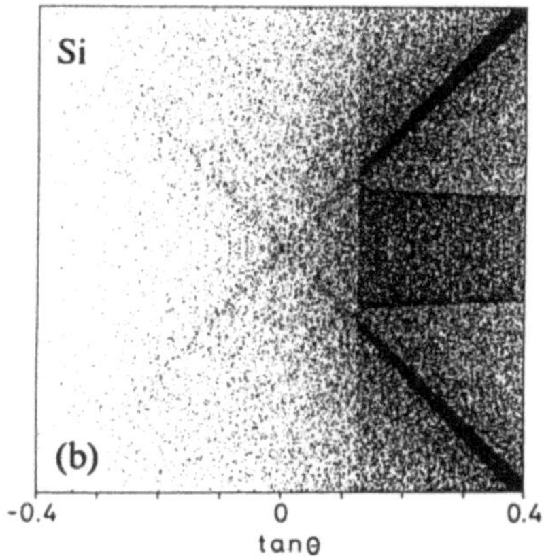

Fig.3 (a) Phonon image for the slotted (100)-Si crystal. (b) Monte Carlo simulation of the phonon intensity with the sample geometry close to (a).

in Eq.(1)) produces only a very weak channeling inconsistent with the experiment, indicating that the phonon focusing itself is not sufficient to explain the strong channeling of elastically scattered phonons.

3. Measurement of elastic scattering rate in Si

The separation of purely scattered phonon flux from the combined scattered and ballistic components by using a slotted crystal gives us a hope to measure quantitatively the elastic scattering rate of phonons. For this purpose we used essentially the same sample geometry as Fig.2(a) but the crystal (thickness is 4.85mm) is now cut with (110) faces for the detector and excitation surfaces[5]. The schematic of the expected intensity distribution along an FT ridge, which corresponds to a single horizontal line perpendicular to the slot, is shown in Fig.2(c). Ideally, for two laser-spot positions displaced infinitesimally to the left and the right of the slot boundary, the scattered flux should be nearly the same. So, we define a "ledge ratio" $R=S/H$ (Fig.4) which indicates the fraction of scattered flux S (observed behind the slot) to the total flux H (measured on the right side of the slot). We find that the ledge ratio is relatively insensitive to adjustable experimental parameters such as power level, gate width, and surface preparation. The dots of Fig.4 indicate the horizontal line scans of T phonon intensity across the slot boundary obtained for a narrow time gate with $0.5t_b$. The analysis procedure consists of making a Monte Carlo calculation of the phonon flux for the experimental geometry for a range of scattering strengths $\Gamma = A\nu^4$, and determining the scattering constant A which best reproduces the experimental ledge ratio

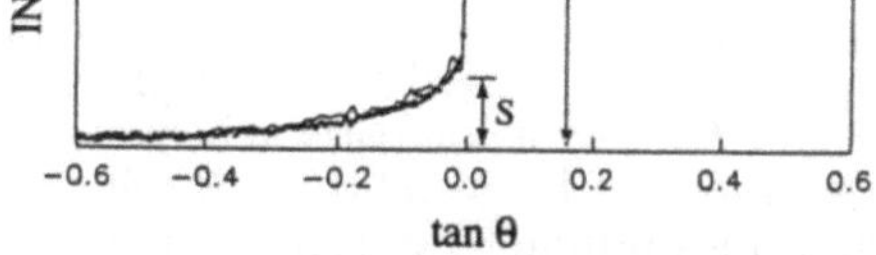

Fig.4 Definition of ledge ratio R and comparison between experimental and theoretical ($A=A_0$) phonon signals.

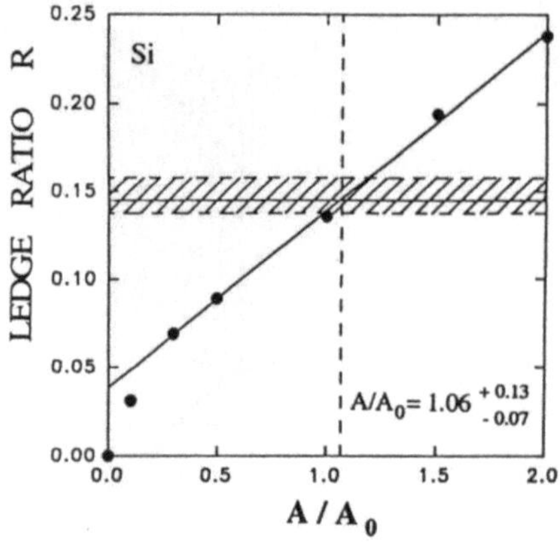

Fig.5 Determination of the scattering constant for Si. Dots are the theoretical ledge ratios and the horizontal line is the experimental value.

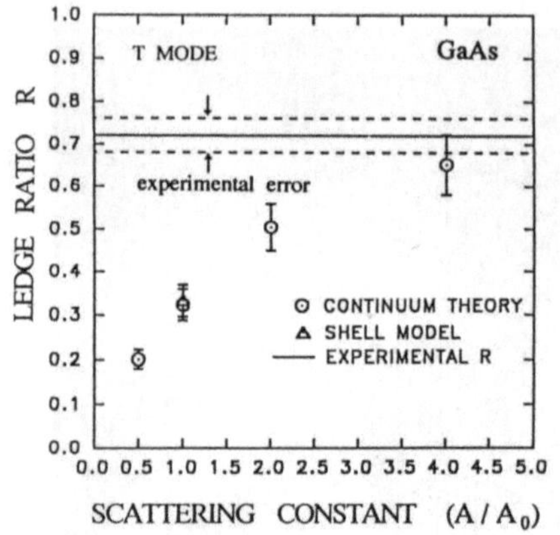

Fig.6 Determination of the T phonon scattering constant in GaAs.

R. The results of calculation with values of *A* ranging from $0.1A_0$ to $2A_0$ ($A_0 = 2.43 \times 10^{-42}$ sec^3 calculated for isotope scattering for Si) have been plotted in Fig.5. The experimentally determined ledge ratio of R=0.145 is shown as a horizontal line on this graph. (The shaded area gives the statistical uncertainty.) Thus the measured scattering-rate constant for T phonons in pure silicon is found to be A=1.06A_0. Also for L phonons the experimental and theoretical (with A=A_0) ledge ratios agree quite well, consistent with the results of the analysis for T phonons.

4. Measurement of the scattering rate in GaAs

A similar experiment utilizing the slotted-crystal technique to measure the scattering rate was performed with undoped-GaAs of (110) orientation (sample thickness 1.94 mm)[6]. A laser beam focused on a copper film on a (110) face provides a heat source and a Pb-Bi tunnel junction with the onset frequency 0.7 THz deposited on the opposite face is used to detect phonons. Figure 6 shows a comparison between the experimental ledge ratio of FT phonons and theoretical predictions for four selected values of the scattering constant *A* (A_0=7.38×10^{-42} sec^3 for GaAs). We find that the experimental scattering rates for

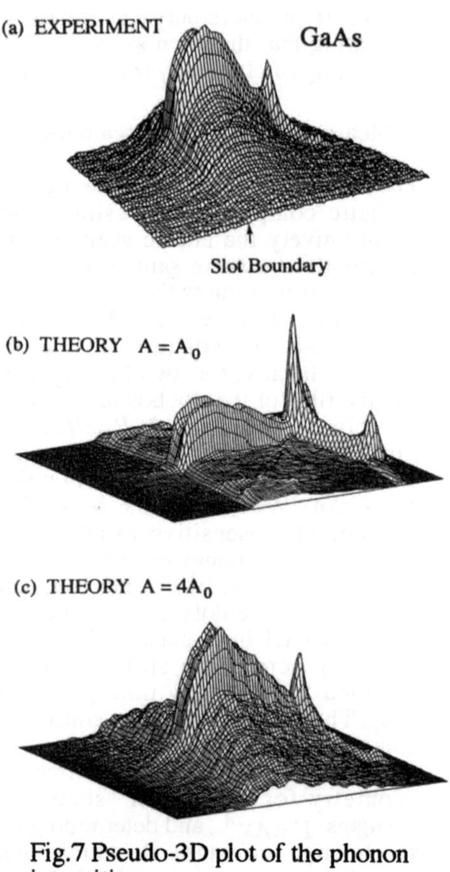

Fig.7 Pseudo-3D plot of the phonon intensities.

T phonons with 0.7 THz in our GaAs sample are approximated by the scattering-rate constant $A=A_T$ with $A_T=(4.0\pm0.7)A_0$. These values are significantly larger than those expected for pure isotope scattering. This result is alternatively seen from the pseudo-3D plot of the phonon flux shown in Fig.7. An analysis for L mode phonons, however, indicates that the scattering-rate constant $A=A_L$, with $A_L=(0.8\pm0.5)A_0$ or $A_L\leq1.3A_0$, which is consistent with the isotope scattering rate.

5. Discussions

Despite our expectations that the measured scattering rate of GaAs would be close to that predicted by isotope-scattering theory, we have found some interesting contrasts to the case of Si. The differential scattering of L and T phonons implies the existence of additional scattering from residual defects in GaAs. Several previous studies have observed the polarization-dependent scattering of phonons from defects in GaAs[7]. Specifically, Culbertson et al. observed a polarization-dependent scattering in undoped GaAs which was diminished after the application of infrared radiation[8]. Their results were interpreted in terms of the so-called *EL2* defect which has been associated with photoinduced metastability effects. Our present results may well be due to this defect in GaAs, which is naturally formed by the liquid-encapsulated-Czochralski growth method for semi-insulating GaAs. Evidence for the presence of *EL2* defects in our GaAs sample is shown in Fig.8. A dramatic change in the position of the FT caustics can be seen after irradiating the sample with the Nd:YAG laser at a wavelength of 1.06 μm for the transition from the normal to metastable states of the *EL2* defect.The wider FT ridges suggest that higher frequency phonons are transmitted in the metastable state of *EL2* defects. The other preliminary results we observed after irradiation with the YAG laser are : (1)An increase in both the scattered and ballistic phonon flux. (2)An increase in the ledge ratio. (3)An increase in the FT peak and a decrease of the ST peak in [110] direction, i.e., a mode dependence of the scattering. These results indicate a change in some inelastic scattering process and also a change in the symmetry of defects with irradiation. To understand all these features observed experimentally we need to develop theoretical studies on phonon-defect scattering, which should be much more complicated than the phonon-isotope scattering.

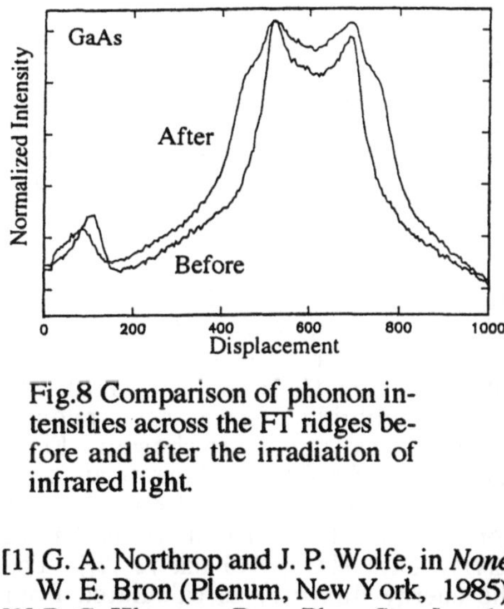

Fig.8 Comparison of phonon intensities across the FT ridges before and after the irradiation of infrared light.

[1] G. A. Northrop and J. P. Wolfe, in *Nonequilibrium Phonon Dynamics*, edited by W. E. Bron (Plenum, New York, 1985), p.165.
[2] P. G. Klemens, Proc. Phys. Soc. London, Sec. A **68**, 1113 (1955).
[3] M. T. Ramsbey, J. P. Wolfe, and S. Tamura, Z. Phys. B **73**, 167 (1988).
[4] J. A. Shields, J. P. Wolfe, and S. Tamura, Z. Phys. B **76**, 295 (1989).
[5] J. A. Shields, S. Tamura, and J. P. Wolfe, Phys. Rev. B **44**, 3001 (1991).
[6] M. T. Ramsbey, S. Tamura, and J. P. Wolfe, Phys. Rev. B (in press).
[7] V. Narayanamurti, Science, **213**, 717 (1981).
[8] J. C. Culbertson, U. Strom, and S. A. Wolf, Phys. Rev. **36**, 2962 (1987).

Study of Phonon Pulse Propagation in Silicon and the Effect of N-Processes

M. Obry, J. Tate, P. Berberich, and H. Kinder*

Physik-Department E10, Technische Universität München, W-8046 Garching, Fed. Rep. of Germany

*Present address: Oregon State University, Department of Physics, Corvallis, OR 97331-6507, USA

We have studied the propagation of heat pulses in Si (111) between 18 K and 32 K. In this transition region from ballistic to diffusive propagation the ballistic peaks are decreasing rapidly with increasing temperature. However, there is no pronounced maximum of diffusive phonons but a slow monotonous decay. This behavior is described by a Monte Carlo simulation assuming N-processes ($1/\tau_{NL} \propto \omega^2 T^3$ and $1/\tau_{NT} \propto \omega T^4$) in the presence of strong isotope scattering ($1/\tau_i \propto \omega^4$).

It is well known that elastic scattering by isotopic impurities acts as a low pass filter for ballistic phonons independent of temperature. Therefore these processes do not block ballistic phonon propagation very effectively when the substrate temperature is raised. This is done by N-processes which have a weak frequency dependence but strongly depend on temperature. Therefore the transition from ballistic to diffusive phonon propagation in dielectric crystals like Si is governed by N-processes, and an analysis of heat pulses propagation as a function of temperature can give information about the inelastic scattering rates even when elastic scattering dominates at thermal frequencies.

The phonon pulses were generated by a NiCr-thin film heater ($A = 1\ mm^2$, $E = 1\mu J$, $t_p = 100$ ns at 18 K to 27 K, $t_p = 3\mu s$ at 32 K). The temperature difference between heater and substrate is estimated to be smaller than 2 K. On the far side of the 3 mm thick Si (111) substrate (p-type, $\rho = 150\ \Omega cm$) a $YBa_2Cu_3O_7$ microbridge served as a phonon detector [1]. The time of flight spectra of the heat pulses are shown in Fig.1 for 5 different temperatures. The pulse shape of diffusive phonons at 32 K was compared with that expected from Fourier's equation of heat conduction (dashed line). The diffusion constant $D = 670\ cm^2/s$ used is in fair agreement with thermal conductivity data ($D = 840\ cm^2/s$ [2]). The solid line on the 32 K-trace shows the energy flux q(t) as calculated from Levinson's theory of nonlocal heat transport [3]. The characteristic times $\tau_i(T) = 40$ ns and $\tau_N(T) = 80$ ns were taken from [2]. Levinson's theory is in better agreement with the data. This indicates that local thermal equilibrium is not established at 32 K in silicon.

Ballistic phonons could be observed up to 27 K (see Fig.1.) The transverse phonons are disappearing first. However, the diffusive phonons do not exhibit a pronounced maximum but only a slow monotonous decay in the

Springer Series in Solid-State Sciences, Volume 112
Phonon Scattering in Condensed Matter VII Eds.: M. Meissner and R.O. Pohl

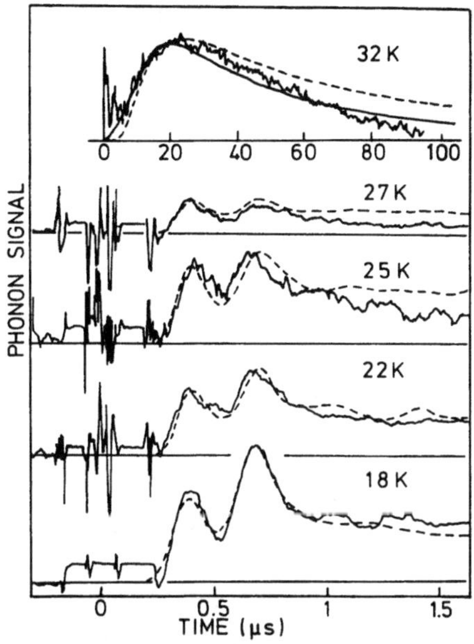

Fig. 1
Heat pulses observed after propagating through a 3 mm Si(111) crystal at 5 different temperatures.

time interval investigated. These results resemble heat pulse measurements of von Gutfeld et al.[4] in sapphire. As Levinson's theory does not apply for ballistic phonon propagation we analyzed our data by a Monte Carlo simulation. Phonon focussing as well as (diffusive) scattering at crystal boundaries was taken into account. The angular average of the L/T mode ratio as fitted to the data at 4 K was 0.27. It is rather large and reflects a preferential emission (or detection) of longitudinal phonons. The scattering rates were approximated by $1/\tau_i = A_i\omega^4$ (isotope scattering), $1/\tau_{NL} = A_{NL}\omega^2T^3$ and $1/\tau_{NT} = A_{NT}\omega T^4$ (normal processes). The resulting pulse shapes are shown in Fig.1 by dashed lines. (They have been smoothed to account for a detector time constant of 120 ns). The constants $A_i = 1.3\cdot10^{-45}s^3$ and $A_{NT} = 9.3\cdot10^{-13}K^{-4}$ were taken from Holland's analysis of thermal conductivity [2], $A_{NL} = 4.3\cdot10^{-25}sK^3$ is a factor of 5 smaller than given in [2]. In summary, the simulation describes the ratio of the ballistic to diffusive contributions reasonably well but not thc monotonous decrease of the diffusive tail at 25 K and 27 K.

References

1. M. Obry, J. Tate, P. Berberich, H. Kinder: In Phonons 89 ed. by S. Hunklinger et al. (World Scientific Singapure 1990), p.328.
2. M. G. Holland: Phys. Rev. 132, 2461 (1963).
3. I. B. Levinson: Sov. Phys. JETP 52, 704 (1980).
4. R. J. von Gutfeld, A. H. Nethercot, Jr.: Phys. Rev. Lett. 17, 868 (1966).

Imaging the Current Density in Superconducting Thin Films Using Nonequilibrium Phonons

S. Chapman and A.J. Kent

Department of Physics, University of Nottingham,
University Park, Nottingham, NG7 2RD, UK

There is considerable interest in the superconducting properties of thin films because certain granular films exhibit a 2D phase transition at the Kosterlitz-Thouless transition temperature T_{KT} [1]. The conventional method of studying such systems is to make detailed measurements of the current-voltage characteristics of the films close to T_{KT} and compare the results with theory [2]. It is usually assumed that the film is narrower than the in plane magnetic penetration depth, $\lambda_\perp$, which is expected to be of the order of millimetres at T_{KT}. However, there are few if any measurements of $\lambda_\perp$ which enable the validity of this assumption to be checked. One way to find $\lambda_\perp$ is to probe the variation in current density, J, across the sample. It is expected to be largest within a region about $\lambda_\perp$ wide at the film edges. In this paper we describe a technique in which a movable beam of nonequilibrium phonons is used to probe the current distribution in thin tin films.

Figure 1 shows the sample arrangement. A granular tin film was vacuum evaporated onto one side of a 380μm thick (001) silicon wafer which was polished on both sides. A set of four electrical contacts (2 current and 2 voltage) were also deposited defining an active area of 4x4mm. Pulses from a Nd-YAG laser were thermalised in a 150nm thick constantan film deposited on the opposite face of the wafer. At liquid helium temperatures the resulting burst of nonequilibrium phonons travel ballistically across the wafer. Upon reaching the tin film some of the phonons were thermalised thus warming it slightly. If the film is biased close to its transition temperature, the warming leads to a change in the electrical resistance giving rise to a voltage pulse at constant film current. By raster (x-y) scanning the laser an image of the intensity of the voltage signal was made. If the film is macroscopically homogeneous with regard to the transition temperature then the size of the voltage signal $\delta V(x,y)=J(x,y)(\mathrm{d}R/\mathrm{d}T)\delta T$, where $\mathrm{d}R/\mathrm{d}T$ is the temperature sensitivity of the film's resistance at a point and δT the increase in temperature. By exploiting the strong transverse mode phonon focusing close to [001] in Si and by setting a narrow box-car gate to exclude the signal due to phonons that had not travelled the shortest path to reach the film we are able to obtain a spatial resolution down to about 100μm using this technique.

Figure 2 shows the film's R-T characteristics in the neighbourhood of the superconducting transition, T_{KT} is indicated. The I-V characteristics at various

Springer Series in Solid-State Sciences, Volume 112
Phonon Scattering in Condensed Matter VII Eds.: M. Meissner and R.O. Pohl

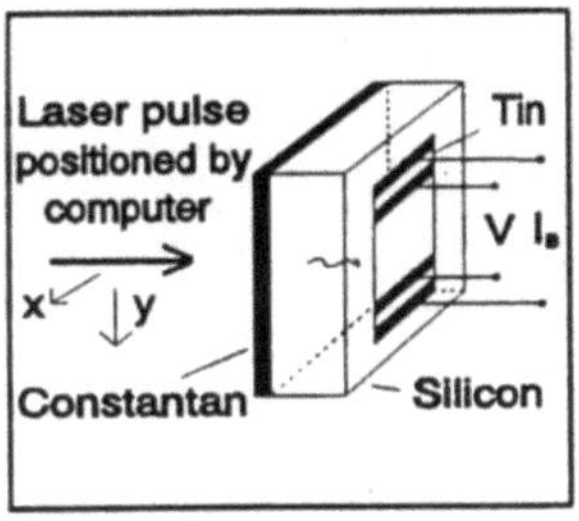

Figure 1

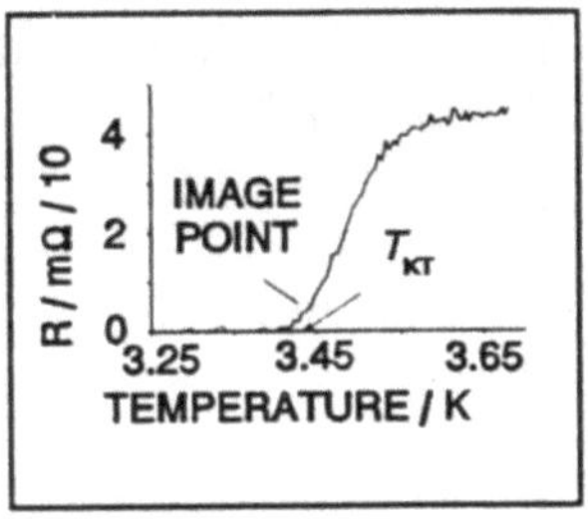

Figure 2

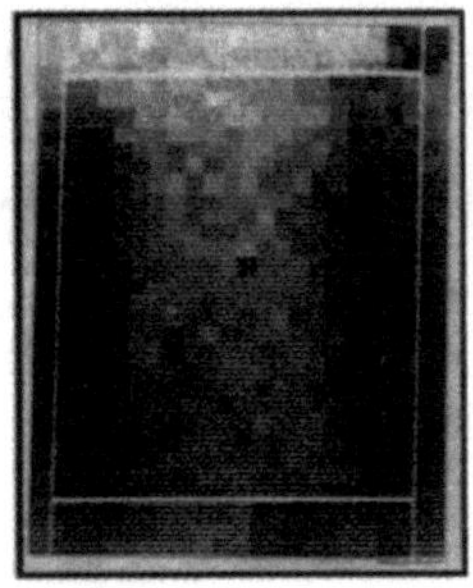

Figure 3

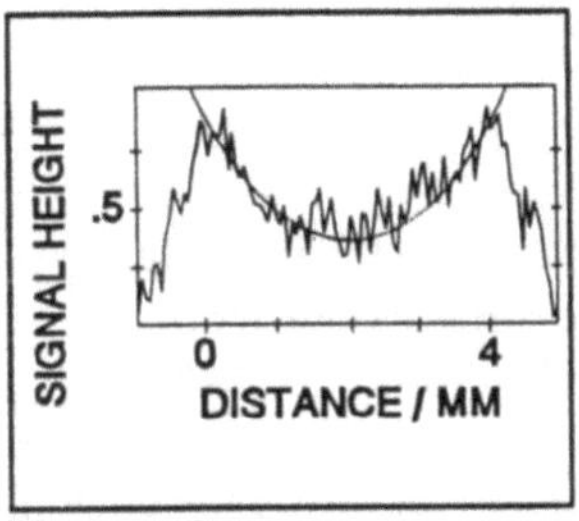

Figure 4

1: Sample geometry.
2: Superconducting transition of 100nm Tin film at .5mA bias current.
3: Full laser image of 100nm Tin film with .5mA bias current at the temperature shown in figure 2.
4: Linescan across Tin film with theoretical fit as discussed in the text.

temperatures suggested that the film could possibly be treated as a granular 'Josephson' medium in which T_{KT} is defined as the temperature at which the current-voltage characteristics follow a characteristic $V \propto I^3$ dependence [2]. Figure 3 shows a full image of the distribution of J, taken with the film biased near T_{KT} and with a total film current of .5mA. It is clear that J is greatest near the edges of the film. A linescan of $J(x)$ taken horizontally through the centre of the image is shown in figure 4. Assuming that $J(x)$ decreases exponentially with characteristic length $\lambda_\perp$ we obtain $J(x)=J_0\{\exp\text{-}x/\lambda_\perp + \exp(x\text{-}w)/\lambda_\perp\}$, where w is the film width. By fitting this to the data in figure 4 we obtain $\lambda_\perp=1.5$mm. If the film were an ideal 2-D Josephson medium we could expect $\lambda_\perp=9.8/T_{KT}=2.85$mm [1]. Although our result is of the correct order of magnitude we suspect that the discrepancy is due to the non-ideal nature of the film.

In conclusion, this technique is able to give a picture of the current distribution in a superconducting thin film. We find that the magnetic penetration length in a granular tin film is large at a temperature close to T_{KT} and is comparable with theoretical predictions. We would like to thank IBM and the Royal Society for financial support for us to attend this conference.

References

[1] Beasley M R, Mooij J E and Orlando T P, (1979) *Phys. Rev. Lett.* **42** p1165
[2] Epstein K, Goldman A M and Kadin A M, (1981) *Phys. Rev. Lett.* **47** p534

Ballistic Phonons Generated by Non-Radiative Decay to the Relaxed Excited State (RES)

M. Dominoni and N. Terzi

Dipartimento di Fisica, Università di Milano,
via Celoria 16, I-20133 Milano, Italy

Abstract. The relaxation around localized centers to the RES in ionic crystals, after femtosecond excitation, is reported in the time range from 10 *fs* up to 30 *ps*, for the case of longlived excited bound electrons (10^{-6}–10^{-3} *s*). The excited electrons, the neighboring ions and the host crystal ions form a complex system, whose non linear dynamics is resolved numerically with a method which allows to study centers with a rich electronic structure. One of the output is the ballistic propagation here reported of the initial pulse (as generated by the femtosecond light) from the impurity space to the crystal.

1. Introduction

The physical scheme of the fotoacoustic effect is the following: a light beam of suitable frequency excites absorbing centers inside a portion of a doped crystal. After the electron excitation, the centers relax non radiatively to the RES, generating elastic pulses which travel from the shined volume to the whole crystal, each starting from a different excited center at a different time. The travelling signal has therefore an initial composition formed by a statistical superposition of the pulses singularly generated at the centers. The subsequent propagation is not free but determined by the crystal characteristics (temperature, purity, perfection, etc.). However, for times smaller than the smaller scattering time (i.e. for the first transient of the relaxation process) the pulse propagation is ballistic and characteristic of the vibrational properties of the imperfect crystal. Here the ballistic propagation of one of the elementary uncorrelated phonon pulses composing the initial signal, is numerically studied using a method recently proposed. [1,2]

2. The System

The pulse is assumed to be generated by a femtosecond light beam. The electron excitation is so fast that the lattice can be considered static. The phonon emission process then follows the light absorption, and the phonons are re-

Springer Series in Solid-State Sciences, Volume 112
Phonon Scattering in Condensed Matter VII Eds.: M. Meissner and R.O. Pohl

leased according to the electron-phonon coupling and the vibrational properties of the crystal, in the time interval in which the the system reaches the RES. Three systems (excited electron of the center, local atoms around the center, surrounding crystal) are strongly coupled during this first transient, with the electron directly interacting with its neighbors (via the electron-phonon EP interaction), which are in turn coupled to the lattice (via the lattice force constants, either perturbed or imperturbed). We use a semi-classical approach and assume that the bound electron have a structure rich in levels, in particular that the excited state is threefold degenerate. The dynamics becomes than that of a non linear complex system, where the lattice relaxation causes a splitting of the electron levels and drives the electron dynamics, and vice versa. The details of the theoretical development can be found in refs. [1,2]. Solutions can be obtained only numerically. Depending on the electronic transition symmetry, on the coupling energies and on their relative strenghts, different relaxation regimes have been found. Here we report about the local lattice relaxation around a positive impurity, substitutional of a K^+-ion in a KI harmonic crystal, after it has been optically excited from a a-type to a t-type electronic state. The *Jahn-Teller* (JT) coupling so generated [3], has A (breathing), E (tetragonal) and T (trigonal) components of strenghts a, b and c respectevely. This is a $t \times (A + E + T)$ JT case. Assuming the standard n.n. limited coupling, the n.n. move after the optical excitation at $t = 0$, under the joint influence of the JT interaction and of the lattice force constants. The resulting motion is shown in figs. 1–2, where the time dependent radial displacements of the n.n. (I^- ions) and n.n.n. (K^+ ions) are reported. The chosen values of the JT coupling constants, $b = 2\hbar\bar{\omega}/\text{Å}$, $c = 6\hbar\bar{\omega}/\text{Å}$, $a = 31\hbar\bar{\omega}/\text{Å}$ ($\bar{\omega} = 0.85 \times 10^{13}\ rad\, s^{-1}$), correspond to an intermediate coupling JT case. No impurity induced change of the lattice force constants has been introduced. The damped radial oscillation of the n.n. (fig. 1) shows that the JT energy is flowing from the n.n. cluster towards the crystal with a relaxarion time of about $\tau \approx 5\,ps$. In other words, the cluster *radiates* phonons. Fig. 2 shows the passing of such a phonon wave packet throughout the n.n.n., with an overall passing time of about 10–15 ps. An intersting feature of the n.n.n. motion is that a fraction of the initial energy resides locally for a long time (at least 20 ps) as

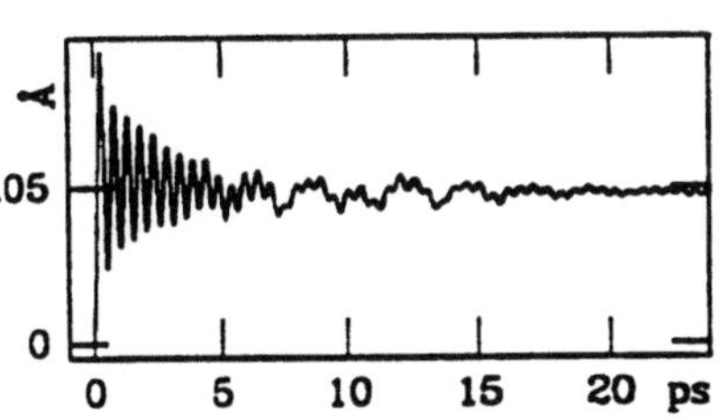

Fig.1. Radial displacement of the (1,0,0) n.n. ion I^- for an initially z-polarized electron state vs. time

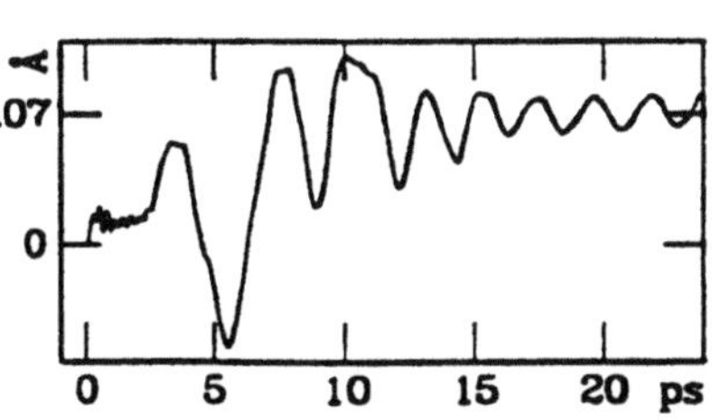

Fig.2. Radial displacement of the (1,1,0) next n.n. ion K^+ for the same case of fig.1

an oscillatory motion of low frequency ($\omega_{low} \approx 0.3 \times 10^{13}\ rad\, s^{-1}$). Notice that in the phonon composition of the initial wave packet this frequency is very low populated. We believe this feature a dynamical signature of the new curvature around the trigonal JT induced distortion of the lattice around the excited impurity.

3. References

[1] M. Dominoni and N. Terzi, Europhys. Lett. **15**, 515 (1991)

[2] M. Dominoni and N. Terzi, *Excited State Lattice Dynamics*, submitted for publication.

[3] see, for instance, *The Dynamical Jahn-Teller Effect in Localized Systems*, ed. Yu.E. Perlin and M. Wagner (North Holland, 1984)

Nonlinear Effects in a Focused Acoustic Beam in Pressurized Superfluid ^{4}He

K. Karaki[1], *T. Saito*[2], *Y. Sasaki*[1], *H. Kishi*[2], *and Y. Okuda*[2]

[1]Olympus Opt. Co. Ltd., Kuboyamacho 2-3, Hachioji, Tokyo 192, Japan
[2]Tokyo Institute of Technology, Meguro-ku, Tokyo 152, Japan

Sound propagation in superfluid ^{4}He can quite easily become highly nonlinear at a moderate sound energy, because its density and sound velocity are both very low. Furthermore, one can change the dispersion relation of phonon in superfluid ^{4}He from anomalous to normal by changing the pressure. The propagation properties of the finite amplitude acoustic wave has been investigated on the basis of these aspects(1). Using a high frequency focused pulse, we studied the propagation of acoustic waves which have much higher energy density than used in previous works, and observed new nonlinear effects(2).

Figure 1 shows a schematic setup of the experiments. A sound pulse of 400MHz is focused by acoustic lens in about a 1μm^2 at 300mK SVP in superfluid ^{4}He. We changed the power density at the focus from 10^3 to 10^7W/m^2 and measured the power of sound pulse reflected by the reflector. Such a high energy density coherent phonon field is easily generated by focused beam.

Typical results are shown in Fig.2. At a low pressure region where the dispersion relation is anomalous, a simple saturation behavior as seen in ordinal

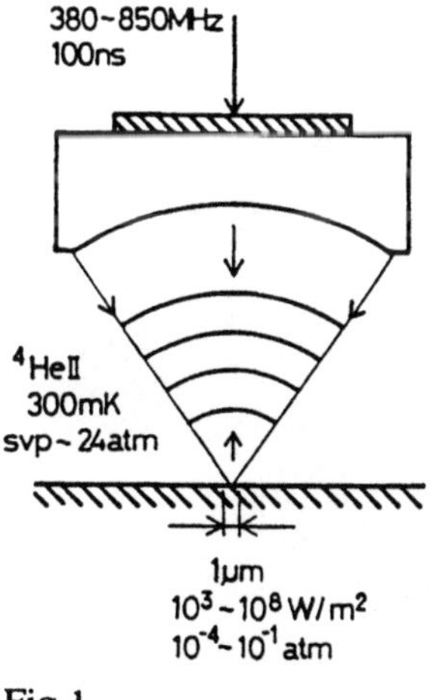

Fig.1.

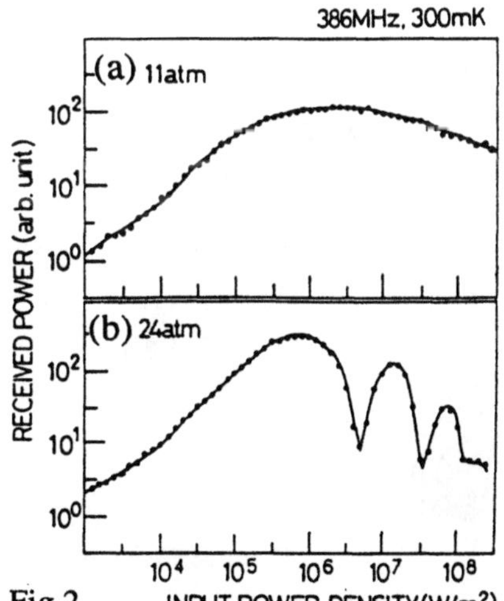

Fig.2.

Springer Series in Solid-State Sciences, Volume 112
Phonon Scattering in Condensed Matter VII Eds.: M. Meissner and R.O. Pohl

liquids is observed; for higher input power, this decreases with increasing the input power. This decrease is explained as the 3-phonon decay process and its parametric amplification[3].

At a high pressure region (normal dispersion), on the other hand, the behavior of received power changes dramatically as seen in Fig.2(b). The received power increases in proportion to the input power at the low input region, is saturated, and then decreases very rapidly ; it then increases, decreases, and increases again. Thus, there are dips or peaks. Second harmonics also have very complicated behavior which is closely related with fundamentals[4]. Since the 3-phonon process becomes forbidden under normal dispersion, the rapid depletion has been explained as the 4-phonon decay and its parametric amplification. Moreover, the second and third peaks have been caused by the waves regenerated from the second and higher harmonics[2]. (Some other process like cavitation may occur, but this cannot explain the re-increase of the received power.) In other words, according to this model, the input power actually disappears at the dips.

Another process, however, can explain these dips. If the focus position is changed by the process like self-focusing or self-defocusing, the acoustic lens would not be able to receive the reflected power, and so dips would appear. To confirm this point, we did the following experiments. We measured the reflected power as a function of input power at several distances (Z) between the acoustic lens and reflector. At a low pressure region, the simple behavior expected from Fig.2.(a) is observed (Fig.3.(a)), while at a high pressure region, results are very complex as seen in Fig.3.(b). At the input power of the dip in Fig.2(b), some peaks exist at the different Z. This means that the dips in Fig.3.(b) are not due to simple disappearance of the input energy, but more complicated phenomena which might include a change of focus or wavefront distortions.

In any event, these nonlinear phenomena in pressurized superfluid ^{4}He are very complicated and are closely related to the phonon dispersion relation, because they are observed only at the normal dispersion region. To understand them, it is essential to measure not only the power but also the phase of reflected waves, now these experiments are ongoing.

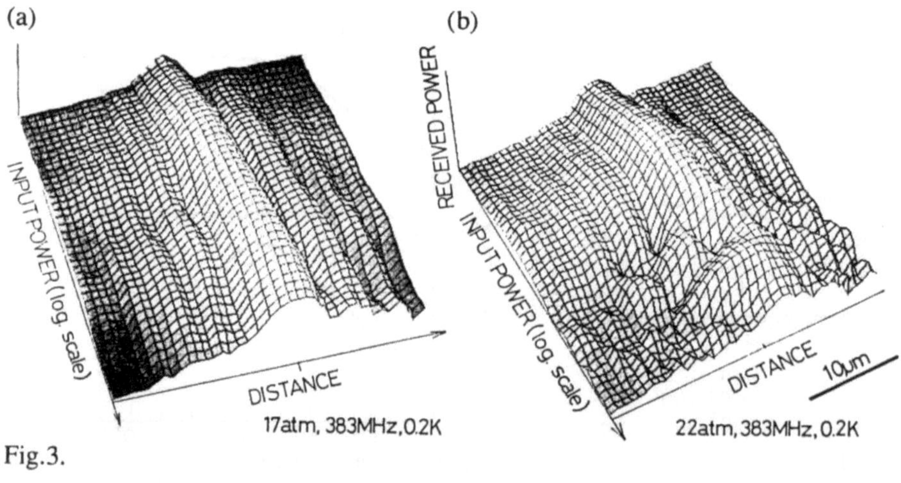

Fig.3.

(1) A.Hikata, H.Kwun, and C.Elbaum ; Phys.Rev.B21('80)3952
J.Joffrin, M.Chapellier, A.Schuhl, and M.Meisel ; J.Low Temp.Phys. 79('90)229
(2) K.Karaki, T.Saito, K.Matsumoto, and Y.Okuda ; Physica B165&166('90)131
K.Karaki, T.Saito, K.Matsumoto, and Y.Okuda ; J.Phys.Soc.Jpn. 60('91)360
A.Moulthrop, M.Muha, B.Hadimioglu, P.Silva, and G.Kozlowski ; IEEE UFF39('92)210
(3) J.Foster and S.Putterman ; Phys.Rev.Lett. 54('85)1810
(4) K.Karaki, T.Saito, K.Matsumoto, and Y.Okuda ; Appl.Phys.Lett. 59('91)908

Monte-Carlo Calculated Nonequilibrium Phonon Pulses in GaAs

D.V. Kazakovtsev[1], *B.A. Danilchenko*[2], *O.M. Kirshon*[2], *and I.A. Obukhov*[2]

[1]Institute of Solid State Physics, Russian Academy of Sciences, 142432 Chernogolovka, Moscow Region, Russia
[2]Institute of Physics, Ukrainian Academy of Sciences, 252008 Kiev, Ukraine

1. Introduction

A number of experiments uses optically excited electron system of the crystal as a source of nonequilibrium phonons. Wide set of these was carried out on GaAs (see [1,2]). The frequencies of the initial acoustic phonons, generated by excited carriers or via fast intermediate stage of optical phonon creation, are up to the boundary acoustic ones. The propagation of these phonons has a complicated character, as they have short mean free paths due both to isotope scattering and anharmonic processes. Thus, there is a problem to calculate the time shapes of registered phonon pulses.

2. Physical Model of the Propagation Process

We take the acoustic phonon spectrum of an isotropic medium, i.e., consisting of one longitudinal (LA) and two degenerate transverse (TA) nondispersive isotropic branches with parameters from [3]. For the isotope scattering time the Rayleigh law with a common value $\tau_I(\nu)=1.35\cdot10^{41}\nu^{-4}$ s is taken. We assume the nonequilibrium phonon occupation numbers n«1, so inelastic scattering reduces to anharmonic three phonon decay. All frequencies of phonons, created by decay, consistent with the energy and quasimomentum conservation laws are taken into account. TA phonon branch is not subject to decay, but intermode conversion due to isotope scattering is present. For LA mode the anharmonic decay time is $\tau(\nu)=A\tau_0(\nu)$, where $\tau_0(\nu)=1\cdot10^{54}\ \nu^{-5}$s is the calculated in [3] value. Constant A for different calculations is changed to reveal the effect of inelastic scattering and is a set of powers of two.

An initial phonon of LA (or TA) mode with frequency ν_0 starts from the center of a sphere at time t=0. In a cascade of scatterings with possible multiplying of number of phonons due to decay all the phonons are tracked until they reach R=1 cm sphere. The electrical signal at the bolometer, proportional to the nonequilibrium phonon energy flux at the detector, is calculated for a set of frequencies ν_0 by summation of a large number of histograms obtained by above outlined procedure.

Springer Series in Solid-State Sciences, Volume 112
Phonon Scattering in Condensed Matter VII Eds.: M. Meissner and R.O. Pohl

3. Results and Discussion

The calculated normalized energy fluxes $\gamma = E^{-1} \cdot dE/dt$ are represented in Fig. 1. For starting LA mode we see a sharp ballistic TA phonon pulse accompanied by quasiballistic tail at $\tau=\tau_0/4$ and quasidiffusive pulse at $\tau=4\tau_0$. For initial TA phonon at the time of the ballistic TA pulse an abrupt increase of phonon flux appears, the step being larger for quasiballistics $\tau=\tau_0/4$ than for quasidiffusion $\tau=4\tau_0$. For $A=\infty$ (no inelastic scattering) typically diffusive spectra are obtained. The changes of pulse shape with starting frequency ν_0 change (Fig. 2) are partly similar to those due to variation of time τ. For ν_0=0.5 Thz the pulses (not shown here) are typically ballistic, their width being about 0.1 μs and peak value 4.5 and 3 for LA and TA modes respectively.

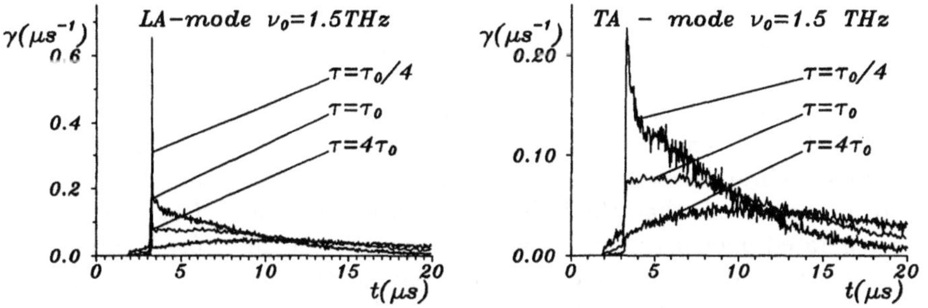

Fig. 1. Calculated phonon energy fluxes for starting at t=0 phonon of LA or TA mode for different anharmonic times τ.

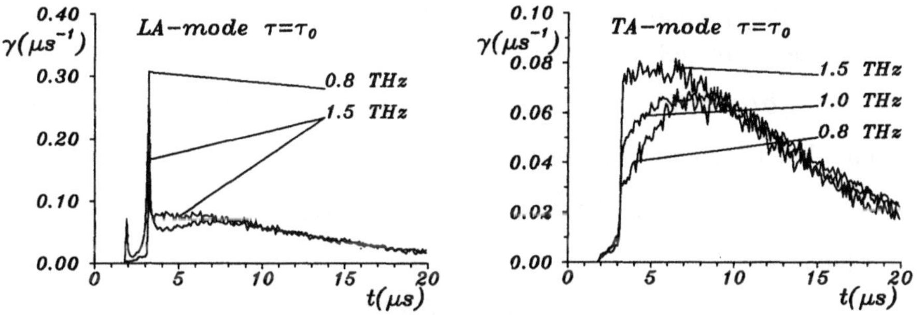

Fig. 2. The same as in Fig. 1, but for starting phonon of different initial frequencies ν_0 and $\tau=\tau_0$.

REFERENCES

1. Ulbrich R.G., in *Nonequilibrium Phonon Dynamics*, edited by W.E.Bron (Plenum, New York, 1985), p. 3.

2. B.A.Danilchenko, V.N.Poroshin, M.I.Slutskii, M.Ashe, phys. stat. sol. (b) **136**, 63 (1986).

3. S Tamura, Phys.Rev.B **31**, 2574 (1985).

Source Terms for the Boltzmann-Peierls Equation for Phonons

Cz. Jasiukiewicz and T. Paszkiewicz

Institute of Theoretical Physics, University of Wroclaw, Pl. Maxa Borna 9, PL-50-204 Wroclaw, Poland

Consider pulses of long-wavelength acoustic phonons impingent upon a detector. These pulses are produced by point or extended sources of monochromatic or Planckian phonons. The sources are located at the origin of a Cartesian coordinate frame with the z-axis perpendicular to the specimen surface and directed into the medium. Assume that the macroscopic sample at temperature T much lower than the Debye temperature θ_D is a perfect crystal, then the generated phonons move ballistically towards the detector. The phonons are characterized by the direction $\hat{\mathbf{q}}$ and the length q of the wave vector, polarization j $(j = 0, 1, 2)$, frequency $\omega(Q)$ $(Q = (\hat{\mathbf{q}}, q, j))$ phase $c(\hat{Q})$ and the group $\mathbf{v}(Q)$ velocities (here $\hat{Q} \equiv (\hat{\mathbf{q}}, j)$). A nonequilibrium state of the phonon gas is described by the distribution function $N(Q; \mathbf{r}, t)$. At the described conditions the deviation $\delta N(Q; \mathbf{r}, t) \equiv N(Q; \mathbf{r}, t) - N_0[\hbar\omega(Q)/k_B T]$ (N_0 is the Planck function) obeys the BPE, which in place of a collision term contains the source term (cf. [1]) $\dot{\zeta}(Q; \mathbf{r}, t)$

$$\partial/\partial t \; \delta \; N(Q; \mathbf{r}, t) \; + \; \mathbf{v}(\hat{Q}) \nabla \; \delta N(Q; \mathbf{r}, t) = \dot{\zeta}(Q; \mathbf{r}, t) \; . \tag{1}$$

Assume that the source term $\dot{\zeta}$ $(Q; \mathbf{r}, t)$ factorizes

$$\dot{\zeta}(Q; \mathbf{r}, t) \; = \; A(Q) F_1(\mathbf{r}) F_2(t) \; . \tag{2}$$

For a point source

$$F_1(\mathbf{r}) \; = \; \delta(\mathbf{r}) \; \equiv \; \delta(x)\delta(y)\delta(z) \; , \tag{3a}$$

where $\delta(x)$ is the Dirac distribution delta. When phonon wave packets are generated by heating with a laser beam a metallic film evaporated on the surface of a dielectric or semiconducting specimen

$$F_1(\mathbf{r}) \; = \; V_h \delta(z) \left(2\pi d^2\right)^{-1} \exp\left[-\mathbf{r}_\perp^2 \Big/ \left(2d^2\right)\right] \; , \tag{3b}$$

where d is a length characterizing the linear dimension of the heated area, $\mathbf{r}_\perp$ is the component of $\mathbf{r}$ perpendicular to the specimen surface. If the shortest characteristic time of the problem is the pulse width one can assume that

Springer Series in Solid-State Sciences, Volume 112
Phonon Scattering in Condensed Matter VII Eds.: M. Meissner and R.O. Pohl

$F_1(t) = \delta(t)$. For the source of monochromatic phonons we taken [1]

$$A(Q) = (4\pi^2/3)\, c^3(\hat{Q})\omega^{-2}(Q)\delta\left[\omega(Q) - \omega_0\right]\theta\left[v_z(\hat{Q})\right], \qquad (4a)$$

where $\theta(x)$ is the Heaviside step function. For the source of Planckian phonons with temperature T_h $(T_h > T)$ the Q dependent part of $\dot{\zeta}$ reads

$$A(Q) = 2V_h\left[c_D^{(s)}/c_D^{(h)}\right]^3 \alpha(h \rightarrow s) N_0\left(\hbar\omega/k_B T_h\right)\theta\left(v_z(\hat{Q})\right). \qquad (4b)$$

where $c_D^{(\sigma)}$ $(\sigma = h, s)$ is the Debye velocity of the heater (h) and specimen (s). The coefficient $\alpha((h \rightarrow s)$ is the transmission probability for phonon scattering at an interface which we assume is diffusive (cf. [2]), and V_h is the heater volume. The solution of BPE is, for (4a) and (4b) respectively, proportional to

$$\theta(t)\delta\left[\mathbf{r} - \mathbf{v}(\hat{Q})t\right], \qquad (5a)$$

$$\theta(t)\delta\left[z - v_z(\hat{Q})t\right]\exp\left\{-\left[\mathbf{r}_\perp - \mathbf{v}_\perp(\hat{Q})t\right]^2/(2d^2)\right\}\Big/\left(2\pi d^2\right). \qquad (5b)$$

With (5) we can calculate components of densities of currents of energy $\mathbf{j}_E(\mathbf{r}, t)$, of the quasimomentum $\pi(\mathbf{r}, t)$ (cf. [1]), the density of energy and a suitable component of the density of quasimomentum falling onto a detector. Here we restrict ourselves to results for $e(\mathbf{r})$ - the total energy density falling onto a detector with a normal $\mathbf{n}$ and located at $\mathbf{r} = (r, \theta_r, \phi_r)$. Phonons which contribute to this density satisfy the conditions: $\theta_r = \theta_v(\hat{Q})$, $\phi_r = \phi_v(\hat{Q})$. We obtained results for: (i) a point source of monochromatic phonons of frequency ω_0 [1], (ii) a point source of Planckian phonons, (iii) a Gaussian source of monochromatic phonons, and (iii) a Gaussian source of Planckian phonons.

Acknowledgements: This work is supported by a grant from the National Committee for Scientific Research of Poland under the contract No 397/2/91.

References

[1] Cz. Jasiukiewicz, D. Lehmann, T. Paszkiewicz, Z. Phys. B–Condensed Matter **84**, 73, 1991

[2] E.T. Swartz, R.O Pohl, Rev. Mod. Phys. **61**, 605, 1989

Part III

Phonon Spectroscopy

Investigations Using Phonon Detection by Exciton Luminescence

A.A. Kaplyanskii, A.V. Akimov, and E.S. Moskalenko

A.F. Ioffe Physico-Technical Institute, 194021 St. Petersburg, Russia

1. Introduction.

Fluorescence detection of nonequilibrium high-frequency acoustic phonons is based on interaction of phonons with different kinds of electronic excitations in solids. For semiconductors the interaction of phonons with intrinsic electronic excitations, excitons, may be used [1,2]. The effects of nonequilibrium phonons on exciton fluorescence are diverse and extremely large in value thus permitting the effective use of them in nonequilibrium phonon physics.

2. Exciton-phonon interaction processes.

1. In event of interaction of acoustic phonon with free exciton (FE) the kinetic energy E and momentum k of the exciton change due to absorption of an acoustic phonon, the total energy and momentum being conserved. Only the phonons with frequencies

$$\hbar\omega \leqslant 2ms^2 + 2(2ms^2E)^{1/2} \qquad (1)$$

are active in the exciton-phonon interaction, m is the exciton mass, s is sound velocity. Phonon-induced intraband FE transitions result: (i) in "heating" of exciton gas and (ii) in drag of FE by phonons if the directed phonon flux exists [3].

2. If several near exciton bands exist, the phonon-induced interband transitions are possible conserving full energy and momentum of quasiparticles.

3. The phonon-induced dissociation of FE requires phonon energy $\hbar\omega > \Delta E$, binding energy of free excitons.

4. The excitons, bound at liquid helium temperature (LHT) at impurities may be released via absorption of phonons with energy $\hbar\omega > E_b$, where E_b is the binding energy of bound excitons (BE). The dissociation BE→FE

Springer Series in Solid-State Sciences, Volume 112
Phonon Scattering in Condensed Matter VII Eds.: M. Meissner and R.O. Pohl

results in redistribution of concentrations of FE and BE in favour of FE.

5. Phonons may induce transitions between localized ground and excited states of BE.

The processes listed above result in phonon-induced changes in exciton luminescence spectra. Below experimental studies with exciton fluorescence detection of nonequilibrium phonons are briefly reviewed.

3. Exciton luminescence and spectrum of phonon hot spot.

The studies of kinetic energy distribution of excitons, interacting with phonons in hot spot (HS) [4] were performed using Cu_2O crystals [5,6]. The optical phonon assisted FE luminescence line 614 nm is due to radiative annihilation of 1s-excitons having different k in band; its spectral shape I(E) directly reflects the kinetic energy distribution f(E) of excitons : $I(E) \sim f(E)\rho(E)$ where $\rho(E) \sim \sqrt{E}$ is density of states in 1s band. Strong SHG pulses of YAG:Nd laser excite simultaneously both phonon HS and luminescence of 1s excitons. The broadening of I(E) shape with increase of pulse power P was observed which indicates the heating of exciton gas. The shape of broadened I(E) cannot be approximated by an equilibrium Maxwell-Boltzman (MB) distribution and hence the distribution of HS phonons, heating the exciton gas is non-Planckian. The effective "temperature" for low energy part of experimental FE distribution function $f(E) \sim I(E)/\sqrt{E}$ is lower, than "temperature" for high energy part. Taking into account (1) this means that phonon distribution has relative deficit of low frequency phonons [5-7].

4. Phonon drag of excitons.

The drag of FE by phonon flux (phonon wind) results in spatial shift of exciton cloud (EC) in sample which may produce different effects in FE fluorescence.

(i) In [2,8] the EC was created by interband optical pumping in thin layer (~1 μm) near the surface of CdS crystal plate at LNT, the phonons being generated simultaneously near the same face (HS). The emission in zero phonon FE line was studied which due to strong reabsorption is observed only from subsurface layer of the plate. At low pumping levels the emission is seen only from pumped face. At higher pumping levels the FE emission appears from opposite face being delayed in time relative to the optical pumping pulse. These observations directly indicate the drift motion of excitons accompanied by smearing of EC through the whole sample (thickness 20 μm) induced by phonon flux, emitted from HS.

(ii) In experiments [9,10] the "ordinary" (oxidized) surface of pure c-Si sample immersed in LH (T=1.7 K) was excited by cw Ar-laser beam, which produced EC near the surface. The heat pulses were injected into the crystal from metal film heater at the opposite face of the sample. The phonon induced temporal changes $\Delta I^F(t)$ in the intensity of FE-LO,TO phonon assisted fluorescence lines were measured. The signal $\Delta I^F(t)$ was positive or negative depending whether the surface was in contact with vacuum or with LH respectively.

Two kinds of exciton-phonon processes (Sect.2) give contributions of opposite sign to the total $\Delta I^F(t)$ signal: (i) dissociation BE→FE which enhances the number of FE ($\Delta I^F>0$), (ii) the exciton drag by phonon flux towards surface, where FE recombine effectively ($\Delta I^F<0$). In the case of vacuum boundary the back-scattered phonon flux compensates incident one, the net flux in EC region becomes negligible and negative FE drag contribution to ΔI^F is cancelled. In the case of LH boundary anomalous Kapitza phonon transmission into LH results in appearance of noncompensated phonon flux and negative FE drag contribution predominates.

5. Reflection and transmission of phonons on crystal-LH boundary.

In [9,10] the heat pulses induced FE-luminescence signals $\Delta I^F(t)$ in c-Si were measured for EC near freshly cleaved in situ atomically pure <111>-surface in contact with LH. The positive signal $\Delta I^F(t)$ was observed indicating the absence of prominent FE drag, which may be attributed to very high phonon reflection at crystal-LH boundary resulting in isotropization of phonon distribution (see Sect.4). After oxidation of the same cleaved surface at the open air the signal $\Delta I^F(t)$ became negative because of Kapitza phonon transmission into LH resulting in appearance of phonon flux and in FE fluorescence drag quenching (Sect. 4). These results correlate well with observations of very high reflection of phonons from perfect cleaved [11] surfaces in contact with LH.

6. Excitonic detector of phonons in c-Si.

The phonon-induced signal $\Delta I^F(t)$ of FE-luminescence, emitted from EC cloud near the surface of c-Si is determined by the balance of two frequency dependent opposite sign contributions (see Sect.2). The positive BE→FE contribution is stimulated by "high-frequency" phonons with $\hbar\omega > \Delta E_b = 3.8$ meV (0.9 THz). The negative FE-drag contribution is stimulated according to (1) by "low-frequency" phonons ω < 0.4 THz. This enables to use the FE and BE fluorescence of c-Si as phonon detector with spectrometric potentialities.

(i) Isotope scattering of THz phonons in c-Si [9,10]. From the shape of FE and BE fluorescence pulses, induced by phonons propagating in <100>,<111> - directions the prominent ballistic component for phonons with ω = 0.9 THz was observed. In special geometry when both the film heater and detector (EC) were at the same face of the sample with a deep cut between them the strong bulk scattering of phonons with ω > 0.9 THz was established. These facts show that the mean free path for THz phonons in pure Si is of the order of few mm.

(ii) Phonon transmission through amorphous films [12]. It was observed that amplitude of negative signal $\Delta I^F(t)$ induced by heat pulses in FE luminescence of EC at "ordinary" surface in contact with LH is strongly increased, when the phonons emitted from film heater pass through the a-Si:H films; the increase was the stronger, the thicker was the film. This showed that the transport of phonons through a-Si:H films is accompanied by effective down-conversion from high-frequency to low-frequency part of phonon spectrum (ω < 0.4 THz), the latter being responsible for negative contribution to ΔI^F. Other measurements revealed the correlation of down-conversion efficiency in a-Si:H film caused by inelastic scattering of phonons with H content. In similar experiments with a-Ge films the time delay of phonon pulse, passed through 0.5 μm film was observed.

(iii) Phonon emission from 2D electron gas [13]. Phonon spectra emitted by two dimensional electron gas (2DEG) in a Si-MOSFET, heated by current pulses were studied by detection of negative BE-luminescence pulses from EC at opposite face of c-Si substrate, which were induced by phonons with ω > 0.9 THz active in BE→FE dissociation. At constant power dissipated in 2DEG the amplitude of $\Delta I^B(t)$ was found to decrease with increase of sheet density of 2DEG. This indicates the shift of emitted phonon distribution to lower (subthreshold) frequencies rather than to higher frequencies as would be expected for emission from heated electrons in a single 2D sub-band. This was explained by the population of the second (excited) sub-band of 2DEG and by increase in the emission of low frequency phonons from low-wavenumber electrons in this sub-band.

7. Other studies using exciton fluorescence.

The dissociation of FE in GaAs at high injected phonon pumping level was observed [14]. The phonon induced quenching of BE luminescence in CdS and GaAs was interpreted as a result of one-phonon transitions from ground to excited BE states decaying via nonradiative Auger-processes [14,15]. The phonon-induced scattering of FE between different exciton bands in c-Si was discussed [16]. In [1,17] the phonon-induced anti-Stokes acoustic phonon wing of zero-phonon BE lines ("vibronic spectrometer") was used to analyze the phonon spectrum of metal film injectors.

References.

1. J. Shah, R.F. Leheny, A.H. Dayem, Phys. Rev. Lett. 33, 818 (1974).
2. N.N. Zinovyev, U. Parmanbekov, I.D. Yaroschetskii, JETP Lett. 33, 584 (1981).
3. L.V. Keldysh, Sov. Phys. JETP Lett. 23, 86 (1976).
4. J.C. Hensel, R.C. Dynes, Phys. Rev. Lett. 39, 969 (1977).
5. A.V. Akimov, A.A. Kaplyanskii, E.S. Moskalenko: in Phonon Physics ed. by J. Kollar, N. Kroo, N. Menyhard, T. Siclos, (World Scientific, Budapest, Hungary, 1985) p. 449.
6. A.V. Akimov, A.A. Kaplyanskii, E.S. Moskalenko Fiz. Tverd. Tela, 29, 509 (1987).
7. E.L. Ivchenko, L.V. Takunov, Fiz. Tverd. Tela 30,1161 (1988).
8. N.N. Zinovyev, L.P. Ivanov, V.I. Kozub, I.D. Yaroschetskii, Sov. Phys. JETP 57, 1027 (1983).
9. A.V. Akimov, A.A. Kaplyanskii, E.S. Moskalenko, R.A. Titov, Zh. Exp. Teor. Fiz. 94, 307 (1988).
10. A.V. Akimov, A.A. Kaplyanskii, E.S. Moskalenko, Physica B. 169, 382 (1991); J. Lum. 45, 135 (1990).
11. J. Weber, W. Sandman. W. Dietsche, H. Kinder, Phys. Rev. Lett., 40, 1469 (1978).
12. A.V. Akimov, A.A. Kaplyanskii, J. Kocka, E.S. Moskalenko, J. Stuchlik, Zh. Exp. Teor. Fiz. 100, 1340(1991).
13. A.V. Akimov, L.J. Challis, J. Cooper, C.J. Mellor, E.S. Moskalenko, Phys. Rev. (1992) in press.
14. A.Yu. Blank, N.N. Zinovyev, L.P. Ivanov, D.I. Kovalev, I.D. Yaroschetskii, Fiz. Techn. Polyprov. 25, 67(1991).
15. B.I. Gelmont, N.N. Zinovyev, D.I. Harchenko, I.D. Yaroschetskii, I.N. Yassievich, Zh. Exp. Teor. Fiz. 94, 322 (1988).
16. N.N. Zinovyev, D.I. Kovalev, I.D. Yaroschetskii, A.Yu. Blank, Pisma Zh. Exp. Teor. Fiz. 53, 147 (1991).
17. N.N. Zinovyev, D.I.Kovalev, V.I. Kozub, I.D. Yaroschetskii, Zh. Exp.Teor. Fiz. 92, 1331 (1987).

Nonequilibrium Acoustic Phonon Influence on Quantum Interference of Free-Exciton Radiative Decay

N.N. Zinov'ev, D.I. Kovalev, and I.D. Yaroshetskii

A.F. Ioffe Physico-Technical Institute, 194021 St. Petersburg, Russia

In this paper we report a new feature of the effect of nonequilibrium acoustic phonons on exciton radiative transitions. Free excitons(FEs) are usually presented with the simple model of the hydrogen atom which is actually accurate only for a model of ideal semiconductor with nondegenerate isotropic and parabolic valence and conduction bands. However all crystals with the diamond or zinc blende symmetry have a degenerate valence band at zero **k**-values. Some of those materials have their absolute minima of the forbidden gap with the nonzero **k**-values. The FE radiative decay in semiconductors with indirect gap proceeds via an intermeduate state. At low temperature the exciton electron emits a near zone edge phonon. It then annihilates a hole in the valence band, emitting a photon. Another possibility for the FE to recombine, is for the exciton hole to be scattered to a high **k**-value state in the valence band allowing the electron to make a vertical transition again emitting a photon. These two recombination paths interfere with each other so that the total probability of the FE radiative decay becomes dependent on the final state energy. Indeed this complicated zone structure does not allow the separation of the center of mass and relative motions. This is considered as a perturbation that lift the degeneracy of the hole state in the FE[1,2]. In the particular case of Si and Ge the FE ground state is split into two Kramers subbands due to the symmetry of this interaction[3]. The optical transitions matrix elements between split valence band states and indirect minima of the conduction band are of different values because of the interference effect which accompanied by the different amplitudes for the hole states involved in these transitions.

The aim of the present work was to investigate the effect of nonequilibrium acoustic phonons on the interference of the FE radiative probabilities and to examine the corresponding quasiresonant scattering of phonons in photoexcited semiconductors.

In the experiments we studied Si single crystals, where the separation of the two subbands is estimated to be 0.35 meV[4] which is within the frequency range of heat generators. A single luminescence line is seen which is believed to contain a component for each subband. The relative population of the two

Springer Series in Solid-State Sciences, Volume 112
Phonon Scattering in Condensed Matter VII Eds.: M. Meissner and R.O. Pohl

subbands can be changed by incident phonons and we have studied the effect of this on the integrated intensity of the FE line. The effect is very sensitive to the ambient temperature of the sample. No effect was seen for $T < 1.9K$ but a marked increase in the intensity due to phonons appeared at higher temperatures. Very different behaviour was seen however in the bound exciton luminescence. In this case phonons reduced the luminescence intensity at all temperatures and no dependence on ambient temperature was found. This confirms previous observation[5], where the effect on BE was interpreted as due to transitions within the BE.

The explanation for the FE luminescence behaviour is attributable to the requirements of energy and momentum conservation: $\hbar(\mathbf{q}) = E_+(\mathbf{k}_f) - E_-(\mathbf{k}_i)$, and $\mathbf{q} = \mathbf{k}_f - \mathbf{k}_i$, where the + and - specify the corresponding exciton subbands, $\mathbf{k}_i$, $\mathbf{k}_f$ and $\mathbf{q}$ are the wave vectors of initial and final states of the exciton and the phonon wave vector correspondingly. These can only be satisfied for excitons with finite **k**-values. So as T increases, the numbers of these also increases until there are sufficient for their transitions to be seen in the luminescence signal. Assuming that the phonon wave vector **q** equals to that of the exciton $\mathbf{k}_i$ to reach the bottom of upper branch at $\mathbf{k}_f = 0$ we found the value $\Delta E = E_{onset}(\mathbf{k}_i) + \hbar \mathbf{q} s_{ph} = 0.3$ meV, where s_{ph} is the sound velocity and $E_{onset}(\mathbf{k}_i) = k_B T/2$ is exciton kinetic energy. Theoretical predictions for the energy splitting of the exciton subbands[3] coincide with the observed value ΔE. A decrease of the net positive signal found at higher temperature could be explained by the process of elastic exciton-exciton collisions which tend to equalise the populations of the subbands. Another experiment that we have done shows the photoinduced scattering of the phonon pulse by FEs. As long as FEs are excited in the path of nonequilibrium acoustic phonon propagation an additional delay occurs in the maximum of time-of-flight pulse and broadening effects are also seen. We believe this is qualitative evidence of phonon quasiresonant scattering effects.

To summarize, a new photoinduced effect of quasiresonant scattering of acoustic phonons has been found. Due to phonon absorbtion FEs undergo transitions to the upper subband of its ground state where quantum interference of partial radiative probabilities becomes constructive one.

References

[1] McLean T P, Loudon R J Phys. Chem. Solids 13, 1, (1960)
[2] Smith D L , McGill T C Phys, Rev.B, 14, 2448, (1976)
[3] Lipari N O, Altarelli M Phys. Rev. B15, 4883, (1977)
[4] Hammond R B, Silver R N Sol. State Commun. 28, 993, (1978)
[5] Zinov'ev N N, Kovalev D I, Yaroshetskii I D this volume.

Processes of Phonon Absorption on Shallow Exciton Impurity Complexes

N.N. Zinov'ev, D.I. Kovalev, and I.D. Yaroshetskii

A.F. Ioffe Phsico-Technical Institute, 194021 St. Petersburg, Russia

In doped semiconductors at low temperatures free excitons(FE) become bound to impurities as exciton impurity complexes (EICs) and no longer exist as separate entities. From the spectroscopic study of EIC it has been established that the energy levels of an EIC are qualitatively similar to that of the impurity to which it is bound[1]. Two problems associated with phonon absorption by the EIC are discussed in this paper. The first could arise if there is an increase of the Auger nonradiative decay due to the transition of the EIC to one of its excited states[2]. The second is concerned with the possible dissociation of the EIC when the phonon energy equals the FE binding energy[3]. To shed light on these problems we have undertaken a study of the EIC excitation by nonequilibrium acoustic phonons.

We have investigated the EIC in Si:P and Si:B. They have almost the same FE binding energy but the configurations of the single electron shells are rather different[1]. The effect of incident phonons on the EIC luminescence is to produce a quenching of the EIC spectral lines. For the EIC in Si:P, contributions due to both nonradiative decay and dissociation were found. In the second case a corresponding increase in FE luminescence occurs allowing us to determine the relative contributions of the two processes. However for the EIC in Si:B only phonon-assisted nonradiative decay was detected.

Effects of the EIC luminescence quenching can be seen by analysing the phonon interaction with the EIC. The matrix element of the transition between the ground state and the continuum due to absorption of acoustic phonons can be described by the expression:

$$|\langle i|H_{e\text{-}ph}|f\rangle|^2 \propto \Xi^2(n_{ph} + ½ \pm ½)|F(q)|^2 \quad ,$$

where n_{ph} is the acoustic phonon occupation number, and Ξ is the effective deformation potential. The factor F(q) is given by

$$F(q) = \frac{1}{[1+(a*q)^2]^2} \quad ,$$

Springer Series in Solid-State Sciences, Volume 112
Phonon Scattering in Condensed Matter VII Eds.: M. Meissner and R.O. Pohl

where a^* is the effective Bohr radius of the bound states with binding energy $E_B = \hbar^2/2m^*(a^*)^2$ and q is the wave vector of the acoustic phonon. The interaction of acoustic phonons with the EIC vanishes when $qa^* > 1$ because of the difficulty of conserving momentum so the interaction is very sensitive to the size of a^*. The value of a^* for the EIC bound to a neutral donor is very much smaller than the effective mass value because of the large wave function amplitude of the electrons in the attractive potential of the central cell. However when the EICs are bound to an acceptor impurity the electrons are repelled by the central cell potential leading to a larger effective Bohr radius a^*. These a^* values are also reflected in the valley-orbit splitting where differences between these two EICs have been confirmed by the negligible valley-orbit splitting observed for Γ_1 - $\Gamma_{3,5}$ states in the EIC in Si:B and the fairly large splitting of the EIC spectrum in Si:P[1].

Another effect of nonequilibrium phonons is in the reduction of the EIC luminescence intensities due to Auger nonradiative decay. In the EIC ground state, the angular momentum is zero (S-state), as is that of the conduction electrons. However holes have non-zero angular momentum. Therefore the Auger decay via $e + e + h \rightarrow e^*$ or $h + h + e \rightarrow h^*$ is strongly forbidden[2]. However incident phonons can excite the EIC to states of nonzero angular momentum allowing the decay to proceed. The resulting decrease in the radiative rates (luminescence) was observed for both Si:P and Si:B.

To summarise, the phonon induced processes of EIC luminescence are due to the combination of two phenomena: dissociation of EIC and nonradiative Auger type decay. Dissociation of the EIC occurs predominantly at donor centres where the effective Bohr radius of the bound carriers is typically less than the phonon wave length.

References

1. M L W Thewalt in "Excitons" ed by E I Rasha and M D Sturge, Series of Modern Problems of Condensed Matter Sciences, v 2.
2. B L Gelmont, N N Zinov'ev, D I Kovalev, V A Kharchenko, I D Yaroshetskii, I N Yassievich, Sov Phys JETP 67, 613, 1988.
3. A V Akimov, A A Kaplyanskii, E S Moskalenko, R A Titov, Sov Phys JETP 67, 2348, 1988.

Femtosecond Studies of Coherent LO-Phonons in GaAs

T. Pfeifer, T. Dekorsy, W. Kütt, and H. Kurz
Institute of Semiconductor Electronics, RWTH Aachen, Sommerfeldstr. 21, W-5100 Aachen, Fed. Rep. of Germany

Abstract. Time-resolved detection of amplitude and phase of coherent LO-phonons, which are impulsively excited in surface-space-charge regions by ultrashort laser pulses, allows detailed studies of the generation and the dynamics of coherent phonon modes in the presence of a dense nonequilibrium electron-hole-plasma.

Optical injection of electron-hole-pairs in GaAs surface-space-charge regions with laser pulses of 50 fs duration and 2 eV photon energy leads to ultrafast screening of electrical surface fields via free carrier transport, thus depolarizing the crystal lattice in a time shorter than one phonon period. The longitudinal electrical field changes associated with the charge separation and the coherent LO-phonons are sampled in a pump-probe set-up with a Reflective Electro-Optic Sampling (REOS) detection scheme.

The REOS-signals of p- and n-doped GaAs (Fig. 1) at an excitation density of $N_{exc} = 1 * 10^{18}cm^{-3}$ reveal a sign reversal due to the reversed band bending. The complete screening of the surface field E_S via the separation of

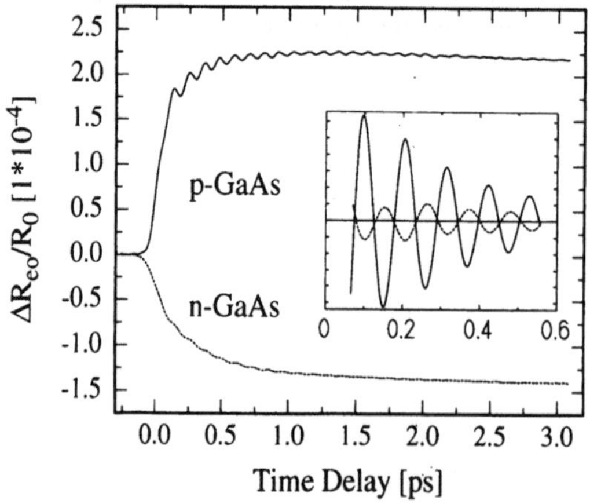

Fig. 1: Time-resolved REOS-signals of (100)-oriented p- and n-doped GaAs, both at a doping density of $N_{dop} = 2 * 10^{18}cm^{-3}$. In the inset the retrieved phonon signal contributions are shown. The measurements were performed at room temperature.

Springer Series in Solid-State Sciences, Volume 112
Phonon Scattering in Condensed Matter VII Eds.: M. Meissner and R.O. Pohl

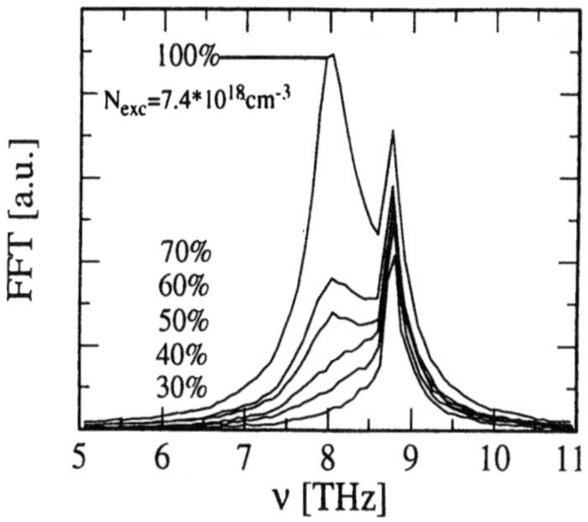

Fig. 2: Fourier Transform of the time-resolved REOS-phonon signals as a function of the excitation density. The data have been taken on a (100)-orientated intrinsic GaAs sample at room temperature.

the photoexcited electrons and holes is accomplished within 2 ps and the saturation values indicate surface fields of comparable magnitude ($E_S \sim \sqrt{N_{dop}}$). A faster initial signal rise is observed in the p-doped GaAs sample, revealing a more effective screening within the pump pulse duration.

The REOS-signals are accompanied by oscillatory features with a frequency of 8.75 THz, which is the Γ_{15}-LO-phonon frequency in GaAs. The coherent phonon amplitude in p-type GaAs is about a factor 5 larger, although the static atomic displacement Q in the surface-space-charge region are of comparable magnitude in both samples ($Q \sim E_S$). This gives evidence for a clear dependence of the phonon amplitude on the velocity of the initial screening process, i.e.$(\partial E_S)/(\partial t)|_{(t=0)}$. The coherent lattice vibration is described by an exponentially decaying $\cos(\omega_{LO}t)$-term, as expected by the model of an oscillator, which is driven by a fast depolarization. The phase shift of π between the coherent oscillations in n- and p-type GaAs is determined by the reversed sign of the driving force.

At even higher excitation densities ($N_{exc} > 1 * 10^{18}cm^{-3}$), in addition to the LO-frequency the TO-frequency occurs, resulting in a mode beating observed in time domain [1]. With increasing excitation density the spectra are dominated by the TO-frequency (Fig. 2). At the high excitation, the dense electron-hole-plasma screens the long range Coulomb interaction of the LO-phonons, thus rendering the LO- into the TO-frequency, which is the bare lattice frequency. This effect occurs at the excitation density, where the average plasma frequency $< \omega_p >^2 \sim < N_{exc}/m^* >$ comes into resonance with the LO-phonon. No plasmon-phonon-coupled modes are visible due to the strong overdamped nature of the hole plasma and large spatial gradients in the carrier distributions.

References

[1] W. Kütt, T. Pfeifer, T. Dekorsy, and H. Kurz, to be published in *Ultrafast Phenomena VIII*, Springer (1992)

Generation of Coherent Phonons via Thermomodulation Induced by Interfering Laser Beams

E.P.N. Damen, P. van der Lek, A.F.M. Arts, and H.W. de Wijn

Faculty of Physics and Astronomy, and Debye Research Institute,
University of Utrecht, P.O. Box 80.000, NL-3508 TA Utrecht, The Netherlands

Abstract. A new method of phonon generation is presented. Thermomodulation of a thin metallic transducer deposited onto a single crystal of $PbMoO_4$ is induced by two interfering cw single-frequency dye lasers. A monochromatic phonon beam is generated with a frequency equal to the beat frequency of the two lasers. The high directionality of the phonons proves their coherency.

In this paper, we introduce a new method of generating nonequilibrium phonons. The heart of the method is thermomodulation (periodic heating) of a thin metallic transducer by two interfering single-frequency cw ring dye lasers. Thermomodulation produces a periodic stress in the transducer, which results in a strain wave in the crystal at a frequency equal to the beat frequency of the lasers. In the present experiment, a gold transducer of 100-nm thickness was deposited onto a single crystal of $PbMoO_4$, and the generated phonons propagated along the c axis.

The detection was accomplished with a Brillouin-scattering spectrometer of conventional design. Both the generated and the thermal phonons reflect a fraction of an argon ion laser beam, and the reflected intensity is analyzed frequency selectively by means of a quintuple-pass Fabry-Perot interferometer. The transmission frequency of the interferometer is actively stabilized, and locked to the position of the anti-Stokes line to permit the detection of the nonequilibrium phonons. Figure 1 shows the intensity of the anti-Stokes line versus the beat frequency of the two lasers. The peak in the intensity occurs at the point where the frequency of the generated phonons best fulfills the Bragg condition. The generated phonons thus are detected on top of the thermal backgound.

To estimate the acoustic power, we first calculate the energy density absorbed in the transducer, equate the generated stress to the energy density on the basis of thermodynamic arguments, and finally evaluate the strain in the crystal. The acoustic power of the strain wave injected into the crystal is then given by $P_{ac} = 16P_0^2\Theta^2/A\Lambda^2\rho v\omega^2$, in which P_0 is the absorbed power per laser, A is the illuminated area of the transducer, Λ is its thickness, Θ is the

Springer Series in Solid-State Sciences, Volume 112
Phonon Scattering in Condensed Matter VII Eds.: M. Meissner and R.O. Pohl

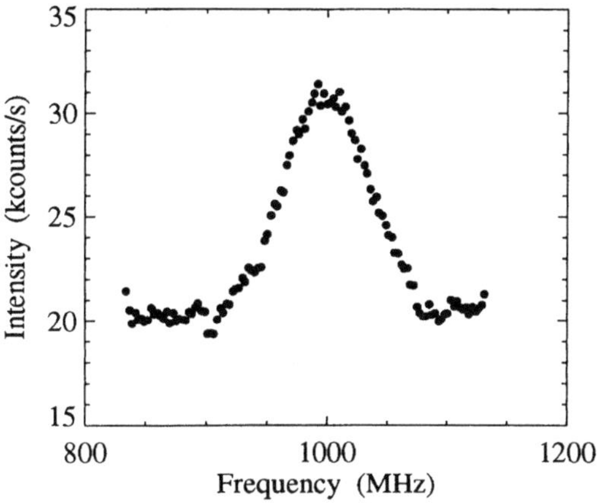

Fig. 1. The intensity of the anti-Stokes Brillouin line vs the beat frequency of the interfering lasers. The peak occurs when the phonons emitted by the transducer match the central frequency (fixed at 1000 MHz) detected by the Brillouin spectrometer. The background is due to thermal phonons.

coefficient of transmission across the interface, and ω is the angular phonon frequency, which equals the angular frequency difference of the interfering lasers. Estimating the acoustic transmittance Θ from the measured thermal conductance across the interface, we arrive at $P_{ac} = 3 \times 10^{-11}$ W. From standard Brillouin scattering theory, the reflected light power is subsequently derived to be 6×10^{-12} W, or 10^4 counts/s at the output of the photomultiplier, in good agreement with the experiment.

A criterion for the coherence of the phonon beam is its divergence in relation to Fraunhofer diffraction of a plain wave emitted by an area having the size of the illuminated area of the transducer. The divergence of the phonons was determined from the frequency width of the Brillouin intensity with allowance for the uncertainty in the Brillouin scattering geometry associated with the finite size of the slits selecting the scattering angle, the profile of the incoming laser beam, and the finite acceptance angle. The angular distribution of the phonon propagation was thus determined to have a full width at half maximum of about 0.01 rad. Within errors, this indeed equals the width of 0.02 rad one expects from Fraunhofer diffraction. Another indication of the small divergence is that the intensity of the phonon beam has not appreciably decreased at a distance of 20 times the effective diameter of the transducer. This leads to the conclusion that the generated phonons leave the illuminated area as a coherent strain wave [3].

Financial support by the Netherlands Foundations FOM and NWO is gratefully acknowledged.

References

[1] A. G. Every, Phys. Rev. B **34**, 2852 (1986).
[2] G. A. Coquin, D. A. Pinnow, and A. W. Warner, J. Appl. Phys. **42**, 2162 (1971).
[3] E. P. N. Damen, A. F. M. Arts, and H. W. de Wijn, to be published.

Stimulated Emission of Phonons in Ruby: Phonon Cavity

P.A. Fokker, J.I. Dijkhuis, and H.W. de Wijn

Faculty of Physics and Astronomy, and Debye Research Institute, University of Utrecht, P.O. Box 80.000, NL-3508 TA Utrecht, The Netherlands

Abstract. Stimulated emission of phonons resonant with the Zeeman-split $\overline{E}(^2E)$ doublet of Cr^{3+} in ruby is observed at 1.5 K after inversion by pulsed optical excitation. Under suitable conditions, the emission resumes after the phonons have completed a round trip through the crystal forming a cavity.

We report on stimulated emission of resonant phonons by the Zeeman-split $\overline{E}(^2E)$ doublet of Cr^{3+} in ruby. The necessary inversion of $\overline{E}(^2E)$ was accomplished by pulsed selective optical excitation into the upper E_+ level. The phonon emission results in a growth of the population N_- of the lower E_- level, which was observed via the relevant component of the R_1 luminescence [1]. The ruby crystal was held at 1.5 K to remove thermal phonons. It contained 700-at.ppm Cr^{3+}, and measured $1.1 \times 0.9 \times 0.4$ cm^3, the c axis being perpendicular to the largest faces. Magnetic fields up to 6 T were applied, and the exciting laser beam, focused to a diameter of 300 μm, ran about parallel with the longest dimension. The luminescence was analyzed with a double monochromator followed by standard photon-counting techniques.

In Fig. 1, representative data are presented for a series of optical pumping powers. Stimulated emission manifests itself in an initial rise of N_- that is at least three orders of magnitude faster than the spin-lattice relaxation time $T_1 = 520$ μs. The most remarkable feature in Fig. 1 is a repetition in the growth of N_- at later times. The delay is 3.3 ± 0.2 μs, which is exactly twice the time it takes for transverse phonons to cover the distance between the end faces of the crystal ($v_t = 6.4 \times 10^5$ cm/s). The phonon pulse generated by stimulated emission thus is seen to cause further enhancement of the phonon emission at the moment all phonons return to their "place of birth" after completing a round trip through the crystal. That is, the crystal forms a phonon cavity with the end faces acting as mirrors. This was confirmed by similar measurements in a crystal having a length of 0.6 cm instead of 1.1 cm, where the delay was reduced to 2.0 ± 0.2 μs. The exponential growth of the traveling phonon pulse ensures its high directionality along the excited zone. The repetition, of course, requires that the phonons become nonresonant after a finite distance. Indeed,

Springer Series in Solid-State Sciences, Volume 112
Phonon Scattering in Condensed Matter VII Eds.: M. Meissner and R.O. Pohl

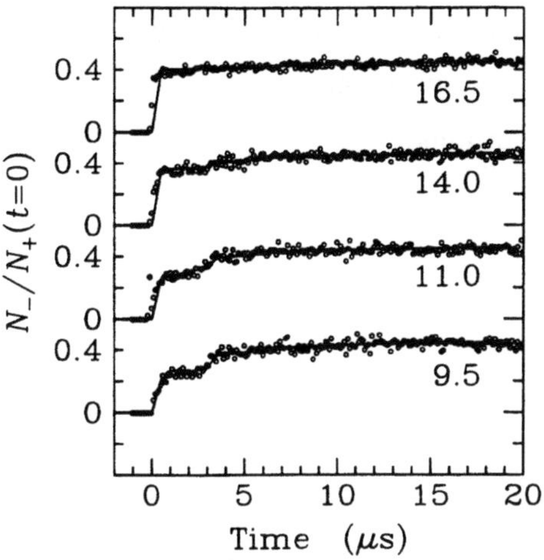

Fig. 1. The population N_- of E_- normalized to the initial population $N_+(t = 0)$ just after pulsed pumping into E_+ for a magnetic field $H = 4$ T at an angle $\theta = 69°$ from the c axis. The $\overline{E}(^2E)$ levels are separated by 1.64 cm^{-1}. The labels denote the parameter $N(x_0, x_1, t)$ at $t = 0$, which is a measure of the initial inversion. Data points were taken every 0.1 μs.

the transition within $\overline{E}(^2E)$ is subject to a sizable macroscopic inhomogeneous broadening associated with the magnetic field and the g factor.

To analyze the data, we developed a model based on position-dependent rate equations for the populations N_- and N_+ and the phonon occupation number p. As it turns out, the natural parameter for the inversion is the dimensionless quantity $N(x_0, x_1, t) = (vT_1\rho\Delta\nu)^{-1}\int_{x_0}^{x_1}(N_+ - N_-)dx$, i.e., the net number of inverted spins a particular phonon interacts with over the distance $x_1 - x_0$ of resonance with $\overline{E}(^2E)$. The quantity $\rho\Delta\nu$ is the density of resonant phonon modes, and v is the phonon velocity. Account is taken of specular reflection by the end faces and the directionality of the amplification by letting only a fraction ε of the phonon modes be active. This fraction is assumed to be determined by the geometry, and to decrease after the first passage of the phonon pulse through the active zone. The development of the phonon pulse as it traverses the active medium is calculated, and the resulting p, N_-, and N_+ are subsequently used as the initial conditions for the next passage. To arrive at acceptable fits, it proved necessary to account for anharmonic loss by upconversion of two resonant phonons to a phonon of double the energy [1]. The fits thus obtained are inserted as the solid curves in Fig. 1. The analysis yields the reasonable results $\varepsilon \sim 0.5$ and 0.01 for the fraction of modes taking part in the first and subsequent passages, $R = 0.15$ for the coefficient of specular reflection, and $\tau_a \sim 60$ ms for the time constant of the upconversion.

Financial support by the Netherlands Foundations FOM and NWO is gratefully acknowledged.

Reference

[1] J. G. M. van Miltenburg, G. J. Jongerden, J. I. Dijkhuis, and H. W. de Wijn, *Phonon Scattering in Condensed Matter*, edited by W. Eisenmenger, K. Laßmann, and S. Döttinger (Springer-Verlag, 1984), p. 130.

Observation of *Quasidiffusive* Phonon Propagation in Silicon

M.E. Msall, M.S. Carroll, J.A. Shield, and J.P. Wolfe

Materials Research Laboratory, University of Illinois
at Urbana-Champaign, Urbana, IL 61801, USA

Photoexcitation of Si is expected to produce a broad range of phonon frequencies as a byproduct of electronic relaxation. These phonons should propagate through the crystal quasidiffusively.[1] In contrast to these expectations, previous experiments on Si immersed in liquid helium observed phonon pulses with no long tail from scattered phonons and sharp phonon caustics.[2] These observations imply that the detected phonons are relatively low frequency and propagate ballistically from a region very near the focused laser source. Our experiments attempt to isolate the factors responsible for this discrepancy between expectation and experiment. In particular, we find that the phonon pulses are very sensitive to the presence of liquid helium at the excitation surface and to the excitation level.

At low excitation levels (less than about 0.5 $\mu J/mm^2$) quasidiffusion is observed in Si, provided that the excitation surface is in vacuum. (Fig 1a) In contrast, a timetrace taken at the same excitation level with helium at the excitation surface does not exhibit the quasidiffusive character. We attribute this difference to the loss of a large portion of the high frequency phonons into the helium bath. A Monte Carlo simulation of the experimental conditions shows the effectiveness of the helium bath in absorbing the high frequency phonons of the diffusive tail. (Fig. 1c)

Even with a vacuum interface, quasidiffusion in Si is observed only at the lowest excitation densities, obtained by defocusing the laser. At higher excitation densities there is a marked decrease in the quasidiffusive tail along with a large increase in early arriving phonons, as seen in Figure 1b.

The quasidiffusive model assumes only spontaneous anharmonic decay processes. At high excitation densities other mechanisms may play a role in the evolution of the excitation energy. In the past, the favored candidate for a density dependent phonon down-conversion mechanism was phonon-phonon scattering and the Levinson 'hot spot' theory.[3] However, we estimate the excitation energies at which the low frequency phonon source appears are roughly three orders of magnitude lower than the predicted phonon hot spot threshold. In addition, the lifetime of this source appear to be at least an order of magnitude longer than the predicted 'hot spot' lifetime.

Springer Series in Solid-State Sciences, Volume 112
Phonon Scattering in Condensed Matter VII Eds.: M. Meissner and R.O. Pohl

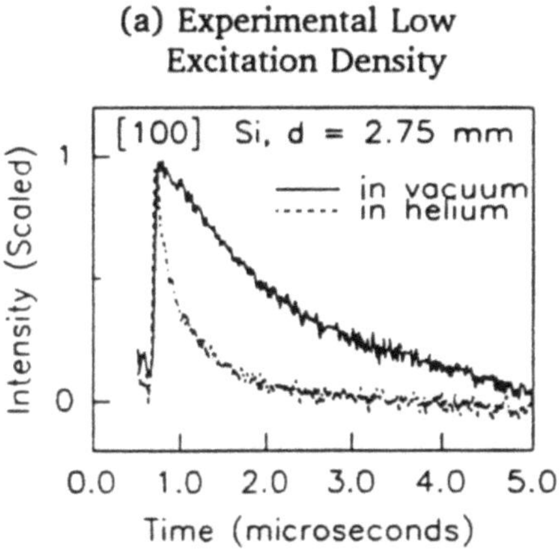

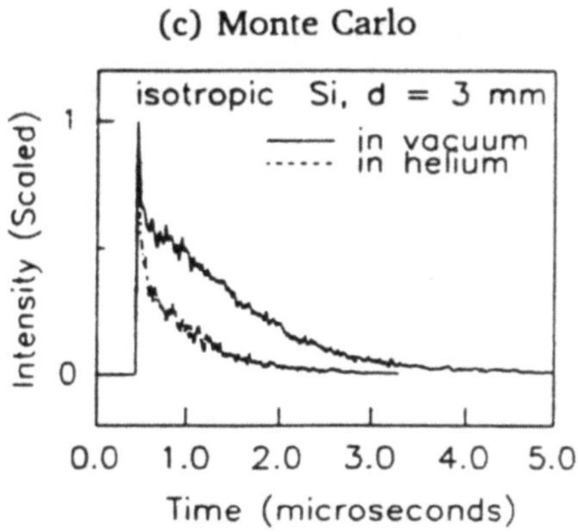

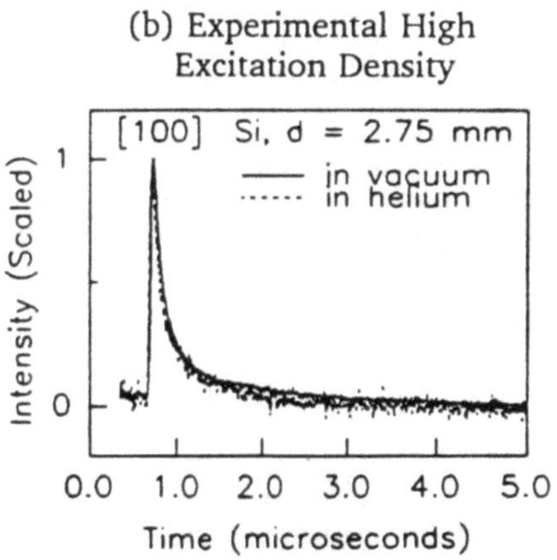

Figure 1 (a) and (b) compare the detected phonon flux following photoexcitation with and without helium at the excitation surface. Quasidiffusive propagation is only observed with the low density vacuum interface excitation. (c) Monte Carlo simulation of the detected phonon pulse after the injection of a number of 4 THz phonons in an isotropic Si sample. The phonons propagate quasidiffusively with a 50% loss factor for each encounter with the helium interface.

We postulate that photo-excited carriers are playing a critical role in the formation of the localized phonon source in photoexcited silicon. At high excitation densities the carrier excess energy can be thermalized by inter-carrier scattering processes which are competitive with (and may surpass) the rate of optical phonon emission. Furthermore, the carriers may condense into droplets of electron-hole liquid which become a source of low frequency acoustic phonons.[4] Supported by MRL grant NSF DMR-89-20538.

References

1. J.A. Shields and J.P. Wolfe, Z. Phys. B75, 11 (1989).
2. H.J. Maris, Phys. Rev. B41, 9736 (1990).
3. D.V. Kazakovtsev and Y.B. Levinson, Sov. Phys. JETP 61, 1318 (1985).
4. Electron Hole Droplets in Semiconductors, C.P. Jeffries and L.V. Keldysh, eds. (North Holland, Amsterdam, 1983).

Monte-Carlo Simulations and the Dynamics of Highly Excited Phonons

J. Wesner, M. Siemon, and W. Grill

Physikalisches Institut der Johan Wolfgang Goethe-Universität,
Robert-Mayer-Str. 2–4, W-6000 Frankfurt, Fed. Rep. of Germany

We have investigated the dynamics of $29cm^{-1}$ phonons in ruby by direct optical excitation of the $2A(^2E)$ (R2) level. A dye laser (t_p = 6ns) generates a column (r = 25 μm) of excited ions and the phonons are detected by a superconducting bolometer, identifying the resonant phonons by their scattering in an additional volume created by an argon laser (inset in fig. (1) or by the luminescence from the $E(^2E)$ (R1) level. We find no strong laser-like nonlinearities as reported by Hu[1], but observe that phonon bottleneck is suppressed for the first 50 ns after the excitation, so that about 70% of the phonons escape from the excited region before strong bottleneck (**b** >=100) sets in [2]. This has been explained qualitatively by "incoherent bleaching" of the acoustically thick medium due to the high phonon occupation numbers (**p** up to 100) following the excitation.

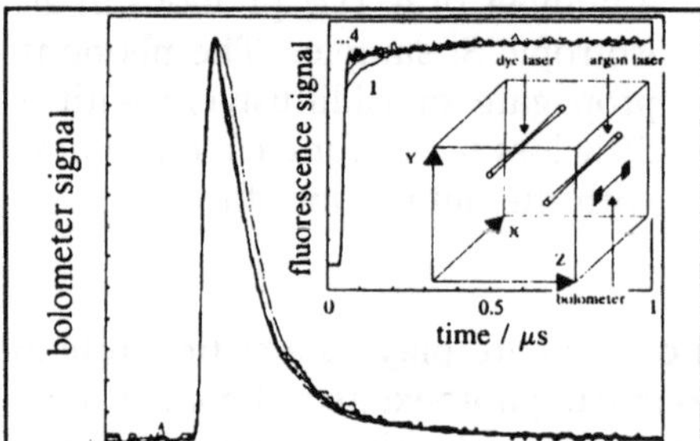

Fig.1: Detected 29 cm^{-1} phonons in ruby and fluorescence of the R1 line after pumping the $2A(^2E)$ level, b>>1. Inset shows setup.

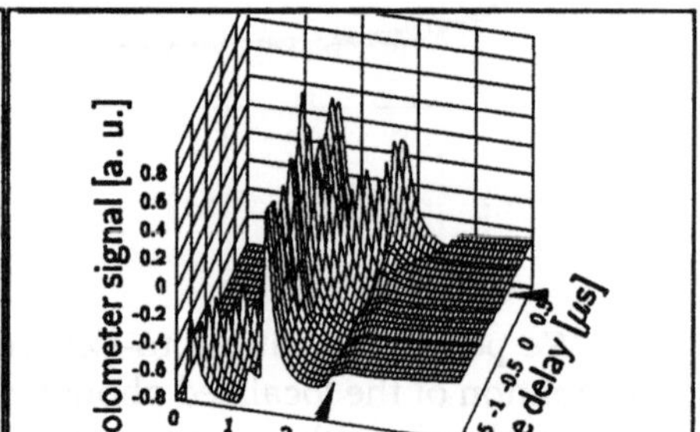

Fig. 2: ´Phonon boxcar´: Bolometer signals of the phonon generating laser (fixed in time) and the ´killer´ laser (shifting in time).

We propose a new method to investigate the shape of the resonant phonon pulses - the ´phonon boxcar´: Using a second dye laser (´killer´) instead of the argon laser, the scattering volume in fig. (1) can be populated in few ns and its onset can be scanned in time relative to the firing of the first laser (fig. (2)). The suppressed phonon signal (lower arrow) reappears if the laser is fired after the phonons have already past the focus position (upper arrow). Using a double differential scheme, the modulation caused by the ´killer´ laser can be separated (fig. (3)) and the time integrated phonon pulse can be mapped (fig. (4)). Again it is found that about 80% of the phonons escape nearly untrapped.

Springer Series in Solid-State Sciences, Volume 112
Phonon Scattering in Condensed Matter VII Eds.: M. Meissner and R.O. Pohl

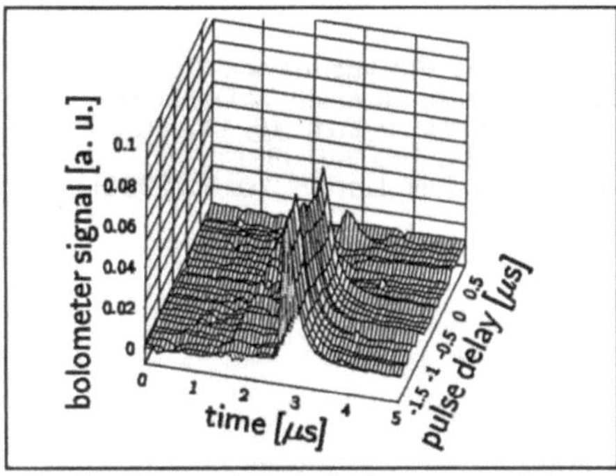

Fig. 3: 'Phonon boxcar': Preprocessed data, showing only the modulated part of the bolometer signal.

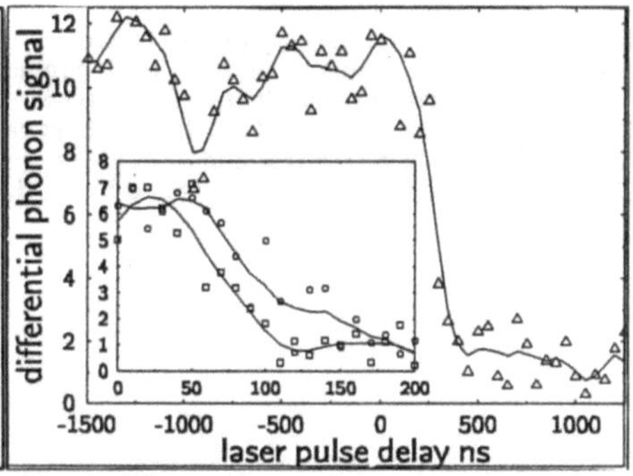

Fig. 4: 'Phonon boxcar' traces, representing the time integrated $29cm^{-1}$ signal.

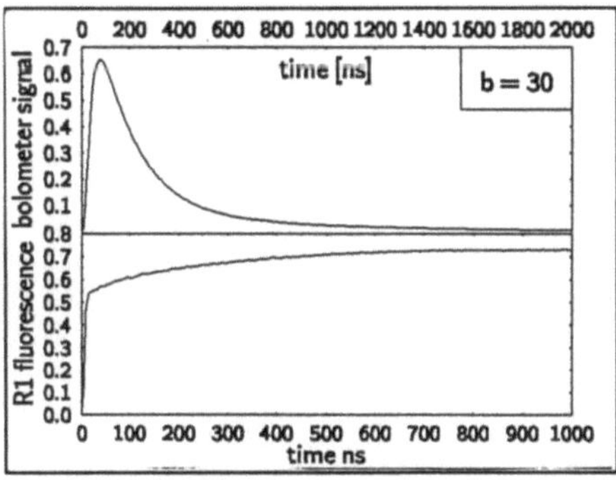

Fig. 5: Simulated bolometer signal and fluorescence intensity

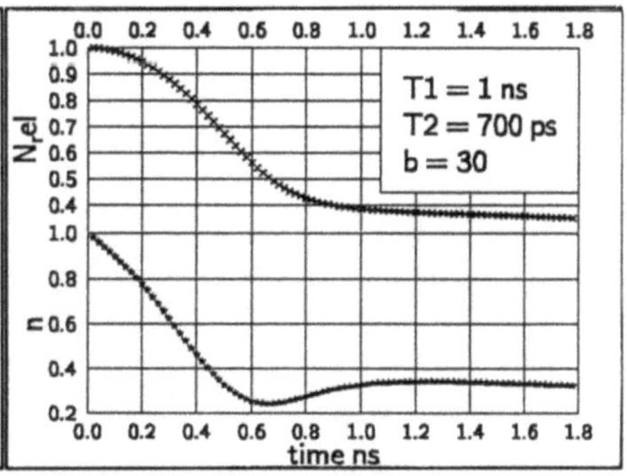

Fig. 6: MC - Simulation of limited superradiance in ruby using the modified rate equations.

If the 'killer' laser is tuned to the R2 level (lower trace in the inset using a delay increment of only 10 ns) instead of the R1 level (upper trace), we find a speed-up of the transition and a shift in delay time, which arises as the phonons can pass the excited volume before the R2 occupation number **n** falls below 0.5.

To describe the results of the experiments, a Monte Carlo simulation takes explicit account of the R2 occupation number **n**, dividing the cylindrical excitation volume in many zones and the simulation time in steps of about 100 ps. After any time step, in any zone, the probability of stimulated emission treated as an effective mean free path for the phonons and a new effective lifetime of the excited ions is re-calculated, considering also the radial profile of the exciting laser and spectral diffusion, using models similar to van Dort [3] or Majetich [4]. Fig. (5) shows a typical result using similar parameters to fig (1).

Recent evidence for superradiant decay in ruby [5] rises the interest to incorporate the effect of a finite polarization lifetime T2 in the simulation. Starting from the Bloch equations with decay:

$$du/dt = -u/T2, \qquad dv/dt = -v/T2 + k\,E\,w \;(1), \qquad dw/dt = -(w+1)/T1 - k\,E\,v \;(2)$$

inserting v from (1) in (2) and by coefficient comparison in the limit of T2 -> 0, one obtains the following set of modified rate equations:

$$dn/dt = -n/T1 + p/T1\,(1-2n) + \sqrt{p}/T1\;dv'/dt \;(3), \qquad dv'/dt = -v'/T2 - \sqrt{p}\,(1-2n) \;(4)$$

The last term in (3) is the "coherent correction" of the usual rate equation, the pseudo-polarization v' (4) has to be accumulated during the simulation for every spatial zone. Fig. (6) shows a result of the modified MC-simulation based on (3) and (4) for parameters similar to the experiments in [5].

[1] P. Hu, Phys. Rev. Lett. 44, 417 (1980)
[2] M. Siemon, J. Wesner and W. Grill, **Proceedings of the 6th International Conference on Phonon Scattering** Heidelberg 1989, and M. Siemon, Thesis, University of Frankfurt
[3] M. J. van Dort, C. R. de Kok, J. I. Dijkhuis and H. W. de Wijn, J. Lum. 45, 112 (1990)
[4] S. Majetich, R. S. Meltzer and J. E. Rives, Phys. Rev. B 38, 11075 (1988)
[5] M. H. F. Overwijk, J. I. Dijkhuis and H. W. de Wijn, Phys. Rev. Lett 65, 2015, (1990)

Investigations of Nonequilibrium Phonons Using Defect-Induced One-Phonon Absorption

W.M. Dennis

Department of Physics and Astronomy, University of Georgia, Athens, GA 30602, USA

Abstract. Monochromatic phonons are generated in $LiYF_4:Pr^{3+}$ using defect induced one phonon absorption (DIOPA) of far infrared radiation. Phonons are detected in a time resolved, frequency selective manner using an absorption vibronic sideband spectrometer. The combination of these techniques enable us to investigate the complete spectral and temporal evolution of the decay products of the monochromatically generated nonequilibrium phonon distributions.

1. Introduction

Phonon decay by anharmonic processes have been investigated in a variety of laser materials. In this paper we present to our knowlege, the first results on nonequilibrium phonon relaxation in $YLiF_4:Pr^{3+}$. In addition we determine for the first time the spectral distribution of the phonon decay products at very early times after monochromatic phonon generation and find good agreement with the predictions of the isotropic dispersionless model.

2. Experimental

High occupation number monochromatic phonons are generated using defect induced one phonon absorption (DIOPA) of far infrared (FIR) radiation [1]. High power FIR pulses are generated using a swept gain super radiant cell which is pumped by a transversely excited atmospheric CO_2 laser. Phonons are detected using an absorption vibronic sideband spectrometer which uses fluorescence to monitor the phonon induced absorption as detailed in reference [2]. The sample is maintained at 10 K using a two stage closed cycle refrigerator. The temporal and spectral resolution is 50 ns and 0.5 cm^{-1} respectively.

3. Results and Discussion

The spectral and temporal evolution of the nonequilibrium phonon population after monochromatic phonon generation at 66 cm^{-1} is shown in fig. 1. The phonon

Springer Series in Solid-State Sciences, Volume 112
Phonon Scattering in Condensed Matter VII Eds.: M. Meissner and R.O. Pohl

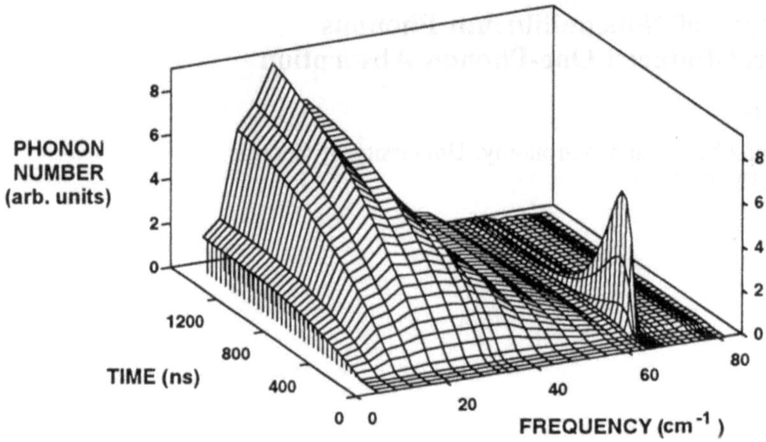

Figure 1. Spectral and temporal evolution of the decay products of 66 cm^{-1} phonons. Occupation numbers can be calculated by dividing the phonon number by the frequency in cm^{-1} squared.

spectrum is given in terms of arbitrary phonon number as opposed to occupation number as this emphasizes the symmetry of the anharmonic decay process. At early times the distribution of the decay products is symmetric about the half frequency 33 cm^{-1} and can be fitted easily to $(\nu_0-\nu)^2\nu^2$ as expected from the isotropic dispersionless model [3].

At later times the broadband decay product distribution down shifts due to further anharmonic processes. This behavior can be accurately modeled using a Monte Carlo simulation that takes into account only spontaneous three phonon processes as expected for low occupation number phonon distributions. Similar behavior and agreement with theory was observed after monochromatic phonon generation at 52 cm^{-1}. Our results indicate that a generation model [4] used by us and others [5] is inappropiate to describe the details of anharmonic decay as such models only allow one decay rate between successive phonon generations.

References

1. W. A. Tolbert, W. M. Dennis and W. M. Yen, Phys. Rev., Lett. **65**, 607 (1990).
2. R. S. Meltzer, J. E. Rives and G. S. Dixon, Phys. Rev., **B28**, 4786 (1983).
3. R. Orbach and L. A. Vredevoe, Physics, **1**, 91 (1964).
4. W. A. Tolbert, W. M. Dennis and W. M. Yen, Phys. Rev., **B44**, 2149 (1991).
5. W. L. Schaich, Solid State Commun., **49**, 55 (1984).

Temperature Dependence of the H^- Local-Mode Line in CaF_2

G.D. Jones[1], *W.J. Knott*[2], *P.T. Lang*[2], *and K.F. Renk*[2]

[1]Department of Physics and Astronomy, University of Canterbury, Christchurch, New Zealand

[2]Insitut für Angewandte Physik, Universität Regensburg, W-8400 Regensburg, Fed. Rep. of Germany

In this paper we present results of a recent study of the temperature dependence of the line center frequency ν and the linewidth Γ for the principal H^- local-mode vibration centred at 965 cm^{-1} in CaF_2 at 10 K. These were measured for temperatures up to 120 K using a tunable carbon dioxide laser, as part of a laser saturation study [1] of this local mode transition. Temperature measurements above 120 K were made using a FTIR interferometer.

Figure 1 gives the linewidth increase $\Gamma(T) - \Gamma(0)$, from the 10 K residual linewidth value $\Gamma(0)$, as a function of temperature. Figure 2 shows the shift $\nu(0) - \nu(T)$ in the line peak frequency $\nu(T)$, from the value $\nu(0)$ at 10 K, as a function of temperature.

The residual linewidth is attributed [2] to an anharmonic decay process in which the local mode excitation decays into several lattice band modes, while conserving energy. For the H^- local mode in CaF_2 occurring at a frequency (965 cm^{-1}) of more than twice the uppermost band phonon frequency (463 cm^{-1}) in CaF_2 [4], at least three band phonons are created. Following the analysis of Elliott et.al. [2], the residual linewidth of 0.29 cm^{-1} [1] can be associated with a quartic anharmonic constant $|C|$ of 3 x 10^{21} J/m^4. The temperature dependence of this three phonon anharmonic decay process is quadratic, but is quite weak giving only a doubled linewidth by 300 K [1].

The dominant temperature dependent linewidth broadening process involves an elastic non-resonant scattering of two band phonons by the local mode. In the Debye approximation for the band phonon modes, the theory gives the linewidth as [2,3]:

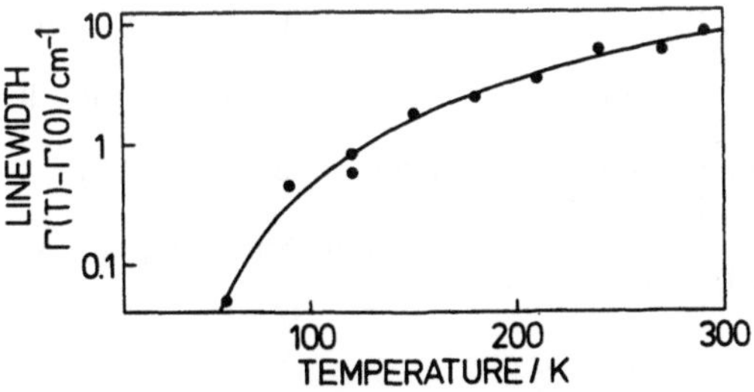

Figure 1: Temperature dependence of the linewidth difference $\Gamma(T) - \Gamma(0)$. The points are the experimental linewidth differences and the solid curve shows a calculated dependence for the parameter values given in the text; $\Gamma(0) = 0.29$ cm^{-1}.

Springer Series in Solid-State Sciences, Volume 112
Phonon Scattering in Condensed Matter VII Eds.: M. Meissner and R.O. Pohl

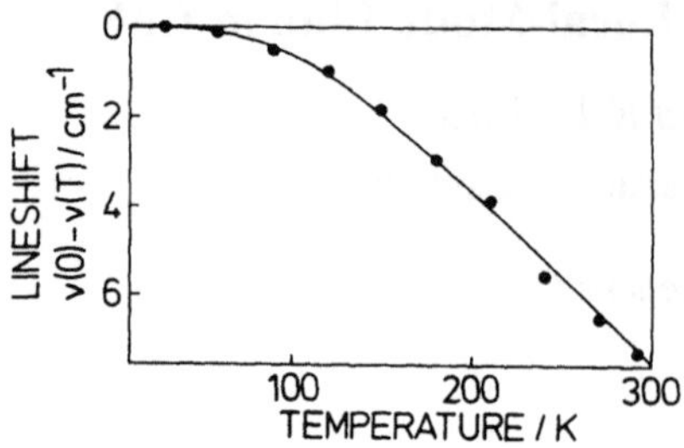

Figure 2: Temperature dependence of the lineshift $\nu(0) - \nu(T)$. The points are experimental and the solid curve shows a calculated lineshift for the parameters given in the text; $\nu(0) = 965\ \mathrm{cm}^{-1}$.

$$\Gamma(T) - \Gamma(0) = \beta\left(\frac{T}{\Theta_c}\right)^7 \cdot \int_0^{\Theta_c/T} \frac{x^6 e^x}{(e^x-1)^2}\, dx \quad ,$$

where $x = \hbar\omega/kT$ and the highest band phonon has a frequency $\omega_c = k\Theta_c/\hbar$. This linewidth increase varies as T^7 at low temperature and as T^2 at high temperature. With the parameters $\Theta_c = 400$ K and $\beta = 84\ \mathrm{cm}^{-1}$, the theoretical curve (Figure 1) describes well the experimental data. The same two phonon scattering process [2,3] gives for the lineshift:

$$\nu(0) - \nu(T) = \delta\left(\frac{T}{\Theta_c}\right)^4 \int_0^{\Theta_c/T} \frac{x^3}{(e^x-1)}\, dx$$

with a dependence as T^4 at low temperature and as T at high temperature. Neglecting the small contribution from thermal expansion of the host lattice [5], the experimental results are well fitted with $\Theta_c = 465$ K and $\delta = 66\ \mathrm{cm}^{-1}$ (Figure 2). The two Θ_c values are somewhat less than than the Debye temperature of 508 K [5]. Similar differences have been found for H^- in alkali halide crystals [3] and have been attributed to preferential coupling to acoustic phonons. We have observed similar behavior of linewidths for two modified H^- centers present in both mixed CaF_2:BaF_2 and CaF_2:SrF_2 crystals.

We acknowledge support by the Ministry of Trade and External Affairs, New Zealand, and the Bayerische Forschungsstiftung through the Bayerisches Verbundprojekt Hochtemperatur-Supraleitung (FORSUPRA).

References

1 P.T. Lang, W.J. Knott, U. Werling, K.F. Renk, J.A. Campbell and G.D. Jones, Phys. Rev. B 44, 6780-4 (1991).

2 R.J. Elliott, W. Hayes, G.D. Jones, H. Macdonald and C.T. Sennett, Proc. R. Soc. London, Ser. A **289**, 1-33 (1965).

3 A.S. Barker Jr. and A.J. Sievers, Rev. Mod. Phys. **47** Suppl. 2, S80 (1975).

4 M.M. Elcombe and A.W. Pryor, J. Phys. C **3**, 492-9 (1970).

5 W. Hayes and A.M. Stoneham in *Crystals with the Fluorite Structure*, ed. W. Hayes (Clarendon Press, Oxford, 1974), pp. 66-67.

Investigation of A^+-States in Si and Ge by Phonon-Induced Conduction Under Uniaxial Stress

P. Gross and K. Lassmann

1. Physikalisches Institut, Universität Stuttgart,
W-7000 Stuttgart 80, Fed. Rep. of Germany

From optical experiments (see for example [1, 2]) it is known that acceptor bound excitons in Si (A^0X) have a ground state multiplet with a splitting in the meV-range. In some models the splitting was ascribed to the Coulomb interaction (jj-coupling) between the holes. In such a model the two $j = \frac{3}{2}$-holes couple to a total angular momentum of $J = 0$ and $J = 2$, respectively (Γ_1 and $\Gamma_3 + \Gamma_5$, respectively, in T_d; $J = 1$ is Pauli-forbidden). In a cubic field the $J = 2$-state can be further split into Γ_3 and Γ_5. The contribution of the electron was neglected. If this model is correct, there should be a comparable splitting in the case of the corresponding A^+-states. By the analysis of the stress dependence of phonon induced conductivity (PIC, [3, 4]) with superconducting Al-tunneling junctions as phonon generators we have now found strong evidence for a split ground state in the cases Si:Ga^+, Si:Al^+ and Ge:Zn^+.

The stress dependence of the A^+-related PIC-thresholds should indicate whether there is a considerable hole-hole coupling or not: if the coupling is very small (independent holes) the stress splitting Δ^A of the A^+-ground state (two-hole state) is linear extrapolating back to zero splitting at zero stress. In this limit the two-hole state can be built up from independent one-hole (Γ_8-) states. A Γ_8-level splits under stress in two levels, which each can take up two holes. The sixfold degenerate $\Gamma_8 \times \Gamma_8$-state then splits under stress into three levels: the lowest two-hole level corresponds to the case of two holes in the lower one-hole level, the fourfold degenerate middle two-hole level corresponds to the case of a hole in the lower and the upper one-hole level, and the highest two-hole level corresponds to the case of two holes in the upper one-hole level. A transition from the lowest to the highest two-hole state would mean a simultaneous excitation of two holes by one phonon and is not allowed for the one-particle phonon interaction.

On the other hand, if the coupling is not negligible, there is a zero stress splitting δ and the stress splitting is nonlinear: $\sqrt{\frac{\delta^2}{4} + (\Delta^A)^2}$. The splitting becomes again linear for stress splittings Δ^A much larger than the hole-hole coupling δ. In this limit, the holes are again decoupled.

Springer Series in Solid-State Sciences, Volume 112
Phonon Scattering in Condensed Matter VII Eds.: M. Meissner and R.O. Pohl

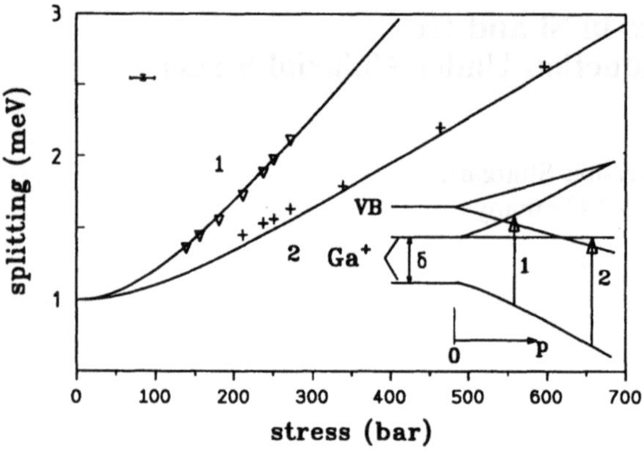

Fig.1: Shift of the signal maxima due to crossings with VB for Si:Ga$^+$. Also shown is the model for the Ga$^+$-ground state.

In our PIC experiments we find as a common feature for the A$^+$-states B$^+$, Al$^+$, Ga$^+$ in Si and Zn$^+$, Be$^+$ in Ge a sharp line emerging at certain stresses at the low energy side of the threshold (see [4]). This line is interpreted as the crossing of one of the upper split A$^+$-states with the valence band. The signal maximum in this situation corresponds to a phonon induced transition within the A$^+$-state. If the upper A$^+$-state is resonant with the valence band, the excited hole can escape to the band and a PIC-signal can be detected. The shift of the maximum of the line corresponds directly to the stress shift of the A$^+$-splitting.

In the cases Si:B$^+$ and Ge:Be$^+$, we find one crossing with the valence band, respectively. In both cases the splitting is linear extrapolating back to zero splitting at zero stress, that is, the hole-hole coupling can be neglected.

In the case Si:Ga$^+$ (Fig. 1) we find two crossings with the valence band and a nonlinear shift of the corresponding signal maxima. This behaviour can be understood, if a split Ga$^+$-ground state is assumed. In this case we can observe transitions from the lowest A$^+$-state to both of the excited states as long as the holes are not decoupled. Hence, the signal due to the transition from the lowest level to the highest level should vanish at high stresses. This is indeed observed experimentally. We can fit the stress shift of the transitions with the above mentioned square root dependence. From the fit we get $\delta = 1meV$.
In a similar way, we get $\delta = 1.2meV$ for Si:Al$^+$ and $\delta = 1.4meV$ for Ge:Zn$^+$.
Financial support from the Deutsche Forschungsgemeinschaft is gratefully acknowledged.

[1] K.R. Elliott, G.C. Osbourn, D.L. Smith, T.C. McGill: Phys. Rev. B**17**, 1808 (1978)
[2] M.A. Vouk, E.C. Lightowlers: J. Lumines. **15**, 357 (1977)
[3] W. Burger, K. Laßmann: Phys. Rev. Lett. **53**, 2035 (1984)
[4] P. Groß, M. Gienger, K. Laßmann: Jap. J. Appl. Phys. **26** (1987) Supplement 26-3 (p.673)

Phonon Spectroscopy of Low-Energy Vibrations of Interstitial Oxygen in Germanium

K. Lassmann, M. Gienger, and M. Glaser

1. Physikalisches Institut, Universität Stuttgart, W-7000 Stuttgart 80, Fed. Rep. of Germany

In oxygen doped germanium we find by phonon spectroscopy with superconducting tunnelling junctions [1] (Fig.1) a series of states from 0.18 meV up to 4.08 meV above the ground state (Fig.3). The sequence can be approximated by a free rotation of the interstitial oxygen atom (Fig.2) corresponding to $E = (\hbar l)^2/(2mr_0^2)$. $\hbar l = 0, \pm 1, \pm 2, \ldots$ is the angular momentum for the rotation around the [111]-axis, m the reduced mass of the system and r_0 the average distance of the oxygen to the [111]-axis.
The 0.18 meV distance between $l = 0$ and $l = \pm 1$ is below the detector threshold; it is seen only as corresponding doublets in the transitions to the excited states. This interpretation of the doublets is confirmed by an activation energy obtained from the temperature dependence of the relative intensity within a doublet. Similarly, transitions from the $l = \pm 2$ level have been identified from the temperature dependence.
In a sample with low oxygen concentration we find that the transitions from the ground state to the $l = \pm 3$ levels are weaker by a factor of 50 than the transitions from the $l = \pm 1$ levels to the $l = \pm 3$ levels.

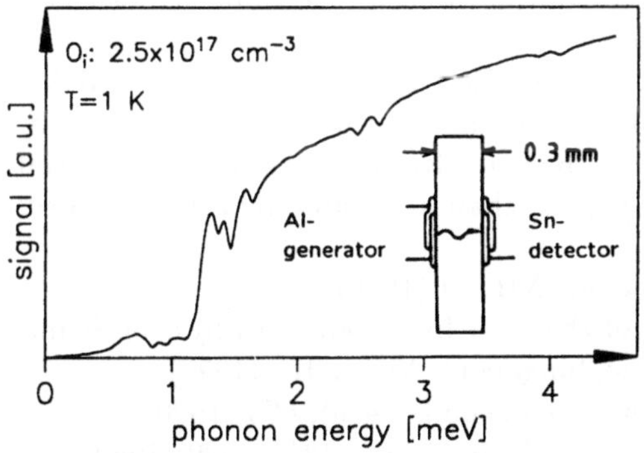

Fig.1: Phonon transmission signal of O_i in Ge versus the phonon energy. Detector threshold at 1.2 meV.

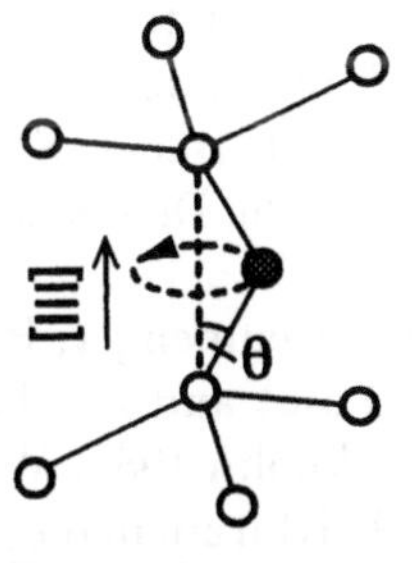

Fig.2: Interstitial site of oxygen in the Ge lattice [2, 3].
$\theta_{Ge-O-Ge} \approx 25°$

Springer Series in Solid-State Sciences, Volume 112
Phonon Scattering in Condensed Matter VII Eds.: M. Meissner and R.O. Pohl

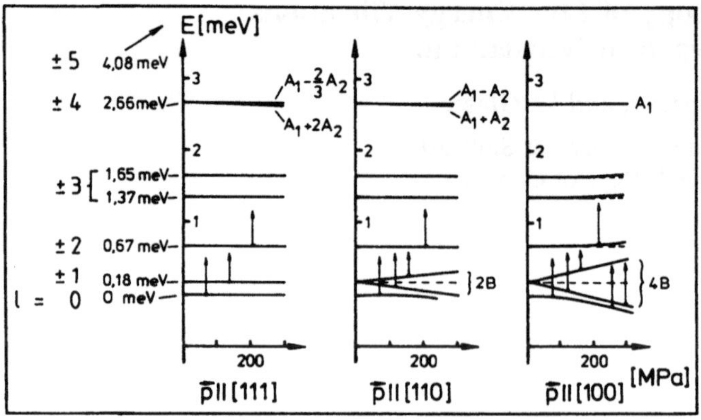

Fig.3: Energy level scheme and stress dependence of the interstitial oxygen in Ge determined by phonon spectroscopy. Phonon transitions from the states labeled $l = 0, \pm1, \pm2$ could be measured.

In this series the states at 1.37 meV and 1.65 meV are interpreted as states belonging to $l = \pm3$ split by the additional six potential barriers from the neighbouring Ge-atoms hindering the rotation [4].
All states show small stress dependence due to the orientational degeneracy, whereas in the sixfold symmetry the $l = \pm1$ level additionally splits for stresses along [100] and [110]. The stress dependence of trigonal centres in cubic crystals is determined by 4 coefficients [5]. We have estimated $A_1 = A_2 = -0.058 \pm 0.02$ meV/GPa, $B = 0.55 \pm 0.05$ meV/GPa, $C = 0 \pm 0.02$ meV/GPa.
We can summarize our results as follows:
Whereas the band of lowest lying mechanical states of interstitial oxygen in Si can be approximated by a two-dimensional oscillator perturbed by a central potential [6, 7], the corresponding states in Ge can be well understood by a hindered rotator with a large central potential opposing a bending oscillation and with a sixfold modulation of the rotation due to the influence of the 6 neighbours to the Ge-O-Ge molecule.
We are obliged to P. Clauws, Gent, V. V. Emtsev, St.Petersburg, and W. Kaiser, Munich, for the supply of samples. Helpful discussions with W. Eisenmenger, H. J. Bauer, Stuttgart, and H. Kaneta, Kawasaki as well as financial support by the Deutsche Forschungsgemeinschaft are gratefully acknowledged.

[1] W. Eisenmenger, Physical Acoustics XII, 79 (1976)
[2] B. Pajot and P. Clauws, Proc. of the 18th Int. Conf. on Phys. of Semic., Stockholm 1986, World Scientific, Singapore, Vol.2, 911 (1997)
[3] L. I. Khirunenko et al., Sov. Phys. Semicond. **24**(6), 663 (1990
[4] H. J. Hrostowski and B. J. Alder, J. Chem. Phys. **33**(4), 980 (1960)
[5] H. J. Hughes and W. A. Runciman, Proc. Phys. Soc. **90**, 827 (1967)
[6] D. R. Bosomworth et al., Proc. Roy. Soc. Lond. **A 317**, 133 (1970)
[7] H. Yamada-Kaneta et al., Phys. Rev. **B 42(15)**, 9650 (1990)

Resonant Phonon Scattering by Calcium Colloids in Electron-Irradiated Calcium Fluoride

C. Wurster, K. Lassmann, and W. Eisenmenger

1. Physikalisches Institut, Universität Stuttgart,
W-7000 Stuttgart 80, Fed. Rep. of Germany

1. Introduction

High frequency phonons with wavelengths in the nm–range should be scattered by mesoscopic defects in single crystals due to excitation of elastic resonances. The technique of thermal conductivity has been applied to investigate such an interaction in the case of Ag–colloids in NaCl [1] and γ–irradiated LiF [2]. Much better spectral resolution ($\sim$ 1 GHz) is possible by phonon spectroscopy with superconducting tunnelling junctions [3] whence more detailed information on the interaction with these defects is to be expected. Here we report on such an investigation of Ca–colloids produced by e^-–irradiation in CaF_2. We find a good correlation of the observed phonon scattering (and its variation with thermal treatment) with a resonance peak in the optical Mie scattering [4].

2. Experimental results

Figure 1 shows the phonon transmission spectra above 285 GHz, the threshold of the Sn–junction, for several differently treated CaF_2 samples. Common for all samples is a scattering dip (3) near 750 GHz of unknown origin. After e^-–irradiation at 120 K two additional dips appear at 398 GHz and 628 GHz which can be annealed by thermal treatment at 720 K. No additional structures were found in the frequency range between $\sim$ 100 GHz and 285 GHz. The appearance of dip (1) as well as its reduction with various annealing steps [Fig. 2] is correlated with an optical absorption line at 550 nm due to Mie scattering at Ca–colloids. A colloid radius of 6 nm is inferred from the position of this line [4]. A small shift of the minimum frequency with annealing temperature as visible in Fig. 2, however, has no observable correspondence in the optical scattering. The annealing behaviour of dip (2) is somewhat different: It deepens at 570 K to 670 K and vanishes also at 720 K. Only a small feature (4) at 377 GHz appears after low temperature e^-–irradiation at 4.2 K. Additional Mie scattering is not observed in this case. γ–irradiation had no effect on the phonon spectra.

Springer Series in Solid-State Sciences, Volume 112
Phonon Scattering in Condensed Matter VII Eds.: M. Meissner and R.O. Pohl

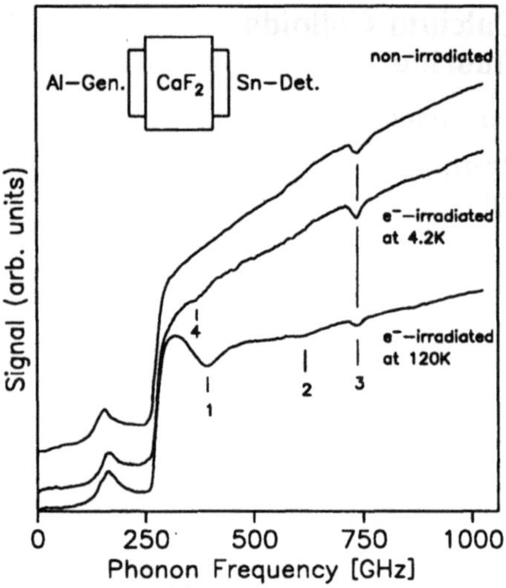

Fig. 1: Phonon transmission of electron irradiated CaF_2.

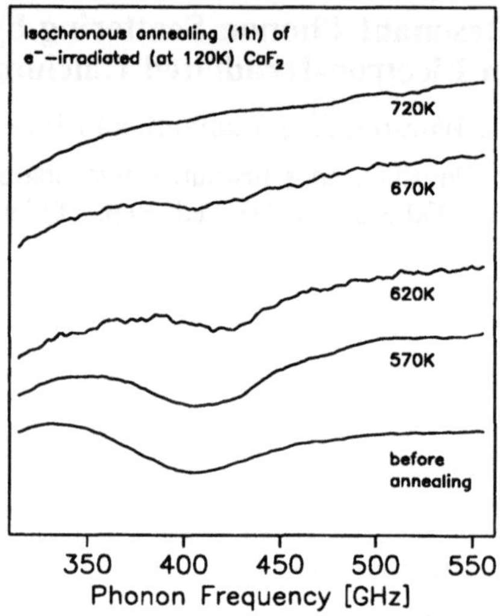

Fig. 2: Reduction of dip (1) at 398 GHz after isochronous (1 h) annealing.

3. Discussion

The correspondence between the annealing behaviour of dip (1) at 398 GHz and the optical absorption line due to Mie scattering at 550 nm led us to assume that dip (1) is due to phonon scattering at Ca–colloids. Evaluating the relative concentration of scatterers from the area below the optical absorption line and from the depth of dip (1) for every annealing temperature both resulting values agree within the error limits only if inelastic phonon scattering is supposed. To estimate the frequency dependence we applied an isotropic continuum model [5] for the total scattering cross section of a sphere for excitation of low lying resonances by transverse acoustic phonons. The resulting frequency pattern consisting of several nearly equidistant sharp peaks is smoothened by convolution with the typical size distribution [4] of the Ca colloids centered at 6 nm. A smooth and broad resonance peak near 400 GHz as experimentally observed is obtained only if a Q–factor as low as 5 is assumed for the individual resonances.

4. References

[1] Worlock J.M., Phys. Rev., **147**, 636 (1966)

[2] Guckelsberger K., Neumaier K., J. Phys. Chem. Sol., **36**, 1353 (1975)

[3] Dittrich E., Scheitler W., Eisenmenger W., Jap. J. Appl. Phys., **26**, 873 (1987), Supplement 26-3

[4] Orera V.M., Alcalá E., phys. stat. sol. (a) **38**, 621 (1976)

[5] Einspruch N.G., Witterholt E.J., Truell R., J. Appl. Phys., **31**, 806 (1960)

New Method for Studying Phonon Thermalization

P.D. Vu and R.O. Pohl

Laboratory of Atomic and Solid State Physics, Cornell University,
Ithaca, NY 14853-2501, USA

1. Introduction

We perform thermal conduction experiments in the boundary scattering regime from 5 K down to 50 mK. The standard method, which uses dc heater power, varies the dominant phonon frequency in the sample by varying the sample temperature. We decouple the dominant phonon frequency from the sample temperature by periodically pulsing the heater power. We vary the dominant phonon frequency injected into the sample by varying the height of the heat pulses. If the height of the pulses is increased, the width is decreased so that the average heat power flowing through the sample and, therefore, the average temperature of the sample as measured by the thermometers, both remain nearly constant.

2. Experimental Matters

We use high–purity, n–type (111) cleaved silicon wafers as samples and measure a rough sandblasted sample, a smooth polished sample, and a polished sample with a 0.5 μm thermal oxide film grown on the surface. The thermal conductance of these samples yields an inverse phonon mean free path which is proportional to the phonon scattering rate. Pulse parameters are as follows: pulse repetition rate of 10 Hz, pulse widths ranging from 50 ms down to 50 μs, pulse peak powers ranging from 1 μW up to 1000 μW, and average sample temperature of about 230-240 mK.

3. Results

Figure 1 shows the inverse phonon mean free path, the scattering of phonons, versus the average sample temperature. The top scale shows the dominant phonon wavelength in silicon. The asterisks are scattering data (using dc heater power) for the rough sandblasted sample, in agreement with the Casimir prediction (solid line). The squares and circles are data for the 0.5 μm thermal oxide film on polished silicon and the smooth polished silicon sample, respectively. Solid squares and solid circles represent data taken with dc heater power while open squares and open circles represent data taken with pulsed heater power. As the pulse height (the peak power) is increased, the scattering rises corresponding to an increase in the dominant frequency of the phonons injected into the sample. The scattering information, then,

Springer Series in Solid-State Sciences, Volume 112
Phonon Scattering in Condensed Matter VII Eds.: M. Meissner and R.O. Pohl

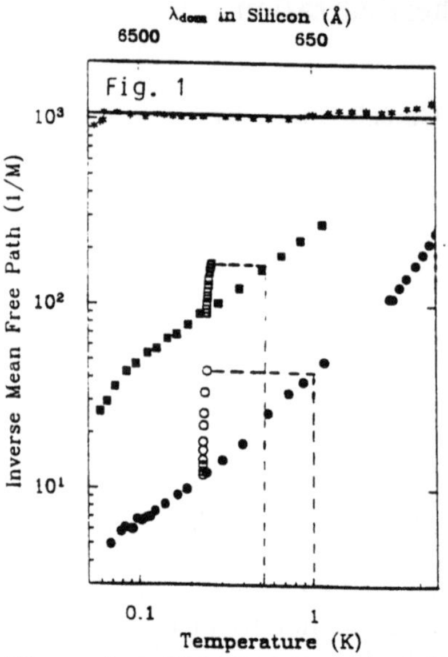

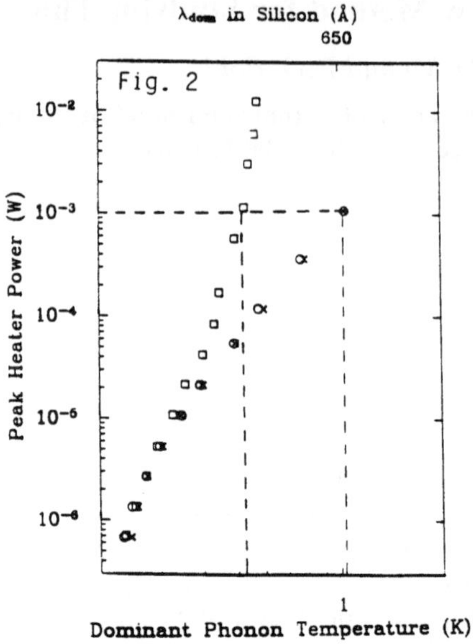

Figure 1 Phonon scattering data for three silicon samples with three different surface treatments. Symbols and results are discussed in the text.

Figure 2 The effect of partial thermalization in a polished silicon sample with 0.5 μm thermal oxide film grown on the surface. See text for discussion.

links the pulse height (peak power) used to the dominant phonon frequency (dominant phonon temperature) in the sample, in the manner shown by the dashed lines. Figure 2 shows the peak heater power used versus the dominant phonon temperature, as determined from the scattering information, for the two samples described. The x's are theoretical calculations of the dominant phonon temperature in the sample given the energy input. These calculations assume adiabatic conditions and are based on the peak heater power, the crystalline specific heat, and the sample to heat sink time constant ($\sim$ 20 μs). The pulse data for the smooth polished sample agree with these adiabatic calculations suggesting that athermal, high–frequency phonons, as high as 1 K, do not thermalize or down–convert in frequency. However, the sample with the 0.5 μm thermal oxide film grown on the surface shows partial thermalization. In particular, 1 K phonons seem to have thermalized down to 0.5 K phonons as shown by the dashed lines. Had complete thermalization taken place, the 1 K phonons would have thermalized down to 230-240 mK, the average sample temperature.

Our new method seems to be a fruitful way of studying phonon thermalization. With further measurements, we hope to be able to determine the elastic versus the inelastic phonon scattering rates.

This research was supported by the NSF, Grant No. DMR–91–15981 and by the U. S. Department of Education, Graduate Fellowship Program in Material Physics, Grant No. P200A10148.

Ultrasonic Properties of Quasi One-Dimensional $(TaSe_4)_2I$ at the Peierls Transition

M. Saint-Paul[1], *R. Britel*[1], *Z. Bihar*[1], *J.C. Lasjaunias*[1], *P. Monceau*[1], and *F. Lèvy*[2]

[1]Centre de Recherches sur les Très Basses Températures, Laboratoire associé à l'Université Joseph Fourier, C.N.R.S., B.P. 166, F-38042 Grenoble Cédex, France

[2]Institut de Physique Appliqué, Ecole Polytechnique Fédérale de Lausanne, CH-1015 Lausanne, Switzerland

$(TaSe_4)_2I$ is a quasi-1D conductor which undergoes a Peierls transition at T_P = 260 K [1]. Phonon softening at $2k_F$ (Fermi momentum) induces lattice modulations associated with an electron charge density wave (CDW). Elastic singularities at the Peierls transition are not yet well understood [2]. The associated singularity in the sound velocity examined theoretically by Nakane [3] varies like $|T-T_P|^{-1/2}$ around T_P.

Possibility of observing critical effects in the elastic properties by ultrasound technique motivates us to measure the elastic constants of $(TaSe_4)_2I$ near the Peierls transition [4]. Ultrasonic measurements have been performed on pure $(TaSe_4)_2I$ crystals and $(Ta_{1-x}Nb_xSe_4)_2I$ crystals doped with x= 0.01 .Shear mode C_{44} was generated along the [001] direction with polarization perpendicular to [001]. Elastic stiffness component C_{44} of pure crystal exhibits a large softening at the Peierls transition as well as a large peak in its imaginary part.Meanwhile the doped crystal does not show any anomalous behaviour in the same temperature range (Fig. 1) which indicates the loss of three dimension ordering between the chains[5].In both cases sound velocity of the C_{44} mode is very small (4.5×10^5 cm/s, $C_{44} = 1.4\times10^{10}$ dyn/cm^2).

The second order transition from quasi-one dimensional metal to incommensurate CDW involves a discontinuity in the longitudinal elastic constant but no discontinuity is expected in the shear constants [8]. For the discussion of the interaction between the strain and the order parameter Q at the Peierls transition, the expansion of the free energy in power of Q and strain components is not easy in the case of $(TaSe_4)_2I$ because of the incommensurate nature of the distortion at T_P [6]. Nevertheless by comparison with theoretical predictions [7] the strong decrease of C_{44} indicates that the strain component e_4 couples with the square of the order parameter. A large critical exponent ρ of about 2 is deduced above and below T_P :

$$\frac{\Delta C_{44}}{C_{44}} = t^{\rho} \quad ,$$

where t is the reduced temperature $t = |T-T_P|/T_P$. Such a critical exponent ρ = 2 cannot be related to the theoretical models [2,3].

In conclusion, the ultrasonic measurements point out a strong softening (> 10^{-2}) of the elastic C_{44} mode at the Peierls transition.Nakane [3]

Springer Series in Solid-State Sciences, Volume 112
Phonon Scattering in Condensed Matter VII Eds.: M. Meissner and R.O. Pohl

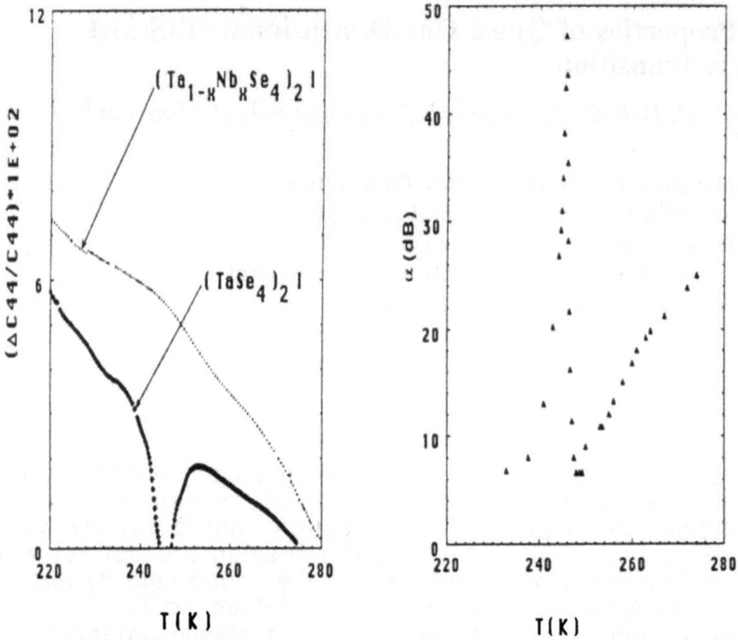

Fig1 Temperature dependence of real part of the elastic constant C_{44} and ultrasonic attenuation α in dB measured at 15 MHz.Inverse of the magnitude of the first echo was measured and taken as a measure for attenuation

predicted a divergence of the sound velocity at T_P but the theoretical parameters are quantitavely inadequate.

[1] For a Reveiw see : P. Monceau, in Electrical Properties of Inorganic Quasi-1D Material, Holland 1985 ; and G. Grüner and A. Zettl, Phys. Rep. **119**, 117 (1985).

[2] J.A. Aronovitz, P. Goldbart and G. Mozurkewicz, Phys. Rev. Lett. **64**, 2799 (1990).

[3] Y. Nakane, J. Phys. Soc. Jpn **55**, 2235 (1986).

[4] M. Saint-Paul, P. Monceau and F. Levy, Solid State Commun. **67**, 581 (1988).

[5] J.E. Lorenzo, R. Currat, P. Monceau *et al.*, to be published.

[6] H. Fujishita, M. Sato and S. Hoshino, Solid State Commun. **49**, 313 (1984).

[7] B. Lüthi and W. Rehwald, in Structural Phase Transitions, ed. K.A. Muller and H. Thomas, Vol. **23** (Springer Verlag 1981) p. 131. L.R.Testardi Phys. Rev. B **12**,3849 (1975).

Ultrasonic Study of the Li-Induced Structural Phase Transition in $K_{1-x}Li_xTaO_3$

K. McEnaney, P. Doussineau, Y. Farssi, C. Frenois, A. Levelut, J. Toulouse, and S. Ziolkiewicz*

Laboratoire d'Acoustique et Optique de la Matière Condensée**,
Université Pierre et Marie Curie, Tour 13, Case 78,
4 place Jussieu, F-75252 Paris Cédex 05, France
*Permanent address: Physics Department, Lehigh University,
Bethlehem, PA 18015, USA
**Associated with the Centre National de la Recherche Scientifique

In the mixed crystal $K_{1-x}Li_xTaO_3$ (KLT), Li^+ ions, when substituted for the larger K^+ ions, occupy off-centre positions thereby creating electric dipole and elastic quadrupole moments. The combination of the highly polarizable $KTaO_3$ host lattice and the system of random-site Li-moments has a profound influence on the low temperature phase of the crystal. Upon cooling, KLT transforms from the cubic paraelectric phase to another phase the nature of which is still controversial. Certain results including X-ray and second harmonic generation (SHG) show a structural change from cubic to possibly tetragonal symmetry. Other results such as the frequency dependence of the dielectric constant and the absence of a macroscopic polarization in zero-field cooled experiments are more easily explained in terms of a progressive freezing of the dipoles and/or quadrupoles. In order to shed light on this problem, we have been looking for signs of a structural phase transition in crystals with different Li concentrations. Our ultrasonic results are interpreted in view of complementary Raman scattering [1] and dielectric constant measurements made on the same samples.

We have measured the attenuation change and the relative phase velocity of longitudinal acoustic waves propagating along the [100] direction in samples with Li-concentrations of 1.5%, 3.5% and 5%. Measurements were made both upon heating and cooling in the range 4 K to 200 K for frequencies between 30 MHz and 1000 MHz. Typical results for each of the samples, shown in figure 1, exhibit two main features;

1) Near 100 K, the dip in sound velocity and a corresponding peak in the attenuation (not shown) are attributed to relaxation of the elastic quadrupole moments associated with the thermally activated Li-system. The shift of the dip to lower temperatures with frequency shows that the relaxation time follows an Arrhenius law with an activation energy of about 1000 K. The observation of a similar relaxation of Li-dipole moments in dielectric experiments suggests that the relaxation is due to $\pi/2$ reorientations of the moments.

2) At lower temperatures and upon heating, a velocity step is seen in all three samples, near 30 K for the 1.5% sample, 55 K for 3.5% and 75 K for 5%. Significantly, the position and amplitude of the steps were found to be independent of frequency. Thermal hysteresis is observed in the more concentrated samples 3.5% and 5%, the width of which is about 3 K and 25 K respectively. In the least concentrated sample, hysteresis, if it exists, must be very small. Similar frequency independent steps and thermal hysteresis have been observed in dielectric experiments on the same samples.

Springer Series in Solid-State Sciences, Volume 112
Phonon Scattering in Condensed Matter VII Eds.: M. Meissner and R.O. Pohl

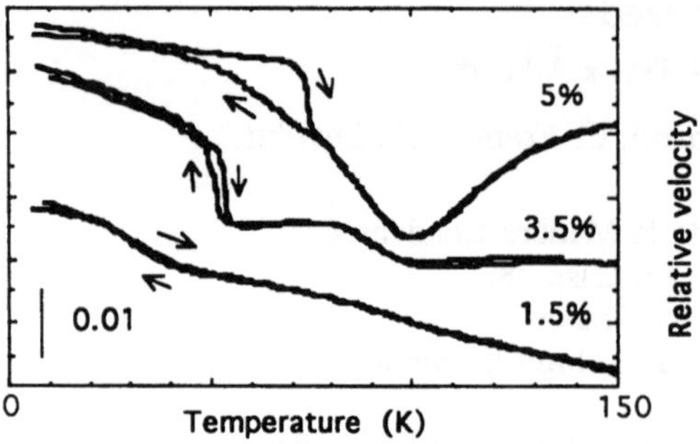

Figure 1. — Relative velocity change of longitudinal acoustic waves propagating along the [100] axis of three different KLT samples as a function of the temperature measured upon both heating and cooling. From top to bottom, sample III : 5% Li, 30 MHz ; sample II : 3.5% Li, 63 MHz ; sample I : 1.5% Li, 216 MHz.

We explain the results taken while heating in terms of an abrupt phase transition from a lower symmetry, probably tetragonal, to the high temperature cubic phase. Upon cooling, the slower variation of velocity in the 5% sample can be interpreted as meaning that the structural change develops from a local level by the growth of lower symmetry domains. Observation of lower symmetry is confirmed by Raman scattering results (obtained upon heating) in the 3.5% sample in which the non-polar TO_3 mode is detected below about 55 K. However, in the 1.5% sample such a mode is not seen below 30 K even though a small step is observed in the acoustic results. In this sample, Raman scattering indicates the absence of a coherent structural distortion over a length scale greater than a phonon wavelength (~ 6000 Å at optical frequency). Nevertheless, the acoustic results, which provide information on the averaged elastic properties, indicate in the same sample the occurrence of a structural change. Combining both sets of results, we conclude that, in the 1.5% sample, the phase change takes place over a restricted length scale. Conversely, for the more concentrated samples, Raman scattering indicates that the domains are larger than 6000 Å in size.

In conclusion, the present ultrasonic study has shown, for Li-concentrations greater than around 2%, the occurrence of a macroscopic structural phase transition. The observation of thermal hysteresis suggests that it is first order while upon cooling the large width of the velocity step may be indicative of a transition via domain growth. Below about 2% concentration, the transition is clearly different in character. There, the results suggest that the structural order is only established over a restricted length scale. The present ultrasonic results and their interpretation are consistent with the results of the previous studies and in particular with the Raman, dielectric and SHG results.

[1] Toulouse J. *et al* *Phy. Rev. Lett* **68**, 232 (1992).

Neutralization of D^--Centers in Si:P by Ballistic Phonons of Various Polarization

B.A. Danil'chenko and S.I. Komirenko

Institute of Physics, Ukrainian Academy of Sciences,
Prospect Nauki 46, 252028 Kiev, Ukraine

1 INTRODUCTION

Impurity atoms in semiconductors at helium temperatures can capture an excess electron to an outer shell, thereby forming D^- centers. By means of phonon induced conductivity method (PIC)[1] it has been shown that D^- centers are neutralized by one-phonon transitions from a bound state into the conduction band. There has been no study of the role played by the polarization of the phonons involved in the impurity-band transitions or in the process of phonon neutralization of D^- centers.

2 EXPERIMENTAL RESULTS and DISCUSSIONS

We are reporting a study of D^-states in *Si:P* crystals with $N_d-N_a=8\cdot10^{14}cm^{-3}$, $N_a=2\cdot10^{13}cm^{-3}$ by means of the heat pulse method and also of the PIC method, described in [2] with the help of a "cooled" preamplifier [3]. The phonon conductivity signal is $\Delta\sigma=e\mu\Delta n \sim N_{D^-}$. The temperature dependence of the PIC-signals $\Delta\sigma(T)$ is exponential with an activation energy $E_1=1.75\pm0.05\cdot10^{-3}$eV (Fig1). It can be seen that at T<2K there is a deviation from this exponential behavior. When temperature is high, the condition $N_a \gg N_{D^-}$ leads to E_1, at lower temperatures E_1 alters by $E_1/2$, which corresponds to $N_{D^-} \approx N_a$.

The result of greatest interest is the difference between the shape of PIC-signals (Fig2b,d) and that of the time-of-flight spectrum of the phonons recorded with the help of the bolometer (Fig2a,c). Specifically, the ratio of phonon amplitudes FTA/LA in the two cases differed by more than an order of magnitude. The ratio FTA/LA of the PIC-signal does not depend on the temperature (Fig1b). However, at the constant value of the phonon flux, under conditions such that the density of D^- centers changes by two orders of magnitude as the temperature is varied (Fig1a), the FTA/LA would necessarily change if this

Springer Series in Solid-State Sciences, Volume 112
Phonon Scattering in Condensed Matter VII Eds.: M. Meissner and R.O. Pohl

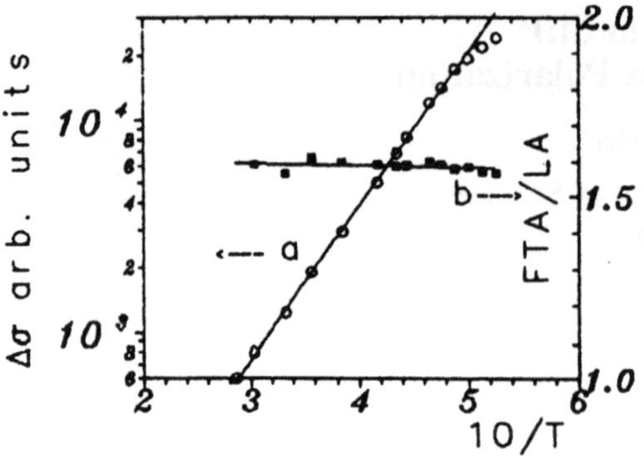

Fig 1 Temperature dependence of the phonon conductivity, measured at the FTA-peak (a). The ratio FTA/LA versus reciprocal temperature (b) Both correspond <110> direction of the phonons propagation.

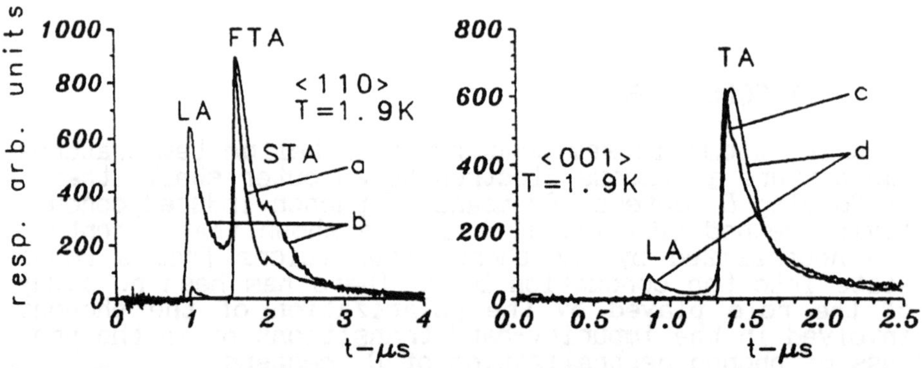

Fig2 Time-of-flight spectra of phonons, recorded with the help of superconductive bolometer (a,c) and PIC-signals (b,d), all normalized on the TA-peak amplitude.

difference were due to a long time scale for the capture to D^- states. The time scale for the restoration of the original population of D^- centers thus does not exceed $5 \cdot 10^{-7}$s under these conditions. This is proved by Fig-2: in <001> direction the flux of LA-phonons is unnoticeable comparatively to TA, but LA-peak in the PIC-signal is significant. The explanation may be the following: the isotopical limit frequency for phonons of the LA mode is $\hbar\omega/k \approx 27K$, while for the FTA is 18K, that causes a difference in the absorption coefficients for the different modes. More over, it would be difficult to expect the absorption coefficients to be equal, since transition matrix elements are different for different modes.

3 REFERENCES

1. W. Burger and K. Lassman, Phys. Rev. Lett. 53,2035,(1984)
2. B. A. Danilchenko and S. I. Komirenko, Sov. JETP Lett., Vol.54,N7,1991.
3. D. L. Daniuk and S. I. Komirenko , will be published in Instr. and Exper. technique, Vol.6,1992.

Part IV

Electron–Phonon Interaction

Novel Spectroscopy of Spatially Confined Acoustic Phonons in Nanometer Scale Structures

M.N. Wybourne and J. Seyler

Department of Physics, University of Oregon, Eugene, OR 97403, USA

In this paper we describe a novel spectroscopy of one-dimensional acoustic modes in narrow Au-Pd wires deposited on silicon substrates. The spectroscopy is based on electron heating measurements in which the low temperature electrical resistance of the wires is measured as a function of an applied electric field.

Three sets of wires were studied with widths, W, of 30, 50, and 90 nm and thickness, d, of 20, 18, and 20 nm, respectively. The samples were delineated using electron beam lithography and were deposited thermally. The wires had contacts suitable for four terminal resistance measurements which were made using an ac bridge, and signal averaging. A fractional resistance change of 1 part in 10^6 could be detected with sensing currents as low as 5 nA per wire. At 4.2 K, the sheet resistance of the wires was $R_{\square} = 12\ \Omega$. With the substrate temperature controlled at 1.5 K, a dc electric field, E, of up to 13 V cm^{-1} was applied across the wires.

The temperature dependence of the equilibrium resistance change, $(R-R_0)/R_0 = \Delta R/R_0$, of all three sets of wires had a minimum value R_0 at $T_0 \approx 8$ K, below which the resistance increased as approximately $T^{-1/2}$, as shown in Fig. 1 for the 30 nm wide wires. The $T^{-1/2}$ dependence below the minimum is consistent with previous reports of electron-electron interactions in Au-Pd wires [1]. The resistance change with E, $\Delta R(E)/R_0$, for the 30 nm wide wire is shown in Fig. 2. At certain values of E, $\Delta R(E)/R_0$ had distinct features, as indicated by the two arrows in Fig. 2. Similar features, but at different values of E were found for all three samples. After thermal cycling the samples to room temperature, the features were reproducible and were independent of the electric field polarity. These experimental facts suggest that universal conductance fluctuations are not the origin of the features.

The acoustic phonon spectrum in the wires is expected to be modified when the dominant phonon wavelength, $\lambda_{ph} \approx V_s h/2k_B T$, becomes comparable with W and d, where V_s is the velocity of sound and T is the temperature. Estimating the longitudinal velocity of sound in Au-Pd to be, $V_l = 4.1\times10^3$ m s^{-1} [2], at 1 K the width of all the wires is less than $\lambda_{ph} \approx 98$ nm. Therefore, the longitudinal phonon spectrum is expected be modified into a series of acoustic subbands, each corresponding to a lateral acoustic waveguide mode of the wire. As E is increased, the electron system will gain energy until the electrons have enough energy to relax by emitting phonons in a higher acoustic subband. Therefore, we

Springer Series in Solid-State Sciences, Volume 112
Phonon Scattering in Condensed Matter VII Eds.: M. Meissner and R.O. Pohl

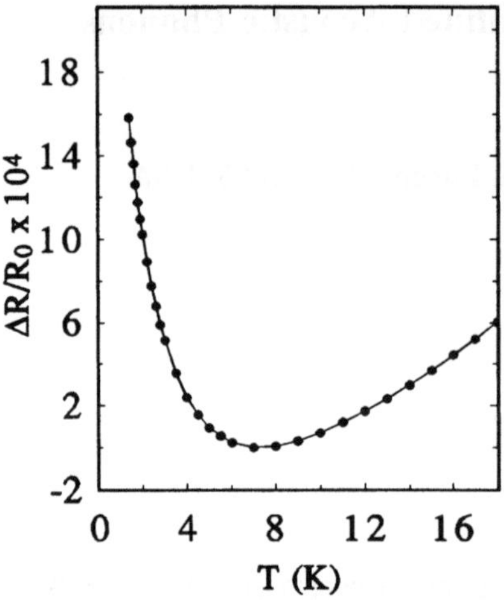

Fig. 1. The equilibrium temperature dependence of $\Delta R/R_0$.

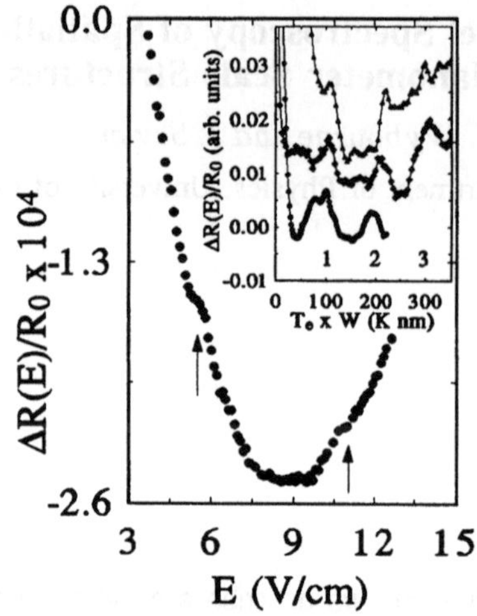

Fig. 2. The electric field dependence of $\Delta R/R_0$ for the 30 nm wires. Inset: Top, middle, bottom curves are for 90, 50, and 30 nm wires respectively.

interpret the features as a differential increase in the electron-phonon scattering rate caused by the increased density of phonon states when additional acoustic subbands are accessed.

The dispersion for uncoupled acoustic modes in a rectangular wire has the form, [3]

$$\omega^2 = V_s^2 \left[q^2 + \left(\frac{n\pi}{W}\right)^2 + \left(\frac{m\pi}{d}\right)^2 \right] ,$$

where q is the wavevector along the wire, and the integers n and m are the subband indices. Therefore, the energy spacing of the subbands scales inversely with W and d. To estimate the energy of the features observed in $\Delta R(E)/R_0$, we calculated dT_e/dE from the equilibrium resistance correction and the slope of the non-equilibrium data. Using this conversion, and removing the steep background, $\Delta R(E)/R_0$ was plotted a function of T_e x W, as shown in the inset of Fig. 2. The features from all three samples were found to scale inversely with W. This result shows that subbands associated with the width of the wire are responsible for the peaks observed in $\Delta R(E)/R_0$. By assigning the value of n to the peaks as shown, we obtain $V_s = 4.2 \pm 0.2\times10^3$ m s^{-1}, which is in close agreement with the value estimated for V_l.

Supported by the National Science Foundation under Grant No. DMR-90-19525.

References

1 J.J. Lin and N. Giordano, Phys. Rev. B35, 545 (1987).
2. W.C. McGinnis and P.M. Chaikin, Phys. Rev. B32, 6319 (1985).
3. B.A. Auld, Acoustic Fields and Waves in Solids (John Wiley, New York 1973) Vol 2, Chapter 10.

The Electron–Phonon Scattering Rate in Thin Free-Standing Metallic Films

J.F. DiTusa[1], *Y.K. Kwong*[2], *K. Lin*[3], *M. Park*[4], *M.S. Isaacson*[4], *and J.M. Parpia*[4]

[1]AT&T Bell Laboratories, Murray Hill, NJ 07974, USA
[2]Solid State Research Laboratory, Tektronix Inc., Beaverton, OR 97077, USA
[3]Department of Physics, SUNY, Stony Brook, NY 11794, USA
[4]Department of Physics, Cornell University, Ithaca, NY 14853-2501, USA

There exists much controversy in the literature regarding the origin of various power laws manifested in the electron-phonon scattering rate (τ_{ep}^{-1}) in metallic films. From theory for bulk systems [1], τ_{ep}^{-1} varies as T^3 or T^4 for the clean or dirty limit, with the power decreasing by T^1 for systems with a 2D phonon excitation spectrum. Experimentally observed power laws for τ_{ep}^{-1} range from 1.4 to 3.7. Typically, the lower values have been attributed to phonons whose excitation spectrum is significantly altered. However, there have been few attempts to systematically explore this hypothesis by varying the sizes.

In this paper, we compare τ_{ep}^{-1} from two experiments. In one, we measure the thermal resistance introduced by the electron-phonon scattering process in thin copper films [2]. In the second, we examine the magnetoresistance of thin aluminum films to determine the inelastic scattering rate which is dominated by τ_{ep}^{-1} [3]. We always compare the rates in free standing and supported films, fabricated simultaneously and with nearly identical physical properties. We expect that the phonon density of states should be 2D for free standing films of thickness d when the thermal phonon wavelength, $\lambda > 2d$. We can express this condition by the relation, $T < hc_s/2d\, k_B$ (c_s is the sound velocity).

The films were deposited on a nitrided wafer, and by a combination of wet and reactive ion etching, the substrate under part of the film was removed, leaving a free standing segment contiguous with an identical supported segment [4].

Electron-heating measurements carried out on copper films show a thermal resistance which varies as T^{-3} for both supported and free-standing films, when the latter are immersed in helium or coated with a few monolayers of helium. The expected bulk dependence for the supported films is T^{-4}. Through a number of checks detailed elsewhere [2], we established that the thermal resistance can be attributed to the electron-phonon scattering rate and not to boundary resistances. The results of these measurements are shown in figure 1a. We also examined thinner and narrower films as well as thicker films which should have exhibited reduced dimensional behavior at different temperatures. In all cases we find that $\tau_{ep}^{-1} = 3 \times 10^8\, T^2\, s^{-1}\, K^{-2}$, was independent of the film thickness and width.

It was known that the inelastic scattering rate of thin aluminum films shows a T^{-3} power law on substrates, consistent with the coupling of the substrate phonons into the thin film [5]. In figure 1b, we plot the results of our experiments illustrating that the inelastic scattering rate ($\tau_{in}^{-1} \approx \tau_{ep}^{-1} = 1.4 \times 10^7\, T^3\, s^{-1}\, K^{-3}$) is unaffected by whether the films are free standing or supported. We also include the results of a calculation to estimate the temperature dependence of τ_{ep}^{-1} in the

Springer Series in Solid-State Sciences, Volume 112
Phonon Scattering in Condensed Matter VII Eds.: M. Meissner and R.O. Pohl

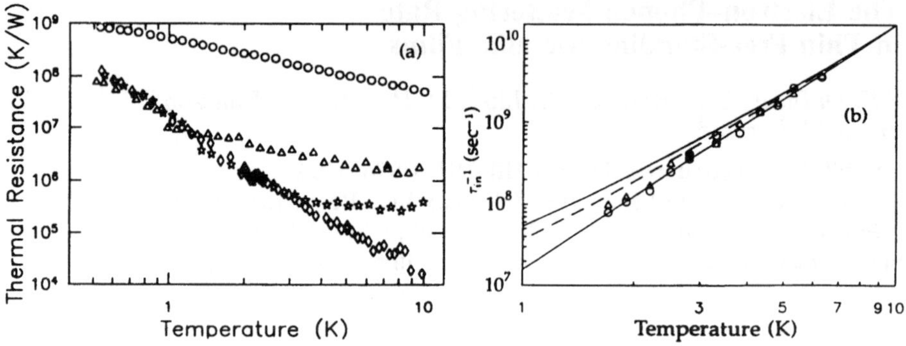

Fig. 1. (a) The thermal resistance for energy transport out of the electron gas in 20nm thick copper films. (◇) Supported film, (◦) free standing films in vacuum, (△) Free standing film covered with 3 to 4 monolayers of ^{4}He, (⋆) free standing films immersed in liquid helium. In (b) we show the inelastic scattering rates for supported (△) and free standing (◦) aluminum films. The upper solid line is the expected result for free standing films accounting for only longitudinal phonons. The dashed line includes transverse as well as longitudinal phonons. Both sets of results in (a) and (b) show no significant crossover effects due to dimensionality constraints.

aluminum films [3], which shows that below 3 K there should be a significant difference between the τ_{ep}^{-1} in free standing and supported films.

From the experiments described in this paper, we draw two conclusions. First, there is no simple explanation for the disparity in the power laws for τ_{ep}^{-1} exhibited by the thin copper and aluminum films. Second, we find that the power law of the electron-phonon scattering rate is unaffected by the constraints imposed by finite size. The observation of the insensitivity to the phonon dimensionality using two different experiments supports the conclusion that the understanding of the electron-phonon interaction in thin films is incomplete.

This research was supported by the NSF through DMR-90-16301 and the Cornell MSC under DMR-88-18588.

References

1. J. Rammer and A. Schmid, Phys. Rev. B, **34**, 1352, (1986).
2. J.F. DiTusa, K. Lin, M. Park, M.S. Isaacson and J.M. Parpia, Phys. Rev. Lett., **68**, 1156, (1992).
3. Y.K. Kwong, K. Lin, M.S. Isaacson and J.M. Parpia, J. Low Temp. Phys., **88**, 1992 to be published.
4. Y.K. Kwong, K. Lin, P. Hakonen, J.M. Parpia and M.S. Isaacson, J. Vac. Sci. Tech. B, **7**, 2020, (1989).
5. P. Santhanam, S. Wind and D.E. Prober, Phys. Rev. B, **35**, 3188, (1987).

Electron–Phonon Interaction in Quasi One-Dimensional Electron Systems

*L.J. Challis and A.Y. Shik**

Physics Department, Nottingham University, Nottingham NG7 2RD, UK
*On leave from A.F. Ioffe Physico-Technical Institute,
194021 St. Petersburg, Russia

ABSTRACT
The processes of interaction between acoustic phonons and electrons in thin quasi-one-dimensional wires in zero and high magnetic fields are considered theoretically.

We discuss the phonon emission from a quasi-one-dimensional electron gas (1DEG) in which the effective electron temperature T_e exceeds the lattice temperature T. Previous theoretical study of the emission [1] concentrated on investigating how numerically computed values of the total energy loss rate Q in zero magnetic field varied with temperature and sheet density. The present work gives particular attention to the form of the angular and frequency distribution of the phonon emission and also extends the calculations to quantizing magnetic fields. We assume the 1DEG is formed from a 2DEG by a parabolic confining potential $V=m\omega^2x^2/2$ with eigenvalues $\epsilon=\hbar\omega(N+1/2)+\hbar^2k^2/2m$ and eigenfunctions $u_N(x)\exp(iky)$ where N=0,1,2,.., $u_N(x)$ is a harmonic oscillator function with characteristic width $l=(\hbar/m\omega)^{1/2}$ and k is the wavevector in the direction of free motion (y-axis). We restrict discussion to $l>a$, the confinement width of the 2DEG (z-axis), and consider different ranges set by the values of $\hbar\omega/k_BT_e$ and so of ql where $q\sim k_BT_e/\hbar s$ is the dominant wavenumber of the phonon emission (s= phonon velocity).

Range (i) $k_BT_e<<(ms^2\hbar\omega)^{1/2}$, $(ql<<1)$
Only intrasubband transitions are allowed in this range and since the electrons are strongly confined in both x-and z-directions, q_x and q_z can have any values up to $q\sim k_BT_e/\hbar s$ while q_y is restricted to $mk_BT_e/\hbar^2k_N$ $<<q$ by momentum conservation (k_N is the Fermi wavenumber in the Nth subband). So the phonons are emitted almost normal to the 1DEG axis and isotropically in the x-z plane, neglecting phonon focussing. The spectrum of phonon frequencies ω_p can be shown to vary as $\omega_p^{\gamma+3}/(\exp(\hbar\omega_p/k_BT_e)-1)$, where $\gamma=+1$ and -1 for deformation potential and piezoelectric coupling respectively, and the intensity peaks when the Fermi energy E_F coincides with the density of states singularity near the bottom of a one-dimensional subband.

Springer Series in Solid-State Sciences, Volume 112
Phonon Scattering in Condensed Matter VII Eds.: M. Meissner and R.O. Pohl

Range (ii) $(ms^2\hbar\omega)^{1/2} << k_BT_e << \hbar\omega, E_F$

In this range phonons can be emitted by both inter- and intrasubband transitions and since now $q >> l^{-1}$, $q_x \sim l^{-1} << q$ and the possible values of q_y: $mk_BT_e/\hbar^2k_N$ (intrasubband transitions) and $|k_N \pm k_M|$ (intersubband transitions) are also $<< q$, the phonons are emitted approximately normal to the 2DEG plane. Inter-and intrasubband transitions make approximately equal contributions to the emission and with a frequency spectrum weaker in power dependence by 1 than that in range (i) since the q_x contribution to the phase volume no longer increases linearly with T_e. The emission intensity oscillates with E_F as in (i).

Range (iii) $k_BT_e \sim \hbar\omega$

"Resonant" phonon emission at $\omega_p = j\omega$ can now occur as a result of transitions between peaks in the one-dimensional density of states at the bottom of the subbands. These are also emitted normal to the 2DEG plane, as are the non-resonant phonons arising from processes discussed in range (ii), though with intensity independent of the position of E_F.

The energy loss rate $Q \propto (T_e^{\gamma+4} - T^{\gamma+4})$ in range (i) and $Q \propto (T_e^{\gamma+3} - T^{\gamma+3})$ in the other two ranges.

Magnetic field does not change the form of the eigenvalues but increases the subband separation to $\hbar\omega_B = \hbar(\omega^2 + \omega_c^2)^{1/2}$ and the effective mass of the electrons to $m_B = m(\omega_B/\omega)^2$. It also reduces the 1D confinement length to $l = (\hbar/m\omega_B)^{1/2}$ and creates spatial separation between the $<k|$ and $<k-q_y|$ eigenstates; this latter has rather small effect on the matrix elements except at very high magnetic fields but should be detectable through suppression of the wings of the angular distribution of the emission. At low temperatures, range (i), as the field increases the intensity Q initially rises $Q \propto m_B$ or $m_B^{1/2}$ for constant E_F and electron sheet density repectively, and the emission becomes more isotropic. Eventually however the intensity reaches a maximum and then falls because the flattening of the subbands resulting from increasing m_B reduces the Fermi velocity below the sound velocity s restricting emission to the exponentially few thermally excited electrons with $v > s$. The field-dependent subband separation also leads to oscillations in the emission intensity with field. At higher temperatures the decrease in confinement length can transfer the system from range (ii) or (iii) to range (i) resulting in changes in both intensity and angular dependence.

References

[1] Bockelmann U and Bastard G, Phys Rev B42,8947 (1990)
[2] Shik A Y and Challis L J ,submitted for publication

Induced Magnetism in Bismuth by Heat Flow Under Conditions of Phonon Hydrodynamics

I.N. Zhilyaev

Institute of Microelectronics Technology and High Purity Materials, Russian Academy of Sciences, 142432 Chernogolovka, Moscow District, Russia

In perfect bismuth cristals, circular electrical currents appear when heat flows through them at helium temperatures [1-3]. These effects can be explained if we assume a phonon gas flow through the sample and the following cnditions are met. The first: the phonon system has viscosity. This implies that the phonon system is in the phonon hydrodynamics regime [4]. The second: phonons which collide with the sidewalls of the sample, do so inelastically. The phonon pulse distribution is, therefore, inhomogeneous in a cross-section. Due to collisions, the distribution of the pulse in the electron-hole plasma is inhomogeneus too. Circular currents will appear in this case.

When a thermal pulse passes trough a bismuth sample in longitudinal magnetic field a diamagnetic wave arises. This wave was registered by number of coils which had been wound on the sample. The electromotive force pulses in coils were measured. Bolometer measurements demonstrated the magnetization wave similar to heat wave. As the thermal wave propagates along the sample it creates in electron-hole plasma an electrically neutral wave of charge-carrier density. Diamagnetic magnetization appeares in electron-hole plasma because of its disturbance is inhomogeneous in a cross-section. It was theoretically shown [5] that under conditions of second sound propagation the heat pulse front has velocity close to the second sound velocity and conserve this value much longer then the pulse maximum do. The heat pulse front velocity measurements gave values which are in agreement with measurements of second sound wave velocity in bismuth [6]. Magnetization wave amplitude depends on inhomogeneity in a cross-section of phonon concetration wave. This inhomogeneity and heatconductivity depend on phonon free path L due to normal phonon-phonon collisions. Measured temperature dependences of electromotive force amplitude A caused by magnetization

Springer Series in Solid-State Sciences, Volume 112
Phonon Scattering in Condensed Matter VII Eds.: M. Meissner and R.O. Pohl

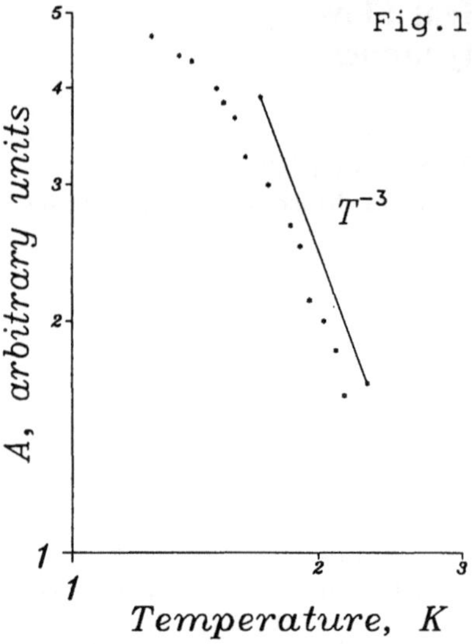

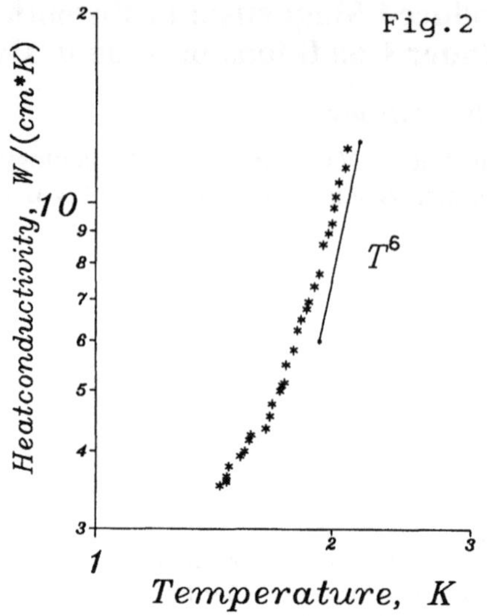

Fig.1. Temperature dependence of a electromotive force amplitude A caused by magnetization wave in coil.

Fig.2. Temperature dependence of heatconductivity.

wave in coil and heatconductivity κ(T) of the same sample are given in Fig.1 and Fig.2 respectively. Curve A(T) is close to T^{-2}-T^{-3} dependence and κ(T) to T^5 -T^6 at T=2K. In bismuth L is proportional to T^{-4} [6]. When T increases L decreases and then boundary layer in which magnetization is exited decreases. Under conditions of phonon hydrodynamics κ(T) is proportional to C/L . As bismuth heat capasity C is proportional to T^3,the difference between powers in dependences A^{-1}(T) and κ(T) is close to 3.

References

1. I.N.Zhilyaev, JETP Lett.40,848 (1984)
2. I.N.Zhilyaev, Sov.J.Low Temp.Phys.14,502 (1988)
3. I.N.Zhilyaev, JETP Lett.49,507 (1989)
4. V.N.Kopylov and L.P.Mezhov-Deglin, Sov.Phys.JETP 38,357 (1974)
5. S.J.Rogers, Phys.Rev.B3,1440 (1971)
6. V.Narayanamurti and R.C.Dynes, Phys.Rev.Lett.28,1461 (1972)

Tunneling as a Probe of Hot Electrons

M. Nahum[1], *P.L. Richards*[1], *and C.A. Mears*[2]

[1]Department of Physics, U.C. Berkeley and Materials Sciences Division, Lawrence Berkeley Laboratory, Berkeley, CA 94720, USA
[2]Laboratory for Experimental Astrophysics, Lawrence Livermore National Laboratory L-401, P.O. Box 808, Livermore, CA 94550, USA

At very low temperatures the inelastic collision processes (energy loss processes) for electrons in a metal are very infrequent. As a result, the electrons become thermally decoupled from the phonons and will heat up relative to the lattice as they absorb power. The subject of hot-electrons at low temperatures is of interest not only from a theoretical point of view, but also because the optimization of many types of devices requires an understanding of this effect. In particular, the effects of hot-electrons can be detrimental to the performance of low temperature SQUIDs [1] and single electron tunneling devices [2]. Alternatively, it may be possible to construct ultra-sensitive bolometric detectors based on this effect [3].

The few direct measurements of electron heating phenomena have commonly employed dc SQUIDs to measure the Johnson noise arising from the resistance of a thin metal strip and the temperature of the electrons. The electrons were heated above the phonon temperature by passing an electric current through the resistive strip [1,4]. We propose an alternative measurement method which uses the temperature dependence of the current-voltage characteristic of a superconductor-insulator-normal metal (SIN) junction to directly measure the temperature of the electrons in the normal metal electrode. The temperature of these electrons can be raised above the lattice by passing a current through that electrode. The advantages of this method are the relative simplicity of the measurements and the very high temperature resolution.

The usefulness of the junction as a thermometer for the electron temperature stems from the exponential dependence of the tunneling current on temperature. It is possible to show that to first order, the quasiparticle current through the junction depends only on the temperature T of the normal electrode and is identical to the case where both electrodes are at the same temperature [5]. If the junction is operated at a temperature such that $kT \ll \Delta$ and for $eV < \Delta$, then the dependence of the quasiparticle current I on the voltage V across the junction is given by $I = (kT/eR)\sinh(eV/kT)$. Here $R = R_N(kT/2\pi\Delta)^{1/2}\exp(\Delta/kT)$ is the dynamic resistance of the junction at low bias, R_N is the normal resistance of the junction, and Δ is the energy gap of the superconducting electrode

Springer Series in Solid-State Sciences, Volume 112
Phonon Scattering in Condensed Matter VII   Eds.: M. Meissner and R.O. Pohl

[5]. For a junction biased at a constant current, the change in the voltage across the junction, due to a change in the electron temperature is given by

$$S = dV/dT = (k/e)\left[(\Delta/kT + 1/2)\tanh(eV/kT) - (eV/kT)\right]. \qquad (1)$$

The ultimate limitation of the temperature sensitivity is due to the intrinsic noise in the junction and can be estimated from the expression for the voltage noise V_n of a biased SIN junction:

$$V_n = [2eI\coth(eV/2kT)R_D^2]^{1/2}, \qquad (2)$$

where $R_D = dV/dI$ is the dynamic resistance of the junction [6]. The minimum detectable temperature change in a 1 Hz noise bandwidth is then given by the ratio V_n/S, whose minimum value determines the optimum bias current. The energy relaxation of the electrons to their environment will be dominated by the electron-phonon interaction only if the energy loss through the leads and through the junction can be made negligible. If the contacts to the strip and to the superconducting electrode of the junction are made of a superconductor with a higher T_c than that of the junction electrode, then the Andreev reflection of quasiparticles at the interfaces will trap the quasiparticles, and hence the energy in the active region.

As a specific example illustrating the potential of this technique we consider an operating temperature of 100 mK, a normal state resistance of R_N=500 Ω, and $\Delta/kT \approx 6$. Under these conditions the optimum operating point for measurement of the electron temperature occurs at a bias current of 1 nA with a resultant sensitivity of 5×10^{-4} V/K, and minimum detectable temperature change of 300nK in a 1Hz noise bandwidth. This is equivalent to a temperature resolution of $\Delta T/T=1.5\times10^{-6}$ in a one second integration time.

Supported by the Director, Office of Energy Research, Office of Basic Energy Sciences, Materials Sciences Division of the U. S. Department of Energy under Contract No. DE-AC03-76SF00098.

References

[1]. F. C. Wellstood, C. Urbina, and J. Clarke, Appl. Phys. Lett. **54**, 2599 (1989).
[2]. H. Pothier, Ph.D. thesis, Docteur de l'Universite Paris 6, 1991 (unpublished).
[3]. M. Nahum, C. A. Mears, and P. L. Richards, (to be published).
[4]. M. L. Roukes, Ph. D. thesis, Cornell University (University Microfilms International, Ann Arbor, MI, 1985).
[5]. L. Solymar, *Superconductive Tunnelling and Applications* (Wiley, New York, 1972).
[6]. D. Rogovin and D. J. Scalapino, Ann. Phys. **86**, 1 (1974).

Evidence for Phase Transitions in the Electron–Phonon System of Aluminum

J.F. Kos and Z. Hou

Department of Physics, University of Regina, Regina, Saskatchewan, Canada S4S 0A2

1. Introduction

The thermal resistivities (W) and the electrical resistivities (ρ) of Au, Cu and Ag[2] exhibit several regions in which the resistivities vary according to specific power laws. ($\rho \propto T^k$ and $W \propto T^n$) with abrupt changes in k and n at the same temperatures for a given metal. We shall refer to this behavior as W-anomalies and they remain unexplained.

2. Results

Measurements of the detailed variation of the thermal resistivity of aluminum as a function of temperature from 2.5 to 48 K.

The low temperature thermal resistivity may be written as:

$$W = W_o + W_i + W_d + W_v + W_{ee} \tag{1}$$

where:

$$W_o = \rho_o \; / \; L_o T \tag{2}$$

$$(W_i + W_d) = (\rho_i + \rho_d) \; / \; L_o T \tag{3}$$

ρ_i is the ideal electrical resistivity, ρ_d is the deviation from Matthiessen's rule, L_o is the Lorenz number (for T $\rightarrow$ 0), W_v represent vertical processes on the Fermi surface (inelastic electron-phonon scattering) and W_{ee} represents the component of thermal resistivity due to electron-electron scattering.

L_o can be obtained by writing eq (1) as:

$$(W - W_i - W_d) \; T = \rho_o / L_o + (W_v + W_{ee}) \; T \tag{4}$$

Values of $\ln[(W - W_o)/T^2]$ as a function of $\ln(T)$ are presented in fig 1 which shows four straight sections. For each of these sections values of $(W - W_o)$ vs T were fitted to an equation $X = ET^P$.

Springer Series in Solid-State Sciences, Volume 112
Phonon Scattering in Condensed Matter VII Eds.: M. Meissner and R.O. Pohl

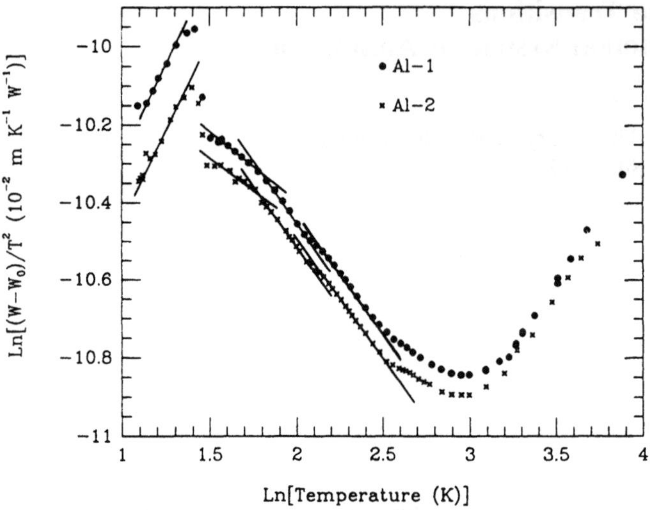

Fig 1 - Graph of Ln[$(W - W_o)]/T^2$] vs Ln(T)

3. Discussion

The determination of $(W - W_o)$ as a function of T exposes the existence of several sections in which $(W - W_o)$ varies as some power of T over a given temperature range with abrupt changes in variations as different temperature ranges are entered (table 3). No physical significance should be associated with the particular values of the powers, however abrupt changes in the powers indicate a chance in phase (or change in scattering mechanisms). Above 12.7 K we have found evidence of bandgap-transitions likely associated with the onset of umklapp electron scattering [4].

References

1. Kos, J.F., Phys. Condens. Matter (1990) 2 4859.
2. Hou, Y.Z. and Kos, J.F., submitted for publication.

Phonon Scattering in the Heavy-Fermion Superconductor URu_2Si_2

K. Behnia, D. Jaccard, and J. Sierro

Département de la Physique de la Matière Condensée,
24 Quai Ernest-Ansermet, CH-1211 Genève 4, Switzerland

Heavy- fermion superconductors have been the subject of intense theoretical and experimental research in recent years. The superconducting instability in these systems appears in a particular environment: a highly correlated Fermi liquid. The strong correlation among electrons in the normal phase has been at the origin of the speculation on a possible unconventional superconductivity in this class of superconductors where, like other exotic superconductors, the mecanism of pairing remains unidentified.

In this paper, we report on the first low-temperature thermal conductivity measurements performed on a single crystal of URu_2Si_2. This moderate heavy-fermion compound goes through an anti-ferromagnetic transition at T_N = 17 K before becoming superconducting at T_C = 1.2 K[1].

Figure 1 shows the temperature dependence of the thermal conductivity of the sample at nominally zero field and at a 3 tesla field along the c-axis. At

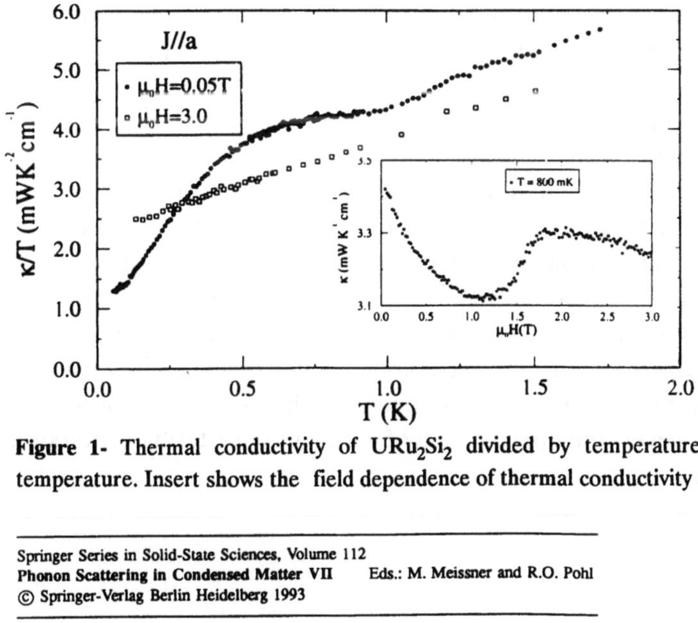

Figure 1- Thermal conductivity of URu_2Si_2 divided by temperature as a function of temperature. Insert shows the field dependence of thermal conductivity at 800 mK.

Springer Series in Solid-State Sciences, Volume 112
Phonon Scattering in Condensed Matter VII Eds.: M. Meissner and R.O. Pohl

1.5 K, the Lorenz number is roughly two times the Sommerfeld value revealing a significant contribution by non-electronic carriers to the thermal conductivity at the onset of superconductivity. Just above T_c, the thermal conductivity of the normal phase can be expressed as $\kappa = aT + bT^2$ with $a = 2.95$ mW K^{-2} cm^{-1} and $b = 1.57$ mW K^{-3} cm^{-1}. One is tempted to attribute the aT term to the conductivity of the electrons and the bT^2 term to the contribution of non-electronic carriers. As expected, the product of a by the residual resistivity (both at zero field and at $\mu_0H = 3T$) is very close to Sommerfeld value, confirming the electronic nature of this term. Comparing the specific heat of URu_2Si_2 and $ThRu_2Si_2$ [2] (which does not order magnetically), the magnon contribution is expected to be important just below T_N and to become negligible in our temperature range of interest ($T<2K$). A T-square phonon conductivity is expected when phonons are mainly scattered by electrons. Our data together with the sound velocity (v_s = 4900 m/s) and the phonon specific heat ($C_{ph} = \beta T^3$ with $\beta = 7.75$ J.K^{-4} m^{-3}) yield $l_{ph} \approx 6$ μm at 2K.

It is instructive to compare the value of the T-square term obtained here with similar terms obtained in the studies on the thermal conductivity of normal state of niobium [3](b = 0.025 mW cm^{-1} K^{-3}) and vanadium [4] (b = 0.012). The large value of b in URu_2Si_2 seems to be essentially a consequence of the inefficiency of the electron scattering of phonons.

The theoretical treatment of electron-phonon interaction in heavy-fermion compounds remains an open question. In metals, in a first approximation [5], this interaction is proportional to the ratio of carrier density to the density of energy states at Fermi level. In the case of URu_2Si_2 below the Néel temperature, the drastic decrease in the density of charge carriers combined with a high density of states explain the smallness of the coupling.

Below T_c , the decrease in the number of electronic excitations leads to a decrease in the electronic contribution on one hand and to an increase in the phonon mean free path on the other hand. These two competing tendencies lead to the plateau of κ/T at temperatures between 0.7 T_c and 0.8 T_c .

The field dependence of the thermal conductivity for a magnetic field along the c-axis at 800 mK is presented in the insert of figure 1. A sudden increase in thermal conductivity marks the upper critical field. For a field in the basal plane the corresponding anomaly is much smaller. In the mixed state, as the field is increased, the enhancement in the electronic conductivity is accompanied by a decrease in the lattice conductivity due to the growing strength of scattering by electronic excitations. At low fields, the presence of a random distribution of vortices is a source of additional scattering of heat carriers. Here, however, the effectiveness of vortices as scattering centers for phonons is highly diminished as the coherence length (≈ 10 nm) is shorter than the phonon wavelength (≈ 230 nm at 1K).

References

1- Palstra et al., Phys. Rev. Lett. **55**(1985) 2727.
2- Schablitz et al., Z. Phys. **B62** (1986) 171.
3- Kes et al., J. Low Temp. phys. **17** (1974) 341.
4-C. L. Tsai et al., Phys. Rev. **B23** (1981), 6430.
5-J. M. Ziman, Electrons and phonons, Clarendon Press, Oxford, 1960.

The Phonon Heat Flux Through a Si–Cu Point Contact

P. Stefanyi[1], *P. Fozooni*[1], *M.J. Lea*[1], *J. Saunders*[1], *A. Feher*[2], *R. Zaboj*[2], *and A.G. Shkorbatov*[3]

[1]Department of Physics, Royal Holloway and Bedford New College, University of London, UK
[2]Department of Physics, Safarik University, 04353 Košice, Czechoslovakia
[3]Institute of Low Temperature Physics and Engineering, 310164 Kharkov, Ukraine

The ballistic transport of phonons through a point contact (PC) can be realized when the characteristic dimension of the contact (d) fulfils two conditions;

(i) $d >> \lambda_{ph}$, where λ_{ph} is the phonon characteristic wavelength.
(ii) $d << l_{ph}$ where l_{ph} is the phonon mean free path [1].

In the present work we concentrate on the regime λ~d in which the wave character of the phonons should be observed. Point contacts were made by mounting a rectangular single crystal of Si on three sharp tips , one of which was made of Cu, the other two of vespel. An NTD doped natural Ge crystal was mounted on the Si as a thermometer; the leads were 50 μm diameter superconducting wires.The measurement technique is described elsewhere [2]. Heat flux curves were measured below 10K. The heat flux is determined from the known heat capacity of silicon and the time dependence of the target temperature. The typical dependence of the heat flux on temperature is shown on Fig.1, for 7 g Si single crystal. It exhibits a plateau above 6K which represents ballistic transport through the PC and a flux enhanced above ballistic transport at lower temperatures with a maximum at 0.7K. The value $Q/T^4 = 2.2$ nW at 10K determines total contact area of 11 μm^2. In order to explain the data we used the following model.

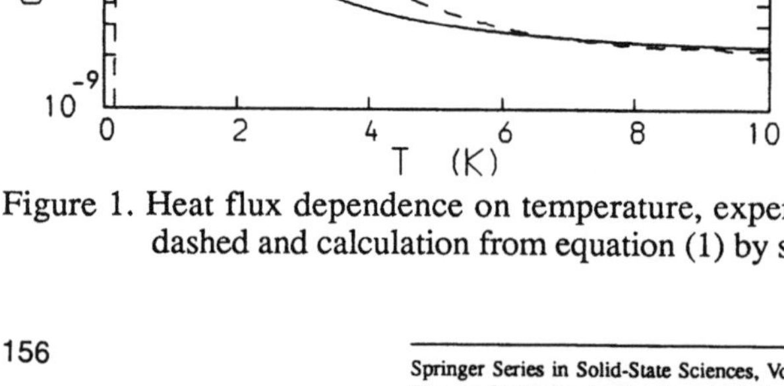

Figure 1. Heat flux dependence on temperature, experimental data is shown by dashed and calculation from equation (1) by solid lines.

Springer Series in Solid-State Sciences, Volume 112
Phonon Scattering in Condensed Matter VII Eds.: M. Meissner and R.O. Pohl

The heat flux through the PC in the ballistic regime can be expressed as [1]:

$$Q(T_2,T_1) = \frac{h}{(2\pi)^3} \int_0^\infty \omega\ A\ S(\omega)\ [N(\omega_1 T_2) - N(\omega_1 T_1)]\ d\omega\ , \quad (1)$$

where $S(\omega)$ is a function related to phonon density of states, $N(\omega,T_i)$ is the Planck distribution function on both sides of the PC and A is the area of the PC. When $\lambda_{ph} << d$ then $A=A_0$ and

$$Q(T_2,T_1) = \frac{\pi^2 A_0\, k_B^4}{120\ h^3} \sum_i v_i^{-2}\, (T_2^4 - T_1^4) \quad . \quad (2)$$

For $\lambda_{ph} \sim d$, A_0 = const. and we can introduce the effective cross section $A(\omega,T)$ for phonons which can be modelled as follows [3]:

$$A(\omega,T)/A_0 = B(d.q)^4\ (1+2C\exp(-0.5)) \qquad d.q < 1.5 \quad (3)$$
$$A(\omega,T)/A_0 = B(1.5)^4\ \{1+C\exp(-d.q/3)\cos^2[(4\pi/3)\ d.q - 2\pi]\} \quad d.q > 1.5$$

Where q is the phonon wave vector, B and C are numerical constants regulating the width and amplitude of the effective cross section. $A(\omega,T)$ is a function which gives Rayleigh scattering cross section $A(\omega,T) \sim \omega^4$ in the long wavelength limit $\omega<v/d$ and geometrical scattering in the short wavelength limit $\omega >>v/d$. In the intermediate region $\omega \sim v/d$, $A(\omega,T)$ is an oscillatory function with the frequency spacing between peaks 1.5 v/d. Since the dominant phonon frequency is related to the temperature by $h/2\pi\omega \sim 3.8 k_B T$ the spacing between peaks is about 4 $k_B T$. Cross section (3) was substituted into (1) and the thermal flux through PC was calculated with various diameters. The result of the calculation is also shown in Fig. 1 together with experimental data. The solid line is the heat flux calculated according to this model with 230 PCs of characteristic diameter d=0.25μm, B=4.5 and C=40. This implies that the measured heat flux is a sum of heat fluxes through a number of small contact areas which seems to be a reasonable picture of the described copper tip-silicon crystal PC. In order to get more confirmation about the PC topology we performed experiments with a 14 g Si single crystal resting on the same arrangement of three tips. The experimental curve shows the same qualitative features as the heat flux through the PC in the previous case.
The heat flux in the region of the plateau Q/T^4=52 nW/K^4 is about 25 times larger than in case of 7 g Si. The maximum is at 0.5K compared to 0.7K for 7 g Si and the maximum value of the heat flux Q/T^4_{max}=85nW/K^4 compared to 35nW/K^4. The enhancement in heat flux reported here at a temperature where the dominant phonon wavelength is of the order of typical point contact dimension demonstrates the wave nature of phonons, however there could be some other explanations. Point contacts of higher quality and with better characterisation will be used in future work. The effect possibly allows the use of the point contacts in phonon spectroscopy and may offer the possibility of selective filtering of non-thermal phonons such as are generated by incoming particles detected by Si target.

References

1.E.N. Bogachek, A.G. Shkorbatov, Sov. J.Low Temp. Phys. **11** (1985) 353.
2.P.Stefanyi, A.Feher, This conference
3.J.W.Schwartz, Ch.T.Walker, Phys. Rev. **155** (1967) 959.

Phonon Spectroscopy of Perovskite Oxides Using Point Contact Technique

H. Srikanth and A.K. Raychaudhuri

Department of Physics, Indian Institute of Science, Bangalore 560012, India

We report the phonon spectroscopy of metallic oxide ReO_3 using the point contact technique. The junction size criterion for ballistic electronic transport and other issues in point contact spectroscopy of oxides have been addressed.

1. Introduction

Role of phonons and electron-phonon interaction in electronic transport in perovskite oxide systems is of current interest. Point contact spectroscopy(PCS) in the Sharvin limit has been used in the past to study phonon spectra in metallic solids and alloys [1,2]. In typical metallic junctions, this happens for $R(V = 0) < 100\ \Omega$. For smaller contact (radius = a) such that $k_f a$=1 (k_f, Fermi wave vector), the Sharvin approximation breaks down and quantum confinement effects make $dR/dV < 0$ much before onset of vacuum tunneling [3].

In perovskite oxides (including high Tc superconductors) one generally sees a decreasing background resistance with bias voltage ($dR/dV < 0$) which can arise from correlation effects [4]. We would then like to ask if it is at all possible to observe proper phonon spectra in perovskite oxides by point contact method in the Sharvin limit ($dR/dV > 0$). In this paper we show that for a single crystal of metallic oxide ReO_3 of relatively high conductivity one can obtain proper Sharvin limit along with the regime $dR/dV > 0$ to do phonon spectroscopy by point contact method.

The point contact experiment was carried out in a bath cryostat where the sample is directly dipped in liquid Helium and the junctions are made by pressing a sharply etched Au wire against the sample. Experimental details are given elsewhere [5].

2. Results

In fig. 1 we show the $|\frac{d^2V}{dI^2}|$ curves for two junction resistances $R_o = 1\Omega$ and $R_o = 165\Omega$. Fig. 2 shows the normalized dV/dI curves as a function of

Springer Series in Solid-State Sciences, Volume 112
Phonon Scattering in Condensed Matter VII Eds.: M. Meissner and R.O. Pohl

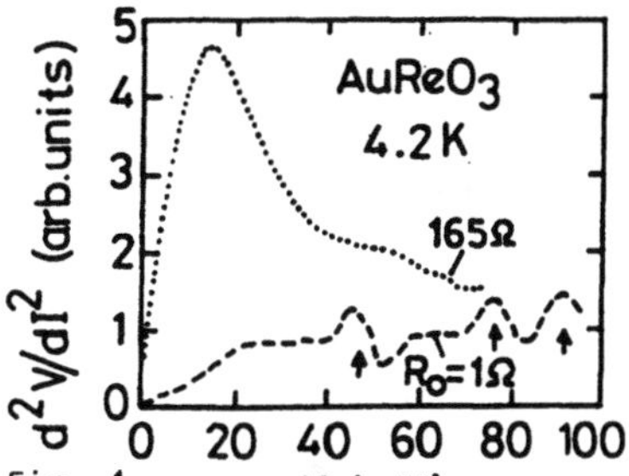

Fig. 1

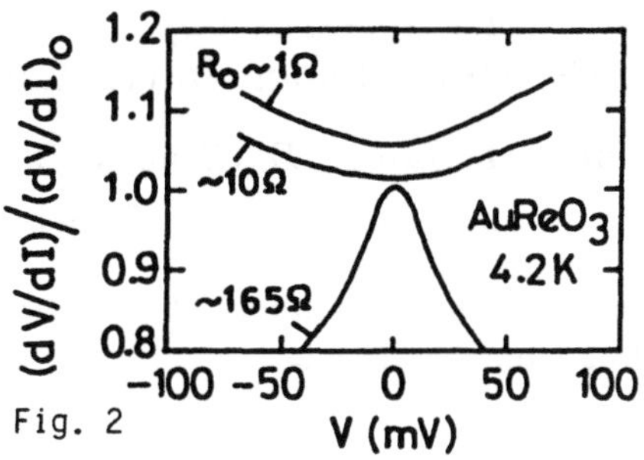

Fig. 2

Fig 1 The $|\frac{d^2V}{dI^2}|$ plots for two junction resistances of $Au - ReO_3$ junctions. $R_o = 1\Omega$ (- - -) and $R_o = 165\Omega$ (. . .). Phonon features are marked by arrows.

Fig 2 The $(\frac{dV}{dI})$ curves for three junction resistances: $R_o = 1\Omega, 10\Omega$ and 165Ω. Note the change in slope as R_o increases.

zero bias resistance (R_o). The change in sign of dR/dV from positive to negative as R_o increases is clearly demonstrated. For the 1 Ω junction we observe reproducible peaks at specific energies marked by arrows (see fig. 1) associated with the phonon energies in ReO_3 which have been confirmed by Raman spectroscopy. Lower energy features are associated with phonons of Au. For the 165 Ω junction ($dR/dV < 0$) no discernible features are seen at the ReO_3 phonon energies. [In these data no background have been subtracted.]

In conclusion we show that it is possible to obtain the phonon spectra in a cubic perovskite oxide ReO_3 by conventional point contact spectroscopy provided the junction resistance is small enough to ensure $dR/dV > 0$. Our observations indicate that in oxides, not only contact size restrictions but material specific properties like strong correlation effects in the system could also play a role in determining whether one can do phonon spectroscopy by the conventional PCS method.

References

[1] A.G.M. Jansen, A.P. van Gelder and P. Wyder, J.Phys. C: Solid St. Phys. Vol. 13, 6073 (1980).

[2] A.M. Duif, A.G.M. Jansen and P. Wyder, J.Phys.:Condens. Matter Vol. 1, 3157 (1989).

[3] H. Srikanth and A.K. Raychaudhuri, Phys. Rev. B (in press).

[4] H. Srikanth, K.P. Rajeev, G.V. Shivashankar and A.K. Raychaudhuri, Physica C Vol. 195, 87 (1992).

[5] H. Srikanth and A.K. Raychaudhuri, Cryogenics Vol. 31, 421 (1991).

Point-Contact Spectroscopy of the Electron–Phonon Interaction in $SmNi_5$

M. Reiffers[1], *P. Szabo*[1], *D. Gignoux*[2], *and D. Schmitt*[2]

[1]Institute of Experimental Physics, Solovjevova 47, 04353 Košice, Czechoslovakia

[2]Laboratoire de Magnetisme L. Neél, CNRS, B.P. 166X, F-38042 Grenoble, France

Point-contact (PC) spectroscopy has been shown to be a very useful experimental method for the direct study of the energy dependence of the interaction between conduction electrons and quasiparticle excitations in metals [1]. For small PC between two metals where both the inelastic l_e and elastic l_i mean free paths of the electrons are larger than PC diameter d, the electron transport through the PC is ballistic. Then the applied voltage directly defines the energy scale for the scattering processes of the accelerated electrons. In this case the second derivative $d^2I/dV^2(eV)$ of the I-V characteristic is directly proportional to the electron-quasiparticle interaction function.

When $l_e, l_i << d$, the Joule overheating of PC region only above ambient temperature arise [1]. Then by the applied voltage one can controles the PC temperature.

$SmNi_5$ crystallizes in the hexagonal $CaCu_5$-type structure. It is an ferromagnet below the Curie temperature T_c = 27.5 K [2]. Its magnetic properties can be described by the crystall electric field (CEF) effects [2]. The situation here is complicated because of the small energy difference between ground CEF multiplet $^6H_{5/2}$ and two excited multiplets $^6H_{7/2}$ and $^6H_{9/2}$. In order to fit the experimental results it is necessary to put into Hamiltonian the spin-orbit coupling and to substitute total angular momentum by its orbital and spin component. Moreover, the direct measurements of the CEF levels doesn't exist because of the neutron absorption by Sm. To study the CEF and phonon role in the electron transport we measured the PC spectra of the oriented single crystal $SmNi_5$.

The PC were made at 1.5 - 2.5 K by pressing a Cu needle to the surface of $SmNi_5$ single crystal (RRR > 20). The magnetic field was applied parallel to the PC axis. The dV/dI and d^2I/dV^2 dependencies were measured using phase-sensitive detection techniques [1].

In Fig.1. the characteristic PC spectra of the $SmNi_5$-Cu heterocontacts in ballistic regime (BR) are shown. We observed characteristic maxima at the energies about 4 meV, 8 meV, 22 meV, 30 meV and a broad peak around 19 meV. Then we found no influence of the magnetic field on the energy position of the characteristic peaks until 10 T. Moreover, taking into account the CEF parameters from [2] one can calculate the self-energies of CEF levels in

Springer Series in Solid-State Sciences, Volume 112
Phonon Scattering in Condensed Matter VII Eds.: M. Meissner and R.O. Pohl

Fig. 1

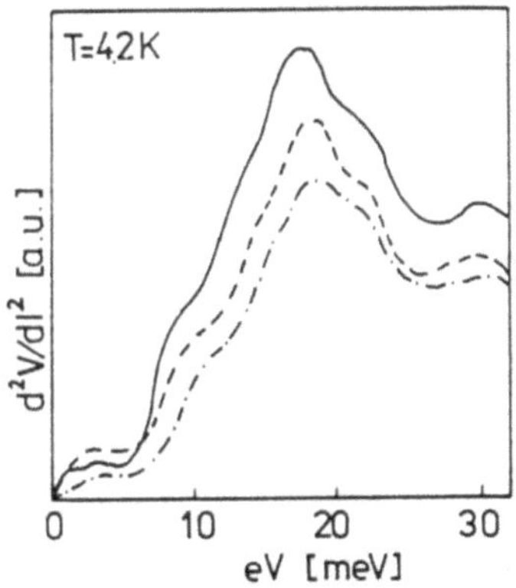

Fig. 1 The characteristic PC spectra of SmNi5-Cu in BR

Fig. 2

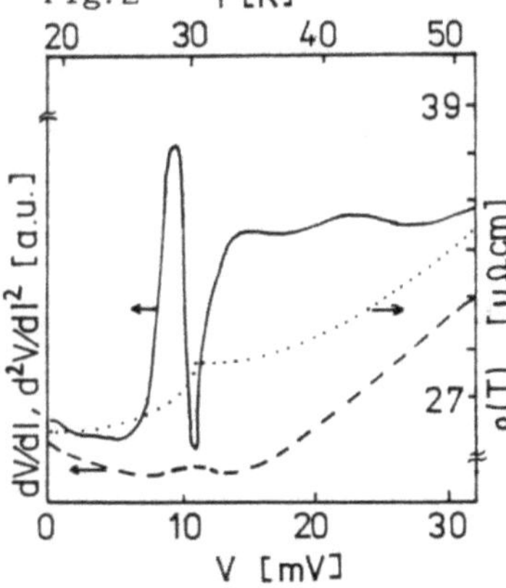

Fig. 2. The comparison of dV/dI(V) —, $d^2I/dV^2(V)$ -- in TR and ρ(T) ... of SmNi5 [2]

SmNi5. From this results only one allowed transition from ground level with energy distance 40 meV as follows from selection rules. Therefore, no magnetic field influence is the confirmation of phonon origin of peaks.

Then the broad peak around 19 meV corresponds to the energy of transversal phonons in Cu. The maximum at 30 meV is ascribed to the longitudinal phonons in Cu. The maxima at 4 meV, 8 meV and 22 meV are connected with the phonons in SmNi5. The phonon part of the PC spectrum coincides well with the spectrum of LaNi5 [3]. One can see the dominating phonon part in PC spectrum of SmNi5.

Except PC spectra in BR, the PC spectra in thermal regime (TR) were measured. In Fig. 2. the dV/dI(V) and $d^2I/dV^2(V)$ dependencies in TR are presented. As follows from the TR theory [1], the dV/dI(V) is similar to the bulk electrical resistivity ρ(T) behaviour [2]. The nonlinearity at V_c around 10 mV on dV/dI is due to the overheating of PC region above the Curie temperature T_c = 27.5 K. Taking into account the TR expression [1] we can estimate the Lorentz function value at T_c as L = 1.8 L_0 which is a reasonable value for such intermetallic compounds.

In conclusion, we observed the dominating phonon part in PC spectra of hetercontact SmNi5-Cu. which coincides well with LaNi5. In TR we obtained the reasonable value of Lorentz function at T_c.

References

[1] A.G.M.Jansen, A.P.van Gelder, P.Wyder; J.Phys. **C13** (1980) 6073
[2] R.Ballou et al., Physica **B149** (1988) 340
[3] A.I.Akimenko et al., Zeits. Phys. **B79** (1990) 191

Nonequilibrium Phonons in Nonlinear Conductivity of the Heterocontact Based on $RENi_5$ (RE = La, Pr, Ce)

M. Reiffers[1], *Y.G. Naidyuk*[2], *and I.K. Yanson*[2]

[1]Institute of Experimental Physics, Solovjevova 47, 04353 Košice, Czechoslovakia

[2]B.I. Verkin Institute for Low Temperature Physics and Engineering, Lenin ave. 47, 310164 Kharkov, Ukraine

Point-contact (PC) spectroscopy is a very useful experimental method for the direct study of the energy dependence of the interaction between conduction electrons and quasiparticle excitations in metals [1]. For small metallic contacts where both the inelastic l_e and elastic l_i mean free paths of the electrons are larger than the contact diameter d, the second derivative $d^2I/dV^2(eV)$ of the I-V characteristic is direct proportional to the electron-quasiparticle interaction function (EQI).

In our experiments PC were made at 1.5 K by pressing a Cu (Fe, Ni) needle to the $RENi_5$ single crystal surface. The dV/dI; d^2I/dV^2 were measured using phase-sensitive detection techniques [1].

The PC spectra of the $LaNi_5$-Cu heterocontacts show only the phonon contribution [2]. The presence of 4f electrons from Pr and Ce caused additional pecularities in PC spectra at low energies. The observed peak at 4 meV in $PrNi_5$ is connected with the excitation of Pr ions to higher CEF levels [3]. In $CeNi_5$ the maximum at about 2 - 4 meV is described to the interaction of conduction electrons with spin fluctuations (paramagnons) [4].

In Fig.1. the comparison of $RENi_5$-Cu (RE = La, Ce, Sm, Pr) PC spectra is shown in order to demonstrate the similarity of the phonon part (5 - 30 meV) of the EQI. The observed background in PC spectra of same intensity as the spectral part) and pronounced asymmetry of the PC resistance dV/dI(V) with respect to the voltage polarity for all compounds are connected with the short elastic mean free path of the phonons in the PC region. Than we have to take into account phonon drag effects. Therefore we measured the heterocontacts $PrNi_5$-Fe (Ni) because Fe and Ni have a different sign of phonon drag. We observed the same asymmetry sign (Fig.2.) in ballistic regime. This testifies the important role of the relaxation of $RENi_5$ nonequilibrium phonons in the PC region.

Springer Series in Solid-State Sciences, Volume 112
Phonon Scattering in Condensed Matter VII   Eds.: M. Meissner and R.O. Pohl

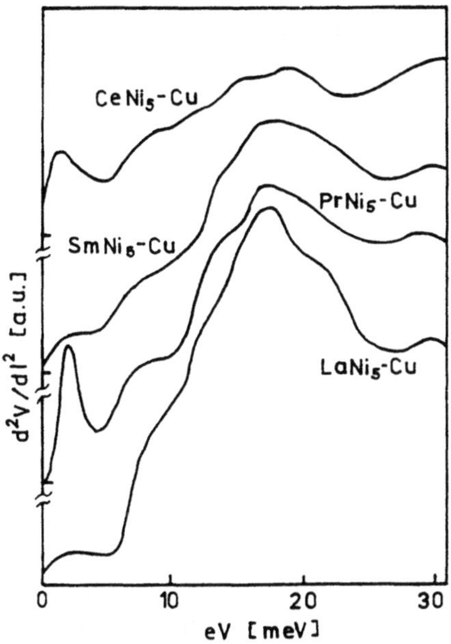

Fig.1. The characteristic PC spectra of the heterocontacts RENi5-Cu (RE = La, Ce, Pr, Sm) in ballistic regime.

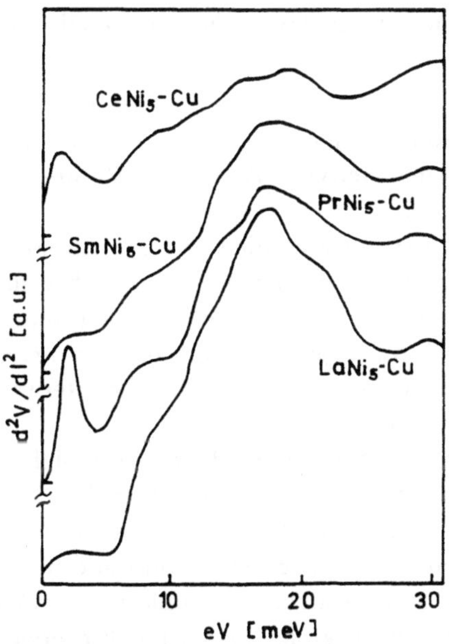

Fig.2. The dV/dI dependences of heterocontacts PrNi5-Fe (Ni) for both voltage polarities up to high voltages.

References

[1] A.G.M.Jansen, A.P.van Gelder, P.Wyder; J.Phys. **C13** (1980) 6073
[2] A.I.Akimenko et al., Zeits. Phys. **B79** (1990) 191
[3] M.Reiffers et al., Phys. Rev. Lett. **B62** (1989) 1560
[4] Y.Naidyuk et al., Proceedings of ICPTM Darmstadt 1992

The Phonon Mechanism of Fluctuator Activation in Ballistic Nanostructures

V.I. Kozub and A.M. Rudin

A.F. Ioffe Physico-Technical Institute, 194021 St. Petersburg, Russia

Abstract. Activation processes for two-level fluctuator (TLF) situated in metallic nanoconstriction and producing telegraph resistance noise are considered for the case of large enough biases. It is shown that these processes are dominated by TLF interaction with non-equilibrium phonons emitted in the constriction rather than with electrons (as it was suggested earlier). The asymmetry of the noise with respect to the sign of the bias observed in experiments is explained to be a result of corresponding asymmetry of the non-equilibrium phonons distribution.

In recent years many experiments have been reported, where the telegraph resistance noise in metal nanometer point contacts has been studied [1,2]. Such a noise is connected with a some "active scatterers" with internal degrees of freedom – *elementary fluctuators.*

At low temperatures these defects can be identified [3,4,5] with well-known two-level systems, typical for disordered atomic structures, while for higher temperatures the generalized concept of soft double-well interatomic potential has been suggested in [6]. Such an object – a sort of *elementary fluctuator* (TLF) – can possess a lot of levels in each well, while the interwell transitions can be due to tunneling or activation.

One of the interesting details of experiments was the dependence of fluctuator's relaxation time τ on the magnitude and sign of the bias V. The authors of [1,2] has explained this behavior as a result of electromigration in course of direct interaction of electrons with TLF introducing some additional fitting parameters, although the nature of very strong asymmetry of $\tau(V)$ was not so clear. Here we would like to note that for high enough energies coupling of TLF with phonons is stronger then one with electrons because of large phase volume. We have studied the TLF behavior in point contact taking into account it's coupling with non-equilibrium phonons (NP) emitted by non-equilibrium electrons (NE). We will show that in this way one can explain the observed $\tau(V)$ behavior without any additional concepts and fitting parameters.

Springer Series in Solid-State Sciences, Volume 112
Phonon Scattering in Condensed Matter VII Eds.: M. Meissner and R.O. Pohl

Starting from the transport equations describing ballistic transport of electrons and phonons one obtains the estimate for the phonon distribution function:

$$N(\omega) \sim \frac{d}{l_e}\frac{eV - \hbar\omega}{\hbar\omega}\theta(eV - \hbar\omega), \quad \nabla N(\omega) \sim \frac{N(\omega)}{d}, \tag{1}$$

where $l_e \approx v_F/\omega_D$ is the electron-phonon mean free path and d is the characteristic contact size. Note, that the spatial distribution $N(r)$ is strongly asymmetric, because phonons are emitted in general in the direction of electron flow.

Considering the TLF-NP interaction within the framework of model [6] one can show that the ratio of TLF-NP and TLF-NE couplings is of the order of $(d/a)(\Delta E^d/\hbar\omega_D)$, where a is the lattice constant while ΔE^d is the characteristic interlevel spacing. For a typical TLF $\Delta E^d/\hbar\omega_D$ is not much smaller than unity. Thus the TLF-NP coupling dominates.

Despite of the anharmonisity of intrawell movement one can estimate the occupation number of n-th level in incoupled TLF-NP system as $\mathcal{N}_n \approx N^n(\omega, r)$, where $\omega \sim \Delta E^d/\hbar$. As a result one obtains for the effective temperature of TLF situated in the contact center:

$$T_e = \frac{\Delta E^d}{log\left(1 + \frac{l_e}{d}\frac{\hbar\omega_D}{\mid eV \mid - \Delta E^d}\right)}. \tag{2}$$

For TLF shifted to other side of the contact having in mind the significant spatial asymmetry of $N(\omega, r)$ depending on the bias sign one can obtain arbitrary strong dependence of τ on bias sign.

References

[1] K.S.Ralls, D.C.Ralph, and R.A.Buhrman, Phys. Rev. B **40**, 11561 (1989).

[2] P.A.M.Holweg, J.Caro, A.H.Verbruggen, S.Radelaar, Phys. Rev. B **45**, 9311 (1992).

[3] A.M.Kogan, and K.E.Nagaev, Solid State Commun., **49**, 387 (1984).

[4] R.Kree, A.Ludviksson, and A.Schmid, Phys. Rev. Lett., **52**, 950 (1984).

[5] V.I.Kozub, Soviet Phys. JETP, **59**, 1303; Soviet Phys. Solid. St., **26**, 1121.

[6] Y.M.Galperin, V.G.Karpov, and V.I.Kozub, Advances in Phys., **38**, 669 (1989).

Phonon–Phonon Drag in Semiconductors with Position-Dependent Band Structure

M. Istrate

Department of Physics, Polytechnic Institute,
Splai Bahlui Stîng, Corp. T, 6600 Iaşi, România

The graded composition of two or more components, the position dependence of the impurity doping, a nonuniform strain or a nonuniform temperature determine the position-dependent band structure of the semiconductor.

In an atomic semiconductor, the electron scattering by lattice vibrations is almost elastic and the electron interact only with long-wave phonons. In semiconductor it is possible thus to exist a two-step drag: at application of an electric field, the electrons drag the electron-phonons and these drag the thermal-phonons; at application of temperature gradient, the electron-phonons aquire an additional drift velocity because of phonon-phonon drag (the thermal-phonons drag the electron-phonons) and these phonons drag the charge carriers [1].

In this paper, we report a theoretical study of the relaxation time for electrons on acoustic phonons in an atomic semiconductor with position-dependent band structure, on taking into account the phonon-phonon drag. We consider the interaction of electrons with the electron-phonons and the charged impurity atoms and the interaction of electron-phonons with electrons, the thermal-phonon, and defects.

The coupled Boltzmann equation for electrons, electron-phonons and thermic-phonons are [1-3]

$$\vec{v}_{\vec{k}}\nabla f(\vec{k},\vec{r}) - \frac{1}{\hbar}(\nabla E_c + \nabla W)\nabla_k f(\vec{k},\vec{r}) = I_{eq}(f,N) + I_{ed}(f) \quad , \qquad (1)$$

$$c_s \nabla T \frac{\partial N(\vec{q},\vec{r})}{\partial T} = I_{qe}(N,f) + I_{qT}(N,F) + I_{qd}(N) \quad , \qquad (2)$$

$$c_s \nabla T \frac{\partial F(\vec{p},\vec{r})}{\partial T} = I_{Tq}(F,N) + I_{TT}(F) + I_{Td}(F) \quad . \qquad (3)$$

At nonequilibrium, the electron or phonon distribution function is a sum of two terms.

Springer Series in Solid-State Sciences, Volume 112
Phonon Scattering in Condensed Matter VII Eds.: M. Meissner and R.O. Pohl

Applying the method of Ref. [4], for a nondegenerate semiconductor, we have found

$$k_a f_1(\vec{k},\vec{r}) = \hbar f_o \tau \vec{k}(1+A)\nabla E_{fn}/m^* k_B T \quad , \tag{4}$$

where

$$\nabla E_{fn} = e\vec{\mathcal{E}} + \vec{C} \quad , \tag{5}$$

$$\vec{C} = -\nabla\kappa + \left(\frac{\partial E_{fn}}{\partial n}\right)\nabla n - \left(\frac{\partial E_{fn}}{\partial n}\right)\frac{1}{4\pi^3}\int d^3k \frac{\partial f_o}{\partial E}\nabla W \quad , \tag{6}$$

$$\tau^{-1} = \tau_a^{-1} + \tau_I^{-1} \quad , \tag{7}$$

$$A = k_a f_o(\vec{k},\vec{r})/x\vec{v}_{\vec{k}}\nabla E_{fn}\frac{\partial f_o(\vec{k},\vec{r})}{\partial E} \quad , \tag{8}$$

$$\frac{1}{x} = -\frac{C_1 m^*}{4\pi\rho\hbar k_B T k^3}\int_0^{2k} N_1(q) q^5 dq \quad . \tag{9}$$

C is determined by the position dependent of band structure. The second term in (4) arises from the anisotropy of the electron-phonon distribution. $N_1(\vec{q},\vec{r})$ is conected with $F_1(\vec{p},\vec{r})$ because of phonon-phonon drag. From (4), (5) and (6) the relaxation time for electrons on acoustic phonons result

$$\tau' = \tau(1+A) + \tau(1+A)\vec{C}/e\vec{\mathcal{E}} \quad . \tag{10}$$

The phonon-phonon drag increases the relaxation time and therefore the carrier mobility. The position dependency of the band structure increases or decreases the relaxation time.

For a degenerate semiconductor, similarity, result the relaxation time.

References

[1] A. A. Belchik, V. A. Kozlov, F. T. P., 20, 53, (1986).
[2] M. Istrate, S. Istrate, phys.stat.sol.(b), 161, 823, (1990).
[3] A. H. Marshak, K. M. van Vliet, Solid State Electronics 21, 417, 429, (1978), 23, 49, (1980).
[4] N. Perrin, H. Budd, Phys.Rev. B9, 3454, (1974).

Acoustoelectric Oscillations in n-InSb

F. Guillon[1], *I. L'Heureux*[1], *S.L. Cyr*[2], *J.H. Page*[2], *R.D. Peters*[2], *and D.A. Reich*[2]

[1]Department of Physics, University of Ottawa, Ottawa, Ontario K1N 6N5, Canada

[2]Department of Physics, University of Manitoba, Winnipeg, Manitoba R3T 2N2, Canada

Recently [1], chaotic acoustoelectric voltage oscillations have been observed in n-InSb subjected to a magnetic field and an external DC current. Under some particular conditions, it was suggested that a direct transition from a steady state to a chaotic regime was often possible. This route is not commonly seen in nature. This fact and previous work on nonlinear acoustic attenuation [2] motivated us to investigate experimentally these voltage oscillations in a InSb sample with electrical characteristics similar to the one used in Ref.1. The linear theory of White [3] is a good model predicting an instability threshold to oscillatory solutions but it is invalid for the large currents considered in Ref.1. On the other hand, we have initiated a proper nonlinear treatment of the acoustoelectric instability [4]. Preliminary results for CdS and zero magnetic field based on a two modes coupling approximation showed that the system may exhibit bistability and hysteresis near the linear threshold.

Following the ideas of Ref.1, we have investigated the acoustoelectric voltage oscillations created by a DC current driven InSb in a transverse magnetic field at 77 K. The sample was immersed directly in liquid nitrogen to reduce sample heating at large electric fields. Power spectra were obtained for various values of the magnetic field B and of the DC current i by analyzing the time series of the voltage oscillations recorded by a 8-bit LeCroy digitizer sampling at a rate of 100 MHz. The present experiments were performed at DC currents well beyond the linear instability threshold where a transition to a chaotic regime is clearly seen.

Examples of typical features are found in the figures. In Fig. 1 the data show the occurrence of period doubling despite the fact that the precursor could not be observed in the current range used. In Fig. 2, a transition to chaos from an oscillatory signal can be seen in the power spectrum as the current increases. Some evidence of hysteresis effects was also observed for particular choices of B and i.

Though a large variety of interesting nonlinear phenomena (including chaos and hysteresis) was detected, no evidence of the direct transition to chaos seen in Ref.1 was observed. Further experimental and theoretical investigations are currently under way.

Springer Series in Solid-State Sciences, Volume 112
Phonon Scattering in Condensed Matter VII Eds.: M. Meissner and R.O. Pohl

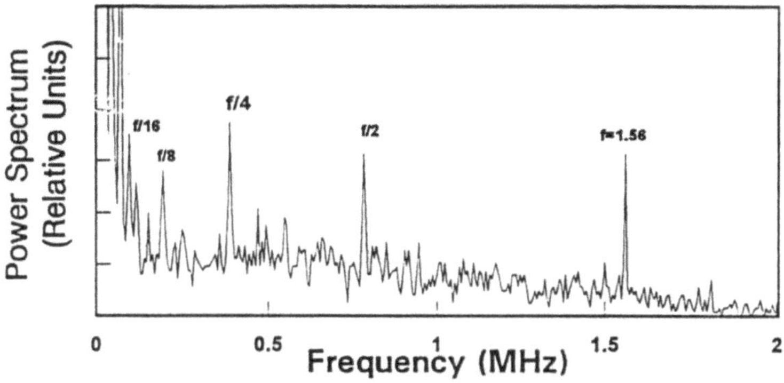

Fig. 1 Power spectrum (linear scale) of the acoustoelectric signal for B=3050 G and i=14.008 mA.

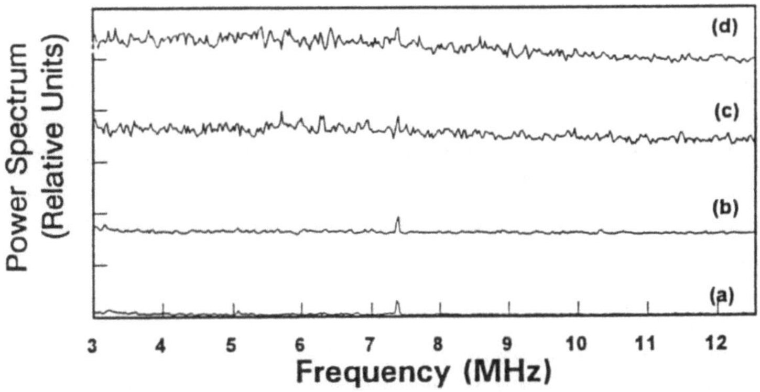

Fig. 2 Same as Fig. 1 for B=5050 G and (a) i=44.083 mA, (b) i=45.032 mA, (c) i=46.020 mA and (d) i= 47.073 mA. For clarity, each curve is shifted vertically by equal increments.

This work was supported by the N.S.E.R.C. of Canada.

References

1. C. W. Skorupka, L. M. Pecora, T. L. Carroll and T. M. Tritt, Phys. Rev. **B 42**, 9252 (1990).
2. F. Guillon, B. Fernandez and J. D. N. Cheeke, Phys. Rev. **B 43**, 4125 (1991).
3. D. L. White, J. Appl. Phys. **33**, 2547 (1962).
4. I. L'Heureux and F. Guillon, Phys. Rev. **A 45**, R5366 (1992).

Phonon Propagation in Doped Compensated Semiconductors: Phonon Magneto-Heat-Conductivity and Possibilities of Phonon Spectroscopy

V.I. Kozub and A.M. Rudin

A.F. Ioffe Physico-Technical Institute, 194021 St. Petersburg, Russia

Abstract. The phonon propagation in doped compensated semiconductors at low temperatures is considered; it is shown to be strongly influenced by their resonance coupling with electron two-level-systems (ETLS) (i.e. close pairs of empty and occupied impurity centers). The phonon absorption (scattering) has a pronounced frequency dependence; it's maximum corresponds to phonons wavelengths of the order of the Bohr radius. External magnetic field modifying the center's wavefunctions leads to changes of the absorption ("magneto-heat-conductance"). The interplay between phonon and microwave resonant absorption by ETLS is analyzed. The possibility of monitoring of the microwave frequency is shown to provide new possibilities of phonon spectroscopy.

As it is known now low-temperature properties of doped compensated semiconductors are influenced by the presence of electronic two-level systems, *i.e.* close pairs of occupied and empty donors with coupled electronic state. In particular as it has been shown recently experimentally [1] as well as theoretically [2] ETLS is a powerful source of sound attenuation. However in [1,2] only low-frequency case has been studied. On the other hand it seems to be of interest to consider ETLS interaction with higher frequency phonons with wave lengths $2\pi/q$ comparable or smaller than ETLS spatial scales r_ω (being the size of the pair resonant with phonons) and a_B (being the Bohr radius). Here we report the results of our theoretical studies of this question.

In a standard way one obtains for the frequency dependence of the resonant attenuation coefficient the following expression:

$$\Gamma(q) \propto \mathcal{A}^2 q r_\omega \mid 1 - \frac{sin(qr_\omega)}{qr_\omega} + \frac{sin(2qr_\omega)}{2qr_\omega} \mid, \quad \mathcal{A} = \frac{1}{\left(1 + \left(\frac{qa_B}{2}\right)^2\right)^2}. \qquad (1)$$

Thus we can see that spatial dispersion leads to rather peculiar frequency behavior. One should specially point at the presence of $\Gamma(\omega)$ peak at $q \sim 1/a_B$ followed by steep decrease ($\propto 1/q^7$).

Springer Series in Solid-State Sciences, Volume 112
Phonon Scattering in Condensed Matter VII Eds.: M. Meissner and R.O. Pohl

It is interesting that $\Gamma(\omega)$ can be significantly influenced by moderate magnetic field, changing the center's wave functions. At weak fields ($\lambda_H \equiv (\hbar c/eH)^{1/2} > a_B$) one obtains the corrections to Γ quadratic in H, while in stronger fields ($\lambda_H < a_B$) $\Gamma(\omega)$ can be changed drastically. E.g for the case $q\lambda_H \gg 1$ and $\mathbf{q} \perp \mathbf{H}$ one obtains:

$$\frac{\Gamma(H,q)}{\Gamma(0,q)} = \frac{[1+(qa_B/2)^2]^4}{exp[-(q\lambda_H)^2]} \left(\frac{a_H}{r_\omega}\right) \left(\frac{a_H}{a_B}\right). \tag{2}$$

Here $a_H \equiv \hbar/(2mE_H)^{1/2}$, $E_H \propto H^{1/3}$ is the binding energy of the electron localized on the impurity. For more detailed analysis see [3]. Thus in particular one can expect magnetic field dependence of phonon heat conductivity.

We have so far considered the ETLS to be in equilibrium with thermal bath. On the other hand the non-equilibrium phonon (NP) pulse can change the ETLS occupation numbers. In particular for the powerful enough pulse the effective temperature of ETLS with energy E matches to the partial temperature of phonons with $\hbar\omega = E$.

As a result one obtains nonlinear behavior of Γ. We want to point also at the possibility to use microwave (MW) irradiation as a probe for a phonon spectroscopy. Indeed ETLS induced MW damping depends on a partial ETLS temperature, modified by NP. On the other hand the resonant saturation of ETLS with a given energy E with MW signal $\omega_{mw} = E/\hbar$ creates MW transparency window for phonons with $\hbar\omega = E = \hbar\omega_{mw}$ thus giving the possibility to measure the contribution of phonons with a given frequency with a help of simple bolometer.

References

[1] Tokumoto H., Mansfield R., Japan. J. Appl. Phys., **22** Suppl.3, 196 (1983).

[2] Yu.M.Galperin, E.Y.Priev, Soviet Phys. Solid St., (1986)

[3] V.I.Kozub, A.M.Rudin, Zh. Eksp. Teor. Phys., **101**, 1073 (1992) (in Russian).

Effect of Nonpolar Optical Phonon Scattering on Free-Carrier Absorption in Quantizing Magnetic Fields

C.C. Wu[1] *and C.J. Lin*[2]

[1]Institute of Electronics, National Chiao Tung University, Hsinchu, Taiwan, Republic of China

[2]Department of Applied Mathematics, National Chiao Tung University, Hsinchu, Taiwan, Republic of China

A quantum theory for the free-carrier absorption of semiconductors to take into account the nonpolar optical phonon scattering in a semiconductor such as germanium has been studied for electrons in a parabolic energy band [1]. In quantizing magnetic fields, different scattering mechanisms give rise to different magnetic-field dependences of the scattering rates [2]. In the present work, we shall study the effect of nonpolar optical phonon scattering on the free-carrier absorption with a nonparabolic energy band of n-type nondegenerate semiconductors such as germanium.

The free-carrier absorption coefficient can be related to the transition probability for an electron to make an intraband transition with absorbing a photon and simultaneously absorbing or emitting an optical phonon. Thus the transition probability is

$$W_i = (2\pi/\hbar) \sum_f [|\langle f|M_+|i\rangle|^2\delta(E_f - E_i - \hbar\Omega - \hbar\omega_o) + |\langle f|M_-|i\rangle|^2 \times \delta(E_f - E_i - \hbar\Omega + \hbar\omega_o)] \ , \tag{1}$$

where $\langle f|M_\pm|i\rangle$ are the transition matrix elements for the interaction among electrons, optical phonons, and the radiation field, E_i and E_f are the initial and final energies of electrons, respectively, $\hbar\Omega$ is the photon energy, and $\hbar\omega_o$ is the optical phonon energy. The free-carrier absorption coefficient is given by

$$\alpha = (\varepsilon^{\frac{1}{2}}/nc) \sum_i W_i f_i \ , \tag{2}$$

where ε is the dielectric constant, n is the index of refraction, and f_i is the free-carrier distribution function.

Using the Landau gauge for the vector potential $\vec{A} = (0,Bx,0)$, the eigenfunctions and eigenvalues for electrons in a nonparabolic energy band with a magnetic field $\vec{B}$ directed along z-axis will be

$$\Psi_{\vec{k}n} = \exp(ik_y y + ik_z z)\Phi_n(x - \frac{\hbar c}{eB}k_y) \tag{3}$$

and

$$E_{\vec{k}n} = -\tfrac{1}{2}E_g\{1 - [1 + (4/E_g)\big((n + \tfrac{1}{2})\hbar\omega_c + \hbar^2k_z^2/2m^*\big)]^{\frac{1}{2}}\} \ , \tag{4}$$

respectively, where ω_c is the cyclotron frequency of electrons, Φ_n

Springer Series in Solid-State Sciences, Volume 112
Phonon Scattering in Condensed Matter VII Eds.: M. Meissner and R.O. Pohl

is the harmonic-oscillator wave function, k_y and k_z are the components of the electron wave vector in the y and z directions, respectively, and m* is the effective mass of electrons. The free-carrier distribution function for the nondegenerate semiconductors can be expressed as

$$f_{\vec{k}n} = \frac{N_o \hbar}{(2\pi m^* k_B T)^{\frac{1}{2}}} \left[\sum_{\ell=0}^{\infty} a_\ell^{\frac{1}{2}} \exp\left(-\frac{E_g a_\ell}{2k_B T}\right) \right]^{-1} \exp\left[-\frac{(E_g a_n + \hbar^2 k_z^2)/m^* a_n}{2k_B T}\right] \quad (5)$$

with

$$a_n = [1 + (4\hbar\omega_c/E_g)(n + \tfrac{1}{2})]^{\frac{1}{2}} \quad , \quad (6)$$

where N_o is the density of free carriers.

The relevant values of physical parameters for n-type Ge are taken to be [3] $N_o = 2.4 \times 10^{13}$ cm^{-3}, $\rho = 5.3267$ gm/cm^3, $\varepsilon = 16$, $m^* = 0.2224m_o$, $E_g = 0.72$ eV, $E_{op} = 19$ eV, $\omega_o = 5.6 \times 10^{13}$ rad/sec, $v_s = 3 \times 10^5$ cm/sec, T = 77 K, and a photon frequency corresponding to that of the 5.6 μm radiation characteristic of a CO_2 laser. In Fig. 1, it shows that both Re(α) and Im(α) oscillate quite considerably with the magnetic field. It can also be seen that the amplitude of oscillations appears quite large. These results are different from those of the parabolic energy band in germanium [1] and of the nonparabolic energy band in gallium arsenide [4].

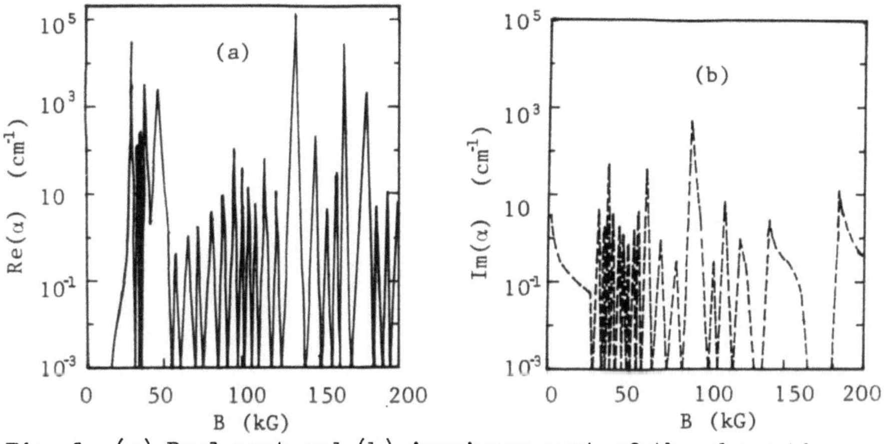

Fig. 1 (a) Real part and (b) imaginary part of the absorption coefficient in n-type Ge as a function of magnetic field B at T = 77 K.

References

[1] T.M. Rynne and H. N. Spector, J. Phys. Chem. Solids, 42, 121 (1981).

[2] L. M. Roth and P. N. Argyres, *Semiconductors and Semimetals* (Academic, New York, 1966), Vol. 1, p 159.

[3] S. M. Sze, *Physics of Semiconductor Devices*, 2nd ed. (Wiley, New York, 1981).

[4] C. C. Wu, J. Tsai, and C. J. Lin, Phys. Rev. B 43, 7328 (1991).

Part V

High-Temperature Superconductors

Some Aspects of Thermal Conductivity in High-T_c Superconductors

C. Uher

Department of Physics, University of Michigan,
Ann Arbor, MI 48109-1120, USA

1. Introduction

Although the key issue of high-temperature superconductivity, namely its microscopic origin, has not yet been satisfactorily resolved, an extensive experimental effort over the past $5\frac{1}{2}$ years has resulted in a wealth of useful data pertaining to physical parameters of the high-T_c perovskites. Studies of the thermal conductivity have proved to be among the most surprising and, unlike galvanomagnetic investigations, they offer important insight into the nature of transport processes over the entire temperature range.

Because of their reduced carrier density, high-T_c perovskites exhibit a thermal conductivity which is entirely different in character than that observed in conventional superconductors. The phonon contribution dominates even in the normal state and the thermal conductivity due to the charge carriers is at most only about 30-40% of the total thermal conductivity. Under such conditions, even a modest carrier-phonon interaction leads to spectacular behavior in the temperature dependence of the heat transport when a sample is cooled through its superconducting transition temperature. An enhancement in the phonon mean-free path as the carriers condense into a Cooper sea leads to a rapid rise in the thermal conductivity and an eventual peak near $T_c/2$. Although these features are observable in all structural forms of Cu-O superconductors, they are particularly well developed in high quality single crystals where interference from defect and grain boundary scattering is small. Under these favorable circumstances the thermal conductivity at its peak is more than twice the value of the conductivity at T_c. Typical data for a sintered sample and for twinned and untwinned single crystals of $YBa_2Cu_3O_{7-\delta}$ are shown in Fig.1.

Detailed thermal transport investigations have been conducted on all major families of high-T_c superconductors over a broad temperature range and for different morphological forms of the structure. In addition, much effort has been channeled towards establishing the influence of oxygen content, atomic substitution and radiation-generated defects on the magnitude and temperature dependence of the thermal conductivity. Comprehensive accounts of the heat transport in high-T_c superconductors are given in several review articles [1-4].

In this paper, I focus on two specific issues which, since their successful study relies on the availability of high quality single crystals (both twinned and untwinned), have not yet been adequately treated in the literature. The

Springer Series in Solid-State Sciences, Volume 112
Phonon Scattering in Condensed Matter VII Eds.: M. Meissner and R.O. Pohl

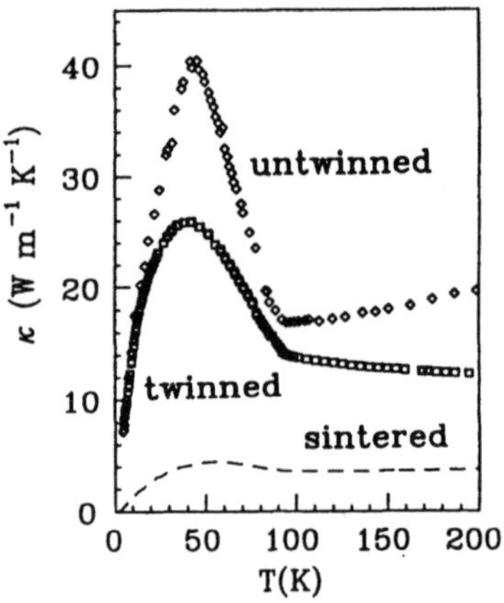

Fig.1. Thermal conductivity of sintered [5], twinned [6] and untwinned (a-direction) [7] samples of $YBa_2Cu_3O_{7-\delta}$.

topics are: i) intrinsic behavior at very low temperatures, and ii) in-plane anisotropy in $YBa_2Cu_3O_{7-\delta}$.

2. Low Temperature Behavior

The early results, generated mostly on sintered samples of $YBa_2Cu_3O_{7-\delta}$ and $La_{2-x}Sr_xCuO_4$, left the question of the dominant low temperature scattering mechanism not just unresolved but in a state of confusion. The approximately quadratic variation of the thermal conductivity observed in a narrow range of liquid helium temperatures, coupled with sound velocity and internal friction measurements that mimic some but not all features of amorphous materials, led several authors [8,9] to propose that thermal transport in high-T_c perovskites has an essentially glassy character. Phonons here are assumed to scatter off of two-level tunneling states (TS) which, although not specified precisely, are hinted to be associated with oxygen vacancies. Although the hypothesis of TS is intriguing, it is based, as far as the thermal conductivity is concerned, on a very narrow temperature range where the temperature dependence is approximately quadratic. In order to establish the limiting temperature dependence of the thermal conductivity, measurements must be extended to subkelvin temperatures.

A compendium of such measurements on sintered $YBa_2Cu_3O_{7-\delta}$ can be found in Refs.1 and 3. Since in this regime $T \ll T_c$, elementary excitations should be condensed out and the heat transport should proceed entirely via phonons. In practice this means that one would expect either a T^3 variation (reflecting phonon scattering on grain boundaries or crystal faces) or, if the scale of scatterers is somewhat shorter than the grain size, some fixed power law dependence (such as T^2 if TS were operative). In reality, the data are at variance with the idea of a single power law dependence and all measurements can be fitted very well by writing [10]

$$\kappa(T) = AT + BT^3. \tag{1}$$

Different explanations have been suggested to account for the crossover from

T^3 to T-linear behavior at very low temperatures. A scenario, first proposed by Gottwick et al., which assumes that a pocket of uncondensed charge carriers is responsible for the T-linear dependence is particularly intriguing. From the magnitude of the coefficient A, Gottwick et al. estimated that about 15% of the carriers do not belong to the condensate. Investigating changes in the low-temperature thermal conductivity of $YBa_2Cu_3O_{7-\delta}$ by cycling a sample between its superconducting and insulating states (using successive vacuum and oxygen annealing treatments), Cohn et al. [11] proved convincingly that the T-linear asymptotic behavior is connected with the superconducting state of $YBa_2Cu_3O_{7-\delta}$ while in its insulating state the thermal conductivity attains its expected T^3 variation. It is important to note that no changes in the grain size or the porosity were detected during these sample cycling procedures. Invoking the Wiedemann-Franz law, the authors estimated that about 5% of the carriers remain uncondensed and that this fraction is large enough to account for the γ-term in the specific heat. Subsequent measurements on $YBa_2Cu_3O_{7-\delta}$ single crystals [12,13] have confirmed the existence of T-linear behavior at very low temperatures. A plot of the data in the form $\kappa(T)/T$ vs. T^2, see Fig.2, leaves no doubt that the T-linear term is an intrinsic feature of the low temperature thermal conductivity of this material. Using an Internal Multilayer Model, Bredl et al. conclude that, while carriers in the Cu-O planes undergo condensation, the T-linear term is connected with unpaired carriers which reside in the chain layers of the structure.

The possible existence of uncondensed carriers well below T_c is certainly a very interesting phenomenon and it could account for several other surprising observations on $YBa_2Cu_3O_{7-\delta}$, e.g., a puzzling residual electronic scattering below 2Δ observed in Raman measurements [14]. Yet, one has to recognize that a T-linear dependence in the thermal conductivity is by no means common to all high-T_c perovskites. In fact, measurements on single crystals of $La_{2-x}Sr_xCuO_4$ [15] do not show any sign of T-linear "pulling", nor has such an effect been detected on sintered samples of (2223) Tl-based compounds [16]. Furthermore, the second most thoroughly studied structure, that of

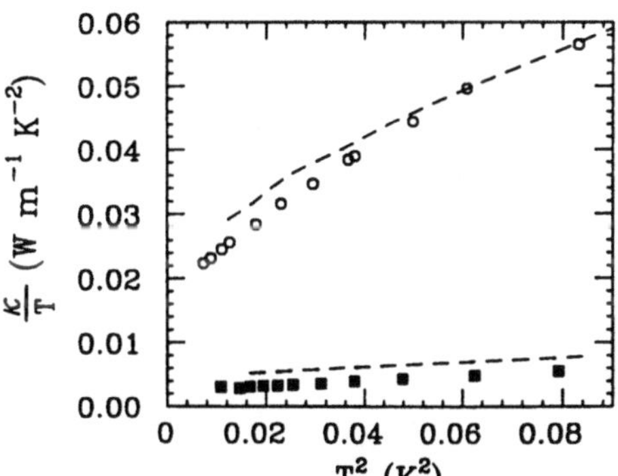

Fig.2. Plot of κ/T vs. T^2 for $YBa_2Cu_3O_{7-\delta}$.
Open circles: single crystal, heat flow parallel to Cu-O planes.
Solid squares: sintered specimen.
Dashed lines indicate the effect of an 8 Tesla field applied perpendicular to the direction of heat flow. The increase in the thermal conductivity is ascribed to the pair breaking effect of the external magnetic field. Adapted from Ref.13.

$Bi_2Sr_2CaCu_2O_8$, displays a very robust quadratic variation which proceeds unabated down to the lowest temperatures (≈30mK) regardless of whether one measures sintered or single crystal specimens [12,17,18]. Although the tunneling state picture might be relevant here, an argument based on the reduced dimensionality of the phonons in this quasi-2dimensional structure is equally persuasive [19]. The existence of unpaired carriers in $YBa_2Cu_3O_{7-\delta}$ must then be viewed strictly as a consequence of its unique atomic structure where chains are supplying carriers which do not participate in superconductivity.

3. In-plane Anisotropy in $YBa_2Cu_3O_{7-\delta}$

It is well known that single crystals of $YBa_2Cu_3O_{7-\delta}$ contain twin boundaries characterized by thin tetragonal, nonsuperconducting regions which impede the transport of charge carriers and phonons. With the development of thermomechanical detwinning techniques [20] one can not only avoid difficulties with twins as scattering centers but can also explore in-plane anisotropy. The anisotropy in this case arises due to the presence of Cu-O chains in the b-direction of crystals. The two existing sets of data [21,22] which explore the in-plane anisotropy in the thermal conductivity display a significant difference between the a and b crystallographic directions, see Fig.3, and thus support the more numerous studies which use the electrical resistivity as an anisotropy probe. However, the problem with the thermal studies is that they show a contradictory trend: the measurements of Cohn et al. indicate $\kappa_a > \kappa_b$ below T_c, while in the most recent data of Yu et al., the anisotropy is reversed, $\kappa_b > \kappa_a$. The former group assumes that phonons dominate and explains their result in terms of enhanced phonon scattering on defects in the Cu-O chains. The latter authors, proposing a fully electronic origin for

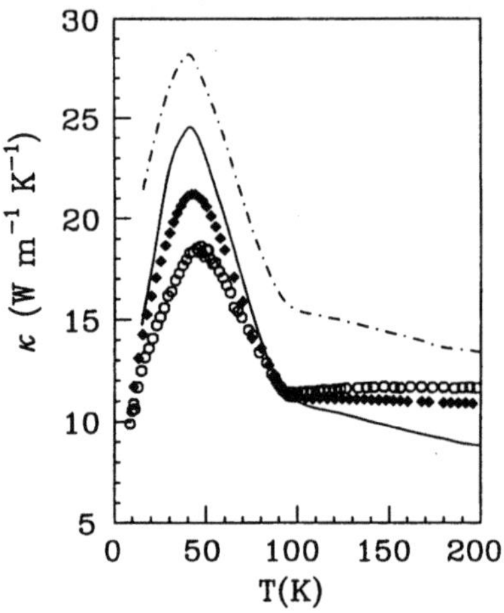

Fig.3. In-plane anisotropy in untwinned $YBa_2Cu_3O_{7-\delta}$ crystals.
Solid diamonds: κ_a from Cohn et al.
Open circles: κ_b from Cohn et al.
Solid line: κ_a from Yu et al.
Dotdashed line: κ_b from Yu et al.
Note contrasting anisotropy below T_c for the two sets of measurements.

the thermal conductivity, attribute their higher thermal conductivity along the chain direction to an additional electronic term associated with the Cu-O chains.

Regardless of whether a purely electronic origin for the thermal conductivity in high-T_c perovskites is plausible or not, the two mutually contradictory sets of measurements call for clarification and invite further study. It should be pointed out that this kind of experiment is very demanding as one must deal with very small samples (typically 1x1x0.2 mm^3). Furthermore, the requirement of measuring the thermal conductivity in two mutually orthogonal directions on the same sample implies a near square sample profile which is not the most favorable aspect ratio for the thermal transport.

The degree of in-plane anisotropy is an important parameter and, in spite of possible experimental difficulties, I expect that a tenacious effort will help to settle this issue.

References

1. C. Uher, in *Physical Properties of High Temperature Superconductors*, ed. D. M. Ginsberg (World Scientific), Vol.3, Ch.3, p.159 (1992).
2. A. Jezowski and J. Klamut, in *Studies of High-Temperature Superconductors*, ed. A. Narlikar (Nova Science Publishers), Vol.4, p.263 (1990).
3. C. Uher, J. Superconductivity **3**, 337 (1990).
4. M. D. Nunez Regueiro and D. Castello, Int. J. Mod. Phys. **5**, 2003 (1991).
5. C. Uher and A. B. Kaiser, Phys. Rev. **B36**, 5680 (1987).
6. S. D. Peacor, J. L. Cohn, and C. Uher, Phys. Rev. **B43**, 8721 (1991).
7. S. D. Peacor, D. Glick, U. Welp, and C. Uher, unpublished.
8. B. Golding, N. O. Birge, W. N. Haemmerle, R. Cava, and E. A. Rietman, Phys. Rev. **B36**, 5606 (1987).
9. M. D. Nunez Regueiro, D. Castello, M. A. Izbizky, D. Esparza, and C. D'Ovidio, Phys. Rev. **B36**, 8813 (1987).
10. U. Gottwick, R. Held, G. Sparn, F. Steglich, H. Rietschel, D. Ewert, B. Renker, W. Bauhoffer, S. von Molnar, M. Wilhelm, and H. E. Hoenig, Europhys. Lett. **4**, 1183 (1987).
11. J. L. Cohn, S. D. Peacor, and C. Uher, Phys. Rev. **B38**, 2892 (1988).
12. G. Sparn, M. Baenitz, S. Horn, F. Steglich, W. Assmus, T. Wolf, A. Kapitulnik, and Z. X. Zhao, Physica **C162-164**, 508 (1989).
13. C. D. Bredl, K. Petersen, S. Arnold, G. Sparn, F. Steglich, T. Takahashi, M. Tachiki, B. Runtsch, and T. Wolf, Z. Physik B - to be published.
14. S. L. Cooper and M. V. Klein, Comments Cond. Mat. Phys. **15**, 99 (1990).
15. D. T. Morelli, G. L. Doll, J. Heremens, M. S. Dresselhaus, A. Cassanho, H. P. Jenssen, S. D. Peacor, and C. Uher, Solid State Commun. **77**, 773 (1991).
16. C. Uher, S. D. Peacor, and J. Shewchun, Physica **C177**, 23 (1991).
17. Da-Ming Zhu, A. C. Anderson, E. D. Bukowski, and D. M. Ginsberg, Phys. Rev. **B40**, 841 (1989).
18. S. D. Peacor and C. Uher, Phys. Rev. **B39**, 11559 (1989).
19. B. A. Merisov, G. Ya. Khadzhai, and A. V. Sologubenko, Sov. J. Low Temp. Phys. **16**, 730 (1991).
20. J. Giapintzakis, D. M. Ginsberg, and P.-D. Han, J. Low Temp. Phys. **77**, 155 (1989).
21. J. L. Cohn, E. F. Skelton, and J. Z. Liu, Phys. Rev. **B45**, 13144 (1992).
22. R. C. Yu, M. B. Salamon, Jian Ping Lu, and W. C. Lee, preprint.

Anisotropy of Phonon Interaction in Y- and Bi-HTSC Monocrystals

V.B. Efimov

Institute of Solid State Physics, Russian Academy of Sciences, Moscow District, 142432 Chernogolovka, Russia

Abstract. For explanation of so big difference of absolute value and temperature dependence of thermal conductivity æ(T) Y- and Bi- monocrystals HTSC in ab-plane and in c-direction we may suggest existence of any dominant scattering mechanism in c-direction, which absence in ab-plane and semitransparent for phonons. On this background phonon-electron interaction, which define hump in thermal conductivity at $T \approx T_c/4$ in ab-plane isn't observed.

In this work we tried to interpreted results of æ(T) of HTSC monocrystals in frame of simple gas kinetic theory, to find phonon free length path λ(T) and to explain these peculiarities æ in *ab*-plane and in perpendicular direction. As was shown in experimental works [1-3] for æ(T) of monocrystals HTSCs the most important properties for our examination are:

a). For $YBa_2Cu_3O_{7-x}$ and $Bi_2Sr_2CaCu_2O_{8-y}$ monocrystals [1] the thermal conductivity has high anisotropy. Absolute value of $æ_{ab}(T)$ is in order higher $æ_c(T)$.

b). The thermal conductivity in *ab*-plane for all monocrystals has a rising below T_c which is connected with disappearance of phonon-electron interaction in superconducting state. At low T æ(T) reduces with cooling of crystal.

Springer Series in Solid-State Sciences, Volume 112
Phonon Scattering in Condensed Matter VII Eds.: M. Meissner and R.O. Pohl

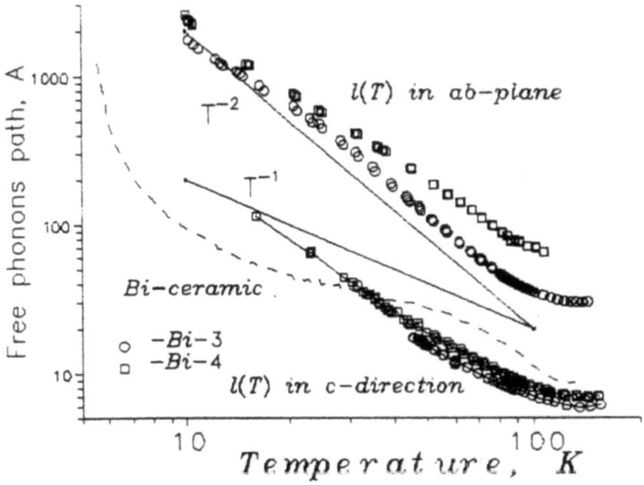

c). In *c*-direction the thermal conductivity monotonic decreases from highest (~ 150 K) to lowest (~10 K) in vestigated T without any peculiarities near T_c.

Simple gas kinetic relationship for æ(T) of phonon system may be written as $æ_{ph} \approx 1/3 C_p v \lambda_{ph}$ It allow us estimate free length path of phonons λ_{ph} which are plotted on Fig. For comparison ibidem plotted date for ceramic. Temperature dependences both $\lambda_{ab}(T)$ and $\lambda_c(T)$ close to $\sim T^2$ at $T < T_c$ and differ from $\lambda(T)$ for ceramics. Absolute values $\lambda_{ab}(T)$ in ten times bigger $\lambda_c(T)$ which may be explain as existence of any phonon scattering mechanism in *c*-direction, which reduce λ (semi-transparence for move phonons perpendicular planes). This suggestion explain difference in temperature dependence of thermal conductivity in and out of plane. On background large thermal resistivity W_c (W=1/æ) (connected with this unknown scattering mechanism) phonon interaction with freezing electrons (as in ***ab***-plane) become invisible.

[1]. V.B.Efimov ***et al*** - Physica C, in press

[2]. M.F.Crommie, A.Zettl - Phys.Rev.B, 1991, 43, 408

[3]. S.J.Hagen ***et al*** - Phys.Rev.B, 1989, 40, 9389

Electron–Phonon Interaction in Thin YBaCuO Films and Fast Detectors

G.N. Gol'tsman, A.D. Semenov, A.V. Sergeev, E.E. Aksaev, I.G. Gogidze. and E.M. Gershenzon

Moscow State Pedagogical University, M. Pirogovskaja str. 1, 119435 Moscow, Russia

Abstract. The thin YBaCuO film response to laser and submillimeter radiation demonstrates the picosecond nonequilibrium peak on the nanosecond bolometric background. Experimental data give an evidence for the spectral dependence of picosecond photoresponse probably due to a poor efficiency of electron multiplication processes. Presented results prove an availability of fast YBaCuO thin film detector.

The photoresponse of HTSC thin films is intensively studied during the past few years. Fast nonequilibrium photoresponse within the time scale of 1-100 ps was found for the first time by the authors [1] at millimeter wavelengths and, more recently, to optical radiation [2,3]. In submillimeter range a time constant attributed to the electron-phonon relaxation time follows the T^{-1} law, giving at liquid nitrogen temperature value of several picoseconds. In this paper we present an evidence for the spectral variation of the fast response.

We carried out the measurements of the fast response dependence on the wavelength with help of two 2-mm backward-wave oscillators (BWO) and Nd:YAG with 20 ps pulses at 1.54 μm and 0.63 μm. We used YBaCuO films of 450-1400 Å thickness deposited by laser ablation onto MgO, ZrO_2 and sapphire subsrates and structured into narrow strips of 5-40 μm width.

The BWO apparatus allows us to measure a film resistance change under influence of a weak radiation power versus a beating frequency. The dependence usually has two knees. First of them appears beside the frequency which is equal to the inverse time of phonon escape, the second one offers the electron-phonon interaction time. Two plateaus prefixing these knees correspond to the bolometric signal and the nonequilibrium signal respectively. Our measurements have showed that a ratio of bolometric to nonequilibrium contribution depends on a film thickness and for d=450 Å on sapphire substrate is about 2-3 at the temperature 76 K. A typical signal trace in laser experiment shows two components .The fast component has a symmetrical shape defined by the instrumental time resolution of the order of 100 ps. This precursor arising from the nonequilibrium electron relaxation process is followed by a

Springer Series in Solid-State Sciences, Volume 112
Phonon Scattering in Condensed Matter VII Eds.: M. Meissner and R.O. Pohl

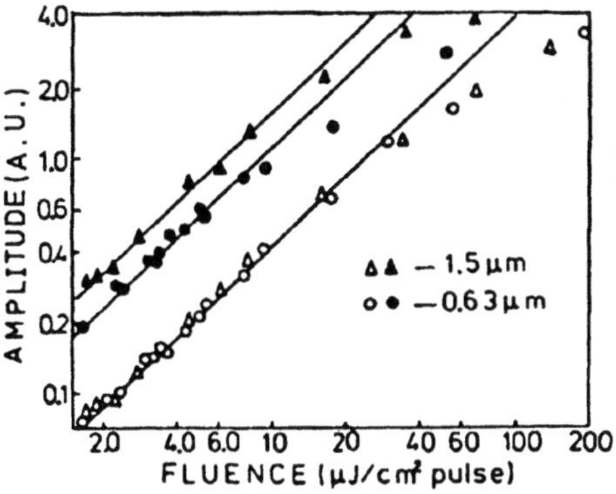

Fig.1. Amplitudes of fast (closed symbols) and slow (opened symbols) components of the response as functions of fluence for two different wavelengths, T=76 K. Solid lines represent a linear fit.

slow exponential bolometric decay with the $\simeq$3 ns time constant. With an increase of a radiation power the saturation of the signal amplitude due to transition of the film into a normal state takes place (Fig.1). The saturation for fast component occurs at lower power level than for the slow one. Another peculiarity is that the curve for $\lambda = 1.54$ μm runs above the curve for 0.63 μm, whereas bolometric curves coincide.

Let us compare the results for submillimeter and visible wavelengths. In former case we are able to get the ratio of amplitudes of fast and slow components directly using stationary parts of frequency dependence. In laser experiment a pulse duration τ_p is longer than the relaxation time of an electron subsystem but much shorter than that of a phonon subsystem τ_b. This fact leads to an additional reduction of the bolometric signal by a factor τ_b/τ_p ~100. Hence, we can conclude that amplitude of nonequilibrium response rises of an order of magnitude towards the long-wave edge of our range.

In our opinion the mentioned spectral dependence reflects a moderately low intensity of the processes of electron multiplication leading to a storage of low energy electrons suppressing a superconductive gap. The further spectral investigation within a wide spectral range are essential for studying the features of electron-phonon interaction in HTSC materials and for improvement of the film-based fast operating detectors [4].

References

[1] E.M.Gershenzon, G.N.Gol'tsman, A.D.Semenov, and A.V.Sergeev, Sov. Phys. JETP Lett. **46**, 285 (1987); and Sol.State Commun. **76**, 493 (1990).
[2] E.M.Gershenzon, G.N.Gol'tsman, I.G.Gogidze, A.D.Semenov, and A.V.Sergeev, Physica C **185-189**, 1371 (1991); and A.D.Semenov, G.N.Gol'tsman, I.G.Gogidze, A.V.Sergeev, E.M.Gershenzon, P.T.Lang, and K.F.Renk, Appl.Phys.Lett. **60**, 903 (1992).
[3] M.Johnson. Appl.Phys.Lett. **59**, 1371 (1991).
[4] G.N.Gol'tsman, B.S.Karasik, A.D.Semenov, E.M.Gershenzon, R.S.Nebosis, R.Steinke, P.T.Lang, W.Schatz, M.A.Heusinger, and K.F.Renk, this issue.

Ultrasonic Study of Three $La_{2-y}Sr_yCuO_{4+x}$ Crystals

J.-Y. Prieur, H. Ji, R. Höhler, and J. Joffrin
Laboratoire de Physique des Solides, Bât. 510,
Université Paris-Sud, F-91405 Orsay, France

Abstract : We present the results of an ultrasonic study of three monocrystals of $La_{2-y}Sr_yCuO_{4+x}$.. Longitudinal and transverse waves have been used along a or c direction at 30 and 90 MHz. For each sample four contributions reproduce the experimental data. The main one is the phonon-phonon interaction, two relaxations are identified and the tetragonal to orthorhombic transition gives an additional critical variation satisfying well defined selection rules.

1.Introduction

La_2CuO_4 has the K_2NiF_4 structure at high temperature and suffers a structural phase transition at T_c=500 K. However, substitution of La by Sr sharply lowers Tc. Stoichiometric La_2CuO_4 is an insulator ; but doping with Strontium or increasing the oxygen content, the compound becomes superconducting. In the present analysis we have used three different monocrystals. Two of them were grown at Riken in Japan. Their composition is $La_{1.92}Sr_{.08}CuO_{4+x}$ with x unknown and La_2CuO_4 with x=.005. These two samples become superconducting respectively at 10 and 40 K. The last monocrystal with composition La_2CuO_4 has been grown at the "Université de Bordeaux". It is oxygen stoichiometric thus avoiding supraconductivity but becoming antiferromagnetic at 310 K.

2. Experimental set-up

The velocity variations were recorded with two automatic systems. In one of them they were determined by monitoring the frequency of the ultrasonic wave in order to keep constant the phase of the output signal (i.e. zero). In the second one the output high frequency signal was digitalised by an oscilloscope. The digitalised signal was then compared to a sine wave with the phase as the fitting parameter.

Both transverse and longitudinal polarisation have been used. They propagated along either a or c axis. The sample dimensions were about 1x2x2 mm^3. A relative accuracy of 5.10^{-6} was achieved.

3.Results

Typical results are displayed on fig.1 where the experimental data points are superimposed on a calculated curve.A quantitative analysis proves that the main contribution to the variations is due to the well-known phonon-phonon interaction.

Springer Series in Solid-State Sciences, Volume 112
Phonon Scattering in Condensed Matter VII Eds.: M. Meissner and R.O. Pohl

Therefore we tried , at first, to calculate a theoretical curve with the following law.where V, γ, T, ρ and C are respectively, the sound velocity, the Grüneisen constant, the temperature, the specific mass and the Debye specific heat.

$$\frac{\Delta V}{V}_{Debye} = -\frac{\gamma^2}{\rho V^2} TC(T) \cdot \quad (1)$$

Two parameters only scale that contribution :γ, Θ_D. Secondly, for all samples, polarisations and directions of propagation, we added a relaxation-type contribution in the temperature range 50-200 K. Thirdly, around 250 K in the x=.005 sample a small relaxation appears in a small temperature interval ; an analog variation appears at 220 K in the strontium sample and at 25K in the stoichiometric one. Finally at temperatures above 300 K a critical variation due to the cristallographic transition is clearly visible. The calculated curve is the sum of four contributions :

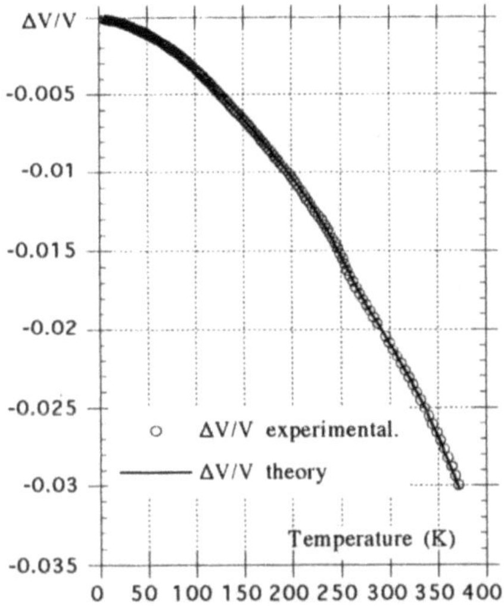

fig. 1 : La_2CuO_{4+x} x=.005 air annealed
Longitudinal Waves (k//plane)
Frequency 30MHz

$$\frac{\Delta V}{V} = \frac{\Delta V}{V}_{Deb} + \frac{\Delta V}{V}_{rel1} + \frac{\Delta V}{V}_{rel2} + \frac{\Delta V}{V}_{crit} \quad (2)$$

$$\frac{\Delta V}{V}_{rel\ 1,2} = -\frac{\Delta V_{1,2}}{T(1+\omega^2\tau^2\exp(\frac{T_0}{T}))} \quad (3) \ ; \qquad \frac{\Delta V}{V}_{crit} = -\frac{\Delta V}{1+\omega^2\tau^2(T)} \frac{T}{T_c - T} \cdot \quad (4)$$

(1) (2) (3) (4) are convenient for all polarisations propagating in the CuO plane ; for longitudinal waves propagating along the c-axis, no critical contribution appears as a consequence of the selection rules.

4.Discussion

The relaxation in the 50-200 K range recalls a similar dip around 25 K in the the sound velocity variations of KTaO3 as measured by H. H. Barrett [1]; there the acoustical waves interacted with an optical mode. In the Lanthanum compounds the large series of modes between 100 and 600 cm-1 might provide an alternative explanation of the dip. A similar one has been used for a PdDx crystal [2]

The 250K relaxation in the x=.005 sample appears at the Néel temperature. However it has been shown [3,4] that a phase separation appears at different temperatures for different Oxygen excess. The Neel temperature and the phase separation temperature coincide in the x=.005 sample. It would be interesting to follow the velocity step with the Oxygen content .and if it follows the Néel temperature or the phase separation . The fact that for the stoichiometric compound the step is at 25 K indicates that the second hypothesis is probably the correct one.

[1] H. H. Barrett, Phys. Rev. **178**, 743 (1969)
[2] Y. Cho and R. G. Leisure, Phys. Rev. B **38**, 5748 (1988)
[3] M. F. Hundley,R. S.Kwok, S. W.Cheong,J. D.Thompson and Z. Fisk, Physica C,**172**, 455
[4] R.K.Kreimer, E.Sigmund, V.Hizhnyakov,F.Hentsch,A. Simon,K.A.Müller and M. Mehring Z. Phys; B- cond. matter **86**, 319 (1992)

Acoustic Surface-Wave Study of Magnetoelastic Effects in a Thin Film of YBCO

R. Höhler, J. Joffrin, and J.Y. Prieur

Laboratoire de Physique des Solides, Bât. 510,
Université Paris-Sud, F-91405 Orsay, France

It has recently been shown that the low temperature acoustic properties of HTC superconductors in the presence of a strong magnetic field can give information on the thermally activated depinning of the vortex lattice [1] .
In this publication we present new experimental and theoretical results that show that acoustic surface wave measurements of thin superconducting films in the presence of a strong magnetic field of variable orientation can provide additional information. In particular the anisotropy of the effective electronic mass can be measured. This aspect has not been exploited by previous surface wave experiments [2,3,4].
We have investigated YBCO layers of a thickness of 300nm on $LiNbO_3$ substrates. The midpoint of the resistive transition of the films was originally at 87K. After the photolithographic procedures we detected a small residual resistance that extends down to 83.5K. The sample holder contains a mechanism that permits a rotation by more than 180° with a relative resolution of 0.5° at low temperatures and in the presence of a magnetic field of 7T. The velocity of the surface waves was measured by a standard phase locked loop technique.The fundamental frequency of the interdigital transducers that we use to excite and detect acoustic surface waves is 100 MHz. Figure 1 shows experimental results at a fixed temperature fo 0.91Tc . Above 300 MHz the angular dependence remains essentially the same down to 65K, but the scale of the effect rapidly increases. At lower temperatures a different angular characteristic sets in which is proportional to the square of the component of the applied field that is parallel to the layer surface. Above Tc, no angular variation was detected.
In analogy to the above mentionned bulk wave data one might think that the anomaly at high temperatures corresponds to the activated depinning of vortices with an anisotropic distribution of barriers. However, this explanation can be ruled out because the angular variation at different temperatures then wouldn't differ just by a scale factor. We have performed a theoretical analysis of the propagation of Rayleigh waves in the presence of a strong magnetic field. In the framework of the anisotropic London model we have shown that there is a coupling of the sound wave with the magnetization component perpendicular to the surface via the strayfield. This magnetization can be predicted in the framework of the anisotropic London model [5] if the

Springer Series in Solid-State Sciences, Volume 112
Phonon Scattering in Condensed Matter VII Eds.: M. Meissner and R.O. Pohl

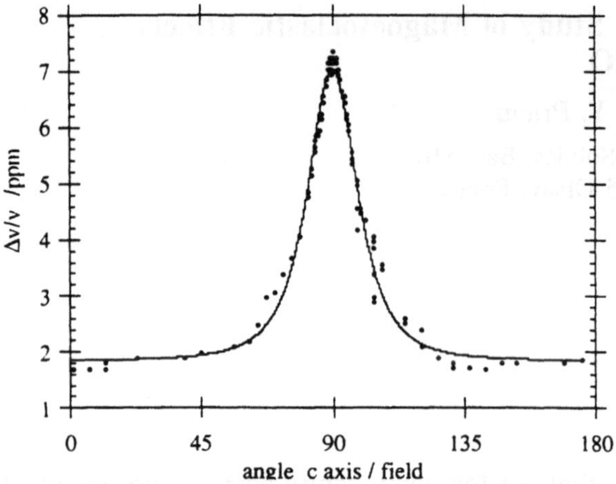

Fig1: Rayleigh velocity anomaly at T=76K, B=7T, f=300MHz. The fit corresponds to a ratio of effective electronic masses perpendicular and parallel to the CuO planes of 20.

ratio of the effective electronic masses parallel and perpendicular to the c axis is known. We have calculated the resulting forces on the superconducting layer by using the Maxwell stress tensor.

To obtain the sound velocity variation we have used the well known numeric techniques of acoustic surface wave physics. Of course the London model is not strictly valid for YBCO at temperatures of 0.9Tc: the superconductor should show twodimensional behaviour. However, it has very recently been shown that the resulting corrections are small [6], and indeed, we observe deviations from an 'ideal fit' that at least qualitatively correspond to these predictions. We would like to thank R. Wördenweber and U. Schneider from the Forschungszentrum Jülich who provided the YBCO samples.

[1] J.Pankert, G. Marbach, A. Comberg, P. Lemmens, P. Frönig, S. Ewert PRL 65 (1990) 3052

[2] H.P. Baum Thesis, University of Wisconsin (1990)

[3] S. Lee, C. Chi ,G. Koren, A. Gupta Phys. Rev. B 43 (1991) 5459

[4] M. Saint-Paul, F. Pourtier, B. Pannetier, J.C. Villegier, R. Nava Physica C 183 (1992) 257

[5] V.G. Kogan Phys. Rev. B 38 (1988) 7049

[6] D. Feinberg Physica C 194 (1992) 126

Phonon Anomaly of $YBa_2Cu_3O_7$ Studied by Inelastic Neutron Scattering

M. Arai[1], *K. Yamada*[2], *Y. Hidaka*[3], *S. Itoh*[4], *Z.A. Bowden*[4], *A.D. Taylor*[4], *and Y. Endoh*[2]

[1]National Laboratory for High Energy Physics, Oho, Tsukuba, Ibaraki 305, Japan
[2]Faculty of Science, Tohoku University, Aramaki, Aoba, Sendai 980, Japan
[3]NTT Applied Electronics Laboratories, Tokai, Ibaraki, Japan
[4]Rutherford Appleton Laboratory, Chilton, Didcot, Oxfordshire OX11 0QX, UK

Abstract. Measurements of S(Q,E) over a wide range of Q-E space were performed on superconducting $YBa_2Cu_3O_7$ using a high energy-resolution neutron spectrometer. We found that S(Q,E) is very sensitive to a local structural evolution and that the temperature dependence of S(Q,E) showed an anomalous behaviour at T=Tc similar to a critical phenomena. Fourier transformation of S(Q,E) of a particular energy also gives direct information of the correlation between specific atoms. From these data we conclude that the apical oxygen is very unstable at Tc and the fluctuation may trigger the distortion associated with a <110> buckling structure.

We have reported in [1] that the dynamical structure factor S(Q,E) is quite sensitive to a local structural evolution and we found an anomalous behaviour at Tc similar to critical phenomena. The Q-dependence of S(Q, E=65meV), for instance, changes drastically at Tc, which is the mode related to apical oxygen (O(4)) mode parallel to c-axis. In the preceding report we suggested that O(4) is very unstable and fluctuating toward <$\bar{1}$10> direction and may induce the <110> buckling local structure of O(4)-Cu(2)-O(2)-O(3) pyramid. In this report we discuss a further analysis of the same data to get more direct information on the atomic pair correlation in the real space.

The dynamical structure factor [2] can be written in the form of

$$S(Q,E) \approx A\left[1+\frac{1}{N}\sum_{i\neq i'}<\eta_i\cdot\eta_{i'}>e^{iQ[i'-i]}\right]_{<Q>} \tag{1}$$

where A is $\frac{h^2Q^2}{2M}e^{-2W}\frac{<n(E)+1>}{E}$. The second term, i.e. distinct term, is correlation between displacement-vectors of the atoms. This situation is quite analogous to spin-spin correlation of a magnetic system. By taking the Fourier transform of the distinct term, we can get so-called Dynamical Pair Correlation Function (DPCF) at an excitation energy E:

$$D(r,E) = \frac{2}{\pi}\int_0^\infty Q\,\{S(Q,E)/A-1\}\,\sin(rQ)\,dQ\,. \tag{2}$$

Liu et al. [3] demonstrated possible vibrational modes of 1:2:3:O7 compound. In Fig. 2 of [3] we can recognize that the modes relevant to O(2), O(3) and O(4) are dominant in high energy region, say above 40meV. Therefore the Q-dependence of S(Q,E) above 40meV mainly reflects the displacement-vector pair

Springer Series in Solid-State Sciences, Volume 112
Phonon Scattering in Condensed Matter VII Eds.: M. Meissner and R.O. Pohl

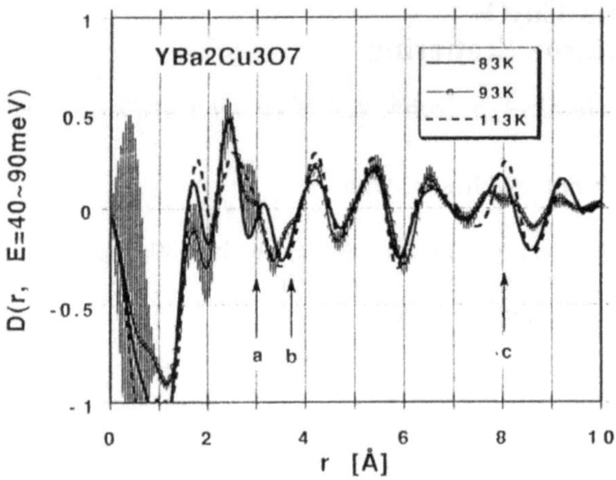

Fig.1 DPCF for 83K(solid line), 93K(circle with error bar), 113K(broken line). Fourier transform of S(2.5Å-1<Q<12Å-1, 40meV<E<90meV).

correlation between those specific oxygen atoms. That is to say, we can extract the correlation of specific atoms without using isotope substitution. In Fig. 1 we show the DPCF below, above and at Tc. S(Q,E) is integrated between 40 and 90meV to improve statistics. The Q-range used for the transformation is 2.5 to 12$Å^{-1}$. The statistical error is drawn only for the data at Tc=93K. The error is enhanced at the low *r* region due to the limited Q-range for the transformation. Therefore we limit on discussion to the region above 2Å. In the figure we can recognize several distinct features.

(1) The peak around 2.7Å separates into two peaks below Tc and those appear at 2.5Å and 3.1Å. (marked a). This evolution may correspond to the buckling of Cu(2)-O(2)-O(3) plane with the movement of 0.1Å for individual oxygen atoms. The buckling creates two different distances between O(2) and O(2) or between O(3) and O(3).

(2) A shoulder appears at 3.6Å only at Tc. (marked b) Possibly the O(4) atom fluctuates and moves toward the <110> direction. The distance between O(4) and O(4) is shortened by 0.3Å. The movement of O(4) to <110> direction is about 0.5Å, which is about 14% of the a or b lattice constant, and this is consistent with the preliminary estimation in [1].

(3) The peak at 8.1Å, which is the O(4)-O(4) correlation over yttrium atom, above Tc becomes a double peak at Tc and shifts to 7.9Å below Tc. (marked c) Possibly the O(4) atom has a strong fluctuation at Tc along the c-axis also. O(4) moves close toward Cu(2) by 0.1Å below Tc. Mustre de Leon et al. [4] reported that there are two possible site for O(4) separated by 0.13Å. Our observation tends to confirm their results. The distance between O(4) and Cu(2) has a very strong relation with the superconductivity [5]. So possibly the shift of O(4) close to Cu(2) below Tc induces a drastic change on the charge transfer from the chain site to the plane site.

[1] M.Arai, et al. Phys. Rev. Lett. 69, 359 (1992)
[2] M.Arai et al., Trans. American Crystall. Assoc. meeting, Toledo, Ohio, 1991.
[3] Liu. et al., Phys. Rev. B38 4992 (1988)
[4] J. Mutre de Leon et al., Phys. Rev. Lett. 65, 1675 (1990)
[5] J.D.Jorgensen et al. Phys. Rev. B41, 1863 (1990)

Temperature Dependence of Bolometric and Non-Bolometric Photoresponse of a Structured $YBa_2Cu_3O_{7-\delta}$ Thin Film

M.A. Heusinger[1], *R.S. Nebosis*[1], *W. Schatz*[1], *K.F. Renk*[1], *G.N. Gol'tsman*[2], *B.S. Karasik*[2], *A.D. Semenov*[2], and *G.M. Gershenzon*[2]

[1]University of Regensburg, Institut für Angewandte Physik III, Universitätsstr. 31, W-8400 Regensburg, Fed. Rep. of Germany
[2]Moscow State Pedagogical University, 119435 Moscow, Russia

Abstract. We investigated the temperature dependence of the transient voltage photoresponse of a current biased structured $YBa_2Cu_3O_{7-\delta}$ thin film in its transition temperature region, around 79 K. Both, picosecond non-bolometric and nanosecond bolometric response to ultrashort far-infrared laser pulses were found for frequencies between 25 cm^{-1} and 215 cm^{-1}. We will discuss optimum conditions for radiation detection and present an analysis of the dynamical behaviour of excited high T_c thin films.

Thin films of several high temperature superconductors have been studied intensely in order to build ultrafast detectors for pulsed radiation. For $YBa_2Cu_3O_{7-\delta}$ thin films bolometric response to radiation from the visible to the far-infrared (FIR) with signal decay times in the order of nanoseconds [1] as well as non-bolometric effects [1-3] with response times in the subnanosecond range have been described. In this paper we report on the temperature dependence of the photoresponse of a current carrying structured $YBa_2Cu_3O_{7-\delta}$ thin film and discuss conditions for ultrafast detection of radiation.

A 70 nm thin $YBa_2Cu_3O_{7-\delta}$ film oriented with the c axis perpendicular to the film surface on a MgO substrate, was structured (inset of Fig. 1) with nine parallel strips of 30 μm width, separated by 30 μm blanks. The strips were connected to two contact pads, that represent together with the

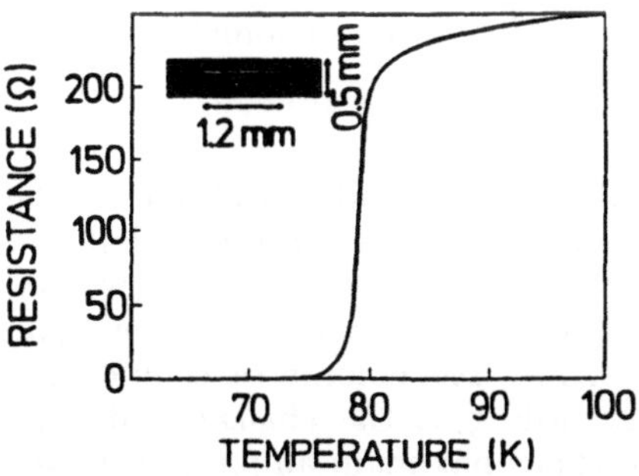

Fig. 1: Resistance of the $YBa_2Cu_3O_{7-\delta}$ film; inset: structure.

Springer Series in Solid-State Sciences, Volume 112
Phonon Scattering in Condensed Matter VII Eds.: M. Meissner and R.O. Pohl

substrate and a conducting ground plate a microstrip line with an impedance of 50 Ω. The film resistance R showed a transition width of 2 K and an onset of superconductivity near 81 K (Fig. 1). The sample was mounted on a holder with a heater for temperature control in a liquid nitrogen bath cryostat with quartz windows for optical access and was biased with a dc current of 20 mA. The photoresponse signals were guided by a wide band (0-18 GHz) co-planar line to a low noise preamplifier (bandwidth 20 GHz, gain 18 dB) and a main amplifier (5 GHz, 42 dB). A single shot oscilloscope with 5 GHz bandwidth and a transient digitizer with 0.8 GHz bandwidth were used to register the amplified signals. The total signal registration system had a time resolution of 120 ps. The FIR laser emission was produced by an optically pumped D_2O-Raman-laser [4], and consisted of a train of ultrashort single pulses. The power incident on the film was regulated by attenuators to be approximately 10 W per pulse.

Fig. 2 shows a photoresponse signal with 120 ps rise and decay time, monitored in the low resistance range (R ≈ 20 Ω) of the film at a temperature of 77.5 K. We attribute this ultrafast signal to quasiparticle heating leading to a change in resistance, followed by a relaxation via electron phonon interaction.

At higher film resistances we observed longer pulses. This is illustrated in Fig. 3 by the photoresponse to a whole FIR pulse train measured near the middle of the transition (79.5 K). The signal can be described as a superposition of different components. First, there is still the ultrafast response. Secondly, a bolometric decay is found, with a decay time of the order of the phonon escape time from the film, attributed to a bolometric effect because of film heating. Additionally a broad background due to a bolometric effect because of substrate heating is observed.

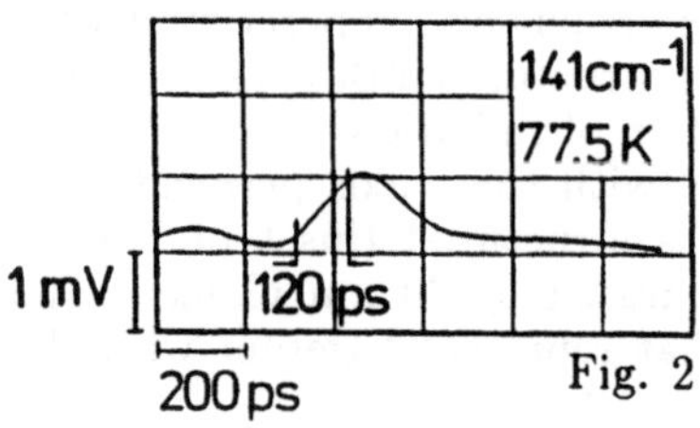

Fig. 2

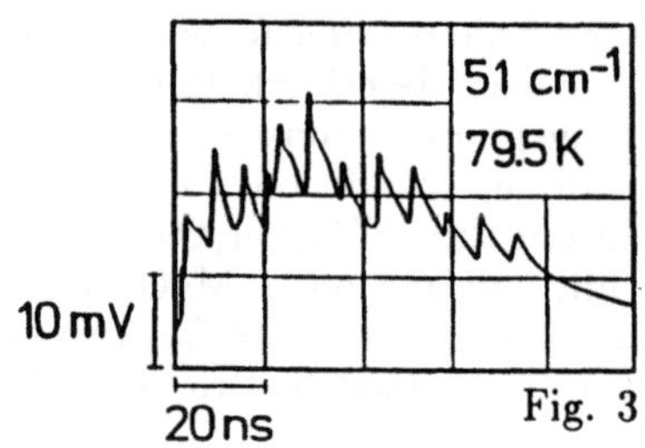

Fig. 3

Fig. 2: Non-bolometric photoresponse for low film resistance, at 77.5 K.

Fig. 3: Photoresponse consisting of a superposition of non-bolometric and bolometric response, at 79.5 K

The temperature dependence of the decay time and amplitude of the bolometric and non-bolometric signal components is shown in Fig. 4 (left and right, respectively). In the low resistance range ultrafast response was observed with an almost constant amplitude. It was superimposed by a bolometric signal at higher temperatures in the linear R(T) regime. The bolometric components show a strong temperature dependence according to the change of dR/dT in the transition temperature region, with their maximum amplitudes in the middle of the transition. The rise in the decay time of the fast bolometric

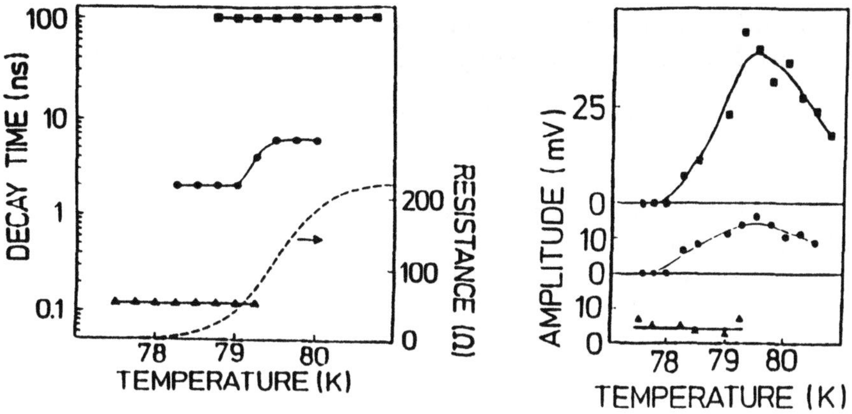

Fig. 4: Temperature dependence of the non-bolometric and bolometric photoresponse decay time (left) and the corresponding amplitude (right) of signal components attributed to quasiparticle heating (▲), sample heating (•) and substrate heating (■).

component is due to impedance mismatch of the electrical circuit and the film, which also causes the ultrafast component not to be observed at the higher temperatures. In the present experiment the amplitude of the bolometric signal due to substrate heating was about an order of magnitude higher than the amplitude of the ultrafast signal.

In conclusion we have reported on the temperature dependence of the photoresponse of a structured $YBa_2Cu_3O_{7-\delta}$ thin film. The sensitivity had the highest value in the middle of the transition and corresponded to bolometric response. On the other hand picosecond response was found in the low resistance range of the film. Thus $YBa_2Cu_3O_{7-\delta}$ thin films used as radiation detectors can be switched from high temporal resolution to high sensitivity only by changing the temperature.

We acknowledge financial support by the Bundesministerium für Forschung und Technologie and the Bayerische Forschungsstiftung through the Bayerisches Verbundprojekt Hochtemperatur-Supraleiter (FORSUPRA). We also thank the DAAD for fellowships (for G.G., B.K., A.S. and E.G.) and we acknowledge financial support by the Scientific Council on the HTS problem (project No. 90462).

References

[1] N. Bluzer, J. Appl. Phys. **71** (3), 1336 (1992).

[2] S.G. Han, Z.V. Vardeny, K.S. Wong, O.G. Symko, and G. Koren, Phys. Rev. Lett. **65** (21), 2708 (1990).

[3] A.D. Semenov, G.N. Gol'tsman, I.G. Gogidze, A.V. Sergeev, E.M. Gershenzon, P.T. Lang, and K.F. Renk, Appl. Phys. Lett. **60**, 903 (1992).

[4] P.T.Lang, W.Schatz, and K.F.Renk, Opt. Commun. **84**, 29 (1991).

Phonon and Luminescence Processes Studied by Raman Spectroscopy in High-T_c Related Oxides

A. de Andrés[1], *S. Taboada*[1], *J.L. Martínez, A. Salinas*[2], *and R. Sàez-Puche*[2]

[1]Instituto de Ciencia de Materiales de Madrid, C.S.I.C., Dep. Física de Materiales C-4, Universidad Autónoma de Madrid, E-28049 Madrid, Spain

[2]Dep. Química Inorgánica, Universidad Complutense de Madrid, E-28020 Madrid, Spain

Abstract. The lattice dynamics of compounds with R_2BaMO_5 formula, where R = rare earth and M = Co, Ni or Cu, are discussed in relation to the 3d metal environement and to the bond distances. Luminescence lines associated with 4f rare earth transitions have been observed and attributed to impurities and components. Broad and weak luminescence bands observed in some compounds are atributed to 3d metal intraband emissions and related to an enhancement of particular phonons.

R_2BaMO_5 compounds present three crystallographic structures depending on the transition metal M and the size of the rare earth. Many of these Ni and Co oxides belong to the Immm space group, which consists of one-dimensional chains of flattened M-O octahedra [1]. Nine Raman active phonons (at $\mathbf{k}=0$: $3A_g+1B_{1g}+2B_{2g}+3B_{3g}$) are expected where only R and in plane oxygen ions vibrate [2]. The cuprates, and the nickelates with the smaller rare earths (Tm-Lu), show a complicated structure with the Pnma space group and Z=4 [3]. M ions are surrounded by five oxygens forming isolated square pyramids. The Cu compounds with the larger rare earth ions (La and Nd) present the tetragonal P4/mbm strcture with Z=2. Here, the coordination of M ions is four and the squares are not connected. In this structure, most of the 11 Raman active modes (at $\mathbf{k}=0$: $3A_{1g}+2B_{1g}+3B_{2g}+3E_g$) are very similar to the Immm phase ones. The detailed normal mode analysis of the Immm structure has been described in [2].

The samples are polycrystalline and were obtained by solid state reaction from stoichiometric amounts of high purity oxides. The X-ray diffraction data reveal that all the samples are single phase. Experiments have been performed with an X-Y Dilor multichannel spectrometer. All spectra have been corrected by the spectral response of the experimental set-up.

Fig. 1 shows the observed frequencies of oxygen modes in the three structures. The highest frequency mode corresponds to a stretching of the four O ions around M ion which is common to the three structures and shows a linear dependence with the M-O distance independently of the M ion or the structure. However the intensity of the peak corresponding to this motion depends on the 3d metal, the maximum intensity is observed for the Cu compounds probably caused by the

Springer Series in Solid-State Sciences, Volume 112
Phonon Scattering in Condensed Matter VII Eds.: M. Meissner and R.O. Pohl

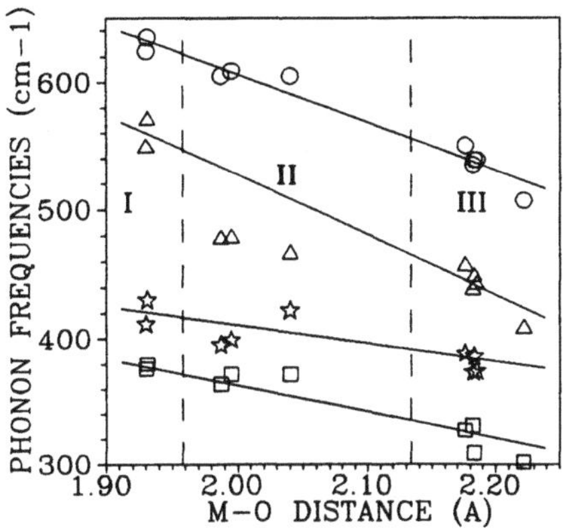

Fig. 1: Oxygen mode frequencies in the three structures versus M-O distances. I: P2/mbm with M=Cu, II: Pnma with M=Cu and Ni, and III: Immm with M=Ni and Co.

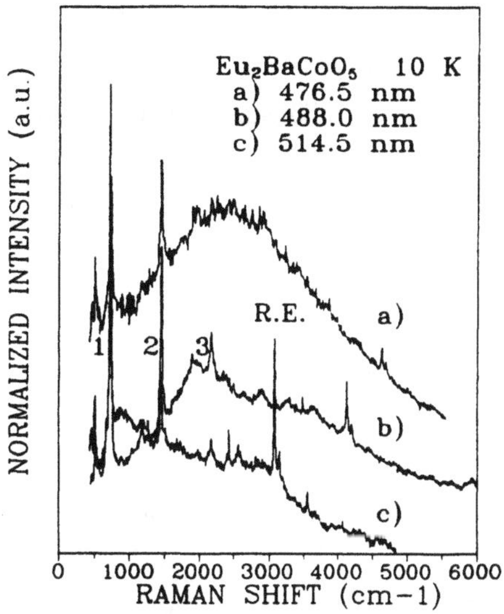

Fig. 2: Phonons and luminescence bands in Eu_2BaCoO_5 at 10 K.

different distribution of the electronic density in the M-O bonds that produces differences in the polarizability derivatives. Copper likes to have shorter "in plane" M-O bonds than the Ni and Co, this is compensated by the longer M-apical oxygen bond or a square coordination. The calculated force constants of the M-O and R-O bonds, obtained from the experimental frequencies, depend with the distances in the expected way ($\propto d^{-3}$) for the three 3d-ions.

Eu_2BaCoO_5 has a complex behavior with the temperature and the incident wavelength (Fig. 2). Opposite resonant effects are observed in the intensities of the sharp rare earth emissions (at 16370 cm^{-1} marked as "R.E." in Fig. 2) and the broad band centered at 18500 cm^{-1} (2.31 eV). The intensity of this latter band is correlated to that of the multiphonon (observed up to order 4). We assign this band to a forbiden crystal field transition in the Co ions partially allowed by its coupling to the strong odd apical oxygen stretching IR phonon (forbidden but observed in the Raman spectra at 720 cm^{-1} and at its multiples marked in Fig.2 as "1,2,3"). A more detailed study will be published elsewhere.

References

1. J. Amador, E. Gutiérrez Puebla, M.A. Monge, I. Rasines, C. Ruiz-Valero, F. Fernández, R. Sáez-Puche and J.A. Campa. Phys. Rev. **B 42**, 7918 (1990).
2. A. de Andrés, J.L. Martínez, R. Saez-Puche and A. Salinas. Sol. St. Comm. **81**, 931 (1992).
3. A. Salinas, J.L. García-Muñoz, J. Rodríguez-Carvajal, R. Saez-Puche and J.L. Martínez J. Sol.St. Chem **100**, xxx (1992).

Raman Phonons in Cuprate Superconductors

O.V. Misochko

Institute of Solid State Physics, Russian Academy of Sciences,
142432 Chernogolovka, Russia; and
Walther-Meissner-Institut für Tieftemperaturforschung,
W-8046 Garching, Fed. Rep. of Germany

Since the new era of high-temperature superconductivity Raman experiments have been widely performed with the aim of investigation of the basic properties of phonons. Since the field is growing so fast, it would be beyond the scope of this paper to give a complete review of all the Raman investigations on cuprate superconductors. The purpose is rather to demonstrate certain peculiarities of lattice dynamics and Raman scattering in these materials.

Experiments with a variety of the cuprates have shown that infrared and Raman spectra normally yield the most information on optical vibrational modes and frequencies.[2,3,4] As all the superconducting cuprates known are either tetragonal or orthorhombic, with a center of inversion, hence Raman allowed modes do not coincide with those infrared ones. The ions in the cuprate structures appear either as ones placed at a center of inversion (and in a mirror plane in the compounds with a single copper-oxygen plane) or as doublets connected by the inversion along z. According to a factor group analysis only the latter are active in Raman scattering. A rough assignment follows from the difference in masses of the vibrating ions, for oxygen modes the frequencies expected to be higher as 300cm-1. The oxygen isotope substitution experiments[4] give an additional evidence for this picture. Now we should ascribe the motion of the individual ions to certain vibrational modes. Taking into account the results of polarization,[3] pressure,[5] rare-earth substitution[6] and resonant studies[7] one can assign the highest frequency phonon in Y123 to displacement of the bridge oxygen with a slight admixture of in-phase vibration of the plane oxygens. The next in frequency phonon consists primarily of in-phase displacements of the plane oxygens with admixture of the bridge oxygen movements. The lowest frequency of oxygen subsystem should be ascribed to out-of-phase movement of the oxygens from the planes. Normal vectors of Tl-1212 are similar to those of the Y-123, but with the difference that the mixing is less pronounced because of larger spacing between the bridge and plane oxygens. In Tl- and Bi-based superconductors with a double layer an intercalation the picture is somewhat complicated. The most strong mixing appears for the oxygens belonging to a distorted octahedron centered at Tl or Bi ions, respectively, while admixture of the plane oxygens is less pronounced. It is thus evident from the Raman data that oxygen subsystem in the cuprates should be treated as a strongly coupled one, all the oxygen phonon eigenvectors (at least for fully symmetric phonons) consist of displacements of different ions because of the large number of particle in the unit cell.

The next point I would like to consider is applicability of the harmonic approximation to the oxygen sublattice. This subject is of importance because of a

Springer Series in Solid-State Sciences, Volume 112
Phonon Scattering in Condensed Matter VII Eds.: M. Meissner and R.O. Pohl

number of theories requires anharmonic phonons to explain high transition temperature. To check it, the pressure and temperature dependence of the oxygen phonons have been measured for the Y-123 and Tl-based crystals. The results obtained suggest that the explicit (phonon occupation driven) and the implicit (volume driven) effects in the temperature dependence of the phonon frequencies are comparable in size, with a tendency for the former to be the larger one. Note that only limiting situation in which the implicit contribution is dominant corresponds to the validity of the harmonic approximation. The reason of such behavior is that for phonon with a complex eigenvector consisting of displacements of different ions is enough only one of the ions moves in anharmonic potential in order to the mode to become anharmonic.

Raman scattering by phonons in the cuprates possesss a number of peculiarities. They are: 1) asymmetric spectral line-shape being dependent on a scattering configuration[11] 2) deviation from the conventional temperature dependence of intensity for some phonons[12,13] 3) appearance of additional phonons in the forbidden polarizations.[13] All these features are common to all cuprates and seem to be explained on the same footing, assuming violation of the translational invariance of crystals caused by the presence of a random potential.

To summarize, we have studied phonon subsystem in three different superconducting families with the use of Raman scattering. Pressure, temperature, polarization and resonant dependencies for all the isotopic substitution of oxygen in the Y123 system allowed to specify the structure of the phonon normal vectors and put forward a hypothesis about the mixing of the oxygen displacements for high-energy phonons in cuprates. Temperature dependence for oxygen phonons indicates anharmonic behavior of the oxygen sublattice.

References

1. A.R. Bishop, in *Earlier and Recent Aspects of Superconductivity* eds. J.G. Bednorz and K.A. Müller, Springer Verlag, Berlin, 1990, p. 482.
2. C. Thomsen, in *Raman Scattering in Solids VI*, ed. M. Cardona, Springer Verlag, Berlin, 1991, p. 285.
3. O.V. Misochko, E.I. Rashba, E.Ya. Sherman, and V.B. Timofeev, Physics Reports **194**, 387 (1990).
4. AV. Bazhenov, L.V. Gasparov, V.D. Kulakovskii, O.V. Misochko, Yu.A. Osip'yan, and V.B. Timofeev, JEPT Lett. **47**, 198 (1988).
5. L.V. Gasparov, O.V. Misochko, M.I. Eremets, A.V. Lomsadze, and V.V. Sruzhkin, Sov. Phys. JETP Lett. **71(5)**, 945 (1990).
6. O.V. Misochko and M.D. Nersisyan, Sov. Phys. Solid State **30(11)**, 2024 (1988).
7. O.V. Misochko and E.Ya. Sherman, JETP **72(1)**, 185 (1991).
8. O.V. Misochko and E.Ya. Sherman, Supercond.:Phys. Chem. Techn. **3**, 2524 (1990).
9. V.D. Kulakovskii, O.V. Misochko, V.B. Timofeev, M.I. Eremets, E.S. Itskevich, and V.V. Struzhkin, JETP Lett. **47**, 626 (1988).
10. O.V. Misochko and V.B. Timofeev, Physica **C162-164**, 1249 (1989).
11. O.V. Misochko and E.Ya. Sherman, Mod. Phys. Lett. **B6**, 281 (1992).
12. O.V. Misochko and E.Ya. Sherman, submitted to Phys. Rev. B.
13. O.V. Misochko, R. Nemetschek, R. Hackl, submitted to Phys. Rev. Lett. B.

Anharmonicity of Bridging Oxygen Vibrations in $YBa_2Cu_3O_{7-x}$ Crystals

L.V. Gasparov, V.D. Kulakovskii, V.B. Timofeev, and E.Ya. Sherman

Institute of Solid State Physics, Russian Academy of Sciences,
142432 Chernogolovka, Moscow District, Russia

Abstract. Raman scattering (RS) of light is used to investigate the behavior of the oxygen modes in $YBa_2Cu_3O_{7-x}$ crystals as a function of exciting radiation power and of temperature.The excitation spectra of these modes are measured. To explain the behavior of the investigated modes it is proposed that the bridging (apex) oxygen ion O4 is located in an anharmonic two-well potential.

1. INTRODUCTION

The $YBa_2Cu_3O_{7-x}$ Raman scattering spectra contain, besides symmetry allowed lines ,some additional lines, e.g.,a line near 590 cm^{-1} which manifests itself most intensely in zz-polarization [1-3]. We shall show in the present paper, on the basis of the entire aggregate of the experimental results, new explanation of this mode.

2. EXPERIMENTS

In our experiments the peak intensity of the investigated lines was normalized to the peak intensity of the 220cm^{-1} line in a BaF_2 crystal. As the pump power to the $YBa_2Cu_3O_{6.5}$ crystal is increased the normalized intensity of the ~ 500cm^{-1} line decreases, Fig.1b, and that of the ~ 590cm^{-1} line increases. The corresponding increase and decrease of the intensities of these modes is reversible. As to the ~330cm^{-1} line (antiphase oscillations of planar oxygen ions (curve 3 in Fig.1a), its intensity remains unchanged.

When $YBa_2Cu_3O_{6.5}$ crystals heated the peak intensity of the ~500cm^{-1} line decreases, that of the ~ 590cm^{-1}

Springer Series in Solid-State Sciences, Volume 112
Phonon Scattering in Condensed Matter VII Eds.: M. Meissner and R.O. Pohl

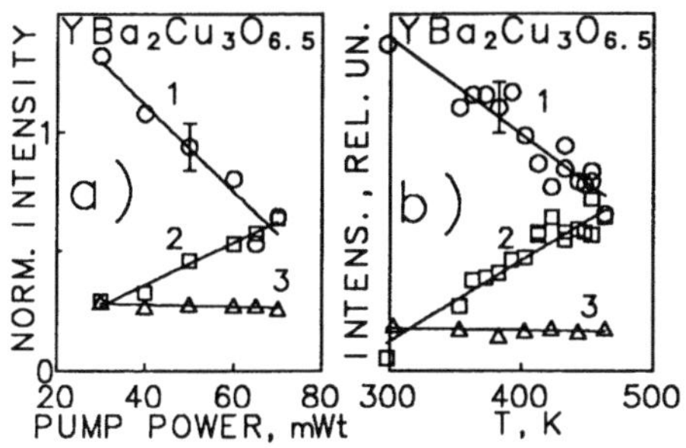

Fig.1.
a) Normalized intensity versus pump power
b) peak intensity versus temperature of crystal sqares-590cm^{-1}, circles- (A_g) 500cm^{-1}, triangles-(B_{1g}) 330cm^{-1}.

line increases, and that of the ~ 330cm^{-1} line remains unchanged,Fig.1b,. As the temperature is raised the lines behave in the same way as when pump power is lowered, and all the changes take place with bridging (apex) and possibly chain oxygen but not with planar oxygen ions (~330cm^{-1} line intensity does not change).

The excitation spectra of ~500cm^{-1} and ~590cm^{-1} modes are evidently identical. For both these modes the scattering intensity in zz-polarization substantially predominate over scattering in xx- and yy-polarizations. All this indicates that the character of the scattering is governed by the electronic properties of the complex Cu1-O4-O1 [1].(i.e., identical hybridized electronic states participate in the scattering).

3. DISCUSSION

We assume that all these experimental data can be attributed to the presence, in the scattering Cu1-O4-O1 complex, of a two-well potential for the O4 ion under conditions of violation of the oxygen stoichiometry. At low temperatures, this oxygen ion is localized mainly in potential well 1 and oscillation with ~500cm^{-1} appears in the system. When the temperature is raised, the particle can land by tunneling or activation in well 2 located closer to the Cu1 ion, and this leads to the appearance of the 590cm^{-1} mode.

References

1. L.V.Gasparov, V.D.Kulakovskii, V.B.Timofeev, E.Ya.Sherman. Sov.Phys.JETP.1991.V.73(5),P.929
2. K.F.McCarty, J.C. Hamilton, R.N. Shelton. at al. Phys.Rev.B.1988. V.38. P.2914.
3. D.Mihailovic,C.M.Foster.Sol.St.Commun.1990.V.74.P.753

High-T_c Superconductivity and Mixed Valence State of Bipolarons

A.N. Kocharian

Yerevan Physics Institute, Alikhanian Br. St. 2, 375036 Yerevan, Armenia

1. Introduction

Strong electron-phonon interaction can produce high T_c superconductivity (HTSC) as a result of electron band narrowing by polarons. The values of T_c, the magnetic field penetration depth λ_2, the thermoelectric power S, the small value of the coherence length and some other properties of HTSC are well explained in the framework of a generalized negative U Hubbard model. In this paper we consider a bipolaron mechanism of HTSC, based on the oxygen model of superconductivity. There is evidence that oxygen ions on square lattices in CuO_2 planes display diamagnetic properties caused by strong intrasite interaction of 2p electrons with local phonons. This means that strong coupling brings about stabilization of two diamagnetic valence configurations O^o and O^{--}. The possibility of competition between valence fluctuations and charge ordering states allows one to apply a mixed valence model approach to HTSC.

2. Results

The effective Hamiltonian taken to order $t^4/|U|^3$ can be reduced to the form[1]

$$H = I_{\|} \sum_{<ij>} (L_i^+ L_j^- + h.c.) + 2I_\perp \sum_{<ij>} L_i^z L_j^z$$
$$-h \sum_i L_i^z + 2t^4/|U|^3 \sum_{n.n.n.} \vec{L}_i \vec{L}_j \quad , \qquad (1)$$

where

$$I_\perp = 2t^2/|U| + 4V - 8t^4/|U|^3 \quad ,$$
$$I_{\|} = 2t^2/|U| - 8t^4/U^3 \quad .$$

In the last term the summation is carried out over all the next nearest neighbors.

The local-pair model using an anisotropic Heisenberg-Ising XXZ model with an effective "magnetic" field $h = |U| - \mu$ and a pseudospin $L = \frac{1}{2}$ trans-

Springer Series in Solid-State Sciences, Volume 112
Phonon Scattering in Condensed Matter VII Eds.: M. Meissner and R.O. Pohl

forms the carrier density into the magnetization in the pseudospin problem. The total magnetization ΣL_i^z corresponds to deviations from half filling. Antiferromagnetic (AF) order in the z-direction corresponds to a charge-density wave (CDW) phase while AF order in the x-y plane corresponds to S-wave superconductivity (SC). The correction to the XXZ model of order $t^4/|U|^3$ describes weak frustration due to next-nearest neighbor interactions between the spins. So the thermodynamic properties can be understood quantitatively at least up to order t^2/U on the basis of known properties of low-dimensional quantum anisotropic AF.

After diagonalization of the effective Hamiltonian, we construct a functional of the thermodynamic potential which depends on the off-diagonal parameter $\Delta =< X_i^{20} >=< \sigma C_r^+ C_r^+ >$ and the non-uniform order parameter η. Solving the self-consistent set of equations linking η and Δ, we obtain a phase diagram of the critical temperature θ versus bipolaron concentration n. It is shown that as well as the pure superconducting state $\Delta \neq 0$ $(\eta = 0)$ and the CDW state $\eta \neq 0$ $(\Delta = 0)$ there is a region of existence of a mixed state (MS) $(\Delta \neq 0, \eta \neq 0)$. At T=0 and g $= I_\perp / I_\parallel > 1$, the boundary between MS and SC takes place at $n = \frac{1}{2}[1 \pm (1 - \frac{1}{g})]^{\frac{1}{2}}$. For a square AF oxygen lattice near $n = \frac{1}{2}$, superconductivity is connected with the motion of clusters consisting of five oxygen atoms in the same electron configuration O^{--} at $n > \frac{1}{2}$ (n-type) and O^o at $n < \frac{1}{2}$ (p-type).

We also report that the analytic expression for the specific heat jump ΔC at the critical temperature θ_c in the pure SC state is

$$\Delta C = -(1-n)/2\theta_c^2 \; (ch^2((1-n)/\theta_c) - 1/2\theta_c)^{-1} \quad , \tag{2}$$

where

$$\theta_c = \Big[arcth(1-n)\Big]/(1-n).$$

We consider doubly occupied and empty sites (hard core bosons) as a two level system which can be mixed by the parameter Δ. There is therefore a close analogy between the interacting hard-core boson model with the two band spinless model (excitonic insulator). The usual mixed valence model is distinguished from this by the assumption that only the effective mass of holes goes to infinity. Due to repulsion of bipolarons at $g > 1$ in the vicinity of $n = \frac{1}{2}$ for quasi two dimensional lattices, a non uniform mixed valence state exists. Frustration of spins on square lattices enlarges the homogeneous mixed valence state. The analysis of a two dimensional XXZ model shows that no real phase transition exits at finite temperatures, however, SC and CDW long range ordering both take place in the ground state. The pair correlation functions show power law decay below the temperature of the Kosterlitz-Thouless transition.

Reference

[1] A.N. Kocharian and P.S. Ovnanian. Z. Phys. B. - Condensed Matter 83, **161** (1991).

Phonon Renormalization of Thermoelectric Power of High-T_c Materials

A.V. Sergeev[1] *and D.V. Livanov*[2]

[1]Moscow State Pedagogical University, M. Pirogovskaja Str. 1, 119435 Moscow, Russia

[2]Moscow Institute of Steel and Alloys, Moscow, Russia

Abstract. Renormalization of thermoelectric power due to Nielsen - Taylor effect (interference between electron-phonon and electron- impurity interactions) is used for the explanation of the temperature dependence and sign of the thermopower in high-T_c materials. In the framework of the model the negative sign of TEP of untwinned $YBa_2Cu_3O_{7-x}$ crystal in *a*-direction observed by Howson et. al. is connected to the strong hole scattering via O-vacancies in chains.

Despite intensive investigations the thermoelectric power of high temperature superconductors continues to be a problem that attracts much attention of both experimentalists and theorists. Resent experiments demonstrate the correlation of the sign with variations in oxygen content: for x$<0.1 \div 0.15$ the *a*-*b*-plane thermopower is negative, when oxygen content is reduced the thermopower decreases to zero and then changes sign to positive. Temperature dependence of thermopower shows the occurrence of extremum (maximum for positive thermopowers and minimum for negative ones) in the temperature range 100-200 K. Measurements on untwinned crystals of $YBaCuO$ have shown that the thermopower in *b*- direction is negative while it is positive in *a*- direction [1].

The observed temperature dependence of thermopower allows a variety of different mechanisms appropriate for ordinary metals [2]. In Ref. 3 this dependence was successfully explained by means of temperature - dependent renormalization due to the strong electron - phonon interaction analogous to electron mass enhancement. The sign of thermopower is still in question, because most experimental works show the same sign of carries for plains and chains.

The phonon renormalization occures is realized not only due to the pure electron-phonon interaction but also due to the interference between electron-phonon and electron-impurity interactions. Both contributions are of the same order and are proportional to the electron - phonon interaction factor (λ). But the first phenomenon results from asymmetry in the properties of particles

Springer Series in Solid-State Sciences, Volume 112
Phonon Scattering in Condensed Matter VII Eds.: M. Meissner and R.O. Pohl

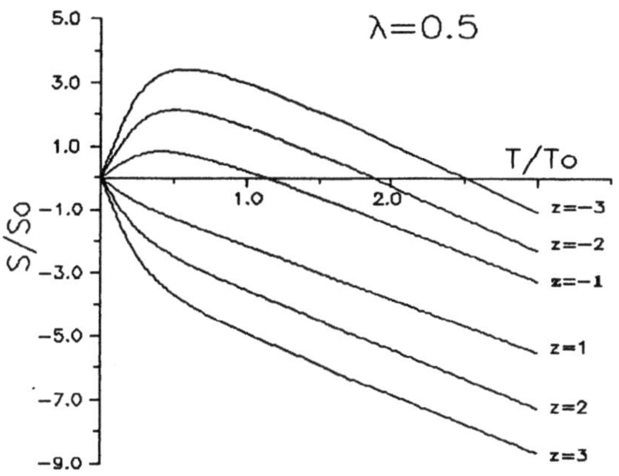

Fig.1. Temperature dependence of thermoelectric power due to interference between electron-phonon and electron-impurity interections.
Z - valence of impurity atoms with respect to the host atom, T_0 - Debay temperature.

near the Fermi surface. The interference contribution is not connected with electron - hole asymmetry, it s magnitude and sign are determined by the non-Born amplitude of the electron-impurity scattering and depend on the valence of the impurity atoms. The interference contribution to thermopower has been calculated in Ref. 4. In this report we presented the results of numerical calculations for temperature dependence of thermoelectric force at different valences of impurity atoms with respect to the valence of the host atom (Z).

As demonstrated by the Figure 1, the model proposed provides an explanation for the experimental temperature dependence described above. Different signs of the effect are attributed to different signs of electron -impurity potentials in plains and chains. It is natural that the electron scattering in chains is due to oxygen vacancies with Z=2, and it leads to negative thermopower. The change of the effect sign with oxygen content decrease may be understood as a result of breaking conductivity along chains and hence to suppression of the negative *b*-axis contribution.

One of us (S.A.V.) woud like to thank Prof. M.Ristig for helpful discussion.

References

[1]. A.J.Lowe, S.Regan, M.A.Howson, Phys.Rev.B 44, 9757 (1991).
[2]. J.H.Cohn, S.A.Wolf, V.Selvamanickam and K.Salama, Phys.Rev.Lett. 66, 1098 (1991).
[3]. A.B.Kaizer and C.Uher, Aust.J.Phys. 41, 597, 1988.
[4]. M.Yu.Reizer and A.V.Sergeev, Sov.Phys.-JETP 66, 1250, (1987)

Part VI

Disordered Systems

Interaction of Soft Modes and Sound Waves in Glasses

A. Bringer[1], *U. Buchenau*[1], *Yu.M. Galperin*[2], *L. Gil*[1], *V.L. Gurevich*[2], *D.A. Parshin*[3], *M.A. Ramos*[1], *and H.R. Schober*[1]

[1]Institut für Festkörperforschung, Forschungszentrum Jülich, Postfach 1913, W-5170 Jülich, Fed. Rep. of Germany
[2]A.F. Ioffe Institute for Technical Physics, Academy of Sciences, 194021 St. Petersburg, Russia
[3]Institut für Angewandte Physik, Universität Heidelberg, Albert-Überle-Str. 3–5, W-6900 Heidelberg, Fed. Rep. of Germany

Abstract. The soft potential model (an extension of the tunneling model to include soft localized vibrations) is shown to describe the anomalous features of the specific heat, the sound absorption and the thermal conductivity of glasses over the entire low temperature range.

We report on a combination of analytical[1] and numerical[2] calculations within the soft potential model[3, 4], aimed at a description of the glassy low-temperature anomalies[5, 6, 7]. Here we give only main results and leave derivations and details to following, more extended publications[1, 2].

The soft potential model postulates soft modes with an effective mass M and a stabilizing fourth order term in the potential[8]

$$V(x) = W\left[D_1\left(\frac{x}{d}\right) + D_2\left(\frac{x}{d}\right)^2 + \left(\frac{x}{d}\right)^4\right] \qquad (1)$$

where the energy W is determined by the equality of a kinetic ground state energy $\hbar^2/2Md^2$ and the potential energy of the purely quartic potential at the distance d. W marks the crossover between tunneling and vibrational states and has been linked[9] to the temperature T_{min} of the minimum in C_p/T^3 (C_p specific heat) at a few Kelvin. The origin of the configurational coordinate x is chosen such that the third order term of the potential vanishes. The coefficients D_1 and D_2 are supposed to be random with a density of states P_s at the origin of the $D_1 - D_2$ plane. For the analytical approximations, that density was assumed to be constant over the whole plane. For the numerical calculations, we assumed a symmetric gaussian distribution of D_1 with $< D_1^2 >= 2.96(k_BT_g/W)^{3/2}$, based on the idea[10] of thermal strains freezing in at the glass temperature T_g, in order to be able to describe the maximum in C_p/T^3 at a temperature T_{max}, three to five times higher than T_{min}.

Both vibrational and tunneling states are assumed to interact with the sound waves, in order to explain the universal anomalies in the acoustic properties and the thermal conductivity of glasses[6, 7]. In the long wavelength limit, the sound waves can be characterized by elastic strains, a uniaxial strain

Springer Series in Solid-State Sciences, Volume 112
Phonon Scattering in Condensed Matter VII Eds.: M. Meissner and R.O. Pohl

ϵ_l in the longitudinal case and a shear strain ϵ_t in the transverse case. The interaction with the soft modes is assumed to be bilinear[1]

$$\delta V_l = \Lambda_l \frac{x}{d} \epsilon_l \qquad \delta V_t = \Lambda_t \frac{x}{d} \epsilon_t \tag{2}$$

with coupling constants Λ_l and Λ_t.

Analytically, the model is difficult to handle. There are three limiting cases where one can use analytical approximations: (i) the tunneling states[4] at slightly negative D_2 and small D_1 (ii) classical relaxation at negative D_2 and higher temperatures and (iii) soft quasiharmonic modes at positive D_2 and/or large D_1. The analytical calculations were done using these three approximations and extrapolating them with appropriate crossover frequencies ($2W/h$) and temperatures (W/k_B) into the crossover region, where the approximations fail. One of the aims of the numerical work was to check these results. The second was to describe the glassy anomalies in the specific heat and the thermal conductivity at higher temperatures.

Below 1 Kelvin, the results of the tunneling model[11] are essentially reproduced. Analytically, one gets the relation $P_s\Lambda_j^2/W = \bar{P}\gamma_j^2$ between the parameters $\bar{P}$, γ_j of the tunneling model and those of the soft potential model ($j = l, t$); numerically, $P_s\Lambda_j^2/W = 0.9\bar{P}\gamma_j^2$.

The classical relaxation at higher temperatures was only studied analytically. One gets a contribution to the mean free path of sound waves

$$l_{rel,class}^{-1} = \frac{2\pi^2 \nu P_s \Lambda_j^2}{\rho v_j^3 W} \left(\frac{T}{W}\right)^{3/4} ln^{-1/4}\left(\frac{1}{2\pi\nu\tau_0}\right) \tag{3}$$

where ρ is the density and v_j is the sound velocity of the glass. ν is the frequency and τ_0 is the inverse of an attempt frequency of the order of THz. The contribution sets in at a comparatively low temperature, W/k_B, and acts on sound waves below the frequency $2W/h$.

At frequencies above $2W/h$, the scattering of sound waves from soft vibrational modes sets in. Analytically, one gets

$$l_{res,vib}^{-1} = \frac{1}{6\sqrt{2}} \frac{2\pi^2 \nu P_s \Lambda_j^2}{\rho v_j^3 W} \left(\frac{h\nu}{W}\right)^3 \tag{4}$$

based[4] on a density of vibrational states $g(\nu)$ proportional to ν^4. With the different assumption of the numerical calculation, one gets a maximum in $g(\nu)/\nu^2$, consistent with the maximum in C_p/T^3, and has to modify the mean free path of Eq. (4) accordingly.

Results for the canonical glass vitreous silica, calculated using P_s=1.5 10^{-6} states per atom, W/k_B=3.8 Kelvin and Λ_l and Λ_t equal to 0.64 and 0.4 eV, respectively (the latter two calculated from tunneling model parameters[12]), are compared to experimental specific heat, thermal conductivity and vibrational density of states data in Fig. 1, showing reasonable agreement over the entire low temperature and low frequency range.

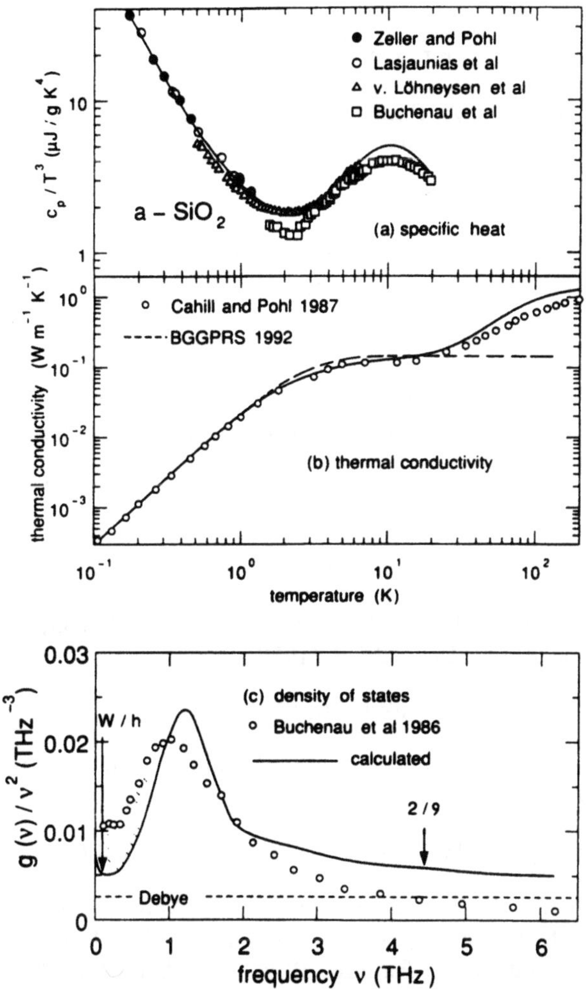

FIG. 1. Comparison of numerical (continuous lines) and analytical (dashed line in (b)) results to experimental data in vitreous silica:

(a) specific heat[5, 13, 14, 15] plotted as C_p/T^3 vs. T

(b) thermal conductivity[16] vs. T

(c) density of states[15], plotted as $g(\nu)/\nu^2$ vs. ν.

References

[1] U. Buchenau, Yu. M. Galperin, V. L. Gurevich, D. A. Parshin, M. A. Ramos and H. R. Schober, Phys. Rev. B (August 1992)

[2] L. Gil, M. A. Ramos, A. Bringer and U. Buchenau (unpublished)

[3] V. G. Karpov, M. I. Klinger and F. N. Ignat'ev, Zh. Eksp. Teor. Fiz. **84**,760 (1983) [Sov. Phys– JETP **57**, 439 (1983)]

[4] M. A. Il'in,V. G. Karpov and D. A. Parshin, Zh. Eksp. Teor. Fiz. **92**, 291 (1987) [Sov. Phys.– JETP **65**, 165 (1987)]

[5] R. C. Zeller and R. O. Pohl, Phys. Rev. B **4**, 2029 (1971)

[6] S. Hunklinger and W. Arnold, in *Physical Acoustics* Vol. XII, ed. by W.P. Mason and R.N. Thurston (New York: Academic Press 1976), p. 155.

[7] J. J. Freeman and A. C. Anderson, Phys. Rev. B **34**, 5684 (1986)

[8] We have chosen to formulate the soft potential model in natural units. The connection to references 1, 3, 4, 9 is given by $d = a\eta_L^{1/2}$ and $P_s = 2P_0\eta_L^{5/2}$.

[9] U. Buchenau, Yu. M. Galperin, V. L. Gurevich and H. R. Schober, Phys. Rev. B **43**, 5039 (1991)

[10] L. Ferrari, W. A. Phillips and G. Russo, Europhys. Lett. **3**, 611 (1987)

[11] W. A. Phillips, J. Low. Temp. Phys. **7**, 351 (1972); P. W. Anderson, B. I. Halperin, C. M. Varma, Phil. Mag. , **25**, 1 (1972)

[12] J. F. Berret and M. Meissner, Z. Physik B **70**, 65 (1988)

[13] J. C. Lasjaunias, A. Ravex, M. Vandorpe and S. Hunklinger, Solid State Commun. **17**, 1045 (1975)

[14] H. v. Löhneysen, H. Rüsing and W. Sander, Z. Physik B **60**, 323 (1985)

[15] U. Buchenau, M. Prager, N. Nücker, A. J. Dianoux, N. Ahmad and W. A. Phillips, Phys. Rev. B **34**, 5665 (1986); U. Buchenau, H. M. Zhou, N. Nücker, K. S. Gilroy, and W. A. Phillips, Phys. Rev. Lett. **60**, 1318 (1988)

[16] D. G. Cahill and R. O. Pohl, Phys. Rev. B **35**, 4067 (1987)

Dynamic Structure Factor and Its Single-Length Scaling for Random Fractal Structures

T. Nakayama and K. Yakubo

Department of Applied Physics, Hokkaido University, Sapporo 060, Japan

We present our numerical results on the dynamic structure factor for large-scale site-percolating networks (site numbers from 93,382 to 110,793), and confirm its scaling properties. It is shown that the dynamic structure factor $S(q,\omega)$ can be scaled by a single-length scale $\Lambda(\omega)$ with the scaling function $h(q\Lambda(\omega))$. This indicates that three characteristic lengths, i.e., a wavelength $\lambda(\omega)$, elastic scattering length $l_{scat}(\omega)$ and localization length $l_{loc}(\omega)$, have the same frequency dependence. Our results are interpreted in terms of the scaling theory recently proposed by Alexander, Courtens and Vacher[1].

The inelastic scattering experiments probe the modulation of the local density due to the lattice vibrations. This implies that the intensity of scattered beam $I(\boldsymbol{q},\omega)$ is proportional to the square of Fourier-transformed density-fluctuation. This viewpoint is more relevant to inelastic neutron scattering than light scattering where the dipole-induced-dipole mechanism becomes important. The intensity is expressed by

$$I(\boldsymbol{q},\omega) = A\left\langle \sum_{\lambda} \delta(\omega-\omega^{\lambda})\rho^{\lambda}(\boldsymbol{q})\rho^{\lambda}(-\boldsymbol{q}) \right\rangle , \qquad (1)$$

where A is the constant independent of $\boldsymbol{q}$ and ω, and $\langle\cdots\rangle$ represents the thermal averages. In the case of small amplitudes, one can write $\rho^{\lambda}(\boldsymbol{q}) = -i\sum_i(\boldsymbol{q}\cdot\boldsymbol{u}_i^{\lambda})e^{-i\boldsymbol{q}\cdot\boldsymbol{R}_i}$, where $\boldsymbol{u}_i^{\lambda}$ and $\boldsymbol{R}_i$ are the vibrational amplitude of the mode λ and the equilibrium position of the i-th atom, respectively. Thus the intensity becomes proportional to the convolution of the Fourier transform of the amplitude-amplitude correlation function. The dynamic structure factor is related to the scattered intensity through the relationship $S(\boldsymbol{q},\omega) = \omega^2 I(\boldsymbol{q},\omega)/k_BT$. We also introduce the dynamic correlation function $C(\boldsymbol{q},\omega)$ defined by

$$C(\boldsymbol{q},\omega) = \langle\langle\tilde{\rho}^{\lambda}(\boldsymbol{q})\tilde{\rho}^{\lambda}(-\boldsymbol{q})\rangle_{\omega}\rangle_{ens} , \qquad (2)$$

where $\langle\cdots\rangle_{\omega}$ denotes the average over all modes with frequencies close to ω, and $\langle\cdots\rangle_{ens}$ means the ensemble average. In Eq.(2), $\tilde{\rho}^{\lambda}(\boldsymbol{q})$ is defined as $\tilde{\rho}^{\lambda}(\boldsymbol{q}) = -i\sum_i(\boldsymbol{q}\cdot\tilde{\boldsymbol{u}}_i^{\lambda})e^{-i\boldsymbol{q}\cdot\boldsymbol{R}_i}$, where $\tilde{\boldsymbol{u}}_i^{\lambda}$ is the displacement of the i-th atom normalized by $\sum_i\tilde{\boldsymbol{u}}_i^{\lambda}\cdot\tilde{\boldsymbol{u}}_i^{\lambda} = 1$. In the definition of Eq.(2), $C(\boldsymbol{q},\omega)$ is related to the dynamic structure factor $S(\boldsymbol{q},\omega)$ with $S(\boldsymbol{q},\omega) = AD(\omega)C(\boldsymbol{q},\omega)$, where $D(\omega)$ is the density of states of vibrational excitations. It should be emphasized that the

Springer Series in Solid-State Sciences, Volume 112
Phonon Scattering in Condensed Matter VII Eds.: M. Meissner and R.O. Pohl

displacements of single fracton mode are distributed anisotropically with fractal correlations and the form of $S(\mathbf{q},\omega)$ deduced by the *averaged* fracton wave functions is misleading as already pointed out by Ref.2.

We consider $d = 2$ site-percolating networks at the percolation threshold ($p_c = 0.593$) with the periodic boundary condition consisting of N particles with mass m and linear springs connecting two nearest-neighbor atoms. The simulated results of q-dependence show that $S(q,\omega)$ for different ω can be scaled by a single characteristic wave number q_0. In particular, it should be emphasized that the wave number dependence obeys the power law: $S(q,\omega) \propto q^{4.0\pm0.1}$ below q_0 and $S(q,\omega) \propto q^{-1.6\pm0.1}$ above q_0. The results of ω-depemdemce indicate that the asymptotic behavior of $S(q,\omega)$ is expressed as $S(q,\omega) \propto \omega^{1.7\pm0.1}$ in the frequency regime of $\omega \ll \omega_0$, and $S(q,\omega) \propto \omega^{-2.25\pm0.1}$ for $\omega \gg \omega_0$.

Let us discuss what our simulations indicate. The relationship between q_0 and ω_0, at which the dynamic structure factors $S(q,\omega)$ have their peaks, represents the frequency dependence of wavelengths (dispersion relation) of oscillating fractons. On the other hand, the frequency dependence of the width of $S(q,\omega)$ in q-space reflects the ω-dependence of the scattering length or the localization length of fractons. The universal behavior of $S(q,\omega)$ shows that these characteristic lengths have the same frequency dependence. Such a remarkable feature of the structure factor has been suggested recently by Alexander, Courtens and Vacher[1]. They have argued for asymptotic behavior of the dynamic structure factor based on the SLSP (single-length scaling postulate).

If the SLSP is valid, the correlation function $S(q,\omega)$ should have the following scaling form:

$$S(q,\omega) = q^{z} h(q\Lambda(\omega)) \ . \tag{4}$$

The asymptotic behaviors of the scaling function $h(x)$ are anticipated to take the forms $h(x) \propto x^{a}$ for $x \ll 1$, and $h(x) \propto x^{-a'}$ for $x \gg 1$, respectively. Here a and a' are some new scaling exponents. Though these power laws are not the deduction from the SLSP, our calculated results indicate clearly the validity of this power law postulate. According to the above scaling behaviors, we have

$$S(q,\omega) \propto \begin{cases} q^{z+a}\omega^{-a\tilde{d}/D} & \text{for } q\Lambda(\omega) \ll 1 \\ q^{z-a'}\omega^{a'\tilde{d}/D} & \text{for } q\Lambda(\omega) \gg 1 \end{cases} \ . \tag{5}$$

Here $\tilde{d}$ and D are the *fracton* dimensionality and the *fractal* dimension, respectively.

The values of the exponents a, a' and z can be extimated from four asymptotic forms (q-dependence for both $q \ll q_0$ and $q \gg q_0$, and the ω-dependence for both $\omega \ll \omega_0$ and $\omega \gg \omega_0$) of $S(q,\omega)$ to be $a = 3.2 \pm 0.1$, $a' = 2.4 \pm 0.1$, and $z = 0.8 \pm 0.1$. These exponents determine the value of exponent σ characterizing the strain due to vibrational modes (fractons) in random fractals. The calculated result of $\sigma = 1.1$ is in accord with the criterion ($\sigma > 1$) proposed by Alexander, Courtens and Vacher[1].

To summarize, we have computed the dynamic structure factors for large-scale percolating networks. The present work has confirmed directly the *SLSP*, i.e., the dynamic structure factor is characterized by a single-length scale $\Lambda(\omega)$. The details were published elsewhere[3].

References

[1] S. Alexander, E. Courtens and R. Vacher: to be published in Phys. Rev. B.

[2] T. Nakayama, K. Yakubo and R. Orbach: J. Phys. Soc. Jpn. **58** (1989) 1891.

[3] T. Nakayama and K. Yakubo, J. Phys. Soc. Jpn. **61** (1992) No.8.

Local Phonons and Nanostructure of Disordered Solids and Glasses

V. Malinovsky

Institute of Automation and Electrometry, Russian Academy of Sciences, Siberian Branch, 630090 Novosibirsk, Russia

As is known, there is a good understanding of the structure of ideal crystals. By contrast to this, the structure of real crystals, amorphous solids and glasses is known only approximately.

The most important differences between elementary excitation spectra of crystalline and amorphous materials manifest themselves in the low-energy range. The low-temperature ($T \leq 1K$) anomalies in specific heat and thermal conductivity of amorphous materials are well known. The phenomenology of these anomalies is available in the framework of the models which involve two-level systems. Less clear is the nature of excess density of vibrational states (DVS) in the energy range 2 to 10 meV. These vibrational states cause the excess specific heat of vitreous and amorphous materials in the temperature range of 10 to 30 K. They are manifested in neutron scattering spectra and IR absorption spectra as well as in Raman scattering spectra in the form of the so-called boson peak at low energies. Raman spectroscopy is the most convenient method for investigation of the vibrational spectra, which also give indirect, model dependent information on nanostructure of non-crystalline solids. The spectrum of the low-frequency Raman spectra can be written in the form

$$\frac{\omega I(\omega)}{n(\omega)+1} = C(\omega)g(\omega)$$

where smooth function $C(\omega)$ describes light to vibrations coupling and $g(\omega)$ is the density of vibrational states (DVS). Position, width and amplitude of boson peak depend on the chemical composition of glass, thermal treatment of the sample, applied pressure and so on. Nevertheless, the shape of the boson peak (i.e. its normalized spectrum, which can be obtained by setting the amplitude and position of maximum equal to unity) is universal in majority of glasses [1]. According to inelastic neutron scattering data the spectrum of low-frequency DVS of glasses has universal shape itself [2]. In [3] it is shown that both spectra of boson peak and the excess low-energy DVS can be fitted by the lognormal function with universal value of standard deviation $\sigma = 0.48 \pm 0.05$ for different glasses. Thus, there is a new universal dimensionless number characteristic for glasses. This universality

Springer Series in Solid-State Sciences, Volume 112
Phonon Scattering in Condensed Matter VII Eds.: M. Meissner and R.O. Pohl

means, that the spectral peculiarities under discussion are a consequence of some universal features of glass structure.

The nature of low-frequency vibrations in glasses which are responsible for the excess DVS and for boson peak has no generally accepted explanation. One of the possible interpretation of low-energy Raman and neutron scattering data in glasses is based on the assumption that in glass there are structure inhomogeneities of size $L \approx 1 \div 2$ nm which lead to the appearance of quasilocalized vibrations with frequencies $\omega \approx \vartheta / L$, here ϑ is the sound velocity. We believe, that despite of chaos which is usually associated with the structure of glasses, there is a universal spatial dimension - an order parameter which is characteristic for glasses of different nature (semiconducting, dielectric, metallic). It can be of the same importance for the theory of the vitreous and liquid states as the elemental cell for the theory of crystals.

Contrary to glasses, the spectral form of boson peak in amorphous semiconductors depends on the chemical composition and thermal treatment of the samples. In this material the high-frequency tail of the boson peak is steeper then in glasses, especially in $a - Si$, $a - Ge$ and to less extent in $a - As$ and $a - As_2S_3$ films. For the structure it means that the width of the size distribution function of nanoinhomogeneities is broader in glasses than in amorphous tetrahedral semiconductors. This result can be understood if we will take into account the existence in glass structure of low-coordinated atoms, which can adjust structure elements more smoothly than it can be done in the case of tetrahedral coordination.

In media with a spatial dispersion of properties the excitation energy is localized on the scales of the order of structure correlation radius. The localization of excitation changes mainly the energy transfer kinetics and determines the mechanism of the structure transformation in real crystals, amorphous solids and glasses under illumination. A single light quantum may transform the structure of the material on the scale of some coordination spheres. Local phonons "loosen" the lattice, provide a random bond breakage, remove steric prohibitions, etc. The local heating model[4] demonstrates a real driving force for structural transformations, gives an opportunity to predict correctly their direction and understand the position of microscopic models which takes into account the specific character of outer atoms shells (broken bonds, valence-alternation pairs, etc). Actually a new physics appears, the physics of cooperative processes.

References

1. V.K. Malinovsky, A.P. Sokolov, Sol. St. Comm., 57 (1986), 757.
2. V.K. Malinovsky, V.N. Novikov, P.P. Parishih, A.P. Sokolov, M.G. Zemlyanov, Europhy. Lett., 11 (1990) 43.
3. V.K. Malinovsky, V.N. Novikov, A.P. Sokolov, Phys. Lett., A153 (1991) 63.
4. V.K. Malinovsky, V.G. Zhdanov, J. Non-Cryst. Sol., 51 (1982) 31.

The Nature of Glass Transition and the Excess Low-Energy Density of Vibrational States in Glasses

V.K. Malinovsky and V.N. Novikov

Institute of Automation & Electrometry, Russian Academy of Sciences, Siberian Branch, 630090 Novosibirsk, Russia

Abstract. It is shown that the excess low-energy (2-10 meV) density of vibrational states (LEDOS) in glasses leads to a significant increase of the mean-square atomic displacements in comparison with the case of crystalline counterparts. The empirical 3/2 rule for relation between glass and melting temperatures is explained. Glass temperature is expressed through the parameters of the excess LEDOS.

1. Introduction

It is well known that the most drastic difference between vibrational spectra of glassy and crystalline modifications of the same compound is in the low-energy part of the spectrum. Here at the energy 2-10 meV there is an excess low-energy vibrational density of states (LEDOS). In its maximum it exceeds the Debye density of states (DOS) by a factor 2-8 for different materials. The excess LEDOS has universal shape in different glasses [1]. In this paper we use the fact that the transfer of a part of vibrational degrees of freedom in glasses from high frequency region of the spectrum to low-frequency one i.e. to the excess LEDOS, leads to an increase of mean-square displacements of atoms $< R^2 >$, and consider some consequences.

2. Results and Discussion

It is easy to express $< R^2(T) >$ through DOS $g(\omega)$. At high temperatures

$$< R^2(T) >= \frac{T}{m}\int \frac{g(\omega)d\omega}{\omega^2}. \tag{1}$$

From equation (1) one can see that the low-frequency part of the spectrum of DOS is reinforced by the factor ω^{-2}. This is the reason of increasing of $< R^2(T) >$ at a certain T in glasses comparing to the crystals of the same composition. We have estimated the value of $< R^2(T) >$ for glasses using the universal properties of the excess LEDOS in glasses found earlier [1,2]. We have found that mean-square atomic displacements in glasses $< R_g^2(T) >$ at $T = T_g$, where T_g is glass temperature, are equal to the value $< R_c^2(T_m) >$, which they have in crystalline counterpart at the melting temperature T_m:

$$< R_g^2(T_g) > \simeq < R_c^2(T_m) > \tag{2}$$

Springer Series in Solid-State Sciences, Volume 112
Phonon Scattering in Condensed Matter VII Eds.: M. Meissner and R.O. Pohl

(within the accuracy of the method). This result can be easily understood, if we will take into account the Lindemann criterion for melting. It is shown also, that at any temperature $T (T \geq T_{Deb})$

$$< R_g^2(T) > / < R_c^2(T) >= (1 + 0.85\beta\omega_0/\omega_D) \qquad (3)$$

where the parameter β is equal to the amplitude of the ratio of the excess LEDOS to the Debye one. ω_0 is the position of the maximum of the excess LEDOS and ω_D is the Debye frequency of the glass.

The behavior of a liquid when the temperature decreases from $T > T_m$ to $T < T_m$ is considered from this point of view. If in the disordered state this material has the excess LEDOS, at decreasing temperature than the excess LEDOS will prevent solidification, so the material will be in the form of undercooled liquid until the temperature is lowered to T_g, where $< R^2(T) >$ is equal to the value given by Lindemann criterion. This is just the case of glassforming materials. If this material has no excess LEDOS, than at $T \approx T_m$ there will be crystallization of the liquid, because the value of $< R^2(T) >$ corresponds in this case to the Lindemann criterion at temperature close to T_m. Thus, there is a correlation between low-energy anomalies of the vibrational spectra of a material and its glassforming ability. The relation between T_g and T_m is expressed through the parameters of the excess LEDOS in good agreement with the empirical rule $T_m/T_g \approx 3/2$.

References

1. V.K. Malinovsky, V.N. Novikov, P.P. Parshin, A.P. Sokolov, and M.G. Zemlyanov, Europhys. Lett. 11 (1990) 43.
2. V.K. Malinovsky, V.N. Novikov, and A.P. Sokolov, Phys. Lett. A 153 (1991) 63.

Nonlinear Dynamics of a Driven Two-Level Tunneling System

A. Würger

Institut für Theoretische Physik, Philosophenweg 19,
W-6900 Heidelberg, Fed. Rep. of Germany

Absorption spectroscopy with electromagnetic and sound waves is an efficient probe to the properties of atomic two-level systems. In many cases however, the external field amplitude is too large to be treated in linear response approximation. With the tunnelling amplitude Δ_0 and $\sigma_z = \pm 1$ labelling two localised states in a double-well potential, the Hamiltonian of a two-level system in an external field reads

$$H = \frac{\Delta_0}{2}\sigma_x + \frac{\tilde{\Delta}}{2}\sigma_z, \qquad \tilde{\Delta}(t) = \Delta + 2\eta \sin(\omega t). \tag{1}$$

Defining $\sigma_\alpha(t) \equiv \Phi_{\alpha\beta}(t,t')\sigma_\beta(t')$ for $t \geq t'$, the von-Neumann-equation can be written as

$$\dot{\Phi}_{x\beta} = -\tilde{\Delta}\Phi_{y\beta} - 2\gamma\Phi_{x\beta} \tag{2a}$$

$$\dot{\Phi}_{y\beta} = \tilde{\Delta}\Phi_{x\beta} - \Delta_0\Phi_{z\beta} - 2\gamma\Phi_{y\beta} \tag{2b}$$

$$\dot{\Phi}_{z\beta} = \Delta_0\Phi_{y\beta} - 2\gamma\Phi_{z\beta}. \tag{2c}$$

Energy dissipation has been accounted for by the damping constant γ. Integrating (2a,c) yields the integro-differential equation

$$\ddot{\Phi}_{y\beta} + 2\gamma\dot{\Phi}_{y\beta} + \epsilon^2\Phi_{y\beta} + \dot{\tilde{\Delta}}(t)\int_{t'}^{t} d\tau e^{-2\gamma(t-\tau)}\tilde{\Delta}(\tau)\Phi_{y\beta}(\tau,t') = \dot{\tilde{\Delta}}\delta_{x\beta}, \tag{3}$$

where $\epsilon = \sqrt{\Delta_0^2 + \Delta^2}$ is the two-level splitting. For the dynamics near resonance the range $\gamma, \eta, |\delta| \ll \omega$ is of primary interest, with $\delta \equiv \epsilon - \omega$. In the Laplace transform of (3),

$$\left(\lambda^2 + 2\gamma\lambda + \epsilon^2\right)\Phi_{y\beta} + \left(\lambda\tilde{\Delta}(\lambda) - \Delta\right) * \frac{\tilde{\Delta} * \Phi_{y\beta}}{\lambda + 2\gamma} = \lambda\delta_{\beta y} + \lambda\tilde{\Delta}\delta_{\beta x} - \Delta_0\delta_{\beta z}, \tag{4}$$

the convolution integral (denoted by an asterisk) shifts the frequency argument by multiples of $i\omega$, yielding $\Phi_{y\beta}(\lambda + in\omega)$ with integer n. This suggests to approximate $\Phi_{y\beta}(\lambda)$ by a sum of functions

Springer Series in Solid-State Sciences, Volume 112
Phonon Scattering in Condensed Matter VII Eds.: M. Meissner and R.O. Pohl

$$\Phi_{y\beta}(\lambda) = \sum_n \psi_n(\lambda - in\omega) \equiv \sum_n \psi_n(\phi), \tag{5}$$

where each ψ_n has four poles close to the origin; this proceeding corresponds to the lowest-order terms of an expansion in terms of η/ω. Retaining the relevant contributions $\psi_0, \psi_{\pm 1}$ only, reduces (4) to three coupled algebraic equations for these functions. Solving this system of equations permits calculation of $\Phi_{y\beta}(\lambda)$ with an accuracy η/ω for both frequencies and residues. Laplace back-transformation gives $\Phi_{y\beta}(t,t')$ and by integrating (2a,c) we obtain the matrix $\Phi_{\alpha\beta}(t,t')$. The damping constant γ gives rise to energy dissipation with a rate $\overline{\eta \cos(\omega t) < \sigma_z(t) >}$.

$\Phi_{\alpha\beta}(t,t')$ contains contributions with frequencies ω, $\omega \pm \Omega$, Ω, where $\Omega \equiv \sqrt{u^2\eta^2 + \delta^2}$ is the Rabi frequency and $u \equiv \Delta_0/\epsilon$; there are two solutions with dissipation rate and forced oscillations given by

$$(I) \qquad \alpha^I = \frac{|\delta|}{\Omega} \frac{u^2\eta^2\gamma}{\gamma^2 + \Omega^2}$$

$$< \sigma_z(t) > = \frac{u^2\eta}{\Omega} \sin(\omega t) + \frac{\Delta}{\epsilon} \frac{\delta}{\Omega}$$

$$(II) \qquad \alpha^{II} = \frac{u\eta}{\Omega} \frac{u^2\eta^2\tilde{\gamma}}{\tilde{\gamma}^2 + \Omega^2}$$

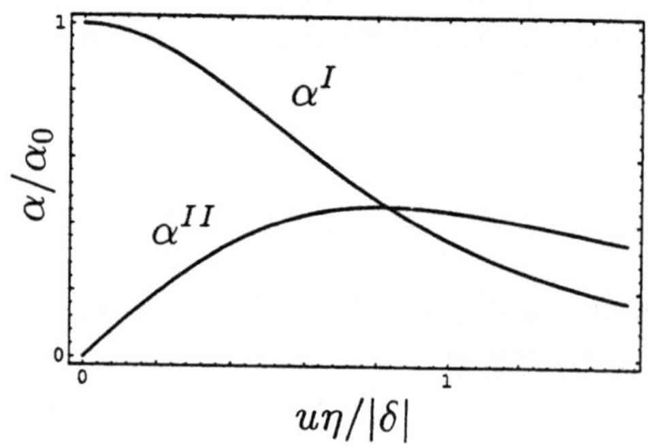

$$< \sigma_z(t) > = \frac{u^2\eta}{\Omega} \left(\cos(\omega t) \sin(\Omega t) + \frac{\delta}{\Omega} \sin(\omega t) \cos(\Omega t) \right) + \frac{\Delta}{\epsilon} \frac{u\eta}{\Omega} \cos(\Omega t), \tag{6}$$

with $\tilde{\gamma} \equiv \gamma(1 + \frac{1}{2}u^2\eta^2/\Omega^2)$. The dissipation rates cross at $u\eta \approx 0.8|\delta|$, whereas the corresponding trajectories in three-dimensional spin space touch at $u\eta = |\delta|$. This may give rise to a hysteresis phenomenon in the parameter $u\eta/|\delta|$.

The second solution is to be encountered for $u\eta > |\delta|$. The residues corresponding to the frequencies $\omega \pm \Omega$ are not equal and they depend on the frequency offset δ. Hence this solution leads to a asymmetric resonance curve as well as to slow Rabi oscillations.

Energy Transport in Disordered Anharmonic Chains

G.S. Zavt*, M. Wagner, and A. Lütze
Institut für Theoretische Physik, Universität Stuttgart,
Pfaffenwaldring 57, W-7000 Stuttgart 80, Fed. Rep. of Germany
*Permanent address: Institute of Physics, Estonian Academy of Sciences, Tartu, Estonia

1. Introduction

The contribution adresses the problem of the combined effect of disorder and nonlinearity on energy transport in oscillatory systems. This problem is relevant for theories of heat conduction in disordered systems and of relaxation processes initiated by optical or other excitation events.

We consider a 1-D-lattice of equal masses M interacting via nearest neighbour harmonic and anharmonic springs (Fermi-Pasta-Ulam system) such that the Hamiltonian in mass-reduced coordinates reads

$$H = \frac{1}{2} \sum_{m=-N}^{N} \left(P_m^2 + \frac{1}{2M} \sum_{m'=m\pm 1} [f_2(m,m')(Q_m - Q_{m'})^2 + \frac{1}{2} f_4 (Q_m - Q_{m'})^4] \right) \tag{1}$$

and displays harmonic *disorder*: $f_2(m,m') = f$ with probability $1-c$ and $f_2(m,m') = f'$ with probability c, whereas the *anharmonicity* is translationally invariant. We consider two types of *classical excitations*, respectively characterized by the initial conditions:
P-Excitation: $P_m(0) = \sqrt{2f/M}A\delta_{m,0} \quad Q_m(0) = 0$
Q-Excitation: $Q_m(0) = A\delta_{m,0}, \quad P_m(0) = 0$
The equations of motion are solved numerically by a 4th-order predictor-corrector method in a self expanding chain. The results are given either as a 3-dim spatio-temporal display of the energy $h_m(\tau)$ per atom (see fig.1), or in terms of momenta $M_\mu = \sum_m m^\mu h_m(\tau)$. E.g. the diffusion coefficient reads $D = \lim_{\tau\to\infty} D(\tau) = \lim_{\tau\to\infty} \tau^{-1}(M_2(\tau)/M_0)$, where dimensionless time units are used: $\tau = \Omega_D t, \quad \Omega_D = \sqrt{4f/M}$. Anharmonic effects are characterized by the anharmonicity parameter $\gamma_4 = (A^2 f_4/f)$.

2. Main Results

(1) A general qualitative feature of all calculations (harmonic, disordered, anharmonic) is the fact that after P-excitation energy is better carried away than after Q-excitation. This phenomenon has already been found for the ideal harmonic chain [1].
(2) In the disordered harmonic chain for P-excitation a part of the energy is trapped near the center (Anderson localization), whereas the wings carry energy away in a ***superdiffusive*** manner ($D(\tau) \sim \tau^{1/2}$). Q-excitation produces stronger localization and the wing transport is less pronounced (***quasi-diffusive***).
(3) Anharmonicity changes the amount of energy confined within the localization lenght. For P-excitation there is a decrease (***delocalization***) accompanied by the rise of ***kink-solitons*** at

Springer Series in Solid-State Sciences, Volume 112
Phonon Scattering in Condensed Matter VII Eds.: M. Meissner and R.O. Pohl

the wings (Fig.1 shows their energy display; $f'/f = 0.5$, $c = 0.2$, $\gamma_4 = 1$). By contrast for Q–excitation the localized part increases (*superlocalization*) and merges into an *immobile soliton* of Takeno–Sievers type.

(4) Disorder destabilizes the kink–solitons (P–excitation). For $\gamma_4 < 1$ their decay follows a *Wadati law* ($\sim \tau^{-1/2}$), for $\gamma_4 > 1$ the decay appears to be *much faster*.

(5)For strong anharmonicity there is always an asymptotically stochastic transport ($D(\tau) = const.$ see Fig.2) This is in agreement with the *Kolmogorov–Arnold–Moser* theorem.

(6)For weakly disordered systems the selftrapped energy portion near the center performs slow spatial oscillations around the excitation site.

[1]M.Wagner, G.Zavt, J.Vazquez-Marquez, A.Lütze, Th.Mougios, G.Viliani, W.Frizzera, O.Pilla, M.Montagna, Phil. Mag. B,1992, Vol.65, 273

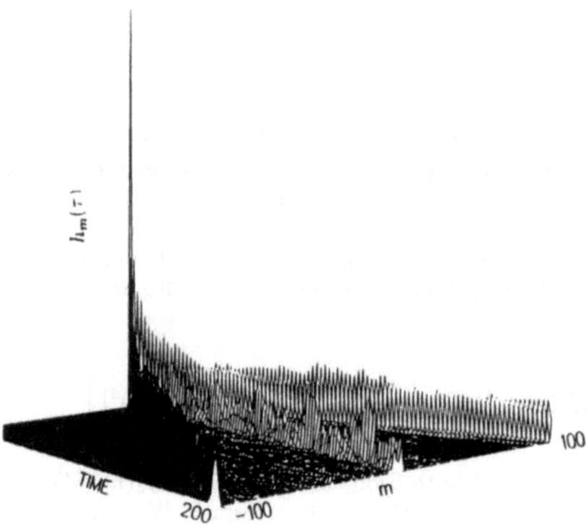

Fig.1 Coexistence of localization and propagating solitons in a disordered anharmonic chain

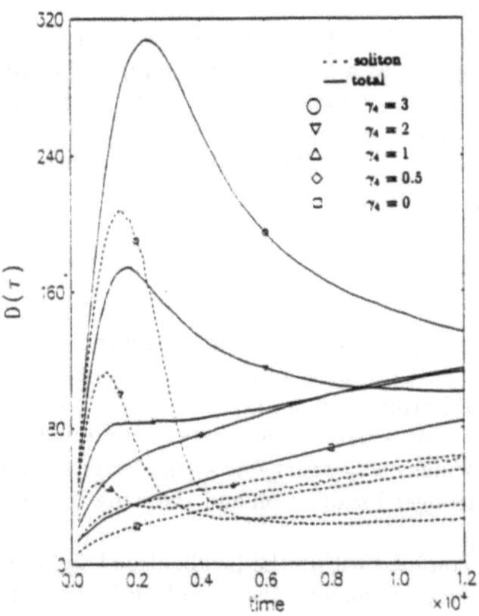

Fig.2 Energy diffusion function $D(\tau)$ for P–excitation in a disordered anharmonic chain

Anomalous Sound Propagation and Thermal Anomalies in Glasses

W. Schirmacher and M. Wagener

Physik-Department E13, Technische Universität München, James-Franck-Str., W-8046 Garching, Fed. Rep. of Germany

We discuss the vibrational properties of glasses in terms of a model which is based on the assumption of statistically fluctuating force constants. In this model a transition from nearly free sound propagation at low frequencies to strongly damped propagation (anomalous sound propagation) at higher frequencies is predicted. We discuss the observed vibrational anomalies in glasses in terms of this transition.

The low-temperature thermal properties and, correspondingly, the low-frequency vibrational properties of glasses and other amorphous solids show a universal behaviour which is completely different from that observed in crystalline materials [1]. Although by now there exists a huge number of attempts [2] for explaining this anomalous behaviour in different temperature/frequency regimes a satisfactory theoretical description is still missing [3].

Recently the present authors have developed a theory of sound propagation in glasses [4, 5, 6, 7] which is based on the mathematical analogy [8, 9] between the equations of motion for a single-particle random walk in a disordered medium and the equations for a set of coupled harmonic oscillators with fluctuating force constants. Solving these equations in effective medium approximation (EMA) a transition from nearly free sound propagation at low frequencies to strongly damped motion (anomalous sound propagation) at higher frequencies is predicted. This transition is analogous to the transition from normal to anomalous diffusion in the equivalent random walk problem [10].

This transition gives a natural explanation and description of the anomalous density of states and the observed hump in the reduced specific heat $C(T)/T^3$.

The other anomalies, namely the $C(T) \propto T$ behaviour below $1\,K$, the excess value of $C(T)/T^3$ near the hump and the behaviour of the thermal conductivity above $10\,K$ cannot be explained within this scheme because it is based on a strictly harmonic Hamiltonian. However, we think that our model might serve as a good starting point for a description of the glassy anomalies. A full description, of course, must include anharmonic interactions [11, 12]. A more thorough description and discussion of our work can be found in [13].

Springer Series in Solid-State Sciences, Volume 112
Phonon Scattering in Condensed Matter VII Eds.: M. Meissner and R.O. Pohl

References

[1] For a review see W.A. Philips (ed.), *Amorphous Solids - Low Temperature Properties*, Springer, Berlin, 1981

[2] For a recent review of model descriptions see S.R. Elliott, Europhys. Lett. **19**, 207 (1992)

[3] J.J. Freeman and A.C. Anderson, Phys. Rev. **B 34**, 5684-90, (1986)

[4] W. Schirmacher and M. Wagener, in: D. Richter, A.J. Dianoux, W. Petry and J. Teixeira (eds.), *Dynamics of Disordered Materials*, 231-4, Springer Proceedings in Physics **37**, Heidelberg, (1989)

[5] W. Schirmacher and M. Wagener, in: S. Hunklinger, W. Ludwig and G. Weiss (eds.), *PHONONS'89*, 531-3, World Scientific, Singapore, (1990)

[6] W. Schirmacher and M. Wagener, Phil. Mag. **65**, 607-13, (1992)

[7] M. Wagener, Doctoral Dissertation, TU München, (1992)

[8] S. Alexander, J. Bernasconi, W.R. Schneider, R. Orbach, Rev. Mod. Phys. **53**, 175 (1981)

[9] W. Schirmacher and M. Wagener, in: U. Rössler (ed.), *Festkörperprobleme/Advances in Solid State Physics* **31**, 39-58, Braunschweig, (1991), Vieweg

[10] These results are similar to those of the phonon-fracton model (A. Aharony, S. Alexander, O. Entin-Wohlman and R. Orbach, Phys. Rev. **B 31**, 2565-7, (1985) and G. Polatsek and O. Entin-Wohlman, Phys. Rev. **B 24**, 7726-30, (1988)) but we think that our fluctuating force constant model is better suited for the description of a glass than invoking fractal topology.

[11] E. Akkermans and R. Maynard, Phys. Rev. **B 32**, 7850-62, (1985)

[12] S. Alexander, O. Entin-Wohlman and R. Orbach, Phys. Rev. **B 34**, 2726-34, (1986)

[13] W. Schirmacher and M. Wagener, to be published, (1992)

Molecular Dynamics Simulations of Thermal Conductivity in Insulating Glasses

M.D. Kluge, J.L. Feldman, and J.Q. Broughton*

Complex Systems Theory Branch, Naval Research Laboratory,
Washington, DC 20375, USA
*NRL-NRC Research Associate

In amorphous solids anharmonicity may either decrease thermal conductivity, as in crystals, due to increased phonon-phonon scattering, or it may increase K by permitting localized vibrational modes to interact among themselves or with extended modes.

According to Green-Kubo Theory thermal conductivity is proportional to the time integral of the ensemble-averaged energy-current autocorrelation function. In the quantum case thermal conductivity may be calculated for a disordered system with harmonic interactions [1]. Classical calculations may be performed for interactions of any form. Their accuracy is limited only by the amount of phase space that can be sampled in a MD simulation.

Following Hardy [2] the energy current is defined as

$$\vec{J}=\sum_{i} E_i\vec{v}_i+\sum_{a}\sum_{i\epsilon a}\sum_{\substack{j\epsilon a\\ j\neq i}} \vec{F}_i^{a}\cdot\vec{v}_i\vec{X}_{ij}\, f_j^{a}. \tag{1}$$

Here E_i is the total energy associated with and located at atom i, and $\vec{X}_{ij}$ is the vector separating atoms i and j. The "a's" are clusters of interacting atoms. F_i^a is the force on atom i due to a cluster "a," of which atom i is an element, and f_i^a is the fraction of the potential energy of cluster a residing on atom i.

The resulting autocorrelation function converges slowly, [3] but can be improved by interpreting the cluster forces, F^a as the difference between the actual force exerted by the cluster and that the cluster would exert at T = 0 [4].

The system studied consisted of a 216 atom model of a-Si using the Stillinger-Weber two and three body potential [5]. The model [6] was created by quenching from the melt under negative pressure. It was confined to cubic boxes, and periodic boundary conditions were employed.

MD calculations were performed to evaluate the autocorrelation function of Eq. (1) and compared with the harmonic calculations of the thermal conductivity of a similar model [1]. The latter calculations gave a thermal conductivity increasing with temperature toward a limiting high-temperature value. The results of the anharmonic calculations, shown in Fig. 1, give a thermal conductivity independent of temperature in agreement with the harmonic calculations. Either

Springer Series in Solid-State Sciences, Volume 112
Phonon Scattering in Condensed Matter VII Eds.: M. Meissner and R.O. Pohl

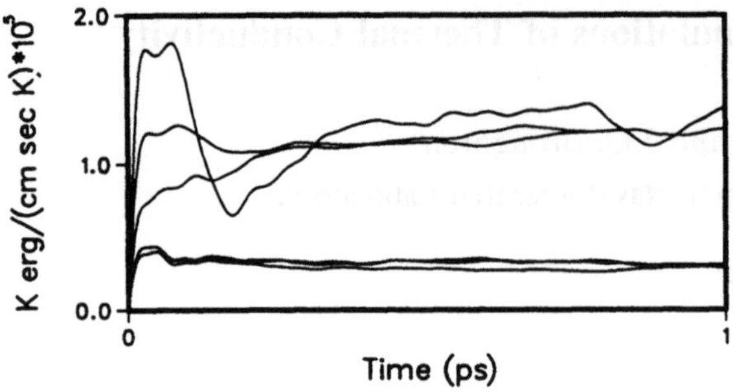

Figure 1. Thermal Conductivity versus cutoff time for a-Si (top) at T = 160 K, 320 K, and 1200 K, and a-SiGe* (bottom) at T = 320 K, 460 K, 810 K.

the SW system does not exhibit large anharmonic effects or there is cancellation between increased phonon-phonon scattering and delocalization.

Similar calculations were performed on a system with 50% Ge*, an imaginary isotopic impurity having twice the mass of Ge, distributed at random throughout the sample, but with the same SW interactions. This system had a much lower thermal conductivity in the harmonic approximation than expected from its high density [6]. This is due to the many localized modes at high frequencies. However, there was no evidence of increased thermal conductivity due to the interaction of these modes. A similar calculation on a 1000 atom model showed slightly higher thermal conductivity due to including longer-wavelength phonons in a larger system.

Acknowledgements

We thank Dr. Philip Allen for his useful discussions. Computations were performed under Grant #DMR900042P from the Pittsburgh Supercomputing Center.

References

1. Allen, P. B. and Feldman, J. L., Phys. Rev. Lett. 62, 645 (1989).
2. Hardy, R. J., Phys. Rev. 132., 168 (1963).
3. Lee, Y. H. et al, Phys. Rev. B43, 6573 (1991).
4. Kluge, M. D. submitted to Phys. Rev. B.
5. Stillinger, F. H., and Weber, T. A., Phys. Rev. B 31 , 5262 (1985).
6. Kluge, M. D., et al, Phys. Rev. B36 , 4234 (1987).

Freezing of Interacting Defects in Glasses

C.C. Yu

Physics Department, University of California, Irvine, CA 92717, USA

In glasses, the temperature range between 3 and 10 K is a crossover region characterized by a plateau in the thermal conductivity and a dramatic drop by a factor of $10^2 - 10^3$ in C/T with decreasing temperature T, where C is the specific heat [1]. In addition the ultrasonic attenuation *at a fixed frequency* decreases by roughly a factor of 30 as the temperature drops from 100 K to 1 K. Below 1 GHz this has been attributed to relaxation and structural rearrangement [1]. However, above 100 GHz, structural relaxation is too slow to contribute and another explanation is needed [2]. In this paper we explore the intriguing possibility that the crossover signals a phase transition in which degrees of freedom freeze out as the system is cooled [3].

We model the glass as a system of randomly placed defects which interact strongly with each other via the elastic strain field [4]. We start by assuming that glasses contain defects that couple linearly to the strain field. Eliminating the strain field yields [4,5]:

$$H_{eff}(\mathbf{r}-\mathbf{r}') = -\sum_{\lambda,\mathbf{k}} \frac{1}{\rho c_\lambda^2} \cos(\mathbf{k}\cdot(\mathbf{r}-\mathbf{r}'))\eta_{\alpha\beta}^{(\lambda)}\eta_{\gamma\delta}^{(\lambda)}\sigma_{\alpha\beta}(\mathbf{r})\sigma_{\gamma\delta}(\mathbf{r}') \qquad (1)$$

where $\sigma_{\alpha\beta}(\mathbf{r})$ is the stress field associated with the defects. The indices α and β range over the real space directions x, y and z. $\eta_{\alpha\beta}^{(\lambda)} = (\hat{\mathbf{k}}_\alpha \hat{\mathbf{e}}_\beta^{(\lambda)} + \hat{\mathbf{k}}_\beta \hat{\mathbf{e}}_\alpha^{(\lambda)})/2$. λ denotes the phonon polarizations and $\hat{\mathbf{e}}$ is the polarization unit vector. ρ is the density and c is the speed of sound. Summing over k results in a dipolar interaction that goes as $1/r^3$ in addition to angular factors.

Numerical simulations involve placing defects randomly on the sites of a simple cubic lattice. A typical defect concentration is 25% and lattice sizes were 4^3, 6^3 and 8^3. We mimic the internal degrees of freedom of the defect by fixing the magnitude of the stress couplings, but allowing the sign of each spatial stress component $\sigma_{\alpha\beta}$ to vary. We average over 50-100 samples; each involving 6000-200,000 Monte Carlo steps per defect, depending on the time needed for equilibration.

As we cool the system down, it undergoes spin glass-like phase transitions which can be studied using finite size scaling [6]. In particular we find that

Springer Series in Solid-State Sciences, Volume 112
Phonon Scattering in Condensed Matter VII Eds.: M. Meissner and R.O. Pohl

the diagonal and off-diagonal components of the defect stress tensor undergo separate phase transitions with T_c(off − diagonal) < T_c(diagonal) and $1/\nu = 1.0 \pm 0.3$ in both cases. Each transition is associated with a divergence of the fourth order elastic susceptibility $\chi_{\alpha\beta\alpha\beta\alpha\beta\alpha\beta}$. The elastic susceptibilities are defined as the coefficients in an expansion of the thermodynamic stress per defect in powers of a small uniform external strain field $\varepsilon_{\alpha\beta}$. It is not clear how these divergences affect the higher order elastic constants since we started with a harmonic Hamiltonian.

Interestingly, there is a tendency for planar antiferroelastic ordering of the off-diagonal stress components. For a 25% concentration of defects the magnitude of the fourier transform of the off-diagonal stress components $|\sigma_{\alpha\beta}(k)|$ has a peak at $ka = \pi$ while the diagonal components do not show any signs of ordering in the quenched ground state.

The specific heat calculated from energy fluctuations of the defects is a broad bump that is linear at low temperatures with a maximum at T_{max}. The specific heat only decreases by a factor of two from T_{max} to $T_{max}/2$ which is much less than that seen experimentally between 3 and 10 K. Thus freezing due to instantaneous $1/r^3$ interactions is too mild to account for the low temperature crossover seen in glasses.

To summarize, we have studied a model of glasses involving randomly placed defects interacting via the elastic strain field. The defects undergo two spin glass phase transitions. The off-diagonal components of the defect stress tensor exhibit antiferroelastic correlations in the quenched ground state. The drop in the specific heat is too mild to account for the crossover seen in glasses.

I would like to thank Eric Grannan and Sue Coppersmith for helpful discussions. This work was supported in part by ONR grant No. N000014-91-J-1502, the Petroleum Research Fund administered by the American Chemical Society, the Alfred P. Sloan Foundation, and an allocation of computer time from the University of California, Irvine.

References

[1] Good reviews are *Amorphous Solids*, edited by W.A. Phillips (Springer, Berlin, 1981); and S. Hunklinger and A.K. Raychaudhuri, Prog. in Low Temp. Phys. **9**, 265 (1986).
[2] T. C. Zhu, H. J. Maris, and J. Tauc, Phys. Rev. B**44**, 4281 (1991).
[3] R.O. Pohl, Am. J. Phys. **55**, 240 (1987); S.N. Coppersmith, Phys. Rev. Lett. **67**, 2315 (1991).
[4] C.C. Yu and A.J. Leggett, Comm. Cond. Mat. Phys. **14**, 231 (1988); C.C. Yu, Phys. Rev. Lett. **63**, 1160 (1989); E.R. Grannan, M. Randeria, J.P. Sethna, Phys. Rev. B**41**, 7784 and 7799 (1990); M.W. Klein, Phys. Rev. Lett. **65**, 3017 (1990).
[5] J. Joffrin and A. Levelut, J. Phys. (Paris) **36**, 811 (1976).
[6] R.N. Bhatt and A.P. Young, Phys. Rev. B**37**, 5606 (1988).

Kinetic Theory of Phonons in Glasses

P. Strehlow[1], *W. Dreyer*[2], *and M. Meissner*[3]

[1]Physikalisch-Technische Bundesanstalt, Abbestraße 2–12, W-1000 Berlin 10, Fed. Rep. of Germany
[2]Technische Universität Berlin, FB 9, Müller-Breslau-Straße 8, W-1000 Berlin 12, Fed. Rep. of Germany
[3]Hahn-Meitner-Institut, Glienicker Straße 100, W-1000 Berlin 39, Fed. Rep. of Germany

1. Introduction

Glasses below 1 K show a qualitative similarity of their thermal properties which can be explained by the widely accepted tunneling model. The most controversially discussed consequence of the tunneling model concerns the interpretation of heat-pulse experiments by introducing a time-dependent specific heat [1]. However, the concept of a time-dependent specific heat is incompatible with thermodynamic principles and reflects rather the fact that an additional variable of state has to be introduced. For that reason, an extended system of field equations is necessary. In addition, the field equations should be symmetric hyperbolic, thus solving the paradox of instantaneous propagation of thermal disturbances and describing wave-like energy transport by phonons.
In the present paper it is shown that the solution of the derived field equations for given boundary and initial values of heat-pulse experiments are in excellent agreement with the experimental data obtained in vitreous silica . Moreover, a new kind of energy propagation in glasses at low temperatures is predicted.

2. Thermodynamics

Based on the coupled Boltzmann equations for phonons and tunneling systems (TS), linearized field equations for disturbances of the phonon energy density e_{Ph} , the TS energy density e_{TS} and the phonon momentum density p_i have been derived and read

$$\begin{aligned} \partial_t e_{Ph} &+ c^2 \partial_k p_k &&= -a(e_{Ph} - b e_{TS}) \\ \partial_t p_i &+ (1/3)\partial_i e_{Ph} &&= -(a+1/\tau)p_i \qquad (1) \\ \partial_t e_{TS} & &&= a(e_{Ph} - b e_{TS}) \quad , \end{aligned}$$

Springer Series in Solid-State Sciences, Volume 112
Phonon Scattering in Condensed Matter VII Eds.: M. Meissner and R.O. Pohl

where τ is the temperature-independent phonon-boundary relaxation time. a and b are given by

$$a=(15/32)\pi^2\Upsilon zkT_R \quad , \quad b=-2\pi^2k^2T_R^{\,2}/(5z\hbar^3c^3i(T_R)) \quad ,$$

$$i(T_R)=-1.645(\Delta_m/kT_R)^{0.0000136}[1+\ln(2.09kT_R/\Delta_m)] .$$

T_R is the reference temperature and the constants z,Υ, Δ_m and c denote the TS density of state, the TS-phonon coupling , the minimal energy splitting and the Debye velocity, respectively.

The temperature disturbance follows from the phonon energy disturbance as $\tilde{T}=T-T_R=(5\hbar^3c^3/2\pi^2k^4T_R^{\,3})e_{Ph}$.

3. Heat-Pulse Experiment

A thin gold film heater (R=33.5 Ω) was evaporated on one side of a glass plate (Suprasil I, 35x35x2 mm^3). In order to measure the temperature deviation $\tilde{T}(L,t)$ a thin pencil-graphite thermometer was prepared on the opposite side at a distance of L=2 mm. The conductance of the thermal link to the bath (T_R= 50 mK) was determined to K=0.605 $ergK^{-1}s^{-1}$. The temperature deviation $\tilde{T}(L,t)$ for a heat pulse of 73 mV and Δt=9.6 μs is shown in Figure 1.

The experimental data are in excellent agreement with the theory from above including the overshoot and the non-exponential relaxation behaviour at later times. In addition, the theory predicts a new pulse at very short times reflecting a wave-like energy transport due to absorption and emission of phonons by fast relaxing TS.

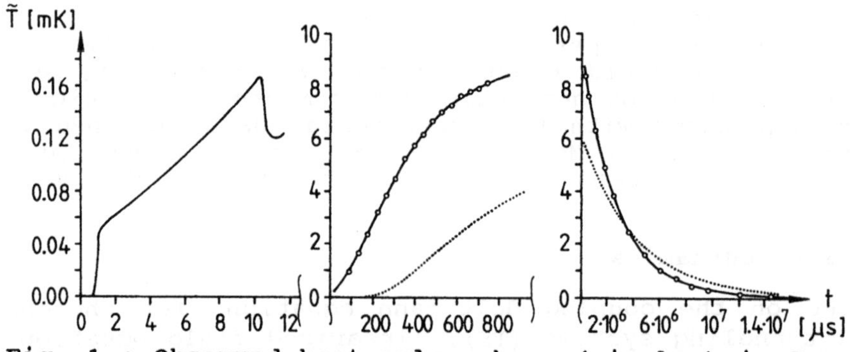

Fig. 1 : Observed heat-pulse shape (circles) in Suprasil I at T_R=50 mK. The theoretical temperature profile (solid lines) is computed from Eq.(1) with $z=1.98\cdot10^{32}$ $erg^{-1}cm^{-3}$, $\Delta_m/k=1.02\cdot10^{-3}$ K and $\Upsilon=2.39\cdot10^{-9}$ cm^3s^{-1}. The dotted line corresponds to the Fourier theory.

4. References

[1] M.T.Loponen et al.,Phys.Rev B 25,1161(1982)

Polarization Properties of Elastic Waves in Anisotropic Disordered Media

L.I. Deich and V.A. Ignatchenko

L.V. Kirensky Institute for Physics, 660036 Krasnoyarsk, Russia

The paper is devoted to polarization properties of elastic waves in anisotropic disordered matter. We consider the anisotropy in the framework of two models.

In the first model a force tensor A_{ijlk} is taken as

$$A_{ijlk}=\lambda\delta_{ij}\delta_{lk}+\mu(\delta_{ik}\delta_{jl}+\delta_{il}\delta_{jk})+\Delta a_{ijlk}(\vec{r}) \tag{1}$$

where λ and μ are constants and $\Delta a_{ijlk}(\vec{r})$ is a random tensor characterized by following properties:

$$<\Delta a_{ijlk}(\vec{r})> = 0, \quad <\Delta a_{ijlk}(\vec{r})\Delta a_{mnrs}(\vec{r}\,')>= N^{mnrs}_{ijkl}K_a(\vec{r}-\vec{r}\,')$$

where <..> means averaging over random functions and the tensor N describes an anisotropy of the correlation properties of the inhomogeneities of the force tensor. The nature of this anisotropy can be various, one particular case was considered in Ref. [1]. Function $K(\vec{r}-\vec{r}\,')$ describes a spatial fall of the correlations.

In the second model the force tensor A is supposed to describe some homogeneous anisotropic material but density of matter G is considered to be random:

$$G=G_0[1+\rho(\vec{r})], \quad <\rho>=0, \quad <\rho(\vec{r})\rho(\vec{r}\,')>=K_\rho(|\vec{r}-\vec{r}\,'|) \tag{3}$$

For both above models equations for averaged displacements have the form of:

$$G_0[\omega_p^2(k)-\omega^2]u_p(k)=G_0^{-1}\sum_t F_{p,t}(k,\omega)u_t(k), \tag{4}$$

where p and t are indexes of polarization, $\omega_p(k)$ is the dispersion law for corresponding homogeneous medium, u_p is the averaged displacement, $\vec{k}$ and ω - wave vector and frequency. The functions F describe the influence of inhomogeneities and in the first order of the perturbation theory has the form:

Springer Series in Solid-State Sciences, Volume 112
Phonon Scattering in Condensed Matter VII Eds.: M. Meissner and R.O. Pohl

$$F_{pt}=N^{mnrs}_{ijkl}e^{i}_{p}(\vec{k})e^{r}_{t}(\vec{k})k_{j}k_{s}\sum_{v}\int\frac{q_{l}q_{n}e^{k}_{v}(\vec{q})e^{m}_{v}(\vec{q})}{\omega^{2}_{v}(q)-\omega^{2}}S_{a}(|\vec{k}-\vec{q}|)d^{3}q$$

for the first model and

$$F_{pt}=\omega^{4}\sum_{v}\int\frac{[\vec{e}_{v}(\vec{q})\vec{e}_{p}(\vec{k})][\vec{e}_{t}(\vec{k})\vec{e}_{v}(\vec{q})]}{\omega^{2}_{v}(q)-\omega^{2}}S_{\rho}(|\vec{k}-\vec{q}|)d^{3}q$$

for second one. Vectors $\vec{e}$ are polarization vectors of waves in homogeneous matter and functions $S(k)$ are the Fourier-transforms of the corresponding functions K.

In the case of uniaxial anisotropy F_{yz} is only non-diagonal component of F for both above models (vector $\vec{k}$ and axis of the anisotropy are in the plane YZ and Z-axis is directed along $\vec{k}$). The directions of polarization are defined by angles φ of their deviation from the corresponding vectors $\vec{e}$:

$$\tan\varphi_{p,v} \propto \pm F_{yz}(\Omega_{p,v},\vec{k})/[\Omega^{2}_{p}(\vec{k})-\Omega^{2}_{v}(\vec{k})] \qquad (5)$$

$\Omega_{p}(\vec{k})$ in Ex. (5) is the dispersion law of wave of p-polarization, modified by inhomogeneities.

It followes from Ex. (5) that: 1). the directions of polarizations depend on wave number; 2). the different polarizations are not ortogonal to each other; 3). Imaginary part of F leads in fact to elliptic polarization of waves.

The dependence of angles φ on k has the form of:

$\tan\varphi \propto a_{1}+ b_{1}k^{2}$ at $k \ll r_{c}^{-1}$ and $\tan\varphi \propto 1/k^{2}$ at $k \gg r_{c}^{-1}$

for the first model and

$\tan\varphi \propto k^{2}$ at $k \ll r_{c}^{-1}$ and $\tan\varphi \propto a_{2} + b_{2}/k^{2}$ at $k \gg r_{c}^{-1}$

for the second one; r_{c} is the correlation radius of the inhomogeneities.

More detailed investigation of these effects should be attached to concrete experimental situations. These effects can be applied to creating new method of experimental investigations of disordered anisotropic systems (for example, such as multilayer materials) based on polarization measurements.

References

1. L. I. Deich, V. A. Ignatchenko. Fis. Tv. Tela. 1987, 29, 825.

On Far-Infrared Attenuation in Glasses

L.I. Deich

L.V. Kirensky Institute for Physics, 660036 Krasnoyarsk, Russia

It is generally accepted that some quasilocal vibrations (QV) coexist with phonons in glasses at *THz* frequencies and it is the excitations that are responsible for the most of unusual properties of these materials. In this paper we study the influence of QV on the far infra-red attenuation $A(\omega)$ of glasses . We assume that QV do not interact with light directly. This assumption has been stimulated by the results of inelastic neutron scattering experiments [1], which pointed to the fact that the modes related to QV in SiO_2 have a symmetry which do not allow them to be active in IR attenuation [2]. The main purpose of the paper is to show that QV can give rise to a bump on the curve $A(\omega)/\omega^2$ similar to one observed in Ref. [2] even if the state density of QV cannot make any contribution to attenuation.

We take into account interaction between QV and phonons, which does allow QV to influence attenuation. This influence appears to depend sufficiently on the correlation properties of the inhomogeneities of a photon-phonon coupling parameter, described by correlation function or its Fourier transform $S(k)$.

Considering QV in the terms of a phenomenological approach, based on the ideas of Ref. [3], we write down the expression for the attenuation in the form of

$$\frac{A(\varepsilon)}{\varepsilon} = Im \frac{1}{\omega} \int \frac{c(k)}{\varepsilon - s^2k^2 - \Sigma(\varepsilon)} d^3k \,, \qquad (1)$$

where $\varepsilon=\omega^2$, s - phonon velocity and Σ is mass operator , caused by QV. It is important to emphasize that Σ depends on frequency, but not on wave number. We neglect the real part of Σ and write down its imaginary part in the form of $i\varepsilon\xi(\varepsilon)$. We suppose ξ to be an increasing function of ε. In Ref. [3] ξ was obtained to be

Springer Series in Solid-State Sciences, Volume 112
Phonon Scattering in Condensed Matter VII Eds.: M. Meissner and R.O. Pohl

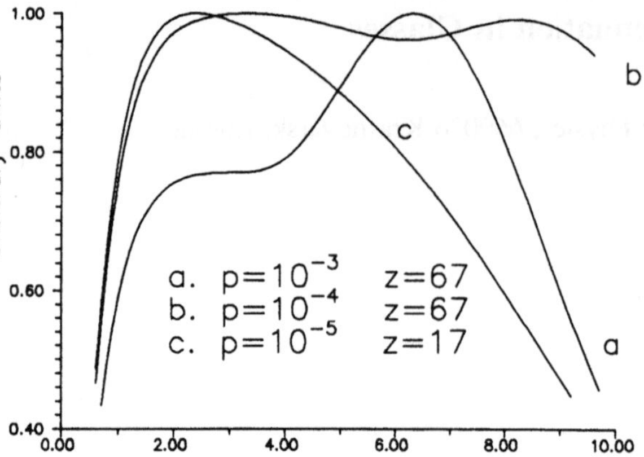

Fig. 1. Attenuation curves for $c(k)$ (2) and $\xi = p\omega^5$

$\propto p\varepsilon^{5/2}$ and we will use this result to make some estimations, though qualitative conclusions do not depend on the concrete form of this function.

Photon-phonon coupling parameter $c(k)$ is taken as

$$c(k) = (1 - z^{-3})^{-1} [z/(k^2 + z^2)^2 - z^{-3}/(k^2+1)^2] \qquad (2)$$

This function is characterized by two correlation radii r_1 and r_2, ($z = r_1/r_2 \geq 1$) and can be regarded as a generalization of Shlemann's function introduced in [4]. Considering $r_2 \neq 0$ we allow the influence of short order.

In fig. 1 we display the plots of expression (1) obtained with $c(k)$ (2) and $\xi = p\omega^5$, where abscissa is the normalized frequency $\omega r_1/s$. The parameter p is proportional to concentration of QV. At $z \leq 20$ (curve c) we have one maximum in the region $\xi \ll 1$, which became sharper with increasing of p. At $z \geq 20$ two maxima appear (curve a) with the first one being in the same region; it becames flatter with p increasing (curve b). This maximum disappears with z approaching infinity, hence taking into account both medium and short order is important. Using the data from [2] and our results we estimated as an example values r_1, r_2 and $\xi(\omega_m)$ for curve a. We found that $r_1 \cong 100\mathring{A}$, $r_2 \cong 1.4\mathring{A}$ and $\xi(\omega_m) \cong 0.025$, which are not at considerable variance with the facts.

REFERENCES

1. U. Buchenau, M. Prager et al. Phys. Rev. 1986, B34, 5665.
2. L. Ghivelder, W. A. Phillips. J. Non-Cryst. Sol. 1989, 109, 280.
3. Yu. M. Galperin, V. G. Karpov, V. N. Solov'ev. JETF. 1988, 94, 373
4. E. Shlemann. Phys. Rev., 1964, 135, A413;

Contribution of Small Clusters of Tunneling Centers to Some Static Properties of Orientational Glasses

T. Kranjc

**Department of Physics, University of Ljubljana,
Jadranska 19, 61000 Ljubljana, Slovenia**

The low temperature properties of glasses, amorphous solids, and of certain impurities in alkali halides (orientational glasses) all arise, according to the tunneling center model, from tunneling states [1,2]. It is therefore important to have some knowledge about the low-lying energy excitations from tunneling units. The tunneling center model, although capable to account for the main anomalous low-temperature properties of these materials, does not give information about the microscopic structure of the tunneling entities. The question about the nature of the tunneling states, however, does not arise in orientational glasses - crystalline solids containing certain impurities capable of tunneling motion - where the microscopic structure of the impurities is well-known [3]. In order to account for the experimentally observed results on the specific heat, dielectric constant and thermal conductivity, a constant density of states of the low-lying energy excitations is needed. In orientational glasses, it seems that the broad distribution in energy spectrum results from a distribution of interactions between tunneling units. It is possible to deduce the relationship between the probability density of the coupling parameters between tunneling units and the probability density of the excitation energies, wherefrom the main static properties of the material may be obtained [4,5].

We concentrate, therefore, on the Hamiltonian of the form

$$H = -\sum_{i\alpha\beta} \Gamma^i_{\alpha\beta}\, s^i_{\alpha\beta} - \sum_{i\alpha} \epsilon^i_\alpha\, p^i_\alpha - \sum_{i<j}\sum_{\alpha\beta} J^{ij}_{\alpha\beta}\, p^i_\alpha\, p^j_\beta \quad , \tag{1}$$

where $p^i_\alpha = |\alpha^i\rangle\langle\alpha^i|$ are projection operators on localized states $|\alpha^i\rangle$ of the i-th center, ϵ^i_α the energy of the i-th center in the state $|\alpha^i\rangle$, $s^i_{\alpha\beta} = |\beta^i\rangle\langle\alpha^i|$ are off-diagonal operators describing the tunneling of the i-th center between the states $|\alpha^i\rangle$ and $|\beta^i\rangle$, $\Gamma^i_{\alpha\beta}$ being the corresponding tunneling matrix elements, and $J^{ij}_{\alpha\beta}$ are interaction energies of the i-th center in the state $|\alpha^i\rangle$ with the j-th center in the state $|\beta^j\rangle$. It is clear that an exact diagonalization of (1) is possible only for some simplified models. To this end, we take into account only tunneling between the nearest-neighbor localized states and the asymmetry of the wells to be zero.

Springer Series in Solid-State Sciences, Volume 112
Phonon Scattering in Condensed Matter VII Eds.: M. Meissner and R.O. Pohl

We have solved the Hamiltonian (1) for some special cases: for symmetrically arranged two, three and four two-level, two three-level, two four-level and two six-level centers. We only present here briefly the case of a cluster of four two-level centers, lying in the corners of a tetrahedron without entering, however, into the question of formation and observability of such clusters [6]. The details of the calculation and other results will be published elsewhere.

We shall write the Hamiltonian (1) for this case as

$$H = -\Gamma \sum_i \sigma_x^i - J \sum_{i<j} \sigma_z^i \sigma_z^j \ , \tag{2}$$

σ_μ^i ($\mu = x, y, z$) being the Pauli spin matrices, which can be represented as a 16x16 matrix. The symmetry group of the Hamiltonian is D_{2h}. Introducing, as a new basis, symmetry-adapted combinations of localized states according to irreducible representations (IR) of the group D_{2h}, splits the Hamiltonian (2) to a block-diagonal form with six 1-D blocks, three 2-D blocks and one 4-D block, which can also be further diagonalized analytically.

The energy eigenvalues classified according to IRs are: IR A_{1g}: $-6J - 2\Gamma^2/3J$, $\pm 2\Gamma - 7\Gamma^2/3J$, $2J + 4\Gamma^2/J$, 0; IR B_{1g}: $2J$; IRs B_{2g} and B_{3g}: 0, IR B_{1u}: $-6J$, 0, $J \pm \sqrt{J^2 + 2\Gamma^2}$; IRs B_{2u} and B_{3u}: $J \pm \sqrt{J^2 + 4\Gamma^2}$. For the A_{1g} symmetry, we have only displayed the energies for $\Gamma/J << 1$.

The energy spectrum of the above system shows an energy gap of 4Γ between the ground state and the first excited state for $J = 0$. As the interaction J increases, excitation energies decrease as $\sim \Gamma^2/J$. It is the strongly interacting clusters that determine predominantly the low-temperature behavior of the tunneling states. The energy spectrum found above enables one to obtain both the partition function of the system as well as energy eigenstates and transition matrix elements between them. Introducing into the Hamiltonian (1) perturbation H' arising from the strain fields ϵ_p due to phonons yields the following matrix element connecting the ground state with the first excited state $M_{12} = -4\gamma/\sqrt{1 + (\Gamma/3J)^2}$, γ being $\partial\epsilon/\partial\epsilon_p$ and we have set $\epsilon_\alpha^i = \epsilon$ for all i, α. These data are needed to determine the specific heat and thermal conductivity due to the tunneling units.

Acknowledgment. The author would like to express gratitude for financial support by SMELT-Ljubljana and Gospodarska zbornica Slovenije.

References

1. *Amorphous Solids, Low Temperature Properties*, ed. W.A. Phillips, Springer, Berlin (1985)
2. *Phase Transitions* **5**, (1985)
3. V. Narayanamurti and R.O. Pohl, Rev. Mod. Phys. **42**, 201 (1970)
4. M.W. Klein, Phys. Rev. **B 29**, 5825 (1984)
5. M.W. Klein, Phys. Rev. **B 31**, 1114 (1985)
6. T. Kranjc, J. Phys. A: Math. Gen. **25**, 3065 (1992)

A Defect Model for Amorphous and Liquid States

A.V. Granato

Department of Physics and Materials Research Laboratory, University of Illinois at Urbana-Champaign, 110 W. Green St., Urbana, IL 61801, USA

The known microscopic configuration and dynamics of interstitials in face-centered cubic metals have been used [1] to construct a model according to which fcc-liquid and amorphous metals are crystals containing a few percent of interstitials. The phenomenological aspects of the model do not depend on the microscopic configuration. The thermodynamic properties will therefore be similar, or universal, for any defect which has similar properties. The essential features are (1) a strong softening of the shear modulus G with defect concentration c, (2) degeneracy in orientation, and (3) low frequency resonance modes ω_R. These properties are known experimentally to be universal for amorphous materials [2,3,4].

We discuss as an example the change in specific heat δC_V to be expected on the basis of the defect model as shown in Fig. 1. We suppose an amorphous or glassy material has been prepared by cooling rapidly from the liquid state. The dashed curve is for a specimen with lower defect concentration as a result of annealing at a temperature T_A just below the glass temperature T_G. In the inset is shown the usual TLS diagram for an isolated defect ($c \rightarrow 0$) and for a finite defect concentration ($c \neq 0$). The formation energy is given by $\alpha G \Omega$, where α is a constant and Ω is the volume per molecule, and is lower and asymmetric for interacting defects. The ground state vibrational level $\hbar\omega_R/2$ is tunnel split with spacing Δ. δC_V for different temperature ranges has many notable features.

(a) For $T < \Delta_o/k$, $\delta C_V \rightarrow 0$ exponentially. This has not yet been observed in measurements $T \geq 10^{-2}$ K, implying that the defect need not be light.

(b) For $T > \Delta_o/k$, δC_V becomes quasi-linear. This is perhaps the best known signature of TLS, and is accounted for by assuming defect interactions [3].

(c) Near the end of the linear region, a depression should be expected when the defect concentration is reduced. This has not yet been observed,

(d) For $T \sim \hbar\omega_R/3k$, a "bump" [4] is obtained which arises from excitations from the ground state to the higher levels of spacing $\hbar\omega_R$. For the interstitalcy model, $\omega_R \sim k\theta/7\hbar$. Although Δ is very sensitive to defect interactions which spread out the specific heat over a wide temperature range (b), the bump should be relatively insensitive to a distribution of barriers and strains and be well approximated by an Einstein model.

(e) For $T <\sim \theta/5$, the T^3 component of the specific heat is enhanced by a lowering of the Debye θ resulting from a reduction of the shear modulus.

(f) For $T \sim \theta$, δC_V should be slightly negative since not all of the defect local modes ω_L will be excited.

Springer Series in Solid-State Sciences, Volume 112
Phonon Scattering in Condensed Matter VII Eds.: M. Meissner and R.O. Pohl

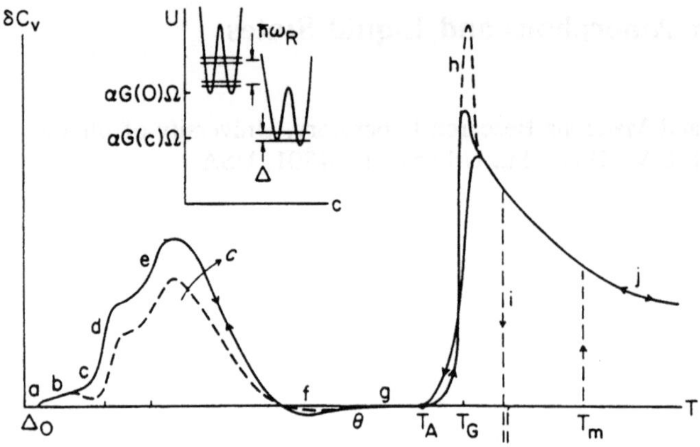

Fig. 1. Schematic sketch of $\delta C_V(T)$ expected from the defect model for a specimen which has been cooled rapidly from the liquid state (not to scale). The dashed curve is for a lower defect concentration obtained by annealing at a temperature T_A just below the glass temperature T_G. The inset indicates the effects of defect interactions at finite defect concentrations.

(g) For $\theta < T < T_A$ $\delta C_V = 0$ since the defect model has the same number of vibrational modes as the perfect crystal, all modes are excited, and the specific heat is the classical Dulong-Petite value 3Nk.

(h) On increasing the temperature through the glass temperature T_G, to the supercooled liquid state, δC_V must jump up to provide the formation energy of the increasing number of equilibrium defects needed with increasing temperature. The temperature dependence in the liquid state is discussed in (j). The peak in the curve near T_G is not a true specific heat, but depends on the time the specimen spent at the annealing temperature T_A. If this is long, then defects have annealed out, and extra energy is required to achieve the equilibrium value. The dashed curve corresponds to a longer anneal at T_A.

(i) Depending on the viscosity and speed of measurement, it may happen that the material crystallizes, releasing energy as indicated by the dashed line.

(j) In the liquid state, the shape of the curve is determined by the defect formation energy, U_F, which is linear in the shear constant, which is not zero at T_m, but is greatly reduced from its crystalline value [5]. As c increases with temperature, G(c) and U_F(c) decrease, so that δC_V decreases.

Some of these universal features are well known and characteristic (b, d, e), while some have not yet been seen (a, c, f). The model is well represented by the standard TLS model, augmented by the softening of the defect formation energy indicated in the inset of Fig. 1.

This work was supported by Dept. of Energy grant no.DEFG02-91ER45439.

References

1. A.V. Granato, Phys. Rev. Lett. **68**, 974 (1992).
2. B. Golding, B.G. Bagley and F.S.L. Hsu, Phys. Rev. Lett. **29**, 68 (1972).
3. M.V. Klein, B. Fischer, A.C. Anderson and P.J. Anthony, Phys. Rev. B**18**, 5887 (1978).
4. C.C. Yu and J.J. Freeman, Phys. Rev. B**36**, 7620 (1987-I).
5. A.V. Granato, Liquid and Amorphous Metals-8, Vienna (1992).

Learning About Phonons with Frequencies Below One kHz

J.D. Maynard

Department of Physics, 104 Davey Laboratory,
The Pennsylvania State University, University Park, PA 16802, USA

Two areas of interest in phonon research are effects in highly disordered solids and nonlinear effects. When phonons propagate in a disordered potential field, they may become Anderson localized, in a fashion similar to electron Schrodinger waves in disordered metals. If there are nonlinear interactions, the question arises as to whether or not the nonlinearity will weaken the Anderson localization. Recently we have been studying such questions using macroscopic acoustic analogs of linear and nonlinear wave propagation in disordered, as well as ordered systems. These systems, designed to precisely simulate certain problems in phonon and electron physics, are operated at frequencies below 1 KHz. We have also used such a system to study the vibrational properties of a 2-D quasicrystal.

The one-dimensional acoustic system used in our Anderson localization experiments is illustrated on the left in in Fig. 1. The wave medium is a long (15 m) 0.178 mm diameter steel wire suspended vertically; tension in the wire is maintained with a weight attached at the lower end. The wave field consists of transverse waves in the wire generated with an electromechanical actuator at one end of the wire. The periodic or nearly periodic potential field V for the wire is provided by small lead masses (with mass m = 249$\pm$ 33 mg) spaced along the wire with an average lattice constant a = 15 cm; a total of 50 masses is used. The masses are sufficiently small so that the potential V may be approximated as a series of delta functions; the system is a precise analog of a Kronig-Penny potential field.

Running parallel to the wire is an aluminum beam which acts as a guide for a trolley carrying a "C"-shaped permanent magnet. The magnet is aligned so that the wire passes between its poles, with the lines of force perpendicular to the wire. When the transverse wave field is present along the wire, the magnetic field induces a current in the wire which is proportional to the velocity of the wire at the position of the magnet. A motor, gear, and chain system is used to translate the trolley and magnet assembly along the aluminum beam. By monitoring the position of the magnet and the current induced in the wire, the entire wave field along the wire, including amplitude and phase, may be recorded.

Springer Series in Solid-State Sciences, Volume 112
Phonon Scattering in Condensed Matter VII Eds.: M. Meissner and R.O. Pohl

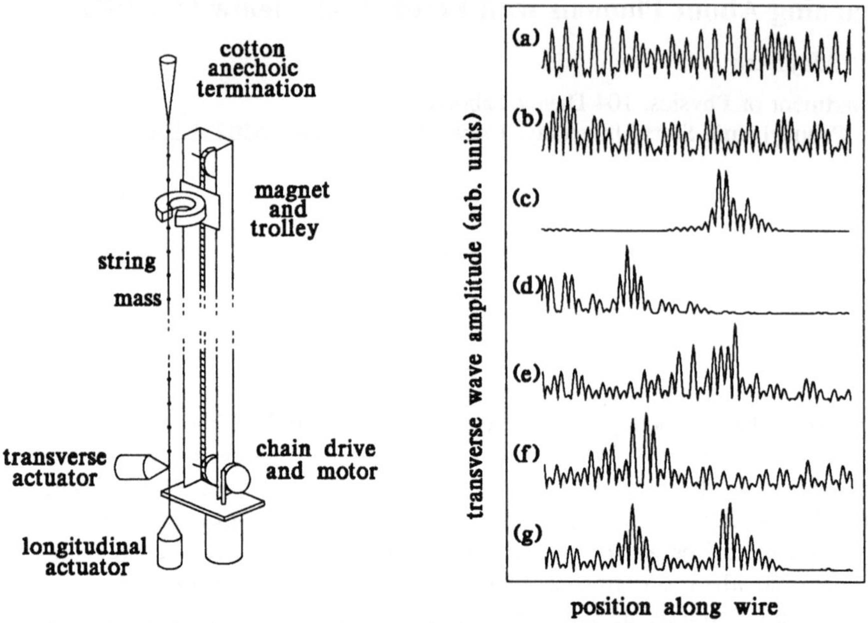

Figure 1. One-dimensional acoustic Anderson localization experiment and the eigenstate amplitude as a function of position along the wire. a) - b) Bloch wave states. c) - f) Eigenstates of the 2% disordered system. g) Phonon assisted hopping.

Before studying the localization effects, we first verified Bloch wave behavior by making measurements with the masses spaced periodically. The frequency response (band structure) of the system was measured by monitoring the transverse wave amplitude near one end of the series of masses while sweeping the frequency of the transverse actuator at the opposite end. By driving the periodic system at one of the eigenfrequencies and translating the magnet, a Bloch wave eigenstate is recorded; two examples are presented in Fig. 1a and b, which show the eigenstate amplitude as a function of position along the wire. These eigenstates are clearly extended, and are in good agreement with theoretical Bloch eigenstates.

Some results of the Anderson localization measurements made with 2% disorder (i.e. with the masses displaced from the periodic lattice sites with a flat random distribution between $\pm 0.02a$) are shown in Fig. 1c through f. The most localized state, Fig. 1c, appeared in the region of the gap of the periodic system.

We have used the wire-mass system to study the effects of a time-dependent potential of the form $V\left[r\left(1+\epsilon\cos\Omega t\right)\right]$ on the Anderson localization. The time dependence is provided by a second electro-mechanical actuator, as shown in Fig. 1, which modulates a strain ϵ in the wire at a low frequency Ω. In a solid state system, this would be analogous to the problem of inelastic electron-phonon scattering and resonant phonon assisted hopping between localized states. The time-average of the mixture for a particular longitudinal strain amplitude is shown in Fig. 1g. If the actuators were stopped, energy would be transferred to the final state with a hopping probability proportional to the ratio of the amplitudes of the two localized states.

While the study of waves in disordered media, including effects such as Anderson localization, and the study of nonlinear systems, including effects such as soliton propagation, have each received considerable attention, the combined field, nonlinear wave propagation in disordered media, is relatively new. A fundamental question is: Does nonlinearity weaken Anderson localization?

The type of nonlinearity of our mass-loaded wire system is particularly suitable for studying effects on localization. A finite amplitude at one localization site changes the instantaneous arc length of the wire, which changes the tension in the wire. The change in tension may then parametrically excite a state which is localized at a different site, resulting in nonlinearly induced hopping.

Our measurements consisted of examining the spectral response (amplitude at a fixed site as a function of frequency), measured at a distance of about four localization lengths from the drive actuator and normalized by dividing by the drive amplitude, for different drive amplitudes. If the system were strictly linear, then the normalized response would not change. If the Anderson localization is weakened by the nonlinearity, then as the drive amplitude is increased, the normalized response at the distant site should increase. We found that the normalized response actually decreased by an average of 30% , indicating an enhanced localization. One might imagine that the nonlinearity causes an Anderson localized state not to parametrically excite a distant site, but rather to "dig a deeper hole" for itself.

One of the original goals of the research with the acoustic simulations was to study two-dimensional Anderson localization. This research goal was diverted with the discovery of quasicrystals. Until a few years ago it was believed that solid matter could exist in two basic forms: crystalline and amorphous (glassy). Crystals are formed by taking a unit cell and periodically repeating it to fill three dimensional space. In order to fill all space, only unit cells of certain shapes are allowed, and consequently only certain rotational symmetries are allowed. In particular, five-fold rotational symmetry is forbidden. However, in 1984 Shechtman at the National Bureau of Standards (now NIST) made an aluminum alloy whose x-ray diffraction pattern indicated five-fold rotational symmetry. While it is possible to have atoms in clusters with five-fold (icosahedral) order, the diffraction patterns had sharp spots, indicating long range order. At first the existence of such materials seemed impossible, but theorists pointed out that while it is impossible to have five-fold rotational symmetry and long range periodic order, it is possible to have five-fold rotational symmetry and another type of long range order which is quasiperiodic.

In order to generate a quasiperiodic pattern, one begins with a periodic lattice in a higher dimension, and traverses this lattice with a lower dimensional surface. A window width is defined, and lattice points falling within that window are projected onto the lower dimensional surface. If the direction cosines describing the orientation of the lower dimensional surface are irrational, then the sites projected onto the surface will form a quasiperiodic pattern. If one begins with a five-dimensional periodic lattice and intersects it with a two-dimensional (plane) surface, then one may obtain a two-dimensional quasiperiodic pattern, or Penrose tiling.

Quasiperiodic patterns are a new symmetry of Nature, and it will be important to understand how this new symmetry effects the properties of materials. A fundamental question is: given a wave equation (for electron Schrodinger waves or phonons) with a potential field having quasiperiodic symmetry, how is this symmetry manifest in the eigenvalues and eigenfunctions. While there are rigorous theorems for quasiperiodic patterns in one dimension, theorems for two and three dimensions, if they exist, have not yet been discovered.

A convenient model for studying the effects of symmetry is the tight-binding model. In this model lattice sites are occupied by local oscillators which, if isolated, would have one or more sharp eigenfrequencies (such as the quantized energy levels of an isolated atom). These local oscillators are allowed to interact through some coupling mechanism to nearest-neighbor lattice sites; the local oscillator eigenfrequencies then broaden out into bands of eigenfrequencies. For the ideal case, the system should have little or no damping, i.e. the local oscillators and the coupling mechanism should have a high quality factor (Q). For the local oscillators in the acoustic simulation ordinary commercial tuning forks (440 Hz) were used. These have the advantage that they can be mounted by the stem and still maintain a high Q oscillation. The tuning forks are epoxied into a heavy aluminum plate in the Penrose pattern at the centers of the rhombuses as shown in Fig. 2. For the nearest neighbor coupling, arcs of 1 mm diameter steel wire are spot-welded from one tine of a tuning fork to that of a nearest neighbor. Using the four sides of each rhombus, four nearest neighbors are identified, and each tine of

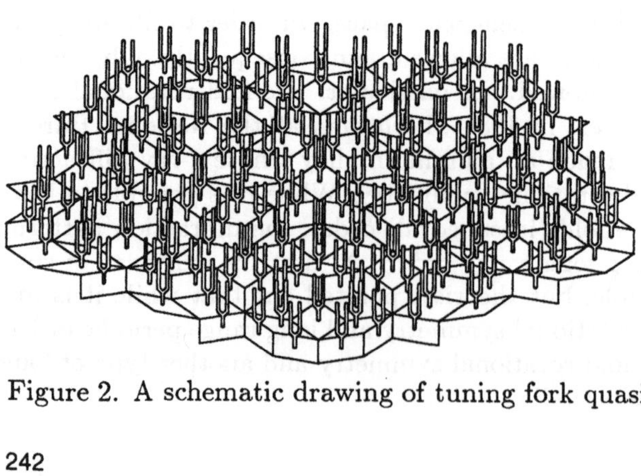

Figure 2. A schematic drawing of tuning fork quasicrystal.

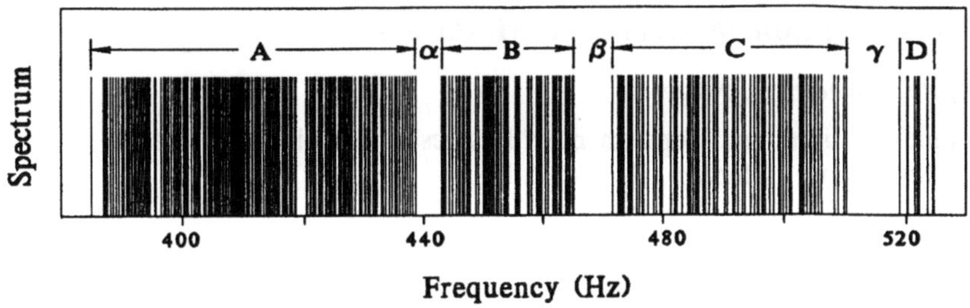

Figure 3. The eigenvalue spectrum of the tuning fork quasicrystal, showing gaps and bands whose widths are in the ratio of the Golden Mean, $(\sqrt{5}+1)/2$.

a tuning fork is coupled to the two nearest tines of the adjacent tuning forks. With this coupling pattern each local oscillator has four nearest neighbors, but the nearest neighbor length varies in a quasiperiodic pattern.

The eigenvalue spectrum, measured in a manner similar to the mass-loaded wire experiment, is presented in Fig. 3. This spectrum shows gaps and bands whose widths are in the ratio of the Golden Mean, $(\sqrt{5}+1)/2$. This result has not been found with theory in two or three dimensions.

For references, see S. He and J. D. Maynard, Phys. Rev. Lett. **62**, 1888 (1990). This work was supported in part by NSF grant DMR 9000549 and the Office of Naval Research.

Sound Propagation in Granular Materials

Chu-heng Liu and S.R. Nagel

James Franck Institute and Department of Physics, University of Chicago, 5640 S. Ellis Avenue, Chicago, IL 60637, USA

What happens when sound propagates in an unconsolidated granular medium? Not only are such "sandpiles" disordered, they also have a very nonlinear response because the contacts between grains, through which the sound travels, is very sensitive to the pressure and hence the amplitude of the sound vibration. The effect of these fragile contacts produces a number of striking effects which are not manifested by the sound in more commonly studied disordered solids such as glasses.

We expect there to be different behaviors depending on the frequency of the vibration, ν. As in other disordered solids at very low frequency, where the wavelength of the sound λ is much larger than the spacing between defects, we expect to be able to treat the medium as homogeneous. At much higher frequency, when λ becomes comparable to the grain diameter, d, the vibrations will be much more sensitive to the inhomogeneities and the geometry of the contacts. Both regimes have distinctive properties.

If we treat the granular material as homogeneous, we can use the fact that the sound velocity depends sensitively on pressure to show that sound will not propagate in the horizontal direction [1]. Clearly at zero pressure, the velocity, c, is zero since the particles do not touch and thus feel no restoring force. As the pressure is increased, the particles are pushed together, distorting their shape and increasing the restoring force as shown in Figure1. Using Hertzian contact theory, one can readily calculate that the velocity depends on the pressure [2,3] as $c \propto P^{1/6}$. (The experimental data of Duffy and Mindlin [3] indicates that the exponent is somewhat less than 1/6. Goddard [4]

Springer Series in Solid-State Sciences, Volume 112
Phonon Scattering in Condensed Matter VII   Eds.: M. Meissner and R.O. Pohl

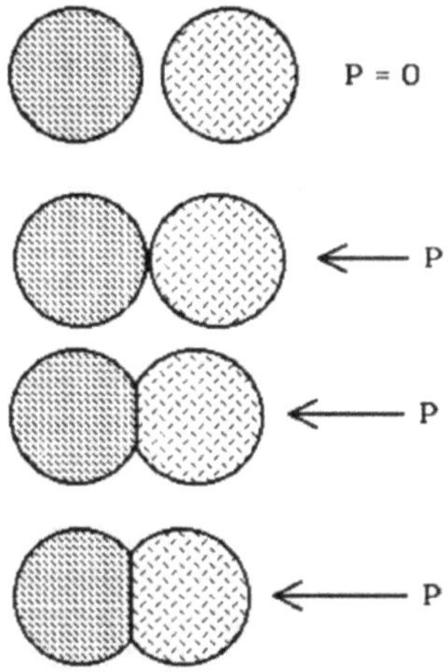

Figure 1 The distortion of two spheres under an external pressure, P. In the top view there is no pressure and the two spheres do not touch each other. As the pressure increases, the distortion becomes larger and the velocity of sound increases due to the increased forces between the grains.

has discussed why this could occur. Our argument does not depend on the exact value of the exponent of h as long as it is positive.) If we ignore the effects of walls and concern ourselves only with a granular medium far from any constraining boundaries, then the pressure within the pile increases linearly with the depth, h, below the top surface so that we have a velocity that depends on the depth, $c \propto h^{1/6}$. If a wave starts to propagate horizontally, it quickly gains a component along the vertical direction and soon reaches the top surface. Thus at low frequencies we expect that horizontal sound propagation is not allowed.

At high frequencies, where the wavelength is comparable to the grain diameter, the sound wave will be very efficiently scattered by the individual grains. Here the precise geometry of packing of the particles will determine the exact path of a wave as it propagates between the source and detector. Such intense scattering can lead to strong interference effects. If even a single particle is displaced, the path can be disrupted so that the amplitude and phase of the detected wave will change by a large amount.

In a series of measurements of sound propagation in a box filled with glass beads [1], we have found a variety of unusual effects. We used 0.5 cm diameter spherical glass

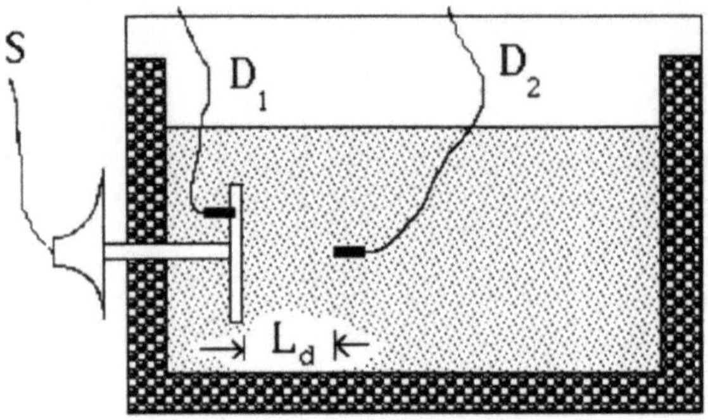

Figure 2. A schematic diagram of the side view of the experimental configuration. D_1 and D_2 correspond respectively to the two detectors used to monitor the source acceleration and the detected signal. S represents the source speaker which was run at a single frequency ν. The walls are padded with 3 cm thick sheets of Styrofoam in order to decrease any effects of sound reflecting from the wall. The entire apparatus was enclosed in additional boxes in order to isolate the system from external vibrations and temperature fluctuations.

beads contained in a square box of side 28 cm with a depth of 8 cm to 15 cm. The configuration is shown in Figure 2. We reduced the reflection of the sound waves from the walls by lining the box with Styrofoam sheets and isolated the box from external vibrations and temperature fluctuations. We connected a 7 cm diameter aluminium disc inside the box to an external speaker in order to transmit the vibrations with a well defined direction of motion. The acceleration of the disc was monitored with an accelerometer attached to its back and controlled with an electronic feedback circuit. The detection accelerometers (diameter 0.7 cm and length 1.2 cm) were comparable to the size of a single grain and placed at distances from 2 cm to 10 cm away from the center of the source. Because they are massive, the accelerometers respond only to vibrations transmitted through their contacts with the solid surroundings and are quite insensitive to the sound pressure in air. We verified that this was the case by performing some of the experiments in a helium atmosphere as well as in air. A measurement of the sound velocity in vacuum also gave the same value as was found in air.

We have found that the vibrations caused by the sound itself was enough to disturb the placement of the particles. Even with a monochromatic source, run at constant frequency ν and amplitude A_s, an enormous amount of noise was observed in the amplitude of the detector, $A_d(t)$. This noise itself had interesting time correlations. A measurement of its power spectrum, S(f), indicated power-law behavior: $S(f) \propto f^{-\alpha}$ with $\alpha \approx 2$. This power law persisted over 5 decades in frequency f. Likewise, if we lowered the amplitude of the vibration and varied the source frequency ν we found a response that fluctuated in a very erratic way. However this response was not noisy in the usual sense: if we repeated the measurement immediately we reproduced all the details of the structure. However if we disturbed the medium and then repeated the measurement we found a very different structure. This indicates that the frequency response is sensitive to the exact placement of all the beads in the container. Similar sensitivity to the microscopic placement of scatterers is common in a variety of interference phenomena such as the "universal conductance fluctuations" seen at low temperatures in mesoscopic metal wires [5].

We have also observed an extreme sensitivity of the sound propagation to temperature fluctuations. A temperature fluctuation of less than a tenth of one degree Kelvin can lead to a hundred percent change in the detected response. Since it is presumably due to the thermal expansion of the beads within the pile, this effect also emphasizes how sensitive the signal is to the exact placement of each of the individual grains of material.

This behavior of the sound propagation in a granular medium indicates only one of the ways in which these materials have novel properties differing from what we find in other systems. Many other properties of sandpiles have equally intriguing properties [6].

We would like to thank H. Jaeger, D. Johnson, J. Knight, D. Sornette, C. Tang and T. Witten for stimulating discussions on various aspects of sound propagation in granular media. This work was supported by NSF Grants DMR-MRL 88-19860 and DMR 91-11733.

References

[1] C.-h. Liu and S. R. Nagel, Phys. Rev. Lett. **68**, 2301 (1992).
[2] L. D. Landau and E. M. Lifshitz, *Theory of Elasticity* (Pergamon , N. Y., 1986) Ch. 1.

[3] J. Duffy and R. D. Mindlin, J. Appl. Mech. (ASME) **24**, 585 (1957).
[4] J. D. Goddard, Proc. R. Soc. Lond. A **430**, 105 (1990).
[5] S. Washburn and R. Webb, Adv. Phys. **35**, 375 (1986); P. A. Lee and A. D. Stone, Phys. Rev. Lett. **55**, 1622 (1985); B. L. Al'tshuler, Pis'ma Zh. Eksp. Teor. Fiz. **41**, 530 (1985).
[6] See for example, H. M. Jaeger and S. R. Nagel, Science **255**, 1523 (1992) and references therein.

Thermal Conduction in Dense Liquids

A.K. Raychaudhuri

Department of Physics, Indian Institute of Science, 560 012 Bangalore, India

We briefly sketch here some of the important aspects of heat conduction in dense liquids near the liquid-solid transition. We point to the similarity of liquid thermal conductivity to glasses, both lying close to minimum thermal concutivity. We point out also the essential difference in liquid and crystal which at melting point have similar phonon mean free path but qualitatively different processes of energy transport.

1. Introduction

Thermal Conductivity λ of a liquid is an old subject[1]. However, it is only recently the physical process involved is being understood. The purpose of the present work is to put the issue of liquid thermal conductivity in the context of heat conduction in solids(glasses and crystals). In this short review I will concentrate only on insulating liquids at or near the melting point (T_m) which we call dense liquids to distinguish it from liquids near the critical point(T_c). In a solid the heat conduction is vibrational in origin. In a gas the heat conduction involves diffusion. In a liquid we have presence of both. However, as $T \rightarrow T_m$, the diffusive process in a liquid is arrested and the local structure develops which persists for time periods longer than the vibrational period. Under these circumstances the heat conduction in a liquid becomes similar to that in a solid and viscosity η (determined by momentum transfer) gets decoupled from λ. In the following I sketch the important features of heat conduction in a liquid.

2. Magnitude of λ and local parameters

In fig.1, we have shown λ of a number of liquids for. We have also shown for comparison λ of few solids. As $T \rightarrow T_m$, the temperature dependence is very small. However, λ starts falling rapidly as $T \rightarrow T_c$. It can be seen that λ of most liquids near $T_m(\lambda_L(T_m))$ lie within a factor of 3 between 0.1 - 0.3 $Wm^{-1}K^{-1}$, and only few liquids have $\lambda_L > 0.5Wm^{-1}K^{-1}$. $\lambda_L(T_m)$ is very close to the classical (high temperature) limit of minimum thermal conductivity and in a way defines a natural scale for λ. Starting from dimensional analysis we obtain a natural unit for λ, which we call λ_{nat}. λ_{nat} is given by,

$$\lambda_{nat} \approx \frac{k_B}{r_s}\nu_a \qquad (1)$$

Springer Series in Solid-State Sciences, Volume 112
Phonon Scattering in Condensed Matter VII Eds.: M. Meissner and R.O. Pohl

where, ν_a = local acoustic vibrational frequency and r_s = interparticle separation. For most substances, $r_s \approx 2.5 - 3.5.10^{-10}m$ *and* $\nu_a \approx 1 - 10THz$. This gives a range of $\lambda_{nat} \sim .05 - 0.5W.m^{-1}K^{-1}$, which is close to $\lambda_L(T_m)$. A number of formulae developed for liquids as well as glasses and disordered solids give $\lambda(T \rightarrow \infty) \approx 2.5\lambda_{nat}$. Since $\lambda_L(T)$ is of the order of λ_{nat}, it is determined by two local parameters r_s *and* ν_a. This can be contrasted with thermal conductivity of a crystalline solid at melting point($\lambda_c(T_m)$), where one still needs the concept of phonon-phonon interactions. The temperature as well as pressure dependence of $\lambda_L(T)$ is determined to a large extent by T and P dependence of r_s and ν_a. At T $\sim T_m$, λ_L follows a linear law with $\lambda_L(T) \approx \lambda_L(T_m)$ - AT. For most liquids A is positive though for few liquids (water, molten B_2O_3, ethylene glycol, molten sulphur etc.) A can be negative. It can be shown[4] from equation 1 that

$$\left(\frac{\delta \ln \lambda_L}{\delta T}\right)_p = -\alpha\left[\tfrac{1}{3} + \nu\right] \tag{2}$$

where ν is the Grüneisen Constant and α is the thermal expansion coefficient. A close correlation of $\left(\frac{\delta \ln \lambda_L}{\delta T}\right)_p$ and α (shown in fig 2) shows that the temperature dependence of λ_L can be explained largely by thermal expansion[2]. A crossover in the sign of $\left(\frac{\delta \ln \lambda}{\delta T}\right)_P$ occuring at $\alpha \approx 5.10^{-4}$ was recently verified with fairly high accuracy measurements under different pressures in glassy and liquid PVC[4]. Similar consideration also applies for the pressure dependence which is measured as g = $\left(\frac{\delta \ln \lambda}{\delta \ln \rho}\right)_T$, where ρ is the density. For most liquids[4] and glasses[5], $g_l \approx g_g \approx 3$. In crystals λ follows a different relation arising

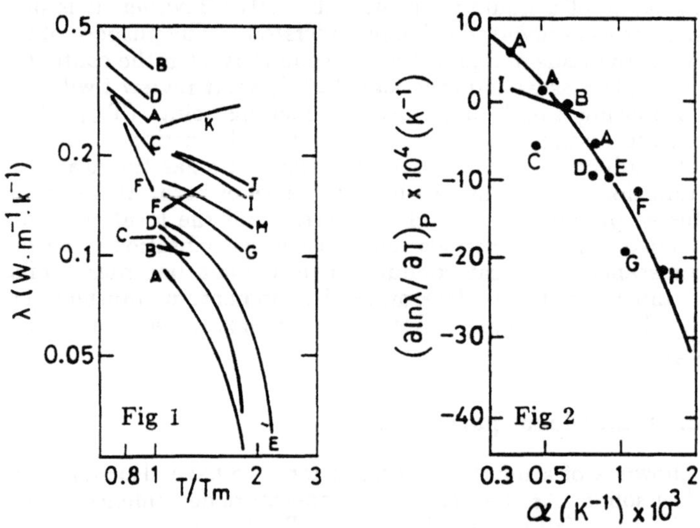

Fig 1 λ of various liquids as a function of T/T_m, A- Kr, B - I_2, C - Ne, D - Ar, E - Br_2, G - Benzene, H - Toluene, I - Acetone, J - Ethyl Alcohol, K - Ethylene Glycol[Data mostly from ref 2 and 3]

Fig 2 Temperature dependence of λ of liquid as a function of α, A - CS_2, B - m terphenyl, C - glycerol, D - biphenyl, E - p terphenyl, F - CCl_4, G - Toluene, H - Cyclohexane, I - PVC. [Data mostly from ref 1 and 4]

from phonon-phonon inteactions[5] and $g_c \approx 6$ - 10. The value of g is a good pointer to the mechanism of heat transport and the difference in g_c and g_L and similarity of g_L and g_g show the qualitatively different mechanism of heat transport occuring in crystals on one hand and liquids and glasses on the other.

3. Comparison of crystal, liquid and glass

How does the thermal conductivity of a liquid at T_m compare with that of solid, like a crystal at T_m or a glass at T_g (glass transition temperature). We show the example of glycerol[4,6] in fig 3(a). It can be seen that the thermal concutivity of the crystal at $T_m, \lambda_c(T_m,$ is about a factor of 1.5 larger compared to $\lambda_L(T_m)$. This difference is largely due to the different densities in two phases. In comparison, however, there is little change in going from liquid to glassy state. Magnitude of λ_c though not too different from $\lambda_L \, and \, \lambda_g$ arises from a qualitatively different process. In fig 3(b) we have shown specific heat C_p of glycerol in crystalline, liquid(supercooled) and glassy state[7]. One can see that in going from liquid to solid we lose a large number of degrees of freedom. These degrees of freedom, though they determine viscosity of a liquid, play virtually no direct role in determination of λ. All that matters are the vibrational degrees of freedom.

In fig 4 we show the ratio $\lambda_c(T_m)/\lambda_L(T_m)$ for a number of liquids with T_m ranging from 80K - 1200K. The ratio mostly lies within 2. This difference of λ_c and λ_L can be further reduced when one takes care of the densities of two phases. If the thermal conductivity of a fictitious liquid with the same density as a solid is denoted by λ_F then $\lambda_c(T_m)/\lambda_F(T_m)$ should reflect the true difference of λ in liquid and solid state. One can see from fig 4 that $\lambda_c/\lambda_F \, at \, T_m$ is nearly one and λ_c/λ_F for most solids lie between 1.2 and 0.95. This again supports our previous statement that the new degrees of freedom introduced on melting does not affect $\lambda_L(T_m)$. In this context we note that a class of compounds like AgI show glass (or liquid) like λ when extra scattering centres

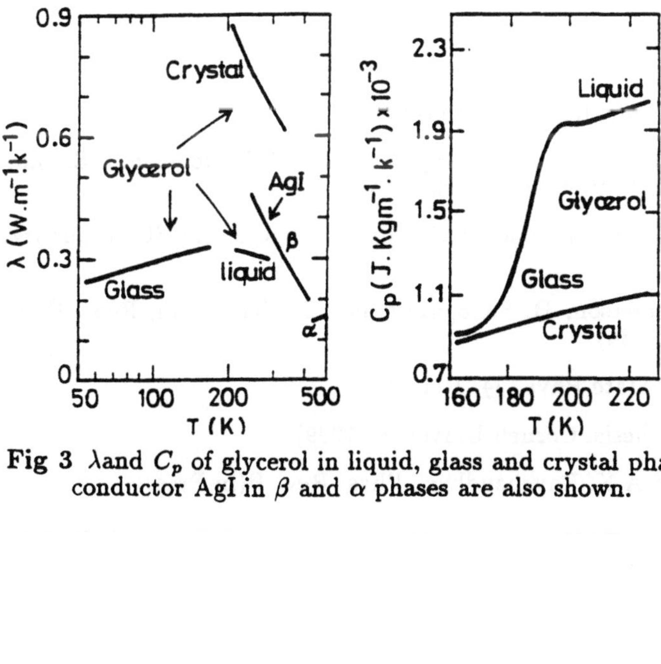

Fig 3 $\lambda$and C_p of glycerol in liquid, glass and crystal phases. λ of superionic conductor AgI in β and α phases are also shown.

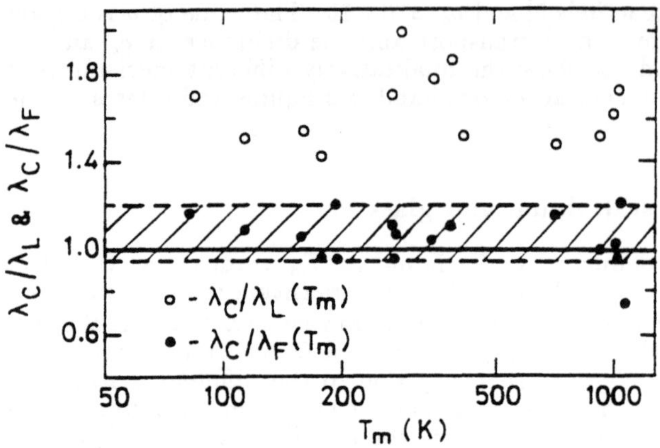

Fig 4 $\lambda_c/\lambda_L(T_m)$ and $\lambda_c/\lambda_F(T_m)$ of various liquids [Data are mostly from ref 2, 3 & 5]

are introduced in the form of charge carriers at $\beta \rightarrow \alpha$ transition[8]. This is shown in Fig.3. Similar observations have been made in clathrate hydrates[4]

4. Ideas, theories and simulations

Recent work on equilibrium and non equilibrium M-D Calculations [9,10] of thermal conduction in dense liquids as well as in glasses[11] added much to our knowledge and showed that heat conduction in dense liquids and glasses should be identical, both lying in the limit given by our natural unit λ_{nat}. We end by asking the question: $\lambda_L(T_m)\, and\, \lambda_c(T_m)$ (if we scale by density) are quite similar implying similar phonon mean free path. But they are qualitatively different. For Crystals inelastic phonon-phonon interactions dominate. For liquids we do not have such process. What then is the difference in the basic physical process that distinguishes these two?

References

1. E. Mclaughlin, Chem. Rev. 64, 390(1964)
2. Thermophysical properties of Matter 3, (Editor Y.S. Touloukian, P.E. Liley and S. Saxena, Plenum, New York, (1970)
3. D.T. Jamieson, Thermal Conductivity data of liquids(HMSO, Edinburg, 1975)
4. R.G. Ross, P. Anderson, B. Sundguvist and G. Băckstrőm, Rep. Prog. Phys 47, 1347(1984).
5. G.A. Slack, Solid State Phys. 34, 1 (1979)
6. D. Cahill, Ph.D Thesis, Cornell University(1989)
7. M. Rajeswari and A.K. Raychaudhuri, Phys. Rev. B(in Press)

8. M.C. Goetz and J.A. Cower, Solid State Comm. 41, 293(1982)

9. R. Vogelsang and C. Hoheisel, Phys. Rev. A35, 3487(1987)

10. G.. Paolini, G.Ciccotti, C. Massorio, Phys. Rev. A34, 1355(1986)

11. J. Michalski, Phys. Rev.B45, 7054(1992)

Temperature Dependence of Short-Range Order and Thermal Conductivity of Inert Liquids

P. Dashora and M.P. Saksena

Dept. of Physics, University of Rajasthan, Jaipur-302004, India

A simple relation has been obtained for the thermal conductivity 'λ' in terms of topological-short-range order parmeter 'x'. The dependence of 'λ' on temperature has thus been studied through the dependence of 'x' on 'T'. The liquid state can be visualized as a system consisting of a large number of atoms which at any instant of time are members of one or the other short-range ordered micro-regions, the size and the mean life of these micro-regions are dependent on temperature. The atomic motions in the liquid state can be described in terms of two distinct motions one, the vibrational motion of the reference atom about its pseudoequilibrium position i.e. its vibrational motion within the ordered micro-region and the second characterizing the change of these pseudoequilibrium positions by diffusion which can also be referred to as the translational motion of the atoms.

Using these aspects, Dashora and Saksena have deduced a relation for the extent of topological short-range order TSRO in the liquids, i.e. the average size of the ordered microregions 'r_m' at the freezing temperature. A parameter 'x_m', representing the extent of the topological short-range order in elements in the liquid state, is defined as the ratio (r_m/a), where 'a' is the radius of the space available per atom in the solid phase just at the melting temperature. Accordingly, TSRO parameter at any temperature 'T' is represented by 'x' = (r/a), where 'r' is the average size of microregions in the liquid at temperature 'T'. Following relations have been deduced /1,2/ for 'x_m' and 'x'

$$x_m = \left(\frac{4\pi}{3}\right)^{1/3} \frac{V_s^{2/3}\,\sigma\, N^{1/3}}{L_m} K \qquad (1)$$

$$\frac{1}{x^3} + \frac{4\sigma\pi K a^2}{9k_B T.x}\left[\left(1-\frac{T_m}{T}\right)+\frac{T_m}{T}\left(\frac{1}{x_m^3}\right)+\frac{4\sigma\pi K a^2}{9k_B .T.x_m}\right]=0 \qquad (2)$$

Springer Series in Solid-State Sciences, Volume 112
Phonon Scattering in Condensed Matter VII Eds.: M. Meissner and R.O. Pohl

Here K is a constant, for spherical surfaces, its value is 3. σ is free surface energy per unit area, L_m is latent heat of fusion, V_s is molar volume in solid phase.

The temperature dependence of λ, in the temperature region considered here, is controlled by the variation of phonon mean free path. For disordered systems, except for long waves, the mean-free path for phonons is constant and is of the order of the short-range order in the system /3/. According to proposed theory, for liquids also we can think of lattice wave being propagated inside the ordered micro-region and being then scattered at the boundaries. Consequently, the expression for thermal conductivity of simple non-metallic liquids can be written as,

$$\lambda = Bx. \tag{3}$$

Where, for a given liquid 'B' is a constant. It depends on the characteristic properties of individual element eg. electronic configuration, strength and space variation on intermolecular forces etc. In absence of knowledge of microscopic structural details, of the system it can be evaluated through a known value of thermal conductivity at a given temperature. The calculated and experimental values for neon, argon, krypton and xenon, listed in table 1, are in good agreement with the experimental data /4/. The present study indicates that the temperature dependence of x successfully accounts for the variation of thermal conductivity with temperature.

Table 1 - Comparison of calcd. and exptl. values of λ of inert liquids

Temp. T K	$10^4 \times \lambda$ **W cm^{-1} K^{-1}** λ_{exp}	λ_{cal}	**Temp. T K**	$10^4 \times \lambda$ **W cm^{-1} K^{-1}** λ_{exp}	λ_{cal}
	Neon			Argon	
24.70	—	12.47	84.4	12.60	13.24
27.59	10.59	10.47	90.0	11.98	11.98
29.81	9.71	9.37	110.0	9.56	9.13
31.15	9.00	8.83	120.0	8.50	8.25
33.00	—	8.20	130.0	7.37	7.57
	Krypton			Xenon	
125	8.79	8.79	170	7.04	7.02
150	7.12	6.87	190	6.20	6.05
175	5.66	5.78	210	5.44	5.33

1. P. Dashora and M.P. Saksena, Natl. Acad. Sci. Lett. 12, 125 (1989)
2. P. Dashora and M.P. Saksena, Ind.J. Pure & Appl. phys. 29 187. (1991)
3. P.G. Klemens, in Solid State physics edited by F. Seitz and D. Turnbull (Academic Press N.Y.) vol 7, ch 1, (1958)
4. L.D. Ikenberry and S.A. Rice, J. Chem. Phys. 39 1561 (1963)

Picosecond Optics Studies of Phonon Propagation in Glasses

H.J. Maris

Department of Physics, Brown University, Providence, RI 02912, USA

1. Introduction

Despite much work a microscopic understanding of the way that heat flows in a glass still does not exist. The thermal conductivity κ has been measured over a broad temperature range for a large number of glasses and other amorphous solids. The results exhibit surprisingly universal behavior [1,2]. Below 1 K κ varies approximately as T^2. Above this temperature there is a temperature–independent region (the "plateau"), followed by a steady rise in κ with increasing temperature. If one analyses these data using the kinetic formula [3]

$$\kappa = \frac{1}{3}C(T)v < \Lambda > \quad , \tag{1}$$

where v is the Debye velocity, $C(T)$ the specific heat, and $< \Lambda >$ the average phonon mean free path, one obtains results of the type shown in Fig. 1 [1]. The mean free path varies approximately as T^{-1} below 1 K, then undergoes a very rapid decrease with increasing T to a value which is only a few lattice spacings. After this drop $< \Lambda >$ remains approximately constant up to room temperature.

No satisfactory explanation of this behavior has been given. However, for the sake of discussion it is convenient to consider the following "standard model". This considers the phonon scattering to have two origins: two- level systems (TLS) and structural disorder. The TLS can scatter phonons via resonant and relaxation processes, but the microscopic origin of TLS has not been established. Resonant scattering from TLS can explain the T^{-1} variation of $< \Lambda >$ below 1 K. The structural disorder leads to elastic Rayleigh scattering of phonons ($\Lambda \propto \nu^{-4}, < \Lambda > \propto T^{-4}$), and thus $< \Lambda(T) >$ has approximately the right temperature dependence to explain the sudden drop in $< \Lambda >$ associated with the plateau. One problem with the standard model is that the disorder in a glass is expected to occur over short length scales, and it is hard to understand how the disorder can be strong enough to make the rapid drop in $< \Lambda >$ occur at a temperature that is a small fraction of Θ_D.

The analysis of thermal conductivity gives information about the average mean free path for thermal phonons at temperature T. More complete information could be obtained from measurements of Λ as a function of ν and T, and there may also be a significant dependence on polarization [4]. Figure 2 indicates qualitatively the regions of ν, T-space that have been explored. The data in the ultrasonic region can be understood reasonably well using the TLS model. At low temperatures the scattering is due to resonance absorption by TLS and varies as

Springer Series in Solid-State Sciences, Volume 112
Phonon Scattering in Condensed Matter VII Eds.: M. Meissner and R.O. Pohl

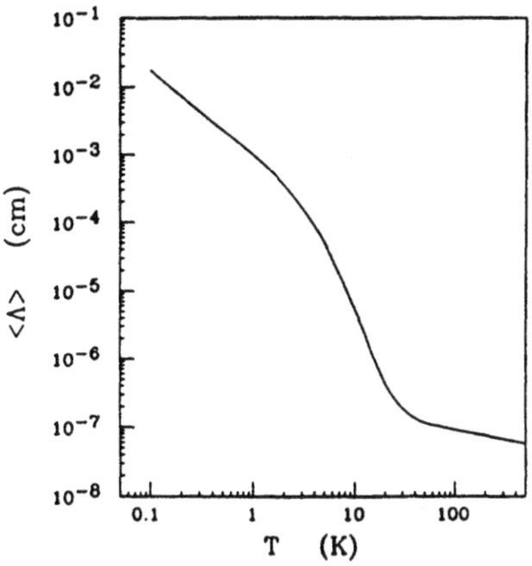

Fig.1. Average phonon mean free path calculated from thermal conductivity.

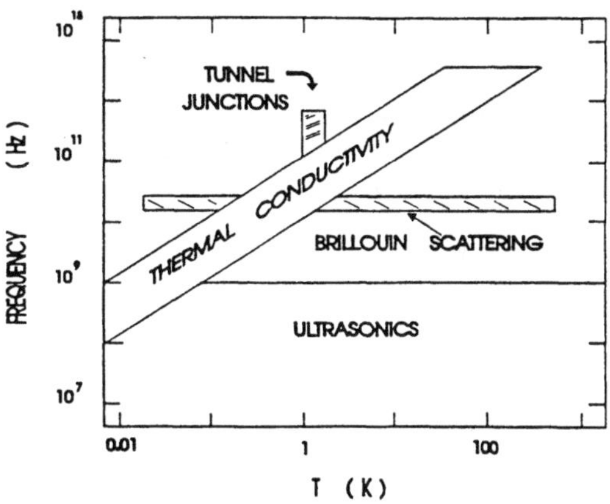

Fig.2. Regions of $\nu - T$ plane explored by various techniques.

T^{-1}, and at high temperatures relaxation processes dominate. The temperature dependence of the scattering in the Brillouin frequency range [5] is qualitatively similar to what is found at lower frequencies. Within the standard model a similar interpretation can probably be given, provided that the distributions for the parameters entering into the TLS are appropriately chosen. The data from tunneling junction experiments are less clear–cut. Results obtained by different groups [6] give significantly different answers. Part of these discrepancies may be due to differences in samples. In addition, the analysis of these experiments is complicated because the detected signal is influenced in different ways by the elastic and the inelastic scattering rates. For a recent report on measurements in this frequency–temperature range, see [4].

From Fig. 2 one can see that there are several regions of the $\nu - T$ plane that have not yet been investigated:

1) Measurements in the tunnel junction frequency range (10^{11} to 10^{12} Hz) have not been made into the temperature range significantly below 1 K. One can speculate that since $h\nu \gg kT$ throughout this range it is unlikely there will be a significant temperature dependence to the scattering, but it would still be worthwhile to check.

2) There are no data between the ultrasonic and Brillouin frequency regimes (frequency range about 10^9 to 3×10^{10} Hz). However since the ultrasonic regime is fairly well investigated and the Brillouin frequency measurements of the scattering rate are accurate, the absence of data in this region is perhaps not of great importance.

3) The third area is a large region of the temperature-frequency plane above the temperature range where tunnel junctions can be used and above the Brillouin frequency ($\nu > 3 \times 10^{11}$ Hz). The only information in this wilderness area comes from the analysis of thermal conductivity data. Studies of the thermal conductivity of glasses containing holes of various size have shown that Λ must be decreasing fairly rapidly with frequency in this regime [7].

Data in the region 3) are of particular interest as a test of the standard model. Even at room temperature the relaxation time τ for a TLS can never be much less than ~ 1 psec since the attempt frequency certainly cannot be more than the Debye frequency. Consequently, the relaxation attenuation due to

TLS (or any other multi–state defect) should saturate by the time a frequency of around $1/(2\pi\tau) \sim 160$ Ghz is reached. As a result, according to the standard model above this frequency the phonon scattering rate should be dominated by temperature-independent scattering arising from structural disorder. Thus, in this frequency regime the prediction is that a) $< \Lambda >$ should be approximately temperature- independent and b) should decrease rapidly with increasing frequency, as ν^{-4} for example. This frequency dependence is required by the thermal conductivity data.

2. Picosecond Ultrasonic Measurements

To make measurements in this frequency range we have used the recently–developed picosecond ultrasonic technique [8]. The sample is in the form of a film deposited onto a substrate. On top of the sample is a thin metal film that serves as a transducer. A light pulse (duration ~ 0.2 ps) is absorbed in the transducer raising its temperature by a few K. The resulting thermal stress launches a longitudinal strain pulse into the sample. After reflection at the sample–substrate interface this pulse returns to the transducer where it causes a small change $\Delta R(t)$ in optical reflectivity. This change is detected by means of a probe pulse that is time–delayed with respect to the light pulse used for generation. Because of the excellent time–resolution of this technique one can perform ultrasonic pulse–echo experiments on films of thickness down to a few 100 Å, and this makes possible the study by ultrasonic means of a wide variety of complex nanostructures, including laterally–patterned structures [8].

An example of ultrasonic echoes obtained by this technique is shown in Fig. 3. The sample is a 750 Å film of a–SiO_2 deposited onto a tungsten substrate. To determine the ultrasonic attenuation (and phonon mean free path) one first measures the shapes of the first and second acoustic echoes $\Delta R_1(t)$ and $\Delta R_2(t)$ in the time domain. The Fourier transforms $\Delta R_1(\omega)$ and $\Delta R_2(\omega)$ are calculated, and the attenuation at frequency ω can then be found from the ratio $\Delta R_1(\omega)/\Delta R_2(\omega)$. Results obtained in this way for a-SiO_2 are shown in Fig. 4 [9].

The results indicate a phonon mean free path that to within experimental error is independent of temperature from room temperature down to 80 K, the lowest temperature at which measurements could reliably be made because of laser heating. The frequency-dependence is approximately quadratic over most

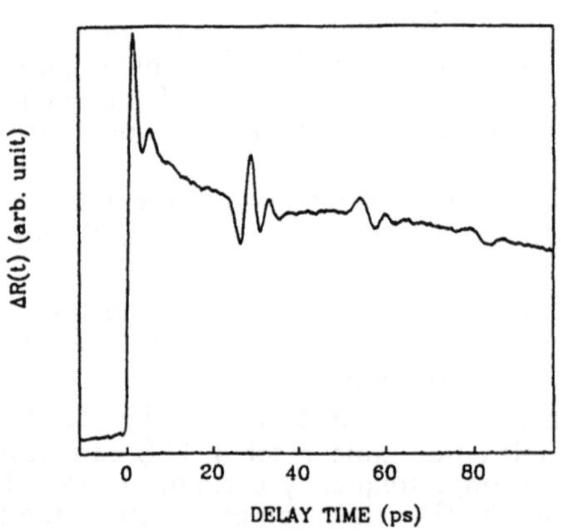

Fig.3. Ultrasonic echoes in a 750 Å film of SiO_2.

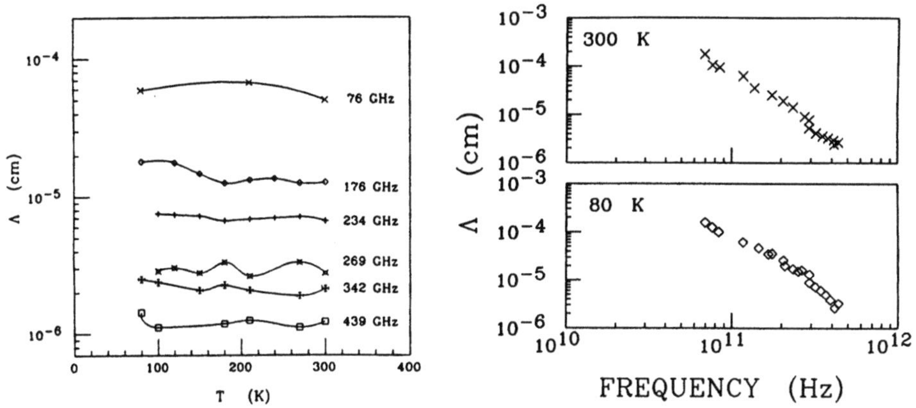

Fig.4. Mean free path of longitudinal phonons in a-SiO_2.

of the range studied, but could be slightly faster than quadratic in the upper part of the frequency range. At the highest frequency studied (440 GHz), $\Lambda \approx 100\mathring{A}$ and the phonon wavelength is 134 Å. In this type of experiment there are a number of possible errors. It is necessary to assume a value for the reflection coefficient of the ultrasonic pulse at the interface between the film and the substrate. In addition, there are possible losses in the transducer film and due to roughness at both surfaces of the sample. To try to reduce the uncertainties due to these effects a series of films of differing thickness was studied.

Since performing these measurements we have begun studies of a polymer system CF_x [10]. Polymers are of interest because the plateau in the thermal conductivity occurs at a lower temperature and so one expects that the rapid decrease in Λ will occur at a lower frequency. The CF_x measurements were made on samples of the type shown in Fig. 5a. A thin layer of CF_x was sandwiched between an Al transducer and a Si substrate. Data obtained for samples with an Al thickness of 130 Å are shown in Fig. 5 b-d. The Al transducer rings at a frequency of around 200 GHz after being excited, and is damped by sound transmission into the CF_x and the Si. For the sample with no CF_x the vibration is damped out almost immediately (Fig. 5b). For a 5 Å CF_x layer the damping is less because the CF_x layer reduces the transmission into the substrate (Fig. 5c). We find that for a 20 Å CF_x layer (Fig. 5d) the rate of damping becomes the same as for a very thick CF_x layer. This indicates that the acoustic wave is substantially attenuated on passing through the CF_x layer; we estimate that Λ can be no more than 20 to 30 Å. This result for Λ is about 40 times smaller than that obtained at SiO_2 at the same frequency. Note that in CF_x the sound velocity is 1.3×10^5 cm s^{-1}, and so at 200 GHz the wavelength is only 65 Å.

3. Discussion

As discussed above, it is unlikely that the large damping rate of phonons in the frequency range above 100 GHz can be explained by a relaxation process. At the highest frequency of 440 GHz, for example, one would require a significant number of defects to have relaxation times less than 0.4 psec. To retain the "standard model" one then has to suppose that the measured damping rate of the phonons is due to the scattering from structural disorder. However, this in

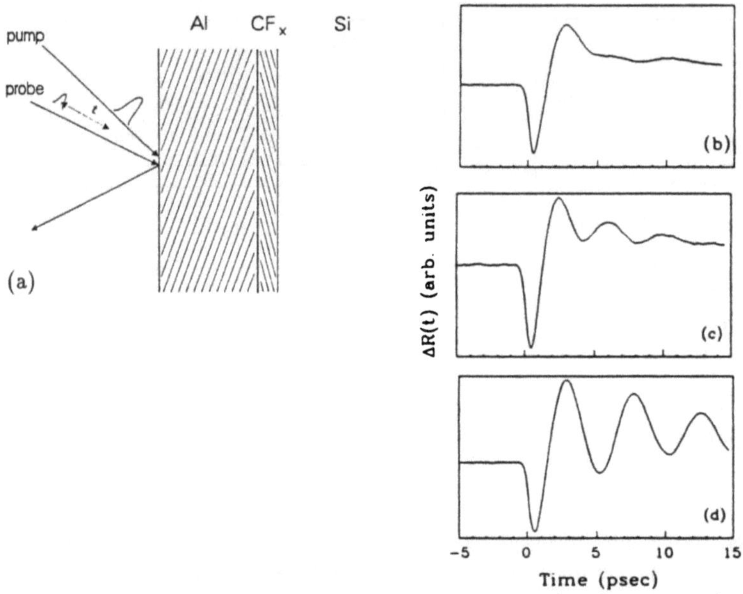

Fig.5. a) Geometry of the CF_x samples. b), c), d) results obtained for samples with 0, 5 and 20 Å of CF_x, respectively.

turn leads to a contradiction. The value of Λ in the studied frequency range (8×10^{10} to 4×10^{11} Hz) would in this case be independent of temperature for the whole range down to $T = 0$ K. This would lead (through Eq. 1) to a thermal conductivity in the range 1 to 5 K that would vary roughly as T, i.e. there would be no plateau, and the conductivity calculated from Eq. 1 would be less than the measured value. To reconcile the thermal conductivity results with the ultrasonic measurements that we have made, it is necessary to suppose that in the temperature range below 80 K the mean free path increases substantially with decreasing temperature. This could be tested via tunnel junction measurements or possibly by picosecond ultrasonics with an especially configured sample to reduce the heating problem.

I thank C.J. Morath, R.J. Stoner, G.I. Tas, J. Tauc, T.C.D. Zhu for their contributions to this research, and J.M. Halbout, G.S. Oehrlein, and G.W. Rubloff for supplying the CF_x samples. This work was supported in part by the Department of Energy through Grant DE-FG02-86ER45267.

References

1. R.C. Zeller and R.O. Pohl, Phys. Rev. B 4, 2029 (1971).
2. For a review see *Amorphous Solids: Low Temperature Properties*, edited by W.A. Phillips (Springer, Berlin, 1981).
3. This approach assumes that phonon are not localized. See, J.E. Graebner, B. Golding, and L.C. Allen, Phys. Rev. **B34**, 5696 (1986).
4. See P. Berberich, G.P. Scherg, R. Gärtner, and H. Kinder, this volume.
5. R. Vacher, J. Pelous, F. Plicque, and A. Zarembowitch, J. Non-Cryst. Solids **45**, 397 (1981).

6. See references listed in ref. 9.
7. See the article by A.C. Anderson in ref. 2.
8. For reviews, see H.T. Grahn, H.J. Maris, and J. Tauc, IEEE J. Quant. Electron. **QE 25**, 2562 (1989); H.N. Lin, R.J. Stoner, and H.J. Maris, J. Non Dest. Eval. **9**, 239 (1990); H.N. Lin, R.J. Stoner, and H.J. Maris, in *Proceedings of the IEEE Ultrasonics Symposium 1990*, editor B.R. McAvoy (IEEE, New Jersey), p. 1301.
9. T.C. Zhu, H.J. Maris, and J. Tauc, Phys. Rev. **B 44**, 4281 (1991).
10. G. Tas, R.J. Stoner, H.J. Maris, G.W.Rubloff, G.S. Oehrlein, and J.M. Halbout, Proc. Matl. Res. Soc., to appear, and to be published.

Waiting-Time Effect on the Specific Heat of Polybutadien

Hong Chen, Kejia Zhu, Xiaoqui Qin, and Xiang Wu
Tongji University, Shanghai 200092, Peoples Republic of China

1. Introduction

Polybutadien (PB) is a typical glass with simple structure and low glass-transition temperature, and interesting metastable states in this material has been observed recently[1]. It was shown by neutron scattering measurements, that the dynamic structure factor of PB follows a Kohlrausch law[2] and that there is a transition to a nonergodic state near the glass-transition temperature[3]. In this paper, we report on our investigations on the responses of the metastable states in PB to different thermal paths by specific-heat measurements.

2. Experimental Details

The sample used in this study was cis-trans-vinyl (46:47:7) PB. The specific heat of a tablet of PB was measured with an adiabatic continuous heating method. After waiting at temperature T_w for a time t_w during the cooling process, we measured the specific heat of the following heating process. From these measurements, we have determined the dependence of the specific heat C_p on the waiting temperature T_w and the waiting time t_w.

3. Results

The specific heat is shown in Fig. 1. A glass transition is observed at the temperature T_g=178 K, comparing well with the result T_g= 181 K of Ref. [2]. It can be seen that there is noticeable dependence on t_w of the specific heat near the glass-transition temperature. The peak becomes sharper and stronger as t_w increases. The dependence of the jump ΔC_p of the specific heat at the glass-transition temperature can be fitted by the Kohlrausch law

$$\Delta C_p(t_w) = \Delta C_p(\infty)\{1 - exp[-(t_w/\tau)^{\beta}]\}$$

with β=0.47± 0.03. The neutron scattering indicated that the relaxation of the structure factor in PB follows a Kohlrausch law with the exponent β=0.45± 0.02 in the time range 9×10^{-3} ns $< t <$ 2.8 ns,[2] much smaller

Springer Series in Solid-State Sciences, Volume 112
Phonon Scattering in Condensed Matter VII Eds.: M. Meissner and R.O. Pohl

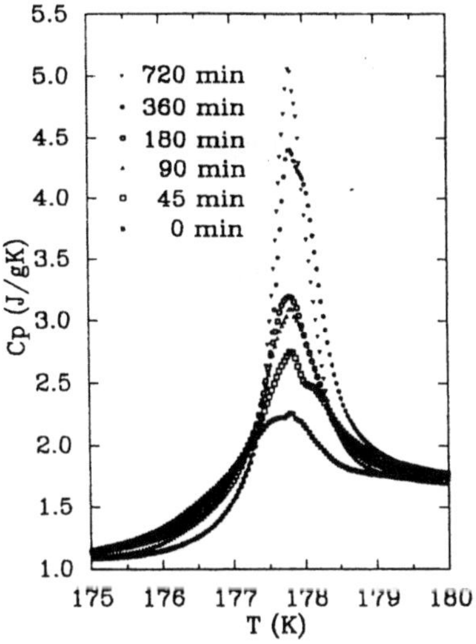

Figure 1. The waiting time t_w effect on the specific heat of polybutadien measured at T_w=169 K and six t_w-values: 0 min, 45 min, 90 min, 180 min, 360 min and 720 min.

than the present waiting-time scale. This means the structural relaxation in Pb can be described by a Kohlrausch law in a much wider time range.

The authors are grateful to Prof. H. von Löhneysen and Dr. J. Wosnitza for stimulating discussions. This work was financed by the Stiftung Volkswagenwerk and by the Chinese National Science Foundation.

[1] H. von Löhneysen et al. Phonon Scattering in Condensed Matter V, A. C. Anderson and J. P. Wolf (eds.) (Springer-verlag, Berlin 1986) p. 35.
[2] D. Richter et al., Phys. Rev. Lett. **61**, 2465 (1988).
[3] B. Frick et al., Phy. Rev. Lett. **64**, 2921 (1990).

Temperature Dependence of the Sound Velocity in a Polar Liquid: SO_2

C. Prieto[1], *A. Chahid*[2], *M. García-Hernández*[2], *F.J. Bermejo*[2], and *J.L. Martínez*[1]

[1]Instituto de Ciencia de Materiales de Madrid (CSIC),
Dep. Física de Materiales (C-4), Universidad Autónoma de Madrid,
E-28049 Madrid, Spain
[2]Instituto de Estructura de la Materia (CSIC), Serrano 123,
E-28006 Madrid, Spain

Abstract. The adiabatic sound velocity and attenuation for liquid SO_2 was measured in a temperature range between the solid and the gas phases (i.e. 200 to 260 K). The results are compared with recent inelastic neutron scattering data on similar temperature range. The extrapolation of these data to Q=0 coincide very well with the present hypersonic velocity from Brillouin experiments.

Sulphur dioxide have been very much studied mostly due to the spontaneous polarization and the piezo and pyro-electric characteristic presented in the solid phase. Recently it was proposed the existence of collective excitations of the polarization function in the fluid state (dipolaronic modes) [1]. In a previous neutron scattering work we were interested in the collective density excitations in a wide Q-range [2]. In that work we observed heavily damped modes and the estimation of the sound velocity from the low Q data. In order to study properly the sound velocity and attenuation in the hidrodynamic region we performed a Brillouin scattering experiment.
Liquid SO_2 was prepared by *in-situ* condensation of gaseous SO_2 into a metallic cylinder with pyrex windows fitted inside a continous flux Helium cryostat with optical access. The excitation light was the 4893Å line coming from a Ar^+ laser. The scattered light was collected at 90 degrees and analyzed by means of a Burleigh DAS10 stabilized triple-pass spectrometer, with a Free Spectral Range of 24 GHz. The temperature was varied between 180K to 260K, with a temperature stability better than 0.1K.
The analysis of the Brillouin spectra was performed with the following function,

$$S(Q,\omega) = \frac{\gamma-1}{\gamma}\frac{2\chi Q^2}{\omega^2+(\chi Q^2)^2} + \frac{1}{\gamma}\left[\frac{\Gamma_B Q^2}{(\omega+v_aQ)^2+(\Gamma_B Q^2)^2} + \frac{\Gamma_B Q^2}{(\omega-v_aQ)^2+(\Gamma_B Q^2)^2}\right] +$$
$$\frac{1}{\gamma}[\Gamma_B + (\gamma-1)\chi]\frac{Q}{v_a}\left[\frac{\omega+v_aQ}{(\omega+v_aQ)^2+(\Gamma_B Q^2)^2} - \frac{\omega-v_aQ}{(\omega-v_aQ)^2+(\Gamma_B Q^2)^2}\right]$$

where γ is the ratio of the heat capacities, v_a is the adiabatic sound velocity, χ is the thermal conductivity, Q is the wave vector of the scattered light and Γ_B is the width of the Brillouin peaks. This model scattering function $S(Q,\omega)$ was convolved with the experimental resolution function and fitted to the experimental data.

Springer Series in Solid-State Sciences, Volume 112
Phonon Scattering in Condensed Matter VII Eds.: M. Meissner and R.O. Pohl

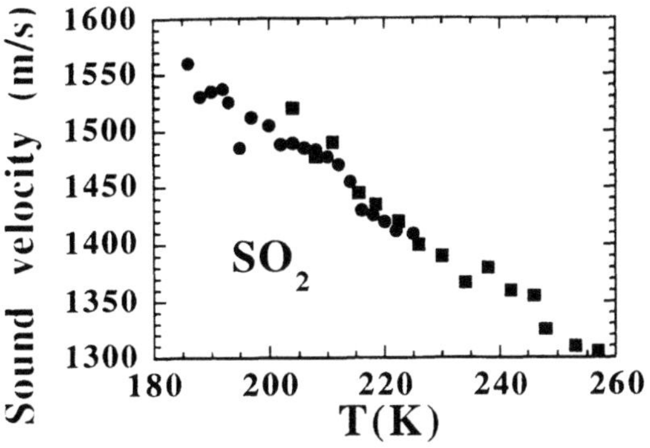

Fig. 1. Sound velocity *vs.* temperature for liquid SO_2. Different symbols indicate different data set.

The temperature dependence of the adiabatic sound velocity is presented in Fig. 1. This adiabatic sound velocity coincides very well with the isothermal sound speed obtained from neutron scattering [2]. It is also interesting to point out that no measurable broadering of the Brillouin peaks was observed at any temperature. This fact can be explained by computing the longitudinal viscosity obtained from neutron data [2], which is the leading contribution to the width of the Brillouin peaks.
A more detailed discussion of the Brillouin scattering data and the comparison with the neutron experiments is done in Ref. 3. No signature of the dipolaronic modes was observed by either neutron nor Brillouin scattering experiments.

REFERENCES

[1] R. Lobo, J.E. Robinson, S. Rodriguez. J. Chem. Phys. **59**, 5922 (1973); and E.L. Pollock, B.J. Adler. Phys. Rev. Lett. **46**, 950 (1981).

[2] J.L. Martínez, F.J. Bermejo, M. García-Hernández, F.J. Mompeán, E. Enciso and D. Martín. J. Phys.: Condens. Matt. **3**, 4075-4087 (1991). See also, F.J. Bermejo, J.L. Martínez, D. Martín, M. García-Hernández, F.J. Mompeán and J. Alonso. J. Chem. Phys. **95**, 5387-5391 (1991).

[3] A. Chahid, M. García-Hernández, C. Prieto, F.J. Bermejo, E. Enciso and J.L. Martínez. Mol. Phys .**78**, xxx (1992). (In press).

Dispersion and Scattering of Monochromatic Phonons in Thin Amorphous PMMA-Films Near the Propagation Cutoff

G.P. Scherg, R. Gärtner, P. Berberich, and H. Kinder

Physik-Department E10, Technische Universität München, W-8046 Garching, Fed. Rep. of Germany

The phonon scattering processes in glasses at frequencies corresponding to the plateau region in thermal conductivity can be studied by monochromatic phonon spectroscopy using thin glass films. A first study on a-SiO_2 was done by Dietsche and Kinder [1]. Inelastic scattering processes were observed, but the role of simultaneous strong elastic scattering was not taken into account in this analysis. Here we study thin PMMA films because the propagation cutoff should occur at lower frequencies. A stress tuned Si:B spectrometer was used which allowed to sort out the elastic and inelastic scattering contributions for longitudinal and transverse phonons separately.

PMMA films of 20 to 60 nm thickness were fabricated on Si (111) crystals by pulling them slowly out of a solution of 10% PMMA/toluene. The Si crystal served as a stress tuned phonon spectrometer which was operated either as monochromator or as analyzer.

For the observation of acoustic thickness resonances, a PbBi junction on top of the PMMA film was biased at $V > 4\Delta/e$ for wide band emission and an Al junction on the far side was used as detector. Fig.1 shows the pulse height for longitudinal (L) and transverse (T) phonons as a function of the spectrometer frequency. The thickness resonances are clearly seen. The maxima and minima are plotted in Fig.2 as a function of $q = n\pi/d$, where n is the order of the resonance and d = 19 nm is the film thickness. The dispersion is linear both for L and T phonons. The slopes, $v_L = 3.0$ km/s and $v_T = 1.59$ km/s (± 5 %) are in good agreement with ultrasonic measurements [2].

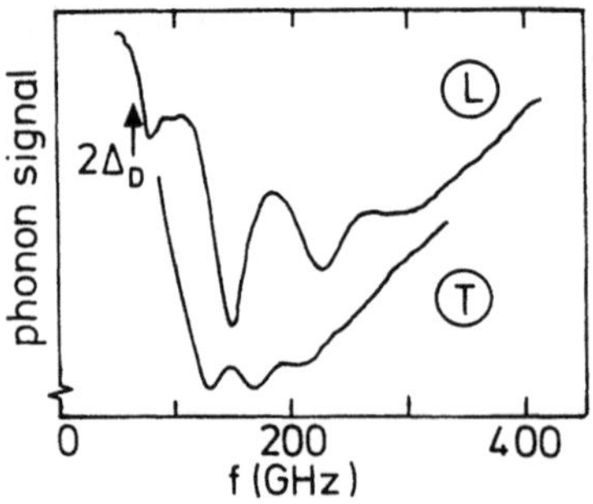

Fig.1 Acoustic thickness resonances

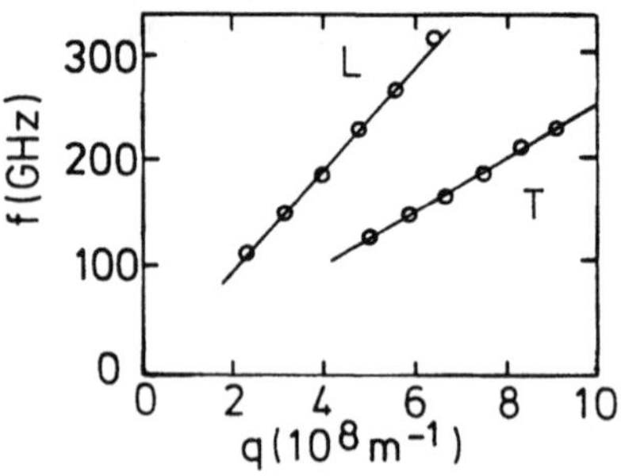

Fig.2 Phonon dispersion relation

Springer Series in Solid-State Sciences, Volume 112
Phonon Scattering in Condensed Matter VII Eds.: M. Meissner and R.O. Pohl

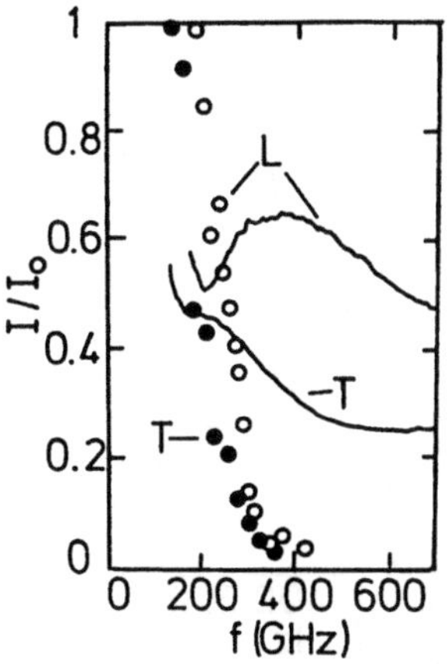

Fig.3 Intensity ratio with and without PMMA film

Phonon scattering was studied by comparing two generator/detector pairs, one with a PMMA film, and one without. In the monochromator mode, a NiCr heater was used as a broad band generator, and the PMMA film was under the detector junction. Then the spectrometer defines the frequency of the incident phonons, and the total transmitted flux is monitored by the detector. The flux ratio I/I_0 with and without PMMA film (42 nm) is plotted in Fig.3 (solid lines). While the T-phonon ratio decreases with frequency, the L ratio increases initially, and remains relatively high.

In analyzer mode, a PbBi junction was used for quasi monochromatic phonon generation. The PMMA film was under the generator, so that the emitted phonon spectrum could be analyzed by the spectrometer. The flux transmitted without frequency shift was measured, and the ratio I/I_0 with and without PMMA film (60 nm) is shown in Fig.3 (symbols). Both L and T phonons are now falling off much more rapidly. Combining the results of Fig.3, we conclude that the L phonons are predominantly inelastically scattered, because they do show up in the total transmitted flux, but not at the original frequency. The phonons produced by inelastic scattering are shifted to lower frequency where they propagate ballistically. So inelastic scattering leaves the total flux unaffected. The total flux of T phonons falls off at high frequencies so that elastic scattering must dominate here. But the decay of the original frequency is much faster, i. e. inelastic scattering is not neglible.

The L phonon data in analyzer mode can therefore be simply analyzed by

$$I/I_0 = \exp(-d/l_i) \quad , \tag{1}$$

where d is the film thickness and l_i the inelastic scattering length. For the T phonons we used a mixed model which is valid for $l_i >> l_e$ where l_e is the elastic scattering length. The fraction of phonons transmitted at the original frequency is:

$$T = x + (1-x)\cdot 2\kappa/N \quad , \tag{2}$$

where $x = \exp(-d/l_e)$ is the fraction of ballistic phonons, $L = (l_e l_i/3)^{1/2}$ the inelastic diffusion length, $\kappa = 4(l_e/3l_i)^{1/2}$, and $N = (1+\kappa)^2\exp(d/L) - (1-\kappa)^2 \exp(-d/L)$. The fraction of phonons inelastically scattered in the film is

$$A = (1-x)\kappa\{(1+\kappa)[(\exp(d/L)-1]-(1-\kappa)[1-\exp(-d/L)]\}/N \,. \tag{3}$$

In the analyzer mode inelastically scattered phonons are sorted out. In the

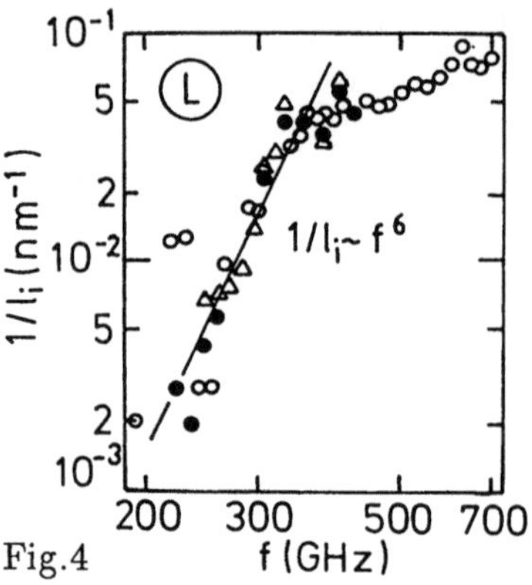

Fig.4

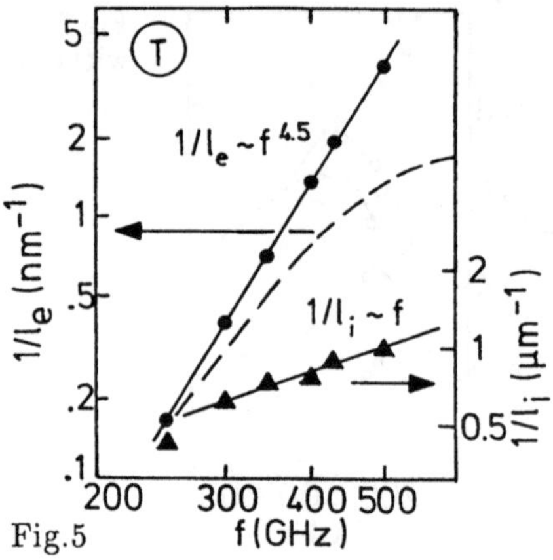

Fig.5

monochromator mode a fraction of 84 % of the inelastically scattered phonons is detected as estimated by a Monte Carlo simulation. So we get:

$$I/I_o)_A = T \qquad (4) \qquad \text{and} \qquad I/I_o)_M = T + 0.84\ A \qquad (5)$$

(4) and (5) are implicit equations for l_e and l_i at a given frequency. (For the actual analysis of the data this model had been refined by taking the average reflection and transmission coefficients at the interfaces into account [3].)

Fig.4 shows the resulting scattering rate of the longitudinal phonons. The open and full circles correspond to film thicknesses of 19 nm and 60 nm, respectively. The data points marked by triangles are deduced from the transmission ratio of the two films: $(I/I_o)_1/(I/I_o)_2 = \exp[-(d_1-d_2)/l_i]$. The scattering rate varies as $l^{-1} \propto f^6$ up to 370 GHz and levels off at higher frequencies. In Fig.5 the elastic and inelastic mean free paths of the transverse phonons are plotted versus frequency (film thickness:19 nm and 42 nm). Note that the elastic scattering rate is by about 3 orders of magnitude stronger than the inelastic rate (see right scale). We find for the elastic scattering rate $l_e^{-1} \propto f^{4.5}$. Graebner et al [4] analyzed the thermal conductivity of PMMA by assuming a mean free path of the form $l = l_{min} + A/f^4$ in our frequency regime, where A is a constant and l_{min} is a minimum phonon mean free path of 0.4 nm. This scattering rate is shown in Fig.5 by a dashed line. Our results compare quite well below 400 GHz, but no levelling off at l_{min} could be found. From our data we would expect a propagation cutoff, ql = 1, at 420 GHz. However, there is no drastic change of the frequency dependence of l_e^{-1} at this frequency.

In summary, the dispersion and the scattering of monochromatic phonons has been studied in PMMA thin films in the plateau regime of thermal conductivity. A linear dispersion was found and no indications of a minimum mean free path have been observed at $(ql)_T \simeq 1$.

References

1. W. Dietsche, H. Kinder: Phys. Rev. Lett. 43, 1413 (1979).
2. C. L. Choy, G. L. Salinger, Y. C. Chiang: J. Appl. Phys. 41, 597 (1970).
3. G. P. Scherg, thesis, TU München (1989) pp. 40 - 57.
4. J. E. Graebner, B. Golding, L. C. Allen: Phys. Rev. B34, 5696 (1986).

Luminescence Study of the Scattering of High-Frequency Phonons in Amorphous Semiconductor Films

A.V. Akimov[1], *A.A. Kaplyanskii*[1], *E.S. Moskalenko*[1], *J. Kocka*[2], and *J. Stuchlik*[2]

[1]A.F. Ioffe Institute, Russian Academy of Sciences, 194021 St. Petersburg, Russia
[2]Institute of Physics, Czechoslovak Academy of Sciences, 180 40 Prague 8, Czechoslavakia

Until now only in few works /1-4/ the methods of nonequilibrium phonon physics are applied to phonon studies in non-crystalline materials. In present work (see also /5/) the luminescent method is used to study the phonon transmittance through thin amorphous semiconductor (Si, Ge) films.

The samples used were parallel-side plates consisting of crystalline (111) silicon (c-Si) substrate 10x5x2 mm^3 and amorphous film (AF) deposited on it (Fig. 1a). Main results were obtained on hydrogenated amorphous silicon (a-Si:H) films and also on amorphous germanium (a-Ge) films. The samples were immersed in pumped liquid helium (T=1.7 K). Heat pulses were injected into AF by means of pulse (duration $2x10^{-7}$ s) current heating of thin constantan film "h", deposited on AF. From "h" into AF a broad phonon spectrum is injected. Phonons transmitted through AF reach c-Si substrate and propagate ballistically to its opposite side, where phonon detection with time resolution 50 ns is performed.

The phonon detection is based on effect of phonons on exciton luminescence of c-Si /6/. Opposite to AF side of c-Si is excited by cw-Ar-laser. As a result excitonic cloud (EC) is formed near the surface of c-Si. Phonon induced changes in the intensities of free (FE) and bound (BE) exciton lines $I^F(t)$ and $I^B(t)$ respectively were measured. These signals are caused by two main processes /6/. First is dissociation BE→FE induced by "high frequency" phonons with frequencies $\Omega > 0.9$ THz. Dissociation BE→FE in EC redistributes FE and BE concentrations in FE's favour, which results in enhancement of FE luminescence lines ($\Delta I^F > 0$) and in the reduction of BE lines ($\Delta I^B < 0$). The second process is FE drag by the flux of "low frequency" acoustic phonons with frequencies $\omega < 0.4$ THz towards samples surface, where FE possess high rate surface recombination. The drag results in decrease of FE number and hence in the reduction of both FE ($\Delta I^F < 0$) and BE ($\Delta I^B < 0$) luminescence. Thus the total effect of phonons on FE and BE luminescence is formed by the sum of contributions of two frequency dependent factors discussed above.

The observed $\Delta I^F(t)$ signal has negative sign (Fig. 1b) and its amplitude increases with AF thickness d. This result shows that transmission of terahertz nonequilibrium phonons through a-Si:H film is accompanied by effective phonon down-conversion. Actually such conversion leads to the enhancement of FE drag

Springer Series in Solid-State Sciences, Volume 112
Phonon Scattering in Condensed Matter VII Eds.: M. Meissner and R.O. Pohl

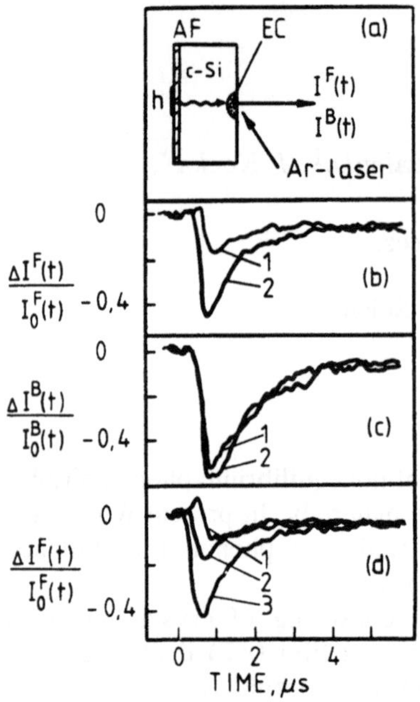

Fig.1. Experimental scheme (a) and relative intensities of FE (b,d) and BE (c) luminescence pulses for two thicknesses of AF a-Si:16% H (b,c) (1 - d=0.2 μm; 2 - d=1.0 μm) and for different concentrations N_H (d) in AF with d=1.0μm (N_H: 6% -1; 10% - 2; 16% - 3). I_0^F, I_0^B- stationary intensities of FE and BE luminescence without phonons. $\Delta I = I - I_0$

by phonons with $\omega < 0.4$ THz and to the reduction of dissociation rate BE→FE induced by phonons with $\Omega > 0.9$ THz. As a result the $\Delta I^F(t)$ negative signal amplitude (Fig.1a) has to increase while the thickness increases. Fig.1d shows luminescence pulses measured for a-Si:H films with different hydrogen concentration N_H(6%, 10%, 16%). It is seen that amplitude of negative $\Delta I^F(t)$ signal increases with N_H. These results show that centers responsible for inelastic scattering includes H. Our conclusion about effective phonon down-conversion in a-Si:H is in agreement with experiments /1-4/.

Using the same technique we observed a delay t= 300-400 ns of the $\Delta I^B(t)$ front in case of phonon transmission through a-Ge (d=0.5 μm) film. This result leads to the assumption of strong localization of THz phonons in a-Ge.

References

1. W.Dietsche and H.Kinder: Phys.Rev.Lett. 43, 1413 (1979).
2. A.S.Alekseev, M.M.Bonch-Osmolovskii, T.I.Galkina, I.B.Levinson and D.P.Utkin-Edin: Pis'ma Zh.Eksp.Teor.Fiz. 37, 490 (1983).
3. T.I.Galkina, A.Yu.Blinov, M.M.Bonch-Osmolovskii, O.Koblinger, K.Lassmann and W.Eisenmenger: Phys.Stat.Sol.(b) 144, K87 (1987).
4. J.Mebert, B.Maile and W.Eisenmenger: Phonons 89, ed. by S.Hunklinger, W.Ludwig and G.Weiss (World Scientific, Singapore, 1989) p.495.
5. A.V.Akimov, A.A.Kaplyanskii, J.Kocka, E.S.Moskalenko and J.Stuchlik : Zh.Eksp.Teor.Fiz. 100, 1340 (1991).
6. A.V.Akimov, A.A.Kaplyanskii and E.S.Moskalenko: Physica B 169, 382 (1991).

Brillouin Scattering of Vitreous Silica Under High Pressure

G. Weiss, K. Döring, R. Ehrenfels, S. Rau, and S. Hunklinger

Institut für Angewandte Physik, Universität Heidelberg,
Albert-Überle-Str. 3–5, W-6900 Heidelberg, Fed. Rep. of Germany

Ultrasonic absorption in vitreous silica (a-SiO_2) exhibits a prominent maximum, the so-called 50 K-peak. It is caused by thermally activated relaxation processes as can be inferred from the position of the maximum which shifts from 30 K at a measuring frequency of 1 kHz to above 100 K at Brillouin frequencies. The microscopic nature of the relaxing 'particles' is unknown but they are believed to originate intrinsically from the random network of the glass. Nevertheless, in the framework of phenomenological models, like the tunneling model or the soft potential model, it is possible to describe the experiments very well in the full range of measured frequencies and temperatures. Both models are founded on the assumption that the particles reside in double well potentials and that their dynamics are based on tunneling between the two equilibrium positions at low temperatures, whereas thermal activation over the barriers is dominating at temperatures exceeding a few Kelvin [1,2].

Owing to the disordered structure the parameters of the double well potentials are widely distributed. The most important parameters for a successful description of the ultrasonic absorption are the ground state energy of a single well, which determines the transition temperature between the tunneling and the thermal activated region, and an upper bound in the distribution function of barrier heights V which limits the absorption at higher temperatures and determines the position of the maximum. A Gaussian distribution in V, centered at zero, yields very good agreement with experiment. The width of the Gaussian is found to be approximately 450 K [1] or 530 K [3], depending on the model.

An extension of these ideas and further implications on possible microscopic models evolved from Brillouin scattering under high pressures up to 2.5 GPa [1]. It was found that the room temperature absorption, dominated by barrier heights of about 1000 K, increases drastically with pressure and the maximum shifts from 100 K to higher temperatures – to more than 300 K at pressures higher than 2 GPa. This was explained by a change of the distribution function $P(V)$ of barrier heights, with $P(V)$ becoming broader quadratically with pressure. Numerical fits yielded a value of about 1400 K for the Gaussian width at a pressure of 2 GPa. At the same time the density of states at small barrier heights appeared to remain unchanged. This is seen in the fact that the pressure affects the absorption only a little below 100 K (see fig.1). Microscopically, one might imagine that existing double well potentials are squeezed in a way which increases their barriers and, additionally, other bistable configurations with small barriers are newly formed.

The effect of pressure on sound velocity is twofold. Firstly, at all temperatures, the sound velocity becomes smaller with pressure and passes through

Springer Series in Solid-State Sciences, Volume 112
Phonon Scattering in Condensed Matter VII Eds.: M. Meissner and R.O. Pohl

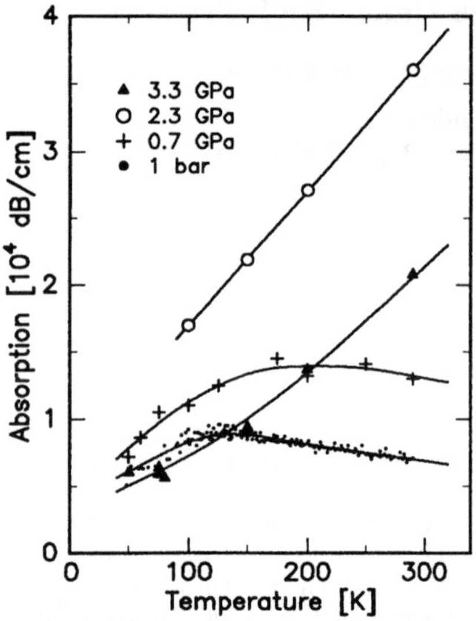

Fig. 1: Temperature dependence of the absorption of 36 GHz longitudinal phonons in vitreous silica at various pressures. Lines are guides to the eye.

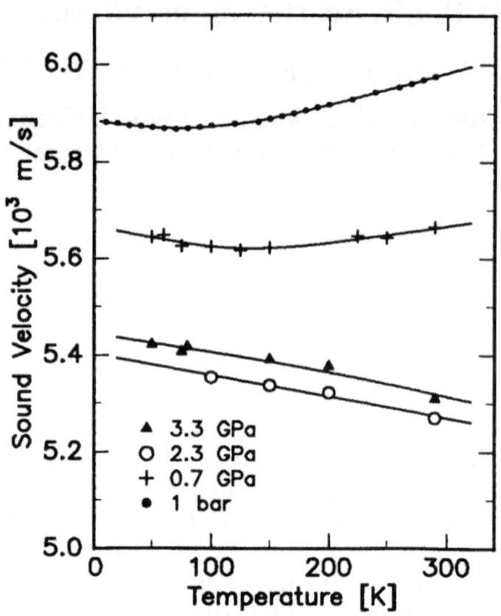

Fig. 2: Temperature dependence of the velocity of longitudinal phonons in vitreous silica measured at pressures up to 3.3 GPa. Lines are guides to the eye.

a minimum at 2.5 GPa in agreement with earlier experiments [4]. Secondly, the region of negative temperature dependence extends to higher temperatures with an unchanged slope. This can be explained consistently by a contribution of the relaxation process discussed above.

In the new experiments we have extended our range of pressure to 3.5 GPa and find a strong reduction of the absorption at all temperatures (fig.1), the maximum absorption occuring around 2.5 GPa. This immediately implies that the number of double well systems with $V/k_B \simeq 1000\,K$ is getting reduced on pressures above 2.5 GPa. It can be modelled by just reducing the height of the Gaussian $P(V)$ or by shifting its center to energies much higher than 1000 K, both giving satisfactory agreement with the absorption data. With the sound velocity, however, there appears a yet unsolved problem. A reduced density of states would result in a diminished slope of the temperature variation in contradiction to the measurements shown in fig.2. Qualitatively it is possible to reduce this contradiction by including a new relaxation mechanism caused by the delayed equilibration of soft harmonic (single well) potentials which may contribute to about 30 % to the negative temperature variation of the sound velocity but not to the absorption [5].

References

[1] D. Tielbürger, R. Merz, R. Ehrenfels, S. Hunklinger: Phys. Rev. B **45**, 2750 (1992)

[2] U. Buchenau, Yu. M. Galperin, V. L. Gurevich, D. A. Parshin, M. A. Ramos, H. R. Schober: to be published in Phys. Rev. B, August 1992

[3] D. A. Parshin: unpublished

[4] M. Grimsditch: Phys. Rev. Lett. **52**, 2379 (1984)

[5] D. A. Parshin: submitted to Phys. Rev. B

Elastic Properties of Amorphous Thin Films

B.E. White, Jr. and R.O. Pohl

Laboratory of Atomic and Solid State Physics, Cornell University,
Ithaca, NY 14853-2501, USA

1. Introduction

The anomalous, low temperature elastic and thermal properties found in all bulk amorphous solids have been well described in terms of a phenomenological two level state (TLS) model, although the microscopic origin of the TLS still remains a mystery. Several workers [1][2][3] have proposed that the mysterious TLS could originate from elastic interactions that arise between stress dipoles found in these solids. In an attempt to determine if these elastic interactions are responsible for the TLS we have developed a technique for measuring the internal friction of amorphous thin films which is sensitive to films with thicknesses as small as 100 Å.

2. Experimental Technique and Preliminary Results

The technique we have developed for measuring the internal friction of thin amorphous films utilizes a silicon double paddle oscillator [4] (Fig. 1) which is driven in the so called anti-symmetric mode [5] in order to minimize non-intrinsic clamping losses. The internal friction of amorphous thin films can be measured by depositing the films on the torsion element connecting the head and wings of the paddle oscillator. The difference in the measured internal friction of the pure Si paddle oscillator and the composite Si/film paddle oscillator is then attributed to the film.

Paddle oscillators are fabricated from (100) Si wafers using photolithography and anisotropic etching techniques. Typical thicknesses of the oscillators range from 200μ to 400μ with corresponding frequencies between 3 kHz and 10 kHz.

The internal friction, Q^{-1}, as a function of temperature, is shown in Fig. 1 for a typical paddle oscillator. As can be seen, Q^{-1} values of 10^{-7} are obtained below temperatures of 10 K where the characteristic "plateau" region in the internal friction of amorphous solids exists. Also in Fig. 1, we show the results of preliminary work obtained on 1μ films of a-Si and a-SiO_2. From the measured internal frictions of these films in

Springer Series in Solid-State Sciences, Volume 112
Phonon Scattering in Condensed Matter VII Eds.: M. Meissner and R.O. Pohl

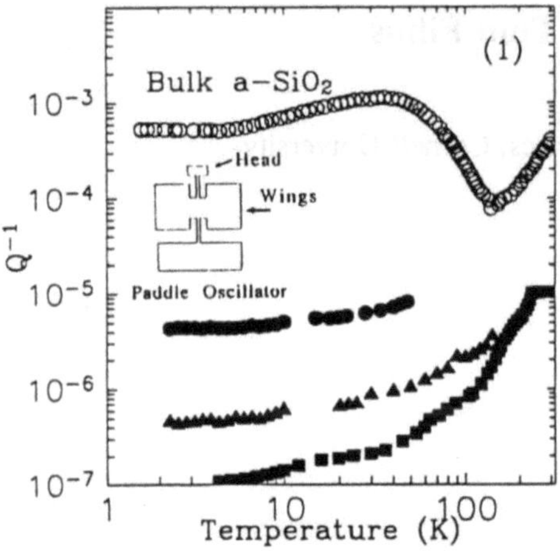

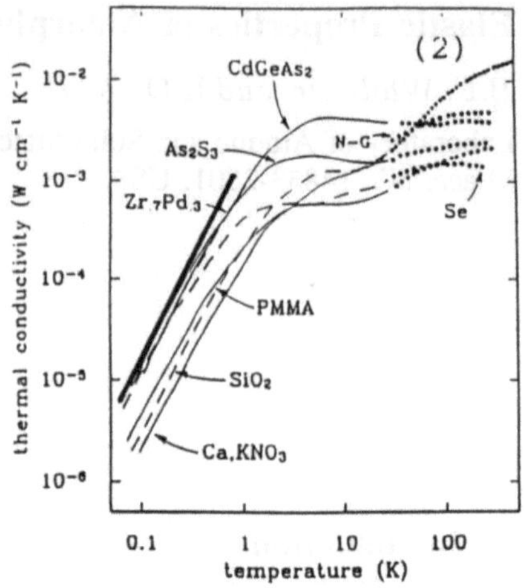

Fig. 1-The measured internal friction of the silicon oscillator-■, the oscillator with 1μ of a-Si-▲, and the oscillator with 1μ of a-SiO_2-●.

Fig. 2-The predicted thermal conductivity of a-Si (thick line) obtained from the measured internal friction of a 1μ film. The thermal conductivity of several other bulk amorphous solids are shown for reference.

the plateau region we find that the calculated $\bar{P}\gamma^2$ for the a-SiO_2 film is in good agreement with that found found in bulk a-SiO_2, while the calculated value of $\bar{P}\gamma^2$ for the a-Si film is a factor of 10 smaller than that found in bulk a-SiO_2. Using these values of $\bar{P}\gamma^2$ and the tunneling model, we are able to predict, for the first time, the low temperature thermal conductivity of a-Si (Fig. 2). We hope that by measuring the internal friction of thin amorphous films in the "plateau" region,as a function of film thickness, we will be able to eliminate, in one dimension, the elastic interactions between stress dipoles in the solid.

This research was supported by the SRC, Grant No. 92/SC/069 and the U.S. Dept. of Education Graduate Fellowship Program in Materials Science, Grant No. P200A10148.

[1] Clare C. Yu and A. J. Legget, Comments on Cond. Mat. Phys. **14**, 231 (1988).
[2] A. J. Legget, Physica B **169**, 322 (1991).
[3] Yaoutian Fu, Phys. Rev. B **40**, 10056 (1989-I).
[4] R. N. Kleiman, G. K. Kaminsky, J. D. Reppy, R. Pindak, and D. J. Bishop, Rev. Sci. Instrum. **56**, 2088 (1985).
[5] J. E. Van Cleve, Ph. D. Thesis, Cornell University, 1991.

Vibrating Reed Experiments on Glasses – New Problems

J. Classen and G. Weiss

Institut für angewandte Physik, Universität Heidelberg,
Albert-Überle-Str. 3–5, W-6900 Heidelberg, Fed. Rep. of Germany

The elastic properties of insulating glasses at low temperatures are determined by the interaction of phonons with localized low energy excitations, which do not exist in pure crystals. In the phenomenological tunneling model, which gives the most successful theoretical approach to the understanding of these states and their interaction with other bulk excitations, they are described as two level systems originating from tunneling transitions of particles between two neighbouring potential minima in the random network of the glass. The energy splitting $E = (\Delta^2 + \Delta_0^2)^{1/2}$ of these systems is determined by the energy difference Δ of the two potential minima and by the tunneling splitting Δ_0 due to the overlap of the wavefunctions of the single wells. Owing to the disordered structure these parameters are assumed to be widely distributed.

The interaction of the tunneling systems with phonons is based on the change of the asymmetry energy Δ by a strain field e and takes place via relaxation and resonant processes. The coupling strength of the interaction is given by the deformation potential $\gamma = 1/2\,\partial\Delta/\partial e$ which is assumed to be independent of Δ. Under the condition of a constant density of states the tunneling model predicts for insulating glasses a logarithmic increase of the sound velocity at low temperatures due to resonant processes and after passing a maximum a logarithmic decrease due to resonant plus relaxation contributions with half the slope of the increase [1]. This prediction also holds for superconducting metallic glasses far below the transition temperature T_c where the electron-tunneling system interaction does not play a crucial role any more. In fact, these predictions have found an experimental confirmation by former measurements [1]. On the other hand, recent measurements of the dielectric constant of several glasses, which should yield similar results as discussed above for the sound velocity, show significant deviations from the expected behaviour [2,3]. Our vibrating reed measurements may be understood as a new precise test of an essential prediction of the tunneling model.

The two figures show the results for the relative change of sound velocity of the very pure vitreous silica Suprasil W and of the superconducting metallic glass $Pd_{30}Zr_{70}$ ($T_c \simeq 2.6\,K$), respectively. The shape of the curves depends on the driving voltage which stimulates electrostatically the reed to forced vibrations. While at higher temperatures for both materials the differences between the various curves are very small, on the low temperature side a strong strain dependence occurs resulting in smaller values of v for larger vibrational amplitudes, as also reported in [4]. The flat part of the curves for $Pd_{30}Zr_{70}$ at high voltages and low temperatures is caused by selfheating of the sample (the internal friction measurements, not presented here, show similar flattening).

Springer Series in Solid-State Sciences, Volume 112
Phonon Scattering in Condensed Matter VII Eds.: M. Meissner and R.O. Pohl

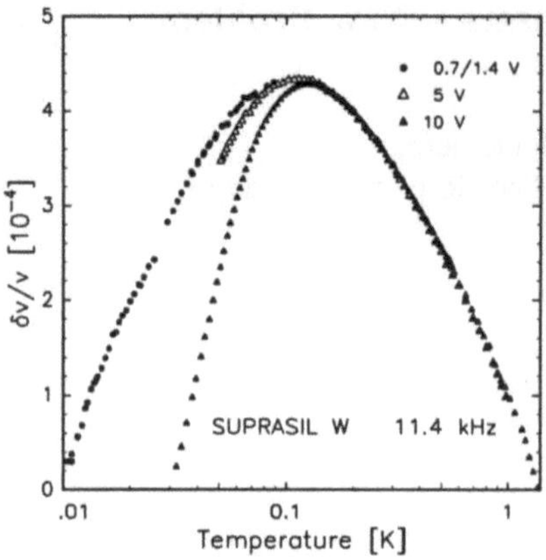

Fig. 1: Relative change of sound velocity of Suprasil W for different driving voltages.

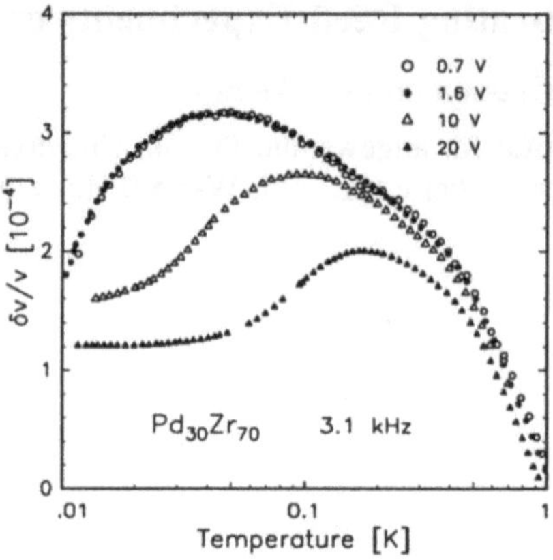

Fig. 2: The same for the superconducting metallic glass $Pd_{30}Zr_{70}$. At high driving voltages selfheating occurs.

The shape of resonance curves indicates strong nonlinear behaviour at low temperatures and high voltages whereas at low voltages and/or high temperatures the normal linear behaviour is found. The nonlinearities occur when the modulation of the asymmetry energy $2\gamma e$ is comparable to or even larger than the thermal energy. In our interpretation the occupation numbers of the tunneling systems can drastically be changed in this case leading to a decrease of sound velocity with increasing strain. These effects will be discussed elsewhere in more detail [5].

In the linear case, i.e. for low driving voltages, where the predictions of the tunneling model should hold we find indeed the sound velocity in both materials to vary nearly logarithmically with T both below and above a maximum, but the ratios of the slopes do not coincide with the expected 2 : –1 behaviour. For Suprasil W we find a ratio of approximately 1 : –1, for the metallic glass of 3 : –1. These deviations are a serious problem for the tunneling model since they cannot simply be explained by modifying the density of states for example. A qualitative explanation of the results can be obtained by the assumption that the deformation potential is not a constant but varies smoothly, at given E, with asymmetry energy Δ in a nonuniversal way. In the case of the dielectric glass the deformation potential for symmetric tunneling systems should be a little smaller (ca. 20 %) than for asymmetric systems, whereas at the metallic glass $Pd_{30}Zr_{70}$ an increase of γ with increasing Δ of about 10 % yields a significant better agreement with our experimental results. A microscopic reason for a dependence of γ on Δ, however, cannot be given.

References

[1] S. Hunklinger, A. K. Raychudhuri: in Progress in Low Temperature Physics IX, Ed.: D. F. Brewer (Elsevier, New York, 1986), p. 265

[2] C. Enss, C. Bechinger, M. v. Schickfus: in Phonons 89, Eds.: S. Hunklinger, W. Ludwig, G. Weiss (World Scient. Publ., Singapore, 1989), p. 474

[3] J. Classen, C. Bechinger, C. Enss, G. Weiss: in preparation

[4] P. Esquinazi, R. König, F. Pobell: Z. Phys. B **87**, 305 (1992)

[5] D. A. Parshin, J. Classen, G. Weiss: to be published

Ultrasonic Attenuation of SiO_2-10%GeO_2 Glass

T. Kosugi[1], *H. Kobayashi*[2], *and Y. Kogure*[3]

[1]Faculty of Science, Hiroshima University, Higashi-Hiroshima 724, Japan
[2]National Research Laboratory of Metrology, Tsukuba 305, Japan
[3]The Nishi Tokyo University, Kitatsuru-gun, Yamanashi 409-01, Japan

Abstract. Precise measurement of ultrasonic absorption on SiO_2-10 mole % GeO_2 glass between 4 - 250 K shows that the most probable origin of the structural relaxation of SiO_2 glass is the transverse motion of the oxygen atom in Si-O-Si bridge among existing models. In addition it is also suggested the similar mechanism is the origin of two level system (TLS) phenomena below 10 K.

The glass state, as often represented by SiO_2 glass, has characteristic thermal and acoustic properties at low temperatures, i.e., the linear dependent specific heat on T below 1 K and T^2 dependence of the thermal conductivity around 1 K, acoustic properties explained by the two level system (TLS) model around or below 1 K, and the large acoustic relaxation peak around the liquid nitrogen temperature [1]. These properties are generally considered to be due to the existence of the double-well potential in some atomic configuration [2 - 4]. However, we do not have the definite solution on the real atomic configuration for it until now.

Our approach for it is very simple but may be appropriate, i.e., we have tried to measure precisely the ultrasonic absorption or the internal friction (IF) of a mixed glass of 90% SiO_2 and 10% GeO_2, where the sites of Si atoms in pure SiO_2 glass are expected to be replaced randomly by Ge atoms [5,6]. The IF is measured by a composite oscillator bar method using - 18.5° X cut quartz transducer (51 kHz)[7].

Figure 1 shows the IF of SiO_2-10% GeO_2 glass. The result of pure SiO_2 glass is also shown, where the absolute value of the IF is reduced to 85 % for the reason stated below. Noted points are that (a) and (b) have very similar values between 4 -

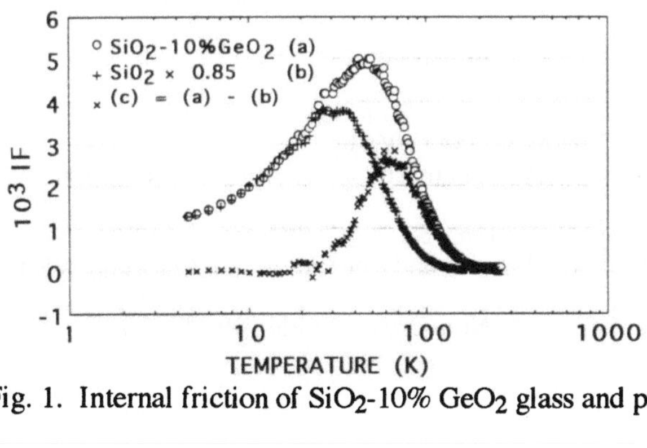

Fig. 1. Internal friction of SiO_2-10% GeO_2 glass and pure SiO_2 glass.

Springer Series in Solid-State Sciences, Volume 112
Phonon Scattering in Condensed Matter VII Eds.: M. Meissner and R.O. Pohl

25 K and the subtracted value (c) has a peak around 65 K. The Young's modulus E of the mixed glass is 95 % of that in SiO_2 glass below 30 K.

There are three plausible structural models proposed for the double-well potential producing the relaxation in SiO_2 glass. First is the transverse motion of bridging oxygen (O) atom across the Si-O-Si bridge (T-model) [2]. Second is the longitudinal motion of O atom along the Si-O-Si bridge (L-model) [3]. Third is the rotation of the relatively rigid SiO_4 tetrahedra (R-model) [4,8]. We expect on acoustic properties the effect of interconnection between SiO_4 and GeO_4 tetrahedra in SiO_2-10 % GeO_2 glass. First let us consider the models T and L. By the mixing, the number of Si-O-Si bridge sites is reduced to 81 %, Ge-O-Ge to 1 % and Si-O-Ge bridge is newly produced by 18 %. Since the magnitude of the IF is shown to be proportional to the density N for the relaxation sites and the relaxation strength Δ ($=D^2/EkT$), where D is the deformation potential and kT has usual meaning [1], for SiO_2-10 % GeO_2 the IF caused by the Si-O-Si bridge is expected to be 85 % of pure SiO_2 glass. Thus we plotted this in Fig. 1. It is reasonable to consider the curve (c) corresponds to the relaxation due to the transverse motion of O atom in Si-O-Ge bridges, because the peak temperature Tp (≈65 K) is reasonably located when we see the fact that in pure SiO_2 and GeO_2 glasses, Tp will come to about 35 K and 100 K for 51 kHz, respectively [2,9]. On the other hand, L-model is denied strongly, because for the Si-O-Ge bridge the double-well potential for longitudinal motion of oxygen atom, even if it would be, must be too largely asymmetric to produce a successive relaxation between its two wells. Next let us consider the recently proposed R-model, i.e., the coupled rotation of five tetrahedra, which consist of one centered tetrahedron and four surrounding tetrahedra [8]. For SiO_2-10 % GeO_2, the number of the group of five SiO_4 is reduced to 59 %, and five GeO_4 to 0.01 %. Instead, the group of the centered SiO_4 with surrounding three SiO_4 and one GeO_4 will be mainly produced (26 %). If we consider the similar subtraction between the IF of SiO_2-10 % GeO_2 and that of pure $SiO_2 \times 0.62$ as we have done for T or L model, we obtain a subtracted component, which is broadly distributed from 4 K to 200 K, having a peak around 60 K. This result does not seem to support R-model, because the secondly major group of R-model should have a similar activation energy to the five SiO_4 tetrahedra, when we suppose its configuration . Thus although R-model is not completely denied, it is not likely.

Further study of the dependence of acoustic spectra on the composition rate in $(SiO_2)_{1-x}(GeO_2)_x$ will lead to a definite confirmation of the real model for the structural relaxation in SiO_2 glass.

References

[1]W.A.Phillips ed.,Amorphous Solids-Low Temperature Properties(Springer-Verlag, 1981).

[2]O.L.Anderson and H.E.Bommel, J. Am. Ceram. Soc. **38**, 125 (1955).

[3]R.E. Strakna, Phys. Rev. **123**, 2020 (1961).

[4]M.R.Vukcevich, J. Non-Cryst. Solids **11**, 25 (1972).

[5]N.F.Borrelli, Phys. Chem. Glasses **10**, 43 (1969).

[6]T.Edahiro, M.Kawachi, S.Sudo and S.Tomaru, Jpn.J. Appl. Phys.**19**,2047 (1980).

[7]T.Kosugi and T.Kino, J. Phys. Soc. Jpn., **53**, 3837 (1984).

[8]U. Buchenau, Proc. Symp. on the Physics and Technology of Amorphous SiO_2 p47. (Plenum, 1988).

[9]K.Sakai, P.K.Choi and K. Takagi, J. Non-Cryst. Solids **109**, 47 (1989).

Scattering of High-Frequency Phonons by Implantation Damage in Silicon

J.K. Wigmore, K.R. Strickland, S.C. Edwards, and R.A. Collins

School of Physics and Materials, University of Lancaster,
Lancaster LA1 4YB, UK

We have studied the diffusive scattering of heat pulse phonons from ion implantation damage at the surface of silicon wafers [1]. By correlating the phonon flux with various stages of thermal annealing, we have been able to identify the presence of two distinct diffusive processes. We found that, firstly, some phonons were scattered from the roughness of the interface between the crystalline wafer and the disordered layer produced by implantation. Other phonons passed straight through this interface into the amorphous layer, where they underwent multiple scattering from the intrinsic two-level systems.

Our sample wafer, kindly donated by Wacker Chemitronic, had (111) orientation and was 480 μm thick, polished on both sides. It was self-implanted on one face with 60 keV Si^+ions to a total dose of 10^{15} cm^{-2} using the SERC Implantation Facility at the University of Surrey. The effect was to produce an amorphous layer extending on average 120 nm into the silicon. The distance over which the structure changed from amorphous to crystalline was believed to be a few atomic spacings, and the distance of the amorphous-crystalline (a-c) interface from the true surface to have a variation of a few nm over a lateral distance scale of 20 - 40 nm. It is known that the roughness of the a-c interface can be modified by annealing the sample at 250^{0}C, whilst a temperature above 550^{0}C is needed to remove the amorphous structure [2]. We report here experiments on different samples which were (A) non-implanted, (B) implanted as described above but not annealed, (C) implanted and subsequently annealed at 250^{0}C for 3 hours, and (D) implanted and annealed at 250^{0}C for 5 hours.

The damage was studied by observing heat pulse phonons reflected from it [3]. The heat pulse generators were excited by 10 ns pulses at power levels varied between 0.1 - 10 mW to change the dominant phonon frequency of the emitted phonon Planck distribution. In order to interpret the data, it was essential to deconvolute the bolometer signal by numerical solution of the bolometer rate equation, as described by Edwards et al. [4], to obtain the phonon flux.

Diffuse reflection at the a-c interface would be expected to show up as a rapidly decaying tail of shape roughly t^{-5} after each specular peak, as was observed. As an indicator of roughness scattering, we took the time for the diffusive tail on the specular transverse mode to fall by 60% of the peak value, t(60%). In addition, there was a second component of diffusive scattering which decayed over a much longer time scale, giving rise to a significant phonon flux at the bolometer at arrival times much later than the specular

Springer Series in Solid-State Sciences, Volume 112
Phonon Scattering in Condensed Matter VII Eds.: M. Meissner and R.O. Pohl

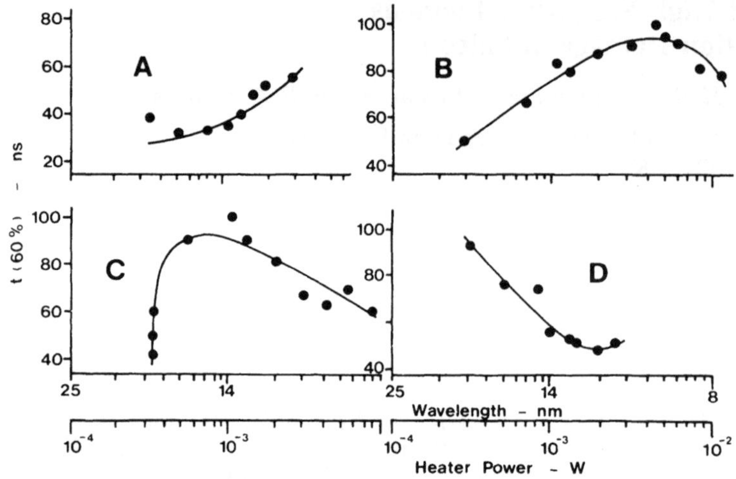

Figure 1 : Variation with heater power and hence dominant phonon wavelength of $t(60\%)$.

peaks. We believe that this corresponded to the trapping of phonons in the amorphous layer, and slow release following multiple absorption and re-emission by the intrinsic two-level systems [5].

In comparison of different samples, it was found that the effect of implantation was to increase both diffusive components, as would be expected. The subsequent annealing stages at 250°C had a significant effect only on the roughness. More detailed information came from varying the power level, by which the dominant phonon wavelength could be scanned through the range between 8 and 25 nm. Figure 1 shows the variation of $t(60\%)$ with dominant phonon wavelength for all four samples. We infer that the maxima in Figure 1 indicated maximum diffuse reflection and hence a correspondence between the phonon wavelength and the roughness scale of the a-c interface. Our data show that the wavelength at which the maximum occurred moved to larger values following successive anneals, implying that shorter wavelength Fourier components of the a-c interface had been removed. From further analysis, we shall infer quantitative parameters for diffusive transport of the phonons in the amorphous layer [5].

We are grateful to Dr. C. Jeynes at the University of Surrey for the implantation. The work is supported by the Science and Engineering Council, to whom KRS is grateful also for a research studentship.

[1] K.R. Strickland, S.C. Edwards, J.K. Wigmore, R.A. Collins and C. Jeynes, Surface and Interface Analysis 18 (1992) to be published.
[2] J. Narayan and O.W. Holland, J. Appl. Phys. 56 2913 (1984).
[3] Hamid bin Rani, S.C. Edwards, J.K. Wigmore and R.A. Collins, Surface and Interface Analysis 14 850 (1989).
[4] S.C. Edwards, Hamid bin Rani and J.K. Wigmore, J. Phys. E22 582 (1989).
[5] J.J. Freeman and A.C. Anderson, Phonon Scattering in Condensed Matter V (ed. A.C. Anderson and J.P. Wolfe) Springer, Berlin, 32 (1986).

Ultrasonic Investigation of Acoustic Phonons at Megahertz Frequencies in Sintered Glass-Bead Networks

H.P. Schriemer and J.H. Page

Department of Physics, University of Manitoba, Winnipeg, Manitoba R3T 2N2, Canada

Three-dimensional percolation networks made from glass microspheres form an interesting physical system in which to study the localization of acoustic phonons due to scattering from structural disorder. In this paper, new ultrasonic velocity and attenuation data are reported to investigate the vibrational properties of this system using samples of improved quality compared with those available in our earlier work [1].

The samples were prepared by first mixing glass microspheres of radius $a \cong 0.8$ μm with Ni particles of a similar grain size, and then removing the Ni particles by etching in acid after the glass beads were sintered together. This enabled the positions of the glass particles and voids to be controlled in a way that corresponds to percolation models. To ensure homogeneous mixing of the glass and Ni particles, the samples were ground, mixed and pressed several times before sintering. This procedure also smoothed the rough surfaces of the Ni particles, leading to more uniform packing and removing the tendency for microcrack formation found in earlier sintered samples. Homogeneity of the powder mixture was further improved by allowing the unsintered mixture to settle in an ultrasonic bath prior to the final pressing. Sintering was performed in a nitrogen atmosphere for one hour at 615° C. After etching, the density ρ of the glass-bead network was found to be less than the value corresponding to the initial occupied volume fraction f of glass, implying that isolated glass clusters were removed from the samples during etching. This is confirmed by the fact that the density (for $\rho < 0.5\rho_{\text{bead}}$) is found to scale as a power law in $f-f_c$, with $f_c = 0.19\pm.01$ as found previously [1] and an exponent $\beta = 0.44\pm.02$ which is in excellent agreement with the accepted value (0.44) for percolation.

The density dependence of the longitudinal ultrasonic velocity for two sets of differently-prepared sinters is shown in Fig. 1(a). In set A, the samples were sintered not in nitrogen but in air so that the Ni particles were allowed to oxidize during sintering, thereby inducing large internal strains in the samples due to the expansion of the Ni particles during oxidation. As a result the bonds between the glass beads were much weaker than in the unstrained samples of set B, giving a corresponding reduction in velocity of approximately 40%. For both sets, the velocity scales with density (for $\rho < 0.5\rho_{\text{bead}}$) as a power law with the *same* exponent, 2.7±.1, implying that the microscopic elasticity in these samples is not modified by large internal strains [2]. Using

Springer Series in Solid-State Sciences, Volume 112
Phonon Scattering in Condensed Matter VII Eds.: M. Meissner and R.O. Pohl

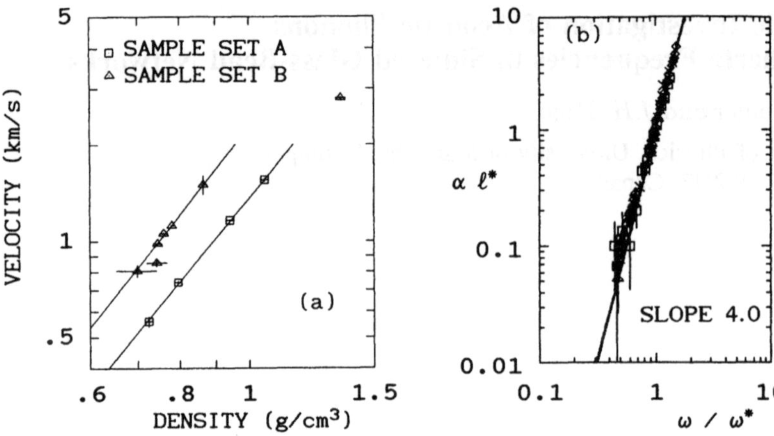

Fig. 1 (a) Longitudinal sound velocity versus density for two sets of glass-bead samples. (b) Attenuation versus frequency for sample set B.

well-known scaling relations for percolation, and the accepted values D = 2.5 for the fractal dimension of the infinite cluster and ν = 0.88 for the correlation length exponent, we find δ = 1.05±.1 for the fracton dimension, close to predictions for bond-bending elasticity [3].

Fig. 1(b) shows the frequency dependence of the ultrasonic attenuation $\alpha(\omega)$, which is inversely proportional to the elastic mean free free path $\ell(\omega)$, for several samples from set B. As in Ref. [1], the data for each sample are scaled by the characteristic length ℓ^* and frequency ω^* defined by $\ell^*=\ell(\omega^*)=\lambda=2\pi v/\omega^*$, ℓ^* being proportional to the correlation length ξ that marks the boundary between the Euclidean and fractal regions of a percolation network. For $\omega > \omega^*$, the mean free path is less than the ultrasonic wavelength λ, indicating the breakdown of wave propagation and an impending crossover to localized vibrational modes. Note that Rayleigh scattering ($\alpha(\omega) \propto \omega^4$) is observed over the entire frequency range investigated (as expected for $k\xi$<1), clearly demonstrating that the anomalous frequency dependence found in our earlier samples [1] was due to long length scale imhomogeneities that have been virtually eliminated in the present samples. This improvement in sample quality is potentially important in that it should enable predictions for $\ell(\omega)$ to be investigated experimentally as the transition to localized (fracton) modes is crossed in a system with δ < 4/3 [4].

References

1. N. Pachet and J.H. Page, in *Phonons 89*, S. Hunklinger, W. Ludwig and G. Weiss, eds. *(World Scientific, Singapore)* p. 820, 1990.
2. S. Alexander, J. Phys. **45**, 1939 (1984).
3. I. Webman and G. Grest, Phys. Rev. B **31**, 1689 (1985).
4. A. Aharony, S Alexander, O. Entin-Wohlman and R. Orbach, Phys. Rev. Lett. **58**, 131 (1987)

Very Fast Relaxation at Low Temperature of Tunneling States in Rare-Earth Doped Glasses

N. Vernier and G. Bellessa

Laboratoire de Physique des Solides*, Bât. 510,
Université Paris-Sud, F-91405 Orsay, France

To explain the low temperature properties of amorphous materials, the existence of Tunneling States (T.S.) has been assumed [1,2]. We report on the existence of a much faster relaxation than usual in a new type of insulating glass: an aluminosilicate glass doped with magnetic rare-earth ions (Dysprosium and Gadolinium). We present here the results of a saturation recovery experiment: we send in our sample a first strong acoustic pulse which saturates the resonant T.S. and then, we look at the saturation rate as a function of the elapsed time (fig.1). We have performed this experiment at about 15mK, for several magnetic ions concentrations from 0% to 10% and for several values of the magnetic field from 0 to 5 Tesla. Our results are shown in Fig.1 for different magnetic fields. The decay of the saturation rate is non exponential. This usual property is due to a distribution of relaxation time T_1. This distribution has a lower boundary, which can be usually evaluated from the

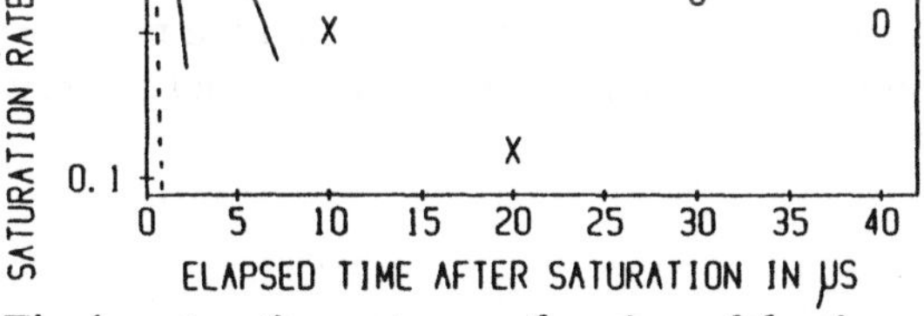

<u>Fig.1</u>: *saturation rate as a function of the time elapsed after the saturation pulse for two glasses doped with 1.5% Dysprosium and 10% Gadolinium. + : Dy glass in 5 Tesla;* o *: Dy glass in zero field; * : Gd glass in 5 Tesla;* x *: Gd glass in 1 Tesla. The dashed line is for the Gd glass in zero field (see text). The straight lines are the initial decays.*

Springer Series in Solid-State Sciences, Volume 112
Phonon Scattering in Condensed Matter VII Eds.: M. Meissner and R.O. Pohl

initial decay. We can see on Fig.1, that for the Dysprosium-doped glass the initial decay is very strong. The lower limit $T_{1,min}$ of the T_1-distribution is less than 5μs. This value is to be compared with $T_{1,min}$=200μs we have found in the non magnetic glass at same frequency and temperature. The magnetic field seems to have a little effect; it seems only to induce a little vertical translation of the curve. The increase of the Dysprosium concentration speeds up the relaxation: for an atomic concentration of 10%, the recovery of the thermal equilibrium is already completed after less than 1μs.

For the Gadolinium doped glass, a second very interesting effect appears. Our results are reported on fig.1. In zero field, the relaxation arises very quickly, since after 1μs the medium has already recovered the thermal equilibrium, like in the most concentrated Dysprosium-doped glass. The behaviour of the Gadolinium-doped glass becomes different from this one as a magnetic field is applied: at 1 Tesla, the saturation rate decreases again quickly with a value $T_{1,min}$ still less than 1μs (but much more slowly than in zero magnetic field). In higher magnetic field, the relaxation continue to slow down and at 5 Tesla the value of $T_{1,min}$ reaches 100μs. Then, we get back at the relaxation rate of the non magnetic glass.

We assign this new relaxation process to the coupling between the T.S. and some magnetic excitations, each of them involving only one ion. The Hamiltonian is [3]:

$$H=-DJ_z^2-g\mu_B\vec{J}.\vec{H} \quad (1)$$

where the first term is the anisotropy energy due to the electrostatic crystalline field and the second one is the Zeeman energy. The first term is generally much greater than the second one (for the Dysprosium ion, the anisotropy energy is about 100K). Hence, at 15mK the magnetic moment can only be aligned along the z-axis. The ions having an anisotropy axis almost orthogonal to the local magnetic field have very low-energy splittings. The reversal of the magnetic moment can only happen through the tunneling effect. These magnetic excitations constitute magnetic tunneling states [3,4]. In the case of Gadolinium, it becomes possible to suppress them by applying a magnetic field. Indeed, for Gadolinium the anisotropy energy is low (about 2K). So, in high field the Zeeman term in the Hamiltonian (1) becomes the strongest and the magnetic moment doesn't stay anymore on its anisotropy axis, but aligns on the magnetic field direction. Thus, there are no more magnetic tunneling states, and the new relaxation process disappears.

REFERENCES

[1] Anderson P.W., Halperin B.I., Varma V.M., Phil. Mag. **25**,1 (1972)
[2] Phillips W.A., J. of Low Temp. Phys. **7**, 351 (1972)
[3] Korenblit I. Ya., Shender E.F., Sov. Phys. JETP **48**, 937 (1978)
[4] Vernier N., Bellessa G., Europhys. Lett.**14**, 349 (1991)
(*) Associated with the Centre National de la Recherche Scientifique.

Low-Temperature History-Dependent Dielectric Constant in Amorphous SiO_2 and $SiO_{1.8}$ Films

B. Tigner, D.J. Salvino, S. Rogge, and D.D. Osheroff

Department of Physics, Stanford University, Stanford, CA 94305, USA

1. Introduction

We report observation of a novel history-dependent dielectric constant in two types of amorphous silicon dioxide, dependent on the history of a DC electric field applied to the samples. When the DC field was held constant, the dielectric constant decayed linearly with the logarithm of the time since the DC field was applied. When the DC field was subsequently swept slowly through a range of values, the dielectric constant showed a hole where the DC field had previously been held. We also report on investigations of the frequency- and AC excitation field-dependence of the temperature of the dielectric constant minimum in these materials.

2. Measurement Technique

The samples were 1μm thick films of reactively rf-sputtered SiO_x ($x \approx 1.8$) [1] and non-reactively rf-sputtered SiO_2. Measurements of the history-dependent dielectric constant were made using an AC capacitance bridge, with AC fields on the samples ranging from 1×10^3 to 5×10^5 V/m at 1 and 10 kHz, at temperatures between 2 and 200 mK. Superimposed on the AC field was a DC bias field which could be swept slowly between -2×10^7 and 2×10^7 V/m to obtain dielectric constant versus bias field "spectra". Investigations of the temperature of the dielectric constant minimum were also carried out with an AC capacitance bridge, with AC fields on the samples ranging from 1.6×10^3 to 1.2×10^5 V/m at frequencies from 10 Hz to 50 kHz.

3. Experimental Findings

When the bias field on a sample was held steady,the measured dielectric constant decayed linearly with the logarithm of the time since the bias field was last changed,

$$C = C_0 \ (1 \ - \ d \log_{10}(t - t_0)\ ),$$

observed in some cases from 10^1 to 10^5 seconds; typical decay rates were $d \approx$ .02%/time-decade for SiO_x samples (figure 1), and $d \approx$ .0004%/time-decade for SiO_2 samples. When the bias field was subsequently swept to obtain "spectra", the dielectric constant showed a hole where the bias field had previously been held (figure 2). As long as the bias field was not allowed to sit at the location of the hole, except momentarily during spectrum sweeps, the hole recovered approximately as the inverse square-root of time,

Springer Series in Solid-State Sciences, Volume 112
Phonon Scattering in Condensed Matter VII   Eds.: M. Meissner and R.O. Pohl

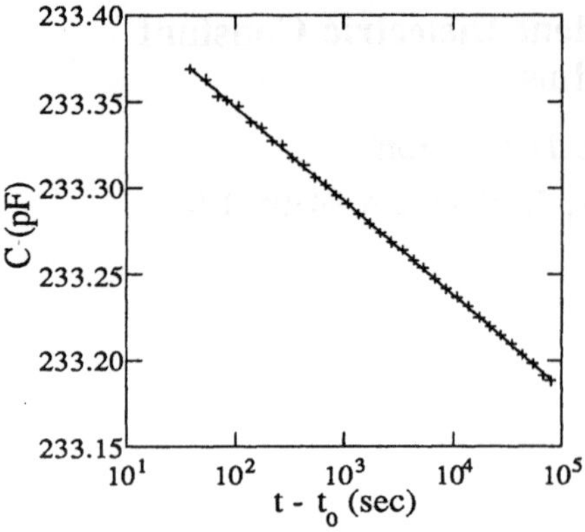

Fig. 1. Logarithmic time dependence of the dielectric constant of an SiO_x sample following the sudden application of a 10^7 V/m DC bias field.

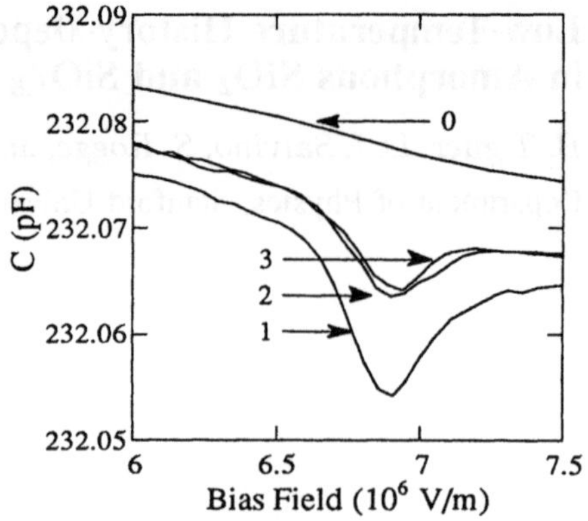

Fig 2. Recovery of a hole in the dielectric constant spectrum of an SiO_x sample. Curve 0 is a background spectrum taken before the hole was created; curves 1, 2, and 3 are the first spectra taken after the hole was created.

$$\Delta C_{depth} = r \ \Delta C_0 \ t^{-1/2},$$

with $r \approx 0.075\%\text{-sec}^{1/2}$ for SiO_x. At the time of this writing, careful measurements of the hole decay in SiO_2 have not been completed.

The frequency-dependence of the temperature of the dielectric constant minimum, which has previously been measured in many glassy materials as obeying a $f^{1/3}$ law, showed a strong dependence on the AC excitation field in both SiO_x and SiO_2. For frequencies above 100 Hz, the temperature of the minimum obeyed a power law, $T_{min} = c\, f^{\alpha}$, but the exponent of the power law varied with the AC field, ranging from $\alpha = 0.34$ at $1.7\text{x}10^3$ V/m to $\alpha = 0.16$ at $1.2\text{x}10^5$ V/m for SiO_2 samples; these data were similar for SiO_x samples. For frequencies below 100 Hz, the temperature of the minimum diverged from the high-frequency power law, increasing as the frequency decreased. Further studies of this unusual behavior are underway.

We gratefully acknowledge the DOE's support of this work through grants DE-FG03-90ER45435 and DE-FG05-89ER75517.

Reference

1. See previous studies of this system by B. Golding, J.E. Graebner, W.H. Haemmerle: Proc. 7th Int. Conf. Amorphous & Liquid Semiconductors (Edinburgh 1977) pp. 367-371

Low-Temperature Heat Release in Polymers

P. Esquinazi, M. Scherl, J. Li, and F. Pobell

Physikalisches Institut, Universität Bayreuth, W-8580 Bayreuth, Fed. Rep. of Germany

Adiabatic time - dependent heat release experiments on disordered systems are of special interest because they may provide additional information on: a) the low energy cut-off u_{min} of the distribution of Tunneling Systems (TS), b) the contribution of thermal activation to the relaxation time of TS, and c) the influence of the electron-TS interaction in the density of states of TS [1]. One of the difficulties in the interpretation of heat release ($\dot{Q}$) data are the not well known contributions from higher-order phonon processes and thermal activation to the relaxation time of TS during the cooling process from T_1 to T_0 ($T_1(T_0)$ is the "charging" (measuring) temperature, usually $T_1 > 2\,\mathrm{K}$ and $T_0 < 1$ K). Rigorously, to calculate $\dot{Q}(T_1, t)$ within the Tunneling Model (TM) and due to the thermal depopulation of the TS during the relatively slow cool down, the cooling process has to be taken into account.

As an example of the non-straightforward interpretation of $\dot{Q}$ - data we present in this work the long-time ($t = 5 \div 200\,\mathrm{h}$) heat release of two polymers with similar thermal and transport properties: Polymethylmethacrylate (PMMA) and Polystyrene (PS). Fig. 1(a) shows the specific heat $C(T, t \rightarrow \infty)$ at 60 mK $< T <$ 200 mK. From the values of the linear terms in C we find that the density of states P of TS is 1.5 times larger for PS than for PMMA, in agreement with optical experiments on the same samples [2] and previous specific heat measurements [3]. The thermal conductivity κ, Fig.1(b), shows the expected T^2 dependence for both polymers due to the resonant interaction of phonons and TS. Since $\kappa \propto 1/P$ we obtain a similar difference for P if we assume that the coupling constants between TS and phonons are equal for both polymers. The heat release $\dot{Q}$ in PMMA shows a $1/t$ dependence, Fig.2, in the measured time and temperature range in agreement with the TM. From these results we obtain an upper limit for the low-energy cut-off, $u_{min} \sim 10^{-10}$. In contrast, $\dot{Q}$ in PS is a factor ten *smaller* than in PMMA and it does not show a $1/t$ dependence; within the TM and taking into account the C and κ data we would have expected $\dot{Q}_{PS}(T_0, t) > \dot{Q}_{PMMA}(T_0, t)$. This apparent contradiction and the time dependence of $\dot{Q}_{PS}$ can be understood by introducing an additional thermal activated relaxation of the TS [4]. Anomalous $t-$dependence of $\dot{Q}$ has been also observed in high-T_c ceramic samples and was interpreted in terms of thermally activated contributions to the relaxation time [5]. Our results indicate that no reliable information about the density of states of TS can be obtained from heat release measurements without the knowledge of the relaxation during the cooling process.

Springer Series in Solid-State Sciences, Volume 112
Phonon Scattering in Condensed Matter VII   Eds.: M. Meissner and R.O. Pohl

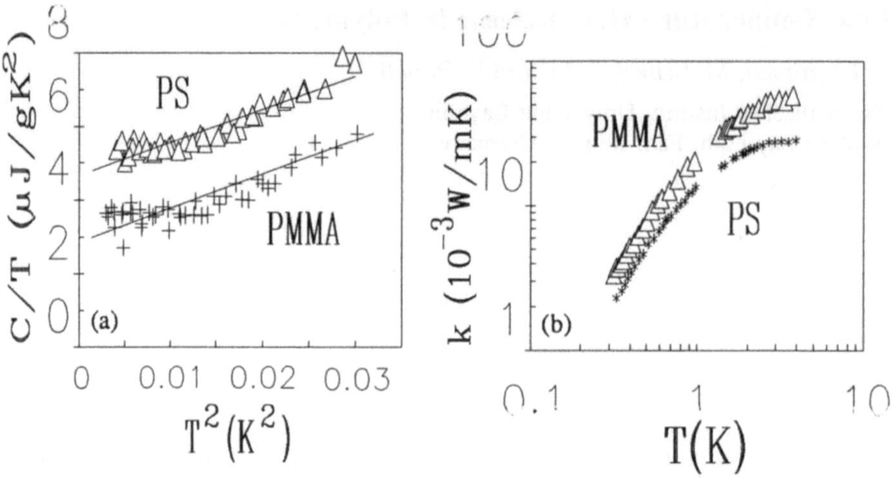

Figure 1: *(a) Specific heat divided by the temperature vs.* T^2 *and (b) thermal conductivity vs.* T *for both polymers.*

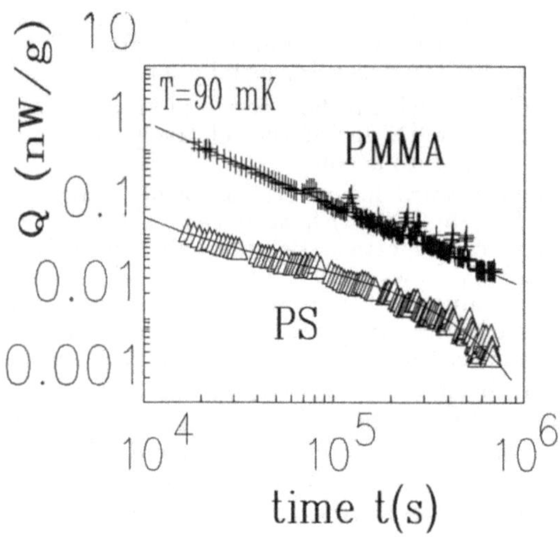

Figure 2: *Heat release vs. time for both polymers after cooling them from 80 K. The full lines are calculated within the TM taking into account the cooling process and for PS an additional thermally activated relaxation process [4]. In the calculations the influence of higher - order processes has been taken into account via an effective "charging" temperature.*

References

[1] M. Deye and P. Esquinazi, Z. Phys. B**76**, 283 (1989), and *Phonons 89*, eds. Hunklinger, Ludwig and Weiss, World Scientific, Singapure 1990, p. 468.

[2] K.-P. Müller and D. Haarer, Phys.Rev.Lett.**66**, 2344 (1991).

[3] R.B. Stephens, Phys.Rev. B**8**, 2896 (1973).

[4] M. Scherl, Diplom. Arbeit, Univ. Bayreuth (1991), unpublished.

[5] S. Sahling and J. Sievert, Solid State Commun. **75**, 237 (1990).

The Maximum Relaxation Time of Low-Energy Tunneling States in Amorphous Solids

S. Sahling

Institut für Tieftemperaturphysik, Technische Universität Dresden, Mommsenstr. 13, O-8027 Dresden, Fed. Rep. of Germany

1. Introduction

Low energy excitation with a broad distribution of their energies and relaxation times are characteristical of amorphous systems. From heat release experiments informations about tunnelling states with very large relaxation times (10^2sec < τ < 10^6sec) can be obtained.

Up to now, in all heat release experiments with amorphous solids (a-SiO_2, epoxy resin, Pentanol-2, amorphous metals - see for example Ref. 1) an upper limit τ_{max} was not found, i.e. τ_{max} of these materials was too large for registration. Since the minimum relaxation time τ_{min} depends strongly on the glass transition temperature T_g (τ_{min} is roughly proportional to $T_g^{1,5}$) /2/, one expects that τ_{max} increases with increasing T_g also. Therefore, the chance to registrate τ_{max} in heat release experiments is much better for amorphous solids with low T_g. In this work experimental results of 3-Methylpentane / 2,3-Dimethylbutane (3MP/2,3DB: T_g = 90K) are presented.

2. Results and Discussion

First the power release of a hollow container (174,7g aluminium, 17,4g copper) was measured. As expected, no long-time thermal relaxation was observed and the time independent heat leak to the sample $\dot{Q}_B$ = 1,7nW with the accuracy $\Delta\dot{Q}_B$ = 0,1nW was determined.

After cooling the sample (container + 16,8g 3MP/2,3DB) from 292K to 4,2K the power release was measured at 4,2K untill the heat leak was constant ($\dot{Q}$<0,1nW). Than the heat release $\dot{Q}$(4,2K,1,6K,t) was measured after cooling the sample from 4,2K to 1,6K.
After about 100hr the power release was again smaller than 0,1nW (see Fig.1). The sample was than heated to T_1 = 2,1K, ...,20K and kept there for at least 100hr. After cooling the sample again to 1,6K the various $\dot{Q}(T_1$,1,6K,t) were measured .

Springer Series in Solid-State Sciences, Volume 112
Phonon Scattering in Condensed Matter VII Eds.: M. Meissner and R.O. Pohl

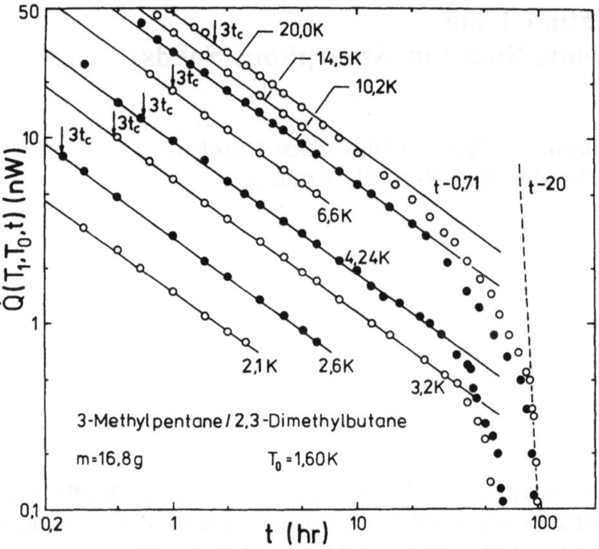

Fig. 1: The heat release $\dot{Q}(T_1, 1{,}6K,t)$ after cooling of 16,8g 3MP/2,3DB from various initial T_1 to 1,6K as a function of time t (t=0: start of cooling, $T(0)=T_1$).
t_c: time neccesary for cooling the sample from T_1 to 1,6K.

In the short time range ($t < 30$hr) the results agree with the data of other noncrystalline organic materials (epoxy resin, pentanol-2), where $\dot{Q}$ is proportional to t^{-a} with $a<1$ (see Fig.1). From the T_1-dependence of $\dot{Q}(T_1,1{,}6K,1hr)$ the distribution parameter of two-level systems $P = 1{,}7\cdot10^{45}/Jm^3$ was obtained. This value is $5 \div 17$ times larger than the corresponding values of a-SiO_2, i.e. P is roughly proportional to T_g^{-1}.

For $t>40$hr the relaxation of the power release is very fast (see Fig.1). This time dependence can be explained by the assumption of an upper limit of the relaxation time spectrum $\tau_{max} \approx (30 \div 50)$hr. τ_{max} depends only weakly on T_1 (see Fig.1) and T_0 ($T_0 \leq 4{,}2K$). The last fact indicates that there is only a small influence of the thermal activation process.

1. S. Sahling, A. Sahling, K. Koláč, Solid State Commun. 65, 1031 (1988).

2. J.F. Berett, M. Meißner, Z. Phys. B70, 65 (1988).

Freezing Phenomena and Heat Release in Glasses

D.A. Parshin[1;*] *and S. Sahling*[2]

[1]Institut für angewandte Physik der Universität Heidelberg, Albert-Überle-Str. 3–5, W-6900 Heidelberg, Fed. Rep. of Germany
[2]Institut für Tieftemperaturphysik, Technische Universität Dresden, Mommsenstr. 13, O-8027 Dresden, Fed. Rep. of Germany

1. Introduction

The usual tunneling model [1,2] predicts that the power $\dot{Q}(T_1, T_0, t)$ released after cooling an amorphous solid from some initial (equilibrium) temperature T_1 to T_0 is proportional to $(T_1^2 - T_0^2)/t$. These time and temperature dependences were observed for the most investigated amorphous and glasslike crystalline materials at temperatures below few kelvin only.

In this work the power release has been calculated in the framework of the soft potential model (SPM). Our calculation show that the most observed differences at higher temperatures are caused by the thermal activation processes.

2. The Soft Potential Model

According to the SPM [3,4], the two-level systems (TLS's) are described by the soft anharmonic oscillator potentials

$$V(x) = \mathcal{E}_0 \left[\eta \left(\frac{x}{a}\right)^2 + \xi \left(\frac{x}{a}\right)^3 + \left(\frac{x}{a}\right)^4 \right] . \tag{1}$$

Here x is the generalized coordinate having the units of length and describing the motion of the tunneling entity, a is the characteristic length of the order of the interatomic spacing ($a \simeq 1$ Å), $\mathcal{E}_0$ is the binding energy of the order of $\overline{M}v^2 \simeq 10$ eV, $\overline{M}$ being the average mass of atoms constituting the glass. The values of the dimensionless parameters η and ξ are distributed due to fluctuations of the structural parameters of a glass. The soft potentials correspond to $|\eta|, |\xi| \ll 1$. In this region the distribution function of these parameters is given by $P(\eta, \xi) = |\eta|\mathcal{P}_0/2$ where $\mathcal{P}_0$ is a constant.

Springer Series in Solid-State Sciences, Volume 112
Phonon Scattering in Condensed Matter VII   Eds.: M. Meissner and R.O. Pohl

3. Results and Discussion

In the tunneling region $T_1 < T_c$ (where T_c is a characteristic crossover temperature from tunneling to activation of the order of few kelvin) $\dot{Q}$ is roughly proportional to $(T_1^2 - T_0^2)/t$ in accordance with the tunneling model.

In the thermal activation region $T_c < T_1$ the heat release is influenced by the freezing behavior of TLS's. Below a freezing temperature T^* a nonequilibrium occupation of a TLS is frozen-in, while at higher temperatures the TLS relaxes during the cooling and does not contribute to the heat release. The freezing temperature is nearly independent of the TLS energy and depends almost linearly on the barrier height. For the time scala of heat release experiments ($10^2\,\mathrm{s} < t < 10^6\,\mathrm{s}$), our calculation yields $T_c < T^* < 1.3\,T_c$ at $T_0 < T_c$ and $T_0 < T^* < 1.3\,T_0$ at $T_0 > T_c$. The power release is independent of T_1 in both cases: $\dot{Q}$ is proportional to $T_0^{9/4} t^{-1} \ln(t/t_0) \approx T_0^{9/4} t^{-0.76}$ (for $t > t_0 \approx 100\,\mathrm{s}$) at $T_0 > T_c$ and proportional to $(T^{*2} - T_0^2)/t$ at $T_0 < T_c$.

The results of our calculation are in a good agreement with the experimental data [5], if we take into account a distribution of T^* or T_c. The distribution function $P(T^*)$ can be determined from the T_1 - dependence of $\dot{Q}$ at $T_0 < T_c$. In agreement with our consideration a sharp peak of $P(T^*)$ was found at $T^* \approx T_c$, where the crossover temperature T_c was calculated from specific heat data. At higher T^* a small excess contribution exists. These TLS's contribute markedly to the heat release because of their large energies.

References

[1] W.A.Phillips, 1987, *Rep. Prog. Phys.*, **50**, 1657.

[2] P.W.Anderson, B.I.Halperin, C.M.Varma, 1972, *Phil Mag*, **25**, 1.

[3] V.G.Karpov, M.I.Klinger, F.N.Ignatiev, 1983, *Sov.Phys.JETP*, **57**, 439.

[4] M.A.Il'in, V.G.Karpow, D.A.Parshin, 1987, *Sov.Phys.JETP*, **65**, 165.

[5] S.Sahling, A.Sahling, M.Koláč, 1988, *Solid State Commun*, **65**, 1031.

*) On leave from: St. Petersburg State Technical University, St. Petersburg, Politechnicheskaya 29, 195251, Russia.

Heat-Pulse Propagation in Vitreous Silica

H. Ikari

Faculty of Education, Shizuoka University, 422 Japan

The universal Plateau in glasses has long resisted even a qualitative explanation[1,2]. This intermediate temperature property is governed by the unknown strong phonon scattering in terahertz frequency range[3,4]. The strong scattering in question occurs for phonons whose wavelength and mean free path are on the order of 100Å. We want to know what kind of propagation mode is possible in such anomalous situation. In this report, we present a brief account of our experimental studies of heat pulse propagation in vitreous silica as a function of radiation power. The heat pulse propagation measurement can easily control the phonon frequency in the THz range by regulating the pulse power, though the phonon spectrum is not monochromatic. The observed phonon diffusion constant at the high power is 1/100 of this value at the low power. This leads to a mean free path $l \sim 100$Å. Thus, our experimental results reveal part of the anomalous diffusion poperties of terahertz phonons transmitting through the bulk glass.

The time of flight measurements of vitreous silica at 2.3K have been performed over the power range 0.1 to 100W/mm^2. The samples were immersed in compressed liquid helium for avoiding the effect of the film boiling [5]. Disk shaped samples of Suprasil W with 40-300μ thicknesses were used. The phonons were generated by passing a 60ns current pulse through a constantan film and were detected by means of a thin-film superconducting bolometer.

In Fig.1, we show detected pulse shapes as a function of time for 0.1mm thick sample at representative energies. As the energy is increasing, the arrival of the thermal pulse is remarkably retarded. This suggests that high energy phonons diffuse across the vitreous silica with the extremely short mean free path. In order to characterize these data on a single curve, it is convenient to determine the leading edge of these pulses and relate it to the diffusion constant. We also obtain the average phonon frequency by using the

Springer Series in Solid-State Sciences, Volume 112
Phonon Scattering in Condensed Matter VII Eds.: M. Meissner and R.O. Pohl

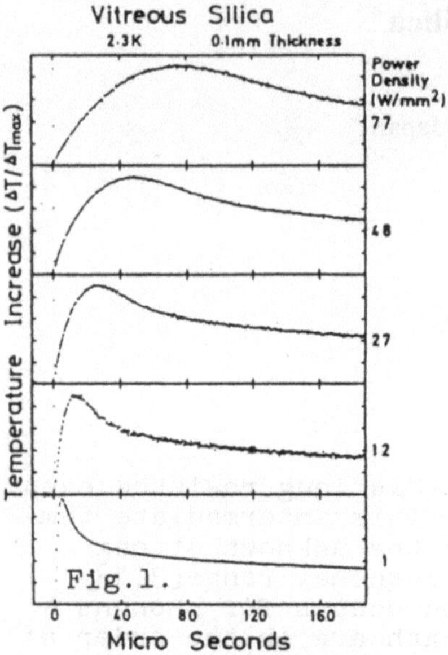

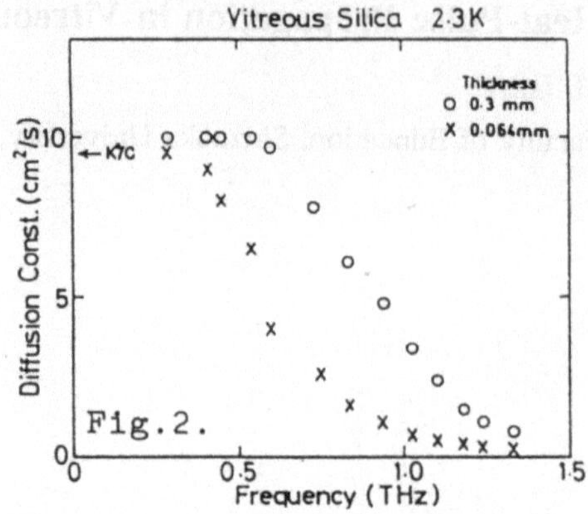

Fig.1(left): Energy dependence of diffusion pulses in vitreous silica at 2.3K.

Fig.2(right): Diffusion constant for various energies.

Stephan-Boltzmann type radiation law. In Fig.2, we now present the phonon diffusion constants as a function of the average phonon frequency. The diffusion constants at the low frequency are equal to the thermal diffusivity K/C. However, when the average phonon frequency exceeds about 500GHz, the diffusion constants are rapidly reduced. The critical frequency also depends on the sample thickness, suggesting phonon diffusion processes with inelastic contributions. This is qualitatively consistent with existing discussions[6] for the plateau problem. The diffusion constant at the high power varies linearly with propagation distance. This suggests a quasidiffusion like propagation mode.

This research has been supported by Iketani Science and Technology Foundation.

1. W.A. Phillips, Amorphous Solids: Low-Temperature Properties (Springer-Verlag, Berlin, 1981).
2. C.C. Yu and A.J. Leggett, Comm. Cond. Mat. Phys. 14, 231 (1988).
3. E.R. Grannan, M. Randeria and J.P. Sethna, Phys. Rev. B41, 7799 (1990).
4. U. Buchenau, H.M. Zhou, N. Nücker, K.S. Gilroy and W.A. Phillips, Phys. Rev. Lett. 60, 1318 (1988).
5. H. Ikari, Phys. Lett. A123, 179 (1987).
6. J. Mebert, B. Maile and W. Eisenmenger, in Phonons89, eds. S. Hunklinger, W. Ludwig and G. Weiss (World Scientific, Singapore, 1989).

Thermal Transport in Superionic Conducting Glasses

Y. Hiki[1], *H. Takahashi*[1], *and Y. Kogure*[2]

[1]Applied Physics Group, Faculty of Engineering, Nakanarusawa, Hitachi 316, Japan

[2]Faculty of Science and Engineering, The Nishi Tokyo University, Uenohara, Yamanashi 409-01, Japan

The thermal conductivity and thermal diffusivity have been measured for superionic conducting glasses AgI-$AgPO_3$ from the room temperature up to the temperatures above the glass transition T_g. The points of interest of ours are: (a) the mechanism of the superionic conduction, especially in glassy materials, (b) the microscopic structures of the glassy superionic conducting glasses, and (c) the mechanism of the thermal transport in superionic conducting materials. These are closely related to each other, and the present study may be useful for the purpose of solving these problems.

The measurment is carried out by using the transient hot-wire method, which is used to determine the thermal conductivity κ and thermal diffusivity D simultaneously in liquids [1] and solids [2]. We have made some improvement of the method, and the principle of our measurment was as the following. A thin wire enclosed in a sample is heated by an electric current, and the wire plays two roles: an electric heating element and a resistance thermometer. The temperature of the wire increases from T_0 to T after time t as $T - T_0 = (q/4\pi\kappa)\ln(4Dt/a^2C)$, where q is the heat generation rate per unit length of the wire, a the radius of the wire, and $C = 1.781$ [3]. Potential leads are attached onto two points separated by a distance l on the wire. The electric resistance of the wire between the points is $R = \alpha T + \beta$ (α, β : constants) when $T - T_0$ is not so large. When the electric current in the wire is I, the generated heat is $q = I^2R/l$. Then, the change of the wire resistance is $R - R_0 = (\alpha I^2 R/4\pi l\kappa)\ln(4Dt/a^2C)$. The change of the voltage $E = IR$ between the potential leads is measured as a function of time t with using a constant current I. In our experiment, several constant current values are used for a measurement. Our expression for processing experimental data is $z_{ij} = R_0 + Ax_iy_j + By_j$. Here, $\ln t = x_i$, $I^2 = y_j$, and $R = z_{ij}$ are variables; R_0 is a constant; and both $A = (\alpha R/4\pi l\kappa)$ and $B = (\alpha R/4\pi l\kappa)\ln(4D/a^2C)$ are also constants with regarding $R \sim R_0$ for small $T - T_0$. By using a number of data points with different i and j, the quantities R_0, A and B can accurately be determined. The thermal conducivity κ is obtained from A value, and the thermal diffusivity D from A and B values.

The experiments are actually carried out as follows. A fused-quartz cylindrical specimen cell 30 mm in length and 12 mm in diameter is used, and a 0.1 mmϕ Pt heating wire with two Pt potential leads is stretched along its axis. A mixture of proper amounts of powdered $AgNO_3$, $NH_4H_2PO_4$,and AgI is melted and kept at 500° C for 2 h, and the melt is cast into the specimen cell through its side-wall window. The cell is then set in an electric furnace. The present measuring system is composed of three parts: (i) A PID temperature

Springer Series in Solid-State Sciences, Volume 112
Phonon Scattering in Condensed Matter VII Eds.: M. Meissner and R.O. Pohl

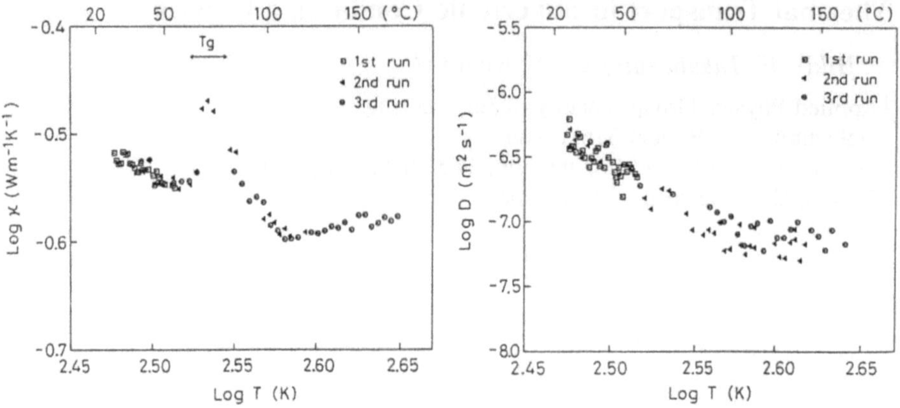

Fig. 1. Temperature dependences of thermal conductivity κ and thermal diffusivity D for $(AgI)_{0.5}(AgPO_3)_{0.5}$ glass.

controlling unit for the furnace. (ii) A constant current power supply for the wire. (iii) A high precision digital voltmeter to measure the voltage between the potential leads. The control of these units and the transfer of the data to the host computer are performed through the IEEE-488 communication line. At an ambient temperature of the specimen, six heating currents, being $I = 0.07 - 0.12$ A with 0.01 A steps, are used. For each of the current, the heating time is 2.52 s, and 101 data points of the voltage are taken during the run. The interval between the runs with different currents is 20 s. After finishing the measurement at a specimen temperature, the PID controller offers another measuring temperature, and the measurement is repeated.

The results for a $(AgI)_{0.5}(AgPO_3)_{0.5}$ specimen are presented in Fig. 1, where the thermal conductivity κ and the thermal diffusivity D are shown as functions of temperature T. The measurement was made by increasing the temperature, and the time reqired for the experimental run was about 24 h. Three runs have been repeated, showing no meaningful discrepancy between the data in these runs. The scatter of data points may mainly be due to the drift and fluctuation of the ambient temperature.

The expeimental results are qualitatively explained by the phonon gas kinetics equation $\kappa = (1/3)Cvl$ and the definition $D = \kappa/\rho C$, where C is the heat capacity per unit mass, ρ the material density, v and l are the velocity and mean free path of phonons. An apparent peak is observed in κ-vs-T near T_g, which reflects the specific heat anomaly in glassy materials. The value of D decreases with T below T_g, and the decrease levels off above T_g. This behavior seems to be very meaningful since D is soley proportional to the phonon mean free path. We are continuing experiments for glasses of $(AgI)_x(AgPO_3)_{1-x}$ ($x = 0 - 0.5$), and a full argument about the phonon transport mechanism in the materials will appear in the near future.

References

[1] Y. Nagasaka and A. Nagashima, Rev. Sci. Instrum. **52**, 229 (1981).
[2] O. Sandberg, P. Andersson and G. Bäckström, J. Phys. E **10**, 474 (1977).
[3] H. S. Carslaw and J. C. Jaeger, *Conduction of Heat in Solids*, 2nd ed. (University Press, Oxford, 1959).

Thermal Diffusivity Threshold in Ge-As-Se Semiconducting Glasses

K. Nandakumar and J. Philip

Department of Physics, Cochin University of Science and Technology, Cochin-682 022, India

Abstract: The thermal diffusivity α of two sets of glasses in the Ge-As-Se system has been measured using the photoacoustic (PA) technique. In both the series the thermal diffusivity shows threshold maximum around the average atomic coordination $\langle r \rangle = 2.67$. The observed behaviour is attributed to the topological changes in the network due to the formation of layered structure in these system.

Recent progress in the explanation of glass forming tendency and the development of new structural models for chalcogenide glasses have evoked renewed interest in the study of the composition dependent variation of various physical properties of these materials. The idea of interpreting physical and structural properties of network glasses in terms of the average coordination number $\langle r \rangle$ is an outcome of the theories by Phillips and Thorpe [1,2,3], based on balancing the constraints acting on the glass network with the available number of degrees of freedom. At $\langle r \rangle = 2.4$, theory predicts that a structural change occurs and the glass transforms to a rigid structure. This value of $\langle r \rangle$ is called the rigidity percolation threshold. This formulation based on short-range interactions has later been modified by Tanaka [4,5]. According to him, chalcogenide glasses possess a layered structure and hence medium-range interactions have to be considered to account for the topological changes in these glasses as $\langle r \rangle$ is varied. He showed that the coordination number of a glass having a stable layer structure is 2.67.

Here we report the results on the composition dependence of thermal diffusivity α of two series of glasses belonging to the Ge-As-Se system using photoacoustic (PA) technique. The compositions studied can be represented as $Ge_xAs_{10}Se_{90-x}$ (x = 5, 10, 15, 20, 25, 28.5, 30, 35 and 40) and $(Ge_{0.5}As_{0.5})Se_{100-x}$ (x = 10, 20, 26.7, 30, 40, 44, 50 and 60). Bulk glasses of various compositions have been prepared by the melt-quenching technique. The amorphous nature has been checked by X-ray diffractometry. The PA spectrometer used for thermal diffusivity measurements is described in detail elsewhere [6]. The

Springer Series in Solid-State Sciences, Volume 112
Phonon Scattering in Condensed Matter VII Eds.: M. Meissner and R.O. Pohl

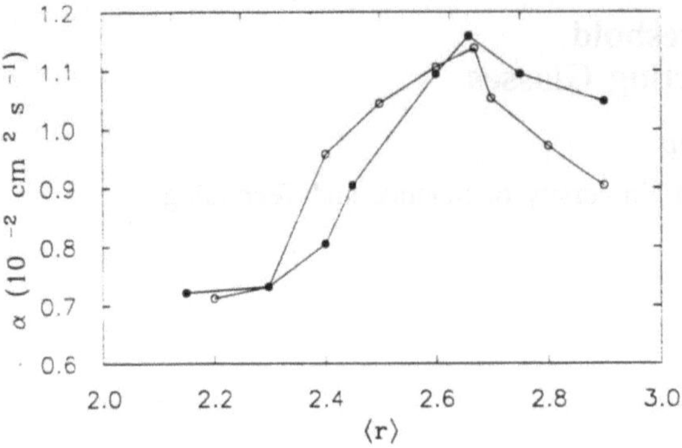

Fig.1: The variation of thermal diffusivity with average coordination number of (o) $Ge_xAs_{10}Se_{90-x}$ and (•) $(Ge_{0.5}As_{0.5})_xSe_{100-x}$ glasses

measurements have been made on samples of appropriate thickness (≈100 μm) mounted on a thermally thick backing material. The variation of PA signal amplitude with chopping frequency is measured to determine the thermal diffusivity of each sample [7].

Fig.1 shows the variation of α with $\langle r \rangle$ for $Ge_xAs_{10}Se_{90-x}$ and $(Ge_{0.5}As_{0.5})_xSe_{100-x}$ glasses. As $\langle r \rangle$ is increased, α increases gradually and shows a maximum around $\langle r \rangle = 2.67$ for both sets of glasses investigated. The thermal diffusivity increases gradually as the 2-d structure is developed through the branching of flexible Se chains as $\langle r \rangle$ is increased and at $\langle r \rangle = 2.67$ the 2-d structure is fully developed and the network goes to a rigid ordered 3-d structure. The decrease in α beyond $\langle r \rangle = 2.67$ may be attributed to the large bonding defects present in the network beyond this critical region. The results indicate that medium range interactions are important in determining the network topology and structure of chalcogenide glasses.

REFERENCES

1. J.C.Phillips, J.Non-Cryst.Solids, **34**, 153 (1979)
2. M.F.Thorpe, J.Non-Cryst.Solids, **57**, 355 (1983)
3. H.He and M.F.Thorpe, Phys.Rev.Lett., **54**, 2107 (1985)
4. K.Tanaka, J.Non-Cryst.Solids, **103**, 149 (1988)
5. K.Tanaka, Phys.Rev. **B39**, 1270 (1989)
6. K.N.Madhusoodanan et al., Phil.Mag., **B58**, 123 (1988)
7. A.Lachaine and P.Poulet, Appl.Phys.Lett., **45**, 953 (1984)

Temperature Dependence of the SeH Vibrational Dephasing Time in Chalcogenide Glasses

R.J. Till and A.J. Sievers

L.A.S.S.P. and M.S.C., Cornell University, Ithaca, NY 14853, USA

The concept of relating the properties of glass-like solids to their microscopic topology rather than to their chemical nature was originally introduced by Phillips [1] and later refined by Thorpe[2]. The key parameter in this model is the mean coordination number $\langle r \rangle$, i. e. the average number of covalent bonds per atom. Constraint counting arguments lead to topological thresholds at $\langle r \rangle = 2.4$ and $\langle r \rangle = 2.67$. Both can be observed in different glasses as extrema or kinks in physical properties as a function of the mean coordination number, such as the glass forming temperature, the mean atomic volume or the dielectric constant[3-5].

The investigation of the low temperature hole relaxation in chalcogenide glass alloys by S. Love et. al.[6] demonstrates that dynamical properties are strongly related to the topology of glasses: the hole relaxation rates increase monotonically with the mean coordination number over three orders of magnitude between $\langle r \rangle = 2$ and 2.8 and are independent of the chemical composition. Some preliminary measurements also suggests that purely geometrical arguments also may play an important role in determining the temperature dependence of dephasing times in glasses at low temperatures[7].

A great deal of effort has gone into characterizing the temperature dependence of dephasing times in glasses[8]. Hole burning or photon echo experiments are used in the optical region to determine the homogeneous linewidths Γ which usually follows a power law

$$\Gamma = \Gamma_0 + a\,T^b. \qquad (1)$$

The temperature exponent b falls roughly into two categories: in organic glasses the experiments yield $b \sim 1.3$, whereas in inorganic amorphous solids $b \sim 2$. Different theories[9] attribute a temperature exponent of 1.3 to the interaction of the optical centers and the glassy two level systems, which have been introduced to explain the thermal anomalies and ultrasonic properties of amorphous solids. The quadratic temperature dependence, on the other hand, can be understood as a two phonon Raman dephasing process, if a small Debye temperature is postulated relative to that of a crystal. To date it has not been possible to map the dephasing time properties between the two different chemical types of glasses.

For the GeAsSe alloy glasses it is possible to vary the mean coordination number from the polymeric chain-like limit of $\langle r \rangle = 2.0$ to $\langle r \rangle \sim 2.8$. For our experiments the chalcogenide glass samples are melt-quenched in a hydrogen

Springer Series in Solid-State Sciences, Volume 112
Phonon Scattering in Condensed Matter VII Eds.: M. Meissner and R.O. Pohl

atmosphere, which produces SeH molecules in the glass with a vibrational stretch mode at roughly 2200 cm^{-1} characterized by an inhomogeneous line width of approximately 100 cm^{-1}. This band shows persistent spectral hole burning because of local structural changes in the glass[6,7]. Using a tunable infrared lead-salt diode laser at a frequency of 2231 cm^{-1}, we measure the homogeneous linewidths by burning holes with burn times between 5 and 160s and then extrapolating to zero burn time at each temperature[10].

The results are that for Se, a purely chainlike structure with a mean coordination number of 2, the temperature exponent $b \sim 1.3$. The alloying of small amounts of As or Ge produces cross linking of these chains and leads to an increase in the b value, which at first increases with alloy concentration but then levels off to a value of $b = 2.2$ for coordination number $<r> > 2.2$.

Clearly, the temperature exponent is not related to the customary topological thresholds at $<r> = 2.4$ or 2.67 but may instead be connected to the dimensionality of the microstructure. According to a model proposed by Phillips[1], the structure consists mainly of one dimensional units for $2.0 < <r> < 2.06$. In the region $2.08 < <r> < 2.18$ the fourfold coordinated Ge and the threefold coordinated As atoms lead to a cross linking of the chains to form new structures, which have two dimensional character. In the third region ($2.2 < <r> < 2.38$) those planar units combine to form three dimensional clusters. His ideas are supported by measurements of the viscosity of As_xSe_{1-x} and Ge_xSe_{1-x} glasses[11].

With this model in mind, we speculate that the change in the value of the measured temperature exponent of the dephasing frequency in chalcogenide glasses is related to the dimensionality of the microscopic structure in the glass. Hence the evolution of the temperature exponent from a small value for polymeric, chain-like organic systems to a large value for inorganic glasses also may be a consequence of the change in dimension of the microscopic structure.

Supported by NSF-DMR-89-18894 and ARO-DAAL03-90-G-0040.
R. J. T. acknowledges partial support from the A. v. Humboldt-Foundation.
We thank B. L. Halfpap and S. P. Love for the supply with samples.

1. J. C. Phillips, J. Non-Cryst. Solids **34**, 153 (1979); **43**, 37 (1981)
2. M. F. Thorpe, J. Non-Cryst. Solids **57**, 355 (1983)
3. S. Mahadevan, A. Giridhar, J. Non-Cryst. Solids **143**, 52 (1991)
4. A. Feltz, H. Aust, A. Blayer, J. Non-Cryst. Solids **55**, 179 (1983)
5. M. Tatsumisago, B. L. Halfpap, J. L. Green, S. M. Lindsay, C. A. Angell, Phys. Rev. Lett. **64**, 1549 (1990)
6. S. P. Love, A. J. Sievers, B. L. Halfpap, S. M. Lindsay, Phys. Rev. Lett. **65**, 1792 (1990)
7. S. P. Love and A. J. Sievers, J. of Lumin. **45**, 58 (1990).
8. R. M. Macfarlaine, R. M. Shelby, J. of Lumin. **36**, 179 (1987)
9. R. Silbey, K. Kassner, J. of Lumin. **36**, 283 (1987)
10. S. Völker, J. of Lumin. **36**, 251 (1987)
11. S. U. Nemilov, Sov. J. Phys. Chem. **37**, 1026 (1964)

Collective Excitations in a Molecular Glass

J.L. Martínez[1], *F.J. Bermejo*[2], *M. García-Hernández*[2], *J. Alonso*[2], *and F.J. Mompeán*

[1]Instituto de Ciencia de Materiales de Madrid (CSIC), Dep. Física de Materiales (C-4), Universidad Autónoma de Madrid, E-28049 Madrid, Spain and Institut Laue-Langevin, B.P. 156X, F-38042 Grenoble Cédex, France

[2]Instituto de Estructura de la Materia (CSIC), Serrano 123, E-28006 Madrid, Spain

Abstract. The excitation spectra of a glass-former system (CD_3OD-6%D_2O) have been studied by neutron scattering techniques. A clear inelastic peak is observed for temperatures above 10K. This behavior is consistent with the temperature dependence of the total vibrational density of states (TDOS), as well as with the single particle dynamics obtained from the equivalent incoherent (hydrogenated) sample. The results are tentatively interpreted as coupling between sound waves and localized excitations.

The thermal and related dynamic properties of glasses well below the glass temperature are now understood on qualitative basis. This is specially true at very low temperatures (<1K). However at intermediate temperatures (1 to 50K) seem clear also that additional excitations are present which are responsible of the anomalous thermal dependence of the specific heat and thermal conductivity.

The experiments were carried out at different instruments of the Institut Laue Langevin (Grenoble, France), specially time of fight (IN6) and triple axis (IN8). The sample container was optimized for each configuration being a plate of 0.6 mm thickness for IN6, or a rectangular box containing different horizontal plates of cadmiun-coated aluminium plates of 0.1mm thickness separated by 1mm, in order to avoid multiple scattering processes (IN8). The sample container with the sample in was quenched inside a liquid nitrogen dewar, which produce the glassy state in this material.

The typical constant-Q spectra from this material present well resolved inelastic peaks at finite low energy in a wide range of Q and temperature values. This inelastic response is located at substantially lower frequency in the glass sample than in the corresponding polycrystal. The Q-dependence of this peak maxima is presented in Fig. 1 for different temperatures (including the polycrystal). This data are complementary to that observed in a previous experiment [1, 2] and closely resembles that observed in metallic glasses [3], where a *rotonlike* minimum is observed for values near the maximum of the structure factor. The data were analyzed in terms of sums of gaussian functions, with the peak corresponding to the low frequency centered around 2meV, with the following expression

$$S(Q,\omega) = \exp(-2W(Q))\,(n(\omega)+1)\,(S^1(Q,\omega) + S^M(Q,\omega))$$

being $W(Q)$ the Debye-Waller term, $n(\omega)$ the detailed balance factor and S^M the multi-phonon term. The one-phonon dynamical structural factor $S^1(Q, \omega)$ is written as,

Springer Series in Solid-State Sciences, Volume 112
Phonon Scattering in Condensed Matter VII Eds.: M. Meissner and R.O. Pohl

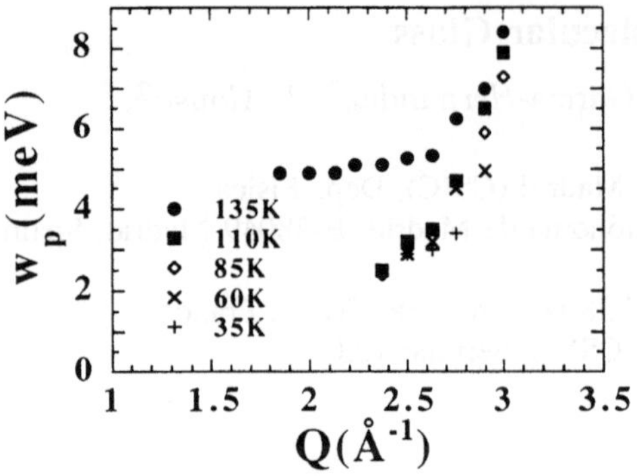

Fig. 1. Low-frequency peak position (ω_p) *vs.* wavevector (Q) at different temperatures. T=135K corresponds to the polycrystalline state, the glass state correspond to temperatures below 135K.

$$S^1(Q,\omega) = \sum_j \frac{2}{\pi \tau_j^2} \exp\left(-(\hbar\omega - \hbar\omega_0) - \frac{\hbar^2 Q^2}{2 M_j})^2 / 2\tau_j^2\right) ,$$

$$\tau_j^2 = \frac{\hbar Q^2}{2 M_{av}} \int Z(\omega)\, n(\omega)\, \omega d\omega \; .$$

The sums runs over the number of components, M_j is the effective mass, ω_0 the recoilless lattice frequency and $\hbar^2Q^2$ the recoil energy, the width factor τ_j behave asymptotically ($Q\rightarrow\infty$). A recoil mass of 1.93 a.m.u. was also calculated from the recoil curves. The data presented in Fig. 1 for the glass state present a very weak dependence on temperature, which may indicate the harmonic nature of the excitations giving rise to this peak.

REFERENCES

[1] F.J. Bermejo, J.L. Martínez, M. García-Hernández, D. Martín, F.J. Mompeán, J. Alonso and W.S. Howells. Europhys. Lett. **15**, 509-514 (1991). See also, F.J. Bermejo, D. Martín-Marero, J.L. Martínez, F. Batallán, M. García-Hernández and F.J. Mompeán. Phys. Lett. **A150**, 201-206 (1990).

[2] F.J. Bermejo, J. Alonso, A. Criado, F.J. Monpeán, J.L. Martínez, M. García-Hernández and A. Chahid. Phys. Rev. **B46**, (1992) (1-September-1992).

[3] J.B. Suck, J. Rudin, H.J. Gunterodt, H. Beck. Phys. Rev. Lett. **50**, 49 (1983); L.J. Lewis, N.W. Ascroft. Phys. Rev. **B34**, 8477 (1986), and references therein.

Vibrations of Disperse Silicas: A Comparative Study

E.F. Sheka[1], *I. Natkaniec*[2], *I.V. Markichev*[1], *and V.D. Khavryutchenko*[3]

[1]Russian Peoples' Friendship University, 3 Ordjonikidze,
117302 Moscow, Russia
[2]Frank Laboratory of Neutron Physics, JINR, 141980 Dubna, Russia
[3]Institute of Surface Chemistry, 252028 Kiev, Ukraine

Abstract. Four families of disperse amorphous silica representing different solid species modifications have been studied by inelastic neutron scattering at 10, 80 and 290K. Doubly amplitude-weighted density of vibrational states (AWDS) have been obtained from the experimental time-of-flight spectra.

Among the species, there are a globular pyrogenic aerosil A380 consisting of spherical particles of 150Å in average diameter; two porous silica gels with average pore size of 20 (SG20) and 100Å (SG100); a polymerlike aerogel AG200 with a mass density of 200 Kg/m^3. Every species family contains three samples: air-dry, air-dried, and deuterated ones. All samples are considered to be complex systems consisting of three components, namely: a silica core, a hydroxil shell and adsorbed (confined) water.

The obtained AWDS spectra are decomposed into the partial basic spectra of the sample components using a new correlation-differential method. The postulate of lack of linear correlation between the basic spectra forms the method's foundation. By making use of the postulate, the construction of a physically significant basis with components meeting the zero correlation coefficient criterion is accomplished by a linear algebra algorithm [1].

The initial composite AWDS spectra of the species were reported earlier [2]. Fig. 1 shows an example of the correlation-differential decomposition procedure of an initial aerosil A380 spectrum. The sum of the presented partial basic spectra fully reproduces the initial spectrum of the air-dry sample [3]. The same decomposition procedure has been applied to the AWDS spectra of the two silica gels [4] and aerogel [5]. All basic spectra thus obtained represent a set of "finger prints" of the samples studied and have allowed one to make their comparative study on a qualitatively new level.

The spectra of the cores, hydroxil and water components of the four studied samples are different. This fact witnesses a principal difference in their core structures that results, in turn, in different hydroxil component structures. The analysis shows [6] that hydroxil components of aerosil and aerogel consist of silanol groups predominantly. In the case of silica gels, this component is made mainly of a combination of silandiol and silantriol groups differing by their weighting factors in different samples. The analysis of such structure allows one to conclude that the basic structural fragments of aerosil and aerogel are rigid and compact enough while that silica gels is more crumble.

The analysis of the water component spectra has revealed the following peculiarities. The first two are related to silica gel spectra. 1. The water components

Springer Series in Solid-State Sciences, Volume 112
Phonon Scattering in Condensed Matter VII Eds.: M. Meissner and R.O. Pohl

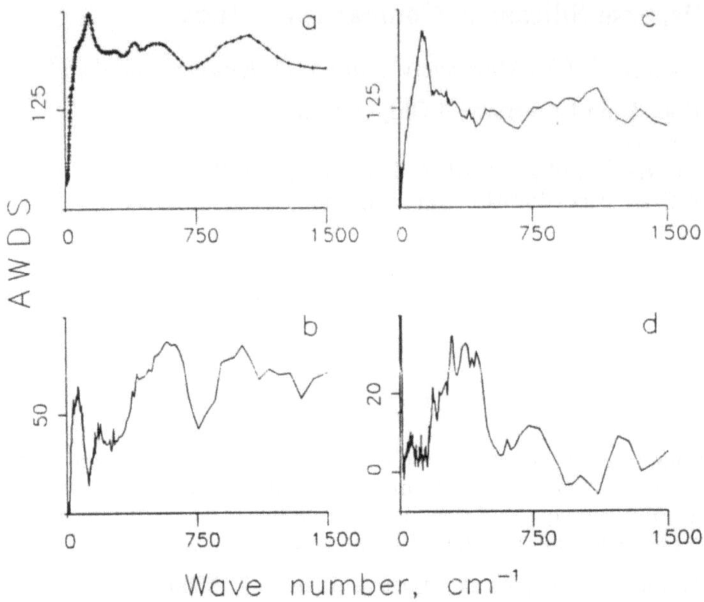

Fig. 1: AWDS spectrum of the air-dry aerosil A380 and partial basic spectra of its components, T=10 K [3].
a. initial aerosil; b. adsorbed water; c. hydroxil component; d. silica core.

of silica gels consist of two constituents, quasifree water (confined water in pores) and adsorbed water. 2. The adsorbed water spectra of silica gels are practically the same without any correlation with the average pore size, contrary to the behavior of the core and hydroxil component spectra. This observation supports a suggestion that the difference in the hydroxil component spectra of silica gels can be explained by its complex content with respect to the silandiol and silantriol groups. At the same time water adsorption occurs through interaction with "the strongest" group that chemically should be the same at every silica gel.

The adsorbed water spectra of aerosil and aerogel are similar and different simultaneously. The difference is connected with the absence of the band at 560 cm^{-1} in the aerogel spectrum. This band is assigned with the hydrogen bond vibrations. Its presence in the aerosil spectrum indicates a collective nature of water adsorbtion. Contrary to this its absence in the aerogel spectrum witnesses an individual molecule mode of adsorbtion. The spectra structure in the region 5-350 cm^{-1} is similar in both spectra, indicating a similar character of water molecule motion with respect to a silanol group.

The spectra analysis presented above in brief has allowed one to bring up a suggestion that the three studied amorphous silica are related to different structure classes of the species. Following Libau's silicate crystal classification [7], they can be attributed to amorphous tectocilica (aerosil), cyclesilica (silica gel) and polysilica (aerogel) [8].

[1] I.V. Markichev, I. Natkaniec, E.F. Sheka: Communication of the Joint Institute for Nuclear Research, P14-92-85 (Dubna 1992).

[2] E.F. Sheka, I. Natkaniec, V.D. Khavryutchenko, P.B. Nechitajlov, A.Yu. Musychka, V.M. Ogenko, I.V. Markichev, J. Brankowski, J. Krawczyk: Journ. Electr. Spectr. Rel. Phenom., 54/55, 855 (1991).

[3] I.V. Markichev, I. Natkaniec, E.F. Sheka: Communication of the Joint Institute for Nuclear Research, P14-92-86 (Dubna 1992).

[4] I.V. Markichev, I. Natkaniec, E.F. Sheka: Communication of the Joint institute for Nuclear Research, P14-92-87 (Dubna 1992).

[5] I.V. Markichev, A.Yu. Musychka, I. Natkaniec, E.F. Sheka: Zh. Struct. Khimii (Russia - J. Struct. Chem.) in press.

[6] E.F. Sheka, I.V. Markichev, V.D. Khavryutchenko, I Natkaniec: Zh. Struct. Khimii (Russia – J. Struct. Chem.) in press.

[7] F. Libau; Structural Chemistry of Silicates: Structure, Bonding and Classification (Springer, Berlin 1985).

[8] V.D. Khavryutchenko, E.F. Sheka: Zh. Fizich, Khimii (Russia – J. Physical Chem.) in press.

Terahertz Phonons in Highly Disordered Crystals

U. Happek[1;2], *W.W. Fischer*[2], *and J.A. Campbell*[3]

[1]Laboratory of Atomic and Solid State Physics, Cornell University, Ithaca, NY 14853-2501, USA

[2]Institut für Angewandte Physik, Universität Regensburg, W-8400 Regensburg, Fed. Rep. of Germany

[3]Department of Physics, University of Canterbury, Christchurch, New Zealand

Abstract. The relaxation of non-equilibrium THz phonons in the mixed-crystal systems $Ca_{1-x}Sr_xF_2$ and $Ca_{1-x}La_xF_{2+x}$ has been studied at low temperature. In the $Ca_{1-x}Sr_xF_2$ system we find only a moderate (up to a factor of two) increase of the relaxation rate, even at high Sr concentrations, and a ν^5 frequency dependence of the relaxation rate, as is characteristic for spontaneous phonon decay due to lattice anharmonicity, similar to that found in pure CaF_2 and SrF_2. In $Ca_{1-x}La_xF_{2+x}$ however, La concentrations in the order of even 1% reduce the lifetime by up to two orders of magnitude. Furthermore, the relaxation rate increases with the cube of the frequency, indicating the existence of a new mechanism dominating the relaxation of the phonons.

Optical techniques for both phonon generation and detection within the crystal volume have been used successfully to study the dynamics of terahertz phonons, including the anharmonic decay [1], elastic and inelastic scattering [2], and surface reflection [3]. Recently, optical techniques have been applied to investigate the relaxation of terahertz phonons in disordered crystals. In these crystals, high frequency phonons are characterized by an extremely short lifetime and mean free path and are not readily accessible by phonon detectors on the crystal surface, thus favoring the use of optical techniques.

In this paper, we report on the study of the relaxation rate of THz phonons at low temperatures in two different mixed-crystal systems, $Ca_{1-x}Sr_xF_2$, where Sr^{2+} ions replace Ca^{2+} ions randomly, and $Ca_{1-x}La_xF_{2+x}$, where, in addition to the replacement of the Ca^{2+} ion by a La^{3+} ion, an interstitial F^- ion is needed for charge compensation. This system shows a gradual change in the behavior of the low temperature thermal conductivity, the internal friction, and the specific heat from a crystalline to glassy regime [4]. Studying the phonon lifetimes in these mixed crystals can test whether inelastic processes, in addition to spontaneous decay due to lattice anharmonicity and to elastic scattering, are present in glass-like systems.

The experimental techniques used here for the generation and detection of the phonons have been applied previously to study the dynamics of high fre-

Springer Series in Solid-State Sciences, Volume 112
Phonon Scattering in Condensed Matter VII Eds.: M. Meissner and R.O. Pohl

quency phonons in CaF_2 and SrF_2. High frequency phonons were generated within the sample volume by multiphonon absorption of CO_2 laser radiation at 28 THz [5]. In this process, three mainly optical phonons are generated, which decay rapidly into high frequency acoustic phonons. Being an intrinsic property of the lattice, the generation of these high frequency phonons does not require the introduction of impurities, as do other optical techniques of phonon generation, such as defect induced one-phonon absorption or nonradiative relaxation of excited electronic impurity levels. In order to detect the phonons, we make use of the phonon-induced anti-Stokes fluorescence from the lowest excited state of Eu^{2+} ions at the appropriate frequency shift ν from the zero phonon line [6]. In the mixed crystals, the zero phonon line is broadened due to the composition of the crystal (for CaF_2 and SrF_2 the Eu^{2+} zero-phonon lines peak at a wavelength $\lambda_0 = 413$ nm and $\lambda_0 = 401$ nm, respectively) and the lowest excited Eu^{2+} level is split due to the reduction in symmetry at the Eu^{2+} site. By selective excitation of a subset of the Eu^{2+} ions in a narrow frequency band at the low frequency wing of the broadened zero phonon line, using a continuous wave dye laser, we can observe anti-Stokes fluorescence and emission from a distribution of the split excited Eu^{2+} levels, which are in equilibrium with the phonons, with a definite relation between phonon frequency and fluorescence wavelength. The Eu^{2+} concentration is kept low (0.01 mol %) to exclude any possible influence of the ions on the phonon decay time.

The crystals (with a typical size of $4 \times 4 \times 4$ mm^3) are immersed in liquid helium at a temperature of 1.6 K and phonons are generated homogeneously in the crystal volume by illuminating it with a Q-switched CO_2 laser (50 ns pulse duration, 10 μJ pulse energy and 10 kHz repetition rate). The phonon-induced fluorescence in the frequency range from 0.3 to 2.0 THz was analyzed in a double monochromator. Using a time-resolved photon-counting technique, signals were integrated for as much as 4×10^7 shots, which is necessary due to the small fraction of Eu^{2+} ions which are in the narrow frequency region of the dye laser radiation and the resulting low sensitivity of the phonon detection technique.

The phonon decay time as a function of frequency is shown in Fig. 1 (left) for pure CaF_2 and for a $Ca_{1-x}Sr_xF_2$ mixed crystal with x = 0.05. The decay time of the mixed crystal is reduced by a factor of 2.4 due to the introduction of the Sr^{2+} ions. We find a ν^{-5} dependence of the lifetime at high frequencies for samples with Sr^{2+} concentrations between x = 0.01 and x = 0.4. This frequency dependence is the signature of spontaneous phonon decay, whereby the phonon spontaneously splits into two phonons of lower frequency, as a result of the lattice anharmonicity. Introducing the decay parameter λ, the anharmonic decay time τ_a is given by $\tau_a^{-1} = \lambda \cdot \nu^5$, where we assume that non-decaying (TA) phonons are in equilibrium with the LA phonons, due to rapid mode mixing resulting from elastic scattering at isotopes and impurities. At low frequencies, the phonon decay time deviates from the power law and becomes almost frequency-independent. In this region, the phonon lifetime

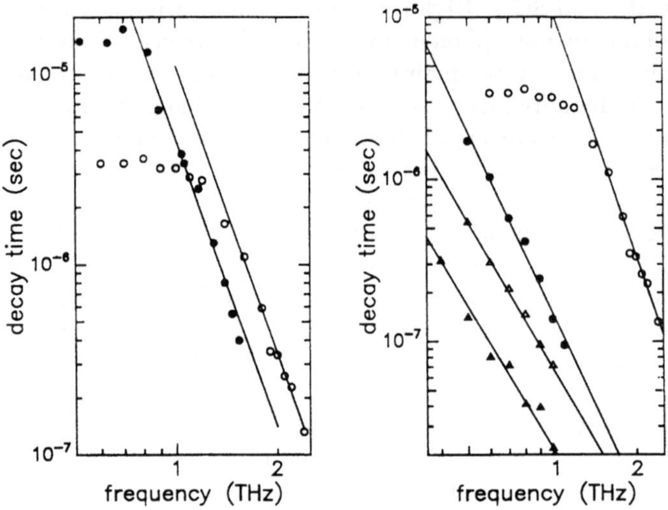

Figure 1. Phonon decay times for pure CaF_2 (open circles) and mixed crystals: note the different frequency dependence in these systems.
Left: $Ca_{1-x}Sr_xF_2$ (solid circles: x = 0.05)
Right: $Ca_{1-x}La_xF_{2+x}$ (solid circles: x = 0.0025; open triangles: x = 0.0075; solid triangles: x = 0.025).

is no longer determined by the anharmonic decay but instead is limited by the diffusive escape of the phonons into the helium bath. In the mixed crystal, phonons are scattered very strongly, leading to an escape time longer than that found for the pure CaF_2 crystal.

The decay parameter λ shows a maximum at at a concentration x = 0.05, indicating that a further increase of the Sr^{2+} concentration in the samples does not lower the lifetime, but rather causes it to increase again, reaching a value that is about a factor of 1.6 shorter than the corresponding time in pure CaF_2. The reason for this behavior is not completely understood. One possibility is that it results from the increase of the lattice anharmonicity due to the Sr^{2+} ions, another is the partial relaxation in the momentum conservation law of the anharmonic decay process, which thereby makes additional decay channels allowed and so reduces the phonon lifetimes.

The effect of the disorder on the phonon lifetimes in $Ca_{1-x}Sr_xF_2$ mixed crystals is, however, remarkably weak: the maximum reduction in the lifetime is less than a factor of 3 and the phonon decay is determined by spontaneous decay due to lattice anharmonicity.

This is in contrast to the strong effect that a small La^{3+} concentration has on the phonon lifetimes in $Ca_{1-x}La_xF_{2+x}$. Figure 1 (right) shows the phonon lifetime as a function of the phonon frequency for different La^{3+} concentrations. For a La^{3+} concentration of $x = 2.5 \times 10^{-3}$ (solid circles) we find a lifetime at $\nu = 1.0$ THz that is a factor of 100 smaller than the

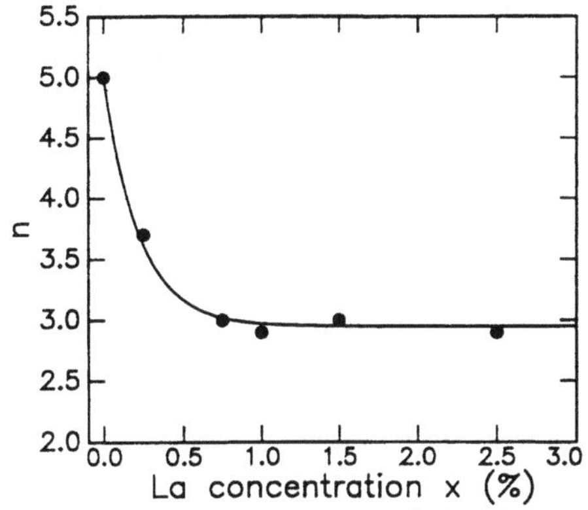

Figure 2. Exponent n of the frequency dependence of the phonon lifetime as a function of La^{3+} concentration, as discussed in the text.

anharmonic phonon lifetime of the pure CaF_2 crystal (open circles). The frequency dependence of the decay time has changed, too; now we find $\tau^{-1} = \lambda \cdot \nu^{3.7}$ instead of a ν^5 dependence of the anharmonic decay. For higher concentrations ($x = 7.5 \times 10^{-3}$, open triangles, and $x = 2.5 \times 10^{-2}$, solid triangles) the lifetime decreases further, reaching a value at 1 THz that is a factor 1000 smaller than the corresponding lifetime in pure CaF_2, and the frequency dependence of the decay time can be described by $\tau^{-1} = \lambda \cdot \nu^3$.

Writing the decay rate as $\tau^{-1}(x) = \lambda(x) \cdot \nu^{-n(x)}$, we plot the exponent n as a function of the La^{3+} concentration in Fig. 2. Starting at n = 5, which describes the spontaneous phonon decay due to lattice anharmonicity, we find a change to n = 3 for x = 1%. Within the accuracy of our experimental data, n(x) seems to be concentration-independent at higher La concentrations.

Dietsche and Kinder [7] have found a similar frequency dependence for the inverse inelastic mean free path l_i^{-1} (l_i corresponds to the phonon lifetime obtained in our experiments multiplied by the speed of sound) in SiO_2 glass in the frequency range between 0.1 and 0.25 THz [7].

The origin of the ν^3 dependence of the decay rate is unclear, but it is evident that inelastic scattering processes, in addition to the spontaneous decay due to lattice anharmonicity, play a dominant role in the relaxational dynamics of phonons in $Ca_{1-x}La_xF_{2+x}$. Elastic scattering at two-level systems, which is traditionally used to describe the thermal conductivity of glasses and also of $Ba_{1-x}La_xF_{2+x}$ mixed crystals [4], cannot explain our results, inelastic processes are needed to interpret our experiments. Possible alternative mechanisms are Raman scattering of the phonons at two-level systems or resonant phonon absorption at three-level or multi-level systems [2], with subsequent relaxation within these systems, leading to a rapid down-conversion of the phonons.

In conclusion, we have demonstrated that optical techniques can be used to study the relaxation of THz phonons in mixed crystals. Our experiments show that the phonon relaxation in glassy mixed crystals is dominated by an inelastic process that is unknown in pure crystals or mixed crystals showing no glassy behavior.

We like to acknowledge helpful discussions with K.F. Renk, S.A. Basun, C.E. Mungan, and P.E. Obermayer. One of us, U.H., wishes to thank the Alexander von Humboldt Foundation and NSF (grant no. DMR-89-18894) for financial support.

References

[1] R. Baumgartner, M. Engelhardt, and K.F. Renk, Phys. Rev. Lett. **47**, 1403 (1981)
[2] U. Happek, T. Holstein, and K.F. Renk, Phys. Rev. Lett. **54**, 2091 (1)
[3] U. Happek, R.O. Pohl, and K.F. Renk in *Phonons 89*, eds. S. Hunklimger, W. Ludwig, and G. Weis (World Scientific, Singapore 1990), 925
[4] D.G. Cahill and R.O. Pohl, Phys. Rev. **B39**, 10477 (1989)
[5] U. Happek, Y. Ayant, R. Buisson, and K.F. Renk, Europhys. Lett. **3**, 1001 (1987)
[6] W.E. Bron and W. Grill, Phys. Rev. **B16**, 5303 (1977)
[7] W. Dietsche and H. Kinder, Phys. Rev. Lett. **43**, 1413 (1979)

Is Silicon a Glass? Measurements of the Mechanical Properties of Boron-Doped and Single-Crystal Silicon

R.E. Mihailovich and J.M. Parpia*

Department of Physics, Cornell University, Ithaca, NY 14853-2501, USA
*Present address: School of Electrical Engineering,
Cornell University, Ithaca, NY 14853, USA

The low-temperature mechanical properties of single-crystal silicon have been studied by Kleiman et al. [1] (KAB). A high-purity (110) wafer, with ρ = 20K Ω-cm and a few ppm non-electronic impurities was used to fabricate the oscillator. They observed a temperature independent dissipation to 100mK, decreasing linearly for T < 100mK. The frequency shift shows softening below 60mK, with a log T dependence near the frequency maximum. Such a mechanical response is qualitatively similar to that of glasses [2]. KAB propose that glassy defects are intrinsic to silicon, and estimate a defect concentration ~1% of that found in vitreous silica. Other explanations have been offered for the mechanical response. [3,4] Keyes proposed that electronic states, split by a small energy, would fit the frequency shift data at a much smaller concentration.

Such electronic states are known to exist in the acceptor hole states of semiconductors, and strongly scatter thermal and ultrasonic phonons. [5,6] We have measured oscillators made from boron doped silicon to determine the importance of the Keyes mechanism. The oscillators were fabricated into the double-paddle geometry of KAB, using the same anisotropic etching technique. We used (100) wafers because this orientation offers geometrically cleaner post-etch corners, believed to yield lower background dissipation. The oscillators had boron concentrations of 6×10^{13} cm^{-3}, 1.4×10^{15} cm^{-3}, 1.1×10^{16} cm^{-3}, and 5.7×10^{16} cm^{-3}, corresponding to resistivities of 200, 7, 1, and 0.2 Ω-cm. The oxygen and carbon content were similar for all samples, at ~10^{18} and 4×10^{16} cm^{-3} respectively.

The low temperature dissipation and resonant period shift at strain levels of 10^{-6} is shown in Figure 1. The dissipation increases with doping level and drops rapidly at low temperatures. The period shift shows crystal softening up to at least 1K and also increases with doping level. Period measurements for the two highest doping samples were taken to a temperature of 20K, both showing a log T dependence up to 1K and softening up to 10K.

For the three highest doped samples, the period shift data is consistent with the analysis of Keyes. The presence of softening to 10K suggests that there is a distribution of energy splittings. The log T dependence indicates the distribution width of about 1K. Keyes estimates that N_a holes with an energy splitting from -U to +U will produce a period shift given by $(P-P_o)/P_o = D^2 (N_a/3U)(1/c_{44})$, where D is an elastic coupling constant and c_{44} the shear elastic constant. Our measured period shift is described well by this relation for a coupling constant of 3 eV, consistent with values measured elsewhere. [6,7]

The dissipation of these three samples also appears to originate in the boron hole states. The scattering of ultrasonic phonons has been calculated by Isawa

Springer Series in Solid-State Sciences, Volume 112
Phonon Scattering in Condensed Matter VII Eds.: M. Meissner and R.O. Pohl

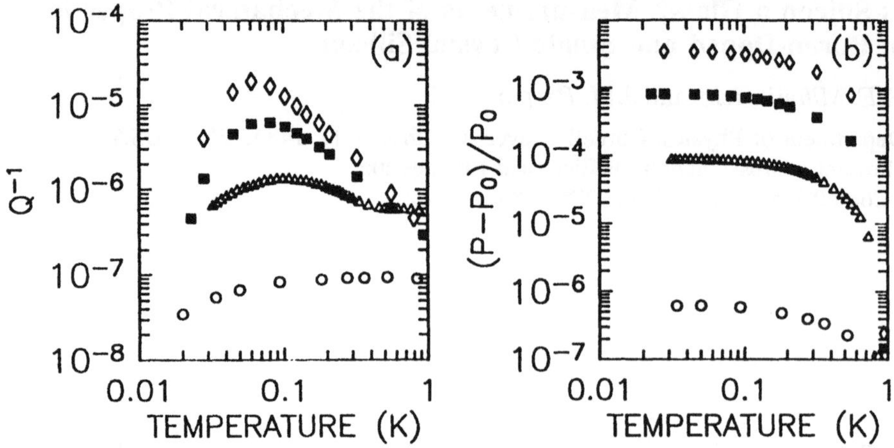

Fig. 1. The dissipation (Fig. 1a) and period change (Fig. 1b) for the four doped silicon oscillators (◇) 0.2 Ω-cm, (□) 1.0 Ω-cm, (△) 7 Ω-cm, (○) 200 Ω-cm.

et al. [8], and should be valid for audio frequency vibrations introduced by the oscillator's motion. The dissipation data can be fit reasonably well using these expressions with a distribution of splitting values. Unfortunately, the dissipation is dominated by the smallest energy splittings; consequently, it is difficult to obtain detailed information on the distribution.

In summary, the mechanical response of our three highest dopant samples is determined by acceptor hole states. The scattering is very strong. The energy splittings appear to arise from elastic introduced by the oxygen impurities.

Our lowest dopant sample's dissipation is qualitatively similar, but an *order of magnitude smaller* than that of the KAB oscillator. In the glass model, the dissipation is proportional to the concentration of glassy defects, implying that our sample has a much lower defect concentration than that of KAB. This is unlikely, since our sample has much higher concentrations of both electronic and nonelectronic impurities. Thus we conclude that glassy defects as proposed by KAB are not intrinsic to silicon. The additional loss observed by KAB may originate in the torsion rod corners from their geometry.

This research was supported by NSF under DMR-88-20170, DMR-91-23857 and the Cornell MSC under DMR-88-18588.

References

1. R.N. Kleiman, G. Agnolet and D.J. Bishop, Phys. Rev. Lett., **59**, 2079, (1987).
2. A.K. Raychaudhuri and S. Hunklinger, Z. Phys. B, **57**, 113, (1984).
3. R.W. Keyes, Phys. Rev. B, **38**, 9806, (1988).
4. W.A. Phillips, Phys. Rev. Lett., **61**, 2632, (1988).
5. M.G. Holland and L.J. Neuringer, in *Proc. of Int. Conf. on the Physics of Semicond., Exeter, July, 1962* (The Inst. of Phys. and the Phys. Soc., London, 1962), p. 474.
6. B. Golding and W.H. Haemmerle, Proc. of the 17th Int. Conf. on Low Temp. Phys. Karlsruhe, Germany, 1984, (North-Holland, Amsterdam, 1984) p. 671.
7. See T. Fjeldly, T. Ishiguro, and C. Elbaum, Phys. Rev. B, **7**, 1392, (1973).
8. Y. Isawa, Y. Takeuti, and N. Mikoshiba, Phys. Rev. B, **15**, 4907, (1977).

Glass-Like Properties of Stoichiometric Crystals

P.A. Medwick[1], *R.O. Pohl*[1], *and T. Tanaka*[2]

[1]Laboratory of Atomic and Solid State Physics, Cornell University, Ithaca, NY 14853-2501, USA

[2]National Institute for Research in Inorganic Materials, 1-1 Namiki, Tsukuba, Ibaraki 305, Japan

1. Introduction

The universal thermal, elastic, and dielectric properties of amorphous solids at low temperatures have been extensively studied for more than twenty years. In an effort to understand this universality, much work has been devoted to the investigation of glass–like properties in certain mixed crystal systems such as KBr:KCN[1]. However, to date, YB_{66} is the only *stoichiometric* crystal reported to display glass–like signatures in its thermal conductivity and specific heat[2]. We have measured the internal friction of YB_{63} as another probe of its glass–like low energy excitations. In addition, we have examined the effect of Gd substitution for Y in the lattice by measuring the thermal conductivity and internal friction of $GdB_{62.5}$.

2. Experimental Matters

Two single crystal samples of YB_{63} and $GdB_{62.5}$ were grown by the indirect heating floating zone method described elsewhere[3]. Crystal quality was monitored by x–ray rocking curve analysis. The internal friction of YB_{63} and $GdB_{62.5}$ was measured using a composite mechanical oscillator technique[4]. The thermal conductivity of $GdB_{62.5}$ was measured using a standard gradient method.

3. Results

Figure 1 shows the internal friction (Q^{-1}) of YB_{63} and $GdB_{62.5}$. Below 10 K, the mechanical loss for YB_{63} displays both a magnitude and temperature dependence characteristic of amorphous solids as illustrated by the data for a–SiO_2. The internal friction of $GdB_{62.5}$ also displays a well–defined plateau in Q^{-1} below $\sim$ 10 K. Note that the rapid rise in Q^{-1} above $\sim$ 100 K for all samples is a measurement artifact.[4]

Thermal conductivity (Λ) of $GdB_{62.5}$ is plotted in Figure 2. The thermal conductivity follows a power law $\Lambda \propto T^2$ below $\sim$ 8 K which then saturates into a plateau as is typical for other amorphous solids such as a–SiO_2. In addition, our results for $GdB_{62.5}$ appear to extrapolate nicely to the high

Springer Series in Solid-State Sciences, Volume 112
Phonon Scattering in Condensed Matter VII Eds.: M. Meissner and R.O. Pohl

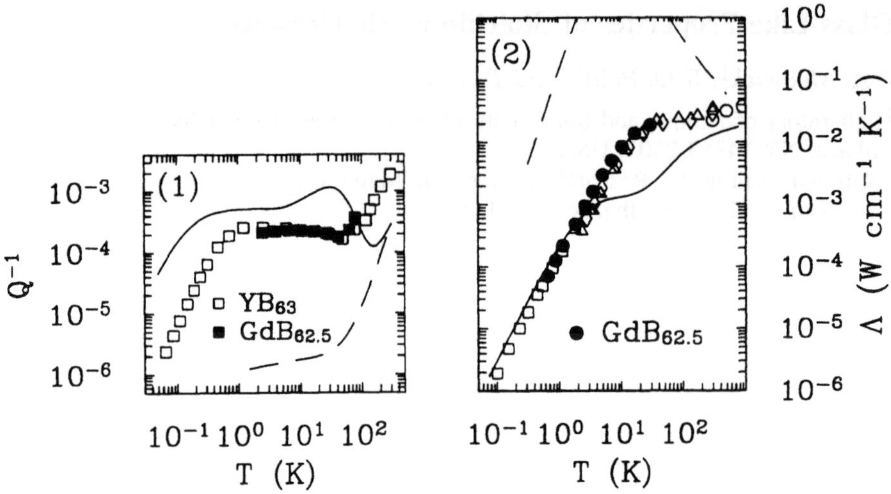

Figure 1 Internal friction (Q^{-1}) of YB_{63} at 164 kHz and $GdB_{62.5}$ at 205 kHz. Shown for comparison are data for a glass (a–SiO_2)[4,5] at 86 kHz (solid line) and a crystal (BaF_2)[4] at 86 kHz (dashed line).

Figure 2 Thermal conductivity (Λ) of $GdB_{62.5}$ along with data for GdB_{66} by Golikova[6] (open circles), $YB_{61.7}$ by Raychaudhuri[2] (open squares), and two samples of YB_{66} by Slack[2] (open triangles, open diamonds). Also shown are data for a–SiO_2 (solid line) and quartz (dashed line), see ref. [7].

temperature measurements for GdB_{66}. Such interpolation is justified in light of the agreement between our data and the measurements for $YB_{61.7}$ and two different samples of YB_{66}.

We see that the substitution of Gd for Y in the crystal structure has no effect on the observed glass–like thermal and elastic signatures. This is a clear indication that these properties are not a result of accidental chemical impurities or different process histories. We conclude that the low energy tunneling states are *intrinsic* to the lattice itself. In particular, these findings demonstrate that glass–like excitations may even be found in crystals with extremely low degrees of disorder (represented by the random site occupancies of the "M" atoms in "MB_{68}") in contrast to the many forms of disorder present in mixed crystals.

This research was supported by the NSF, Grant No. DMR–91–15981.

[1] J. J. DeYoreo et al., Phys. Rev. B **34**, 8828 (1986).
[2] D. G. Cahill et al., Phys. Rev. B **40**, 3254 (1989).
[3] T. Tanaka et al., J. Cryst. Growth **99**, 994 (1990).
[4] D. G. Cahill et al., Rev. Sci. Instrum. **60**, 2706 (1989).
[5] J. E. Van Cleve, Ph.D. dissertation (Cornell Univ., Ithaca, 1991).
[6] O. A. Golikova et al., J. Non–Cryst. Solids **87**, 64 (1986).
[7] D. G. Cahill, Ph.D. dissertation (Cornell Univ., Ithaca, 1989).

Glass-Like Elastic Properties in Polycrystalline Alloys

J.E. Van Cleve[1], *R.O. Pohl*[1], *and A.K. Raychaudhuri*[2]

[1]Laboratory of Atomic and Solid State Physics, Cornell University, Ithaca, NY 14853-2501, USA

[2]Physics Department, Indian Institute of Science, Bangalore 560012, India

1. Introduction

Previous low-temperature elastic measurements on quasicrystals[1] and $Al_{94}Si_6$ [2] show that some highly unstable alloys show crystalline elastic properties below 20K. On the other hand, low-temperature thermal measurements of $Zr_{80}Nb_{20}$ [3] and $Ti_{67}V_{33}$ [4] suggest that alloys formed from columns 4B and 5B of the periodic table are glasslike in compositions near the $\omega-\beta$ phase transition. Here we present precise elastic data on both the temperature variation of the speed of sound $\frac{\delta v}{v}$ and the internal friction Q^{-1}on several 4B-5B alloys using the same techniques used to measure $Al_{94}Si_6$, silica, and a BaF_2 crystal. We used a compound torsional oscillator at frequencies between 60 kHz and 200 kHz and took data between .05K and 100K in order to look for the resonant and relaxational effects of glasslike tunneling systems, and to look for empirical correlations above 2K.

2. Experimental Details

The oscillator method is described elsewhere[5]. The 4B-5B alloys were arc-melted at least 4 times in a Ti-gettered argon atmosphere, then drop-cast into ingots roughly 4.5 mm in diameter. The $Ti_{67}V_{33}$ ingots were annealed at 1175K for 4 hours, and the $Zr_{80}Nb_{20}$ ingots were homogenized at 1275K for 24 hours. The $Nb_{37}Ti_{63}$ ingot was manufactured by Teledyne Wah Chang Albany and was annealed in vacuum at 1175K for two hours. All samples were gas-quenched from their annealing temperatures at $\approx 1Ks^{-1}$. Samples were inspected for cracks visually and mechanically.

3. Results

Although the structurally polycrystalline 4B-5B alloys are cooled relatively slowly from the melt ($\approx 1Ks^{-1}$) compared to melt-spun amorphous metals like Pd-Si alloys ($\approx 10^5Ks^{-1}$), each alloy measured has the same low-temperature elastic properties as silica in both temperature dependence and magnitude, as shown in Figures 1 and 2. Each has a non-diffusive phase transition at or near the measured composition. Alloys with diffusive phase transitions, such as $Al_{94}Si_6$, and alloys with continuous solutions across the phase diagram, such as the 5B-5B alloy system Nb-Ta, do not show glasslike properties[6]. These

Springer Series in Solid-State Sciences, Volume 112
Phonon Scattering in Condensed Matter VII Eds.: M. Meissner and R.O. Pohl

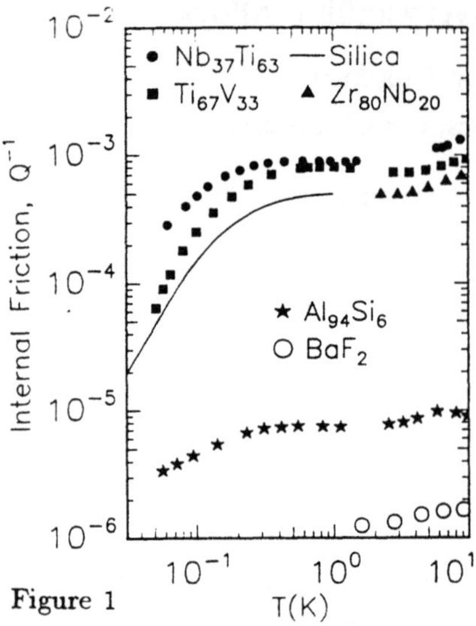

Figure 1

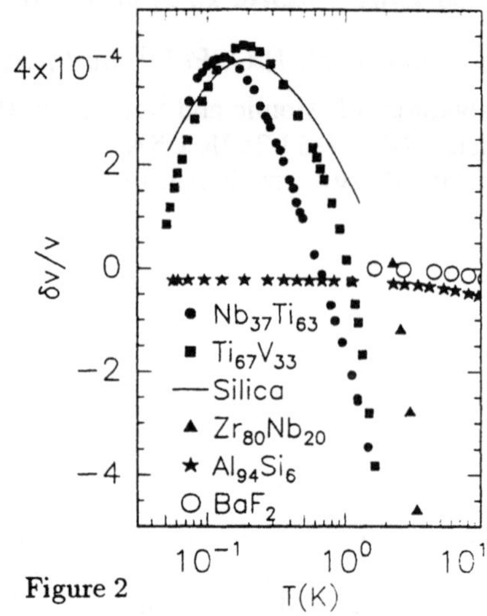

Figure 2

Figure 1. Internal friction Q^{-1}measured on 4B-5B alloys compared to our results on $Al_{94}Si_6$ [2], silica (Suprasil W), and Cahill's measurement of BaF_2 [5]. The 4B-5B alloys show glasslike Q^{-1}while $Al_{94}Si_6$ is crystalline.

Figure 2. Variation of speed of sound with temperature, $\frac{\delta v}{v}$, measured for the same materials. The 4B-5B alloys show glasslike $\frac{\delta v}{v}$ while $Al_{94}Si_6$ is crystalline.

observations suggest that solids in which non-diffusive phase transitions can occur are likely to develop glass-like excitations.

Our detailed measurements on a few 4B-5B alloys permits us to review the literature on these materials[6] and infer whether the samples used in other studies would have glasslike low-temperature properties, even if only either $\frac{\delta v}{v}$ or $Q^{-1}$was measured for $T > 4K$. For the Zr-Nb phase diagram, on which the most information was available, the combined data indicate that glasslike alloys are confined to the Nb-poor side of the phase diagram, which contains the ω phase. The Nb-rich side, characterized by a structural continuity between the Zr-Nb solid solutions and the β phase of pure Nb, does not show glasslike properties.

The research at Cornell was supported by the National Science Foundation, Grant No. DMR 91-15981.

[1] J. E. Van Cleve, N. A. Gershenfeld, K. Knorr, and P. A. Bancel, Phys. Rev. B1 **41**, 980 (1990).

[2] J. E. Van Cleve, J. Chevrier, and R. O. Pohl, *Phonons '89*, S. Hunklinger, W. Ludwig, and G. Weiss, eds., World Scientific, London, p. 579 (1990).

[3] L. F. Lou, Solid State Comm. **19**, 335 (1976).

[4] B. S. Chandrasekhar, H. R. Ott, and H. Rudigier, Solid State Comm. **42**, 419 (1982).

[5] D. G. Cahill and J. E. Van Cleve, Rev. Sci. Instrum. **60**, 2706 (1989).

[6] T. E. Noble, Master's thesis, Cornell University (1990), unpublished.

Glass-Like Low-Temperature Acoustic Properties of Polycrystalline Metals

P. Esquinazi, R. König, D. Valentin, and F. Pobell

Physikalisches Institut, Universität Bayreuth, W-8580 Bayreuth, Fed. Rep. of Germany

Low energy excitations and their interactions with phonons and conductions electrons determine the low temperature acoustic properties (sound velocity v and internal friction Q^{-1}) of amorphous solids. At $T < 1$K and at audio frequencies these properties are characterized by a maximum in v at T_{co}, relatively large values of Q^{-1} and a "plateau" in $Q^{-1}(T > T_{co})$ [1]. Anomalous acoustic properties below 1 K are also found in "special" crystalline solids with some kind of disorder (see Refs. in [1] and also Ref.[2]). It turns out that after more than 20 years of research the "glass-like" acoustic properties appear to be the rule rather than the exception. Until recently [3] *simple* polycrystalline metals were ignored as candidates in the search of the "anomalous glass-like behavior" of the acoustic properties. We have measured the acoustic properties of polycrystalline superconductors NbTi, Nb and Ta and of the normal metals Ag and Cu. Our results strongly support the conclusion of our recently published work [4] namely: there are basically no differences in the low - temperature, low - frequency acoustic properties of polycrystals and amorphous materials.

The acoustical studies at frequencies of 100 Hz $\leq \nu \leq$ 8 kHz and in the temperature range 0.05 mK $\leq T \leq$ 1 K have been performed with the vibrating reed and wire techniques. The sample holders were screwed to the upper flange of our Cu nuclear refrigeration facility [5]. We have investigated five different NbTi wires with diameters 20 $\mu m \leq d \leq 50 \mu m$; one of the wires was treated with acid reducing its diameter from originaly 50 μm to 20 μm to study possible surface contributions. Besides differences in thermal conduction and self heating contributions at $T <$ 10 mK, similar results have been obtained in all the NbTi wires. Typical results of $\Delta v(T)/v$ are plotted in Fig.1(a) where we have also included data of SiO_2 [4]. The slopes of the logarithmic T-dependence of $v(T)$ of NbTi are similar to that of SiO_2. Pure Nb and Ta show the same behavior but the slopes of $v(T)$ are a factor four and ten smaller. Q^{-1} of the three superconductors shows a plateau between 0.1 K and 1 K and its value agrees with the predictions of the Tunneling Model (taking into account the slopes of $v(T)$) as well as the frequency dependence of the onset of the plateau at T_0, see Fig. 1(b). Non-linear behavior in $v(T)$ at $T < T_{co}$ was observed for all the measured samples in agreement with the SiO_2 results [4]. The non-linear behavior can be explained taking into account the change of the population of the tunneling systems (TS) due to the acoustic wave and self heating effects [4].

"Glass-like" behavior in the acoustic properties is also observed in Ag and Cu, see Fig.2. In contrast to SiO_2 and to the polycrystalline superconductors,

Springer Series in Solid-State Sciences, Volume 112
Phonon Scattering in Condensed Matter VII Eds.: M. Meissner and R.O. Pohl

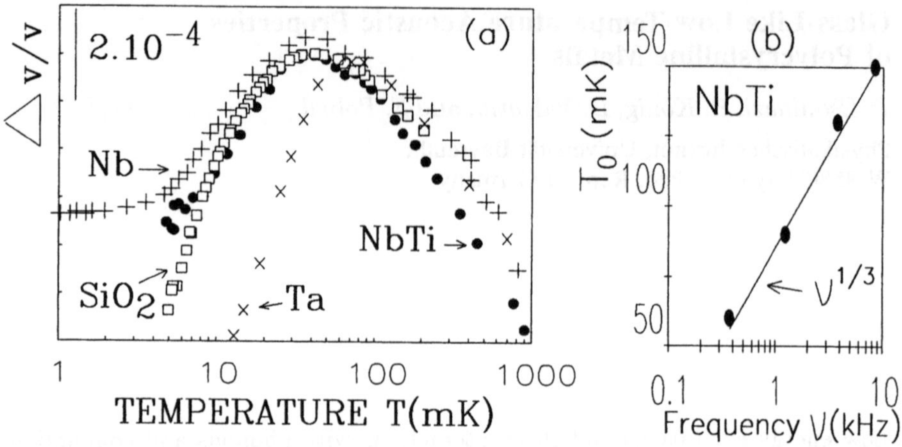

Figure 1: *(a): Relative change of sound velocity against temperature. The frequencies and applied strains (at 10 mK) were:* *SiO_2: 400 Hz,* 1×10^{-7}, *NbTi: 370 Hz,* 1.1×10^{-6}, *Nb: 1.5 kHz,* 2×10^{-6}, *Ta: 5.5kHz,* 1.6×10^{-5}. *The $\Delta v/v$-scale for Nb (Ta) has to be divided by a factor 4 (10). (b) Frequency dependence of the onset T_0 of Q^{-1} for NbTi.*

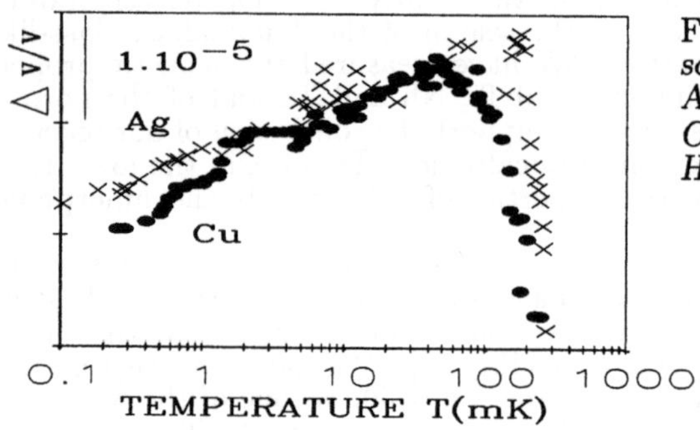

Figure 2: *Relative change of sound velocity of cold rolled Ag (vibrating reed) [4] and Cu (vibrating wire) at 250 Hz and 300 Hz, respectively.*

$\Delta v(T)/v$ is strain independent and its logarithmic slope at $T < T_{co}$ is two order of magnitude smaller which might be partially due to the influence of conduction electrons in the relaxation of the TS.

References

[1] S. Hunklinger and A.K. Raychaudhuri, In: *Progress in low temperature physics.* D.F. Brewer (ed.) Vol IX. Amsterdam:North Holland 1986.

[2] N. Birge, B. Golding, W. Haemmerle, H. Chen and J. Parsey, Jr., Phys.Rev. B**36**, 7685 (1987). H. Hikata and C. Elbaum, Phys.Rev.Lett. **54**, 2418 (1985).

[3] P. Thauer, P. Esquinazi and F. Pobell, Physica B**165/166**, 905 (1990).

[4] P. Esquinazi, R. König and F. Pobell, Z. Phys. B**87**, 305 (1992). R. König, P. Esquinazi and F. Pobell, J. Low Temp. Phys., in press (1992).

[5] K. Gloos, P. Smeibidl, C. Kennedy A. Singsaas, P. Sekowski, R. Mueller and F. Pobell, J. Low Temp. Phys. **73**, 101 (1988).

How do the Two-Level-System Distributions Arise in Amorphous Solids?

S.K. Watson, D.G. Cahill**, and R.O. Pohl*

Laboratory of Atomic and Solid State Physics, Cornell University, Ithaca, NY 14853, USA
*Present address: UC San Diego, La Jolla, CA 92093, USA
**Present address: University of Illinois, IL 61801, USA

The two-level-systems ('TLS') in amorphous solids are characterized by two features: (a) They have a constant density of states with a broad distribution of relaxation times; (b) Their density of states has practically the same magnitude in all amorphous solids. Certain properly disordered mixed crystalline solids share these features. In the most thoroughly studied system, $(KBr)_{1-x}(KCN)_x$, lattice-mediated elastic interactions between tunneling centers (the CN^-) are believed to play a crucial role in the development of the TLS at intermediate cyanide concentrations ($0.2 < x < 0.6$). In the present work we take a novel approach to creating the TLS in a mixed crystal. We show that feature (a) arises when tunneling centers in crystals are subjected to random internal stresses by comparing the behavior of KBr+~0.5 mole% CN^- and $KBr_{.5}Cl_{.5}$ doped with ~0.5 mole% CN^-, where internal stresses are intentionally introduced through the random substitution of 50% of the Br^- for Cl^-.

In Figure 1 the specific heat of KBr doped with 0.8 mole% CN^- (asterisks) is orders of magnitude larger than the same system with 50% of the Br^- replaced by Cl^- (solid triangles). The specific heat of $KBr_{.5}Cl_{.5}$+0.9 mole% CN^- is, in fact, very similar to $(KBr)_{1-x}(KCN)_x$ in the glasslike concentration range, x=0.41 (open circles). In $KBr_{.5}Cl_{.5}$+0.9% CN^-, however, the glasslike specific heat is achieved with nearly 40 times less CN^- by introducing random stresses into the crystal!

In order to quantify these results and determine the energy and volume density of TLS, P, and the transverse coupling of TLS to phonons, γ_t, the internal friction of $KBr_{.5}Cl_{.5}$ + 0.9 mole% CN^- was measured in torsion at ~90 kHz. In Figure 2 the internal friction of $KBr_{.55}Cl_{.45}$ is close to the background (*ie.* the loss due to the mounting arrangement). The data for the mixed crystal doped with cyanide and for the amorphous solid, however, are much larger and show the characteristic behavior of the internal friction expected for a system with TLS. From the Tunneling Model fit to the data for $KBr_{.5}Cl_{.5}$ + 0.9 mole% CN^- (solid line through triangles), we determine that $P=6\times10^{44}$ J^{-1} m^{-3} and γ_t=0.11 eV. By comparison, $P=9\times10^{44}$ J^{-1} m^{-3} and γ_t=0.16 eV for $(KBr)_{.59}(KCN)_{.41}$ while $P=8\times10^{44}$ J^{-1} m^{-3} and γ_t=0.65 eV for a-SiO_2.

To summarize, we have shown that $KBr_{.5}Cl_{.5}$ + 0.9 mole% CN^-, in which random internal stresses are intentionally introduced through the substitution of 50% Cl^- for Br^-, has glasslike thermal and elastic properties below 1 K. Since

Springer Series in Solid-State Sciences, Volume 112
Phonon Scattering in Condensed Matter VII Eds.: M. Meissner and R.O. Pohl

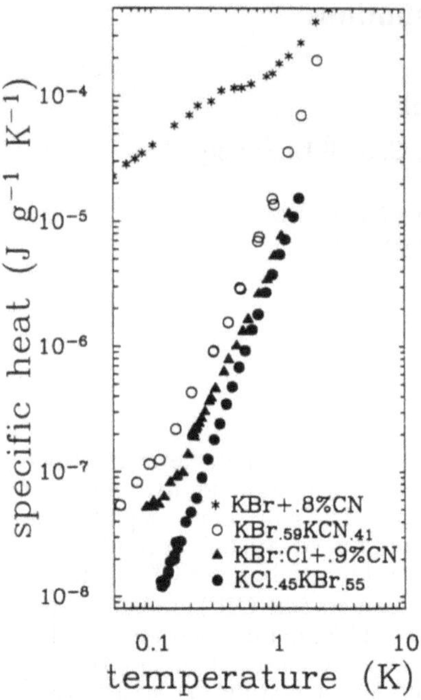

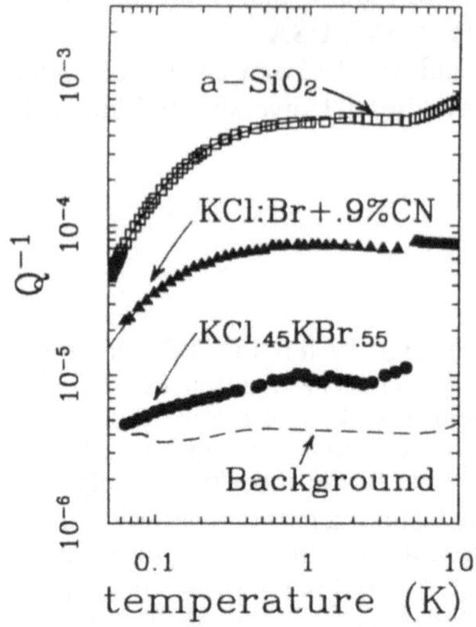

Fig. 1 (left): Long time specific heat of KBr+0.8%CN$^-$ [1], $KBr_{.59}KCN_{.41}$[2], $KCl_{.45}Br_{.55}$ and $KBr_{.5}Cl_{.5}$+0.9%CN$^-$ (this work).

Fig. 2 (right): Internal friction of $KCl_{.45}Br_{.55}$, $KBr_{.5}Cl_{.5}$+0.9%CN$^-$, and an amorphous solid, a-SiO_2.

random internal stresses are likely to be common to all amorphous solids, we conclude that their broad energy distributions and relaxation times have a very natural explanation.

This work has been supported by the U.S. National Science Foundation, Grant No. DMR 911-5981.

References

[1] W. Knaak and M. Meissner, *5th Intl. School on Condensed Matter Physics; Disordered Systems and New Materials; Sept. 19-26, 1988,* eds. A.M. Prohorov, M. Borissov *et al.* (World Scientific, Singapore, 1989).

[2] J. J. De Yoreo, W. Knaak, M. Meissner, and R. O. Pohl, *Phys. Rev. B* **34**, 8828 (1986).

Collective Relaxation of CN^--Tunneling States in $(KBr)_{1-x}(KCN)_x$

C. Enss, D. Arndt, and M.v. Schickfus

Institut für angewandte Physik, Universität Heidelberg,
Albert-Überle-Str. 3–5, W-6900 Heidelberg, Fed. Rep. of Germany

It is well known that the low temperature thermal, acoustic and dielectric properties of CN^- doped KBr crystals are dominated by atomic tunneling systems. To study the dynamic behavior of the CN^- tunneling states we have performed low temperature dielectric experiments in the audio frequency range on KBr crystals with CN^- concentrations between $x = 0.001$ and 0.3.

The simplest model describing atomic tunneling states is a particle moving in a double well potential. This potential is characterized by an asymmetry Δ of the two wells and by the tunnel splitting Δ_0. The total energy of such two level systems is given by $E = (\Delta_0^2 + \Delta^2)^{1/2}$. The CN^- tunneling states couple to an external electric field $\boldsymbol{F} = \boldsymbol{F}_0 \cos(\omega t)$ through their electric dipole moment $\boldsymbol{p}$. Generally this leads to both resonant and relaxational processes. In our experiments, however, because of the low frequencies, only the relaxation contribution is important. Therefore the dielectric loss angle is given by:

$$\tan(\delta) = \iint P(\Delta, \Delta_0) \frac{(\boldsymbol{p} \cdot \widehat{\boldsymbol{F}})^2}{\epsilon\epsilon_0 k_B T} \left(\frac{\Delta}{E}\right)^2 \operatorname{sech}^2\left(\frac{E}{2k_B T}\right) \frac{\omega\tau}{1 + (\omega\tau)^2} \, d\Delta d\Delta_0 \,,$$

where $P(\Delta, \Delta_0)$ is the distribution function of the potential parameters, τ the relaxation time, and $\widehat{\boldsymbol{F}}$ the unit vector of the electrical field. The dependence of $P(\Delta, \Delta_0)$ on the concentration of the tunneling defects is still not well understood and is known only for the very dilute and the highly concentrated case. For samples with concentrations $0.01 \leq x \leq 0.1$ it seems likely that electric and elastic interaction among the CN^- ions will generate a broad distribution of asymmetry energies Δ, while the tunnel splitting Δ_0 remains essentially fixed. From the last term in the above equation one can see that the main contribution to the dielectric loss is due to tunneling systems for which $E \simeq k_B T$ and $\omega\tau \simeq 1$.

The experimental results for the dielectric loss angle of $(KBr)_{0.99}(KCN)_{0.01}$ for different frequencies are shown in fig. 1a. The most interesting features are the high absolute and strongly frequency dependent value of $\tan(\delta)$, and that for all frequencies there is a maximum below 200 mK. The most surprising point is that this maximum shifts to lower temperature with increasing frequency!
This means that the relaxation rate of the tunneling systems increases as the sample cools. This temperature dependence can certainly not be explained

Springer Series in Solid-State Sciences, Volume 112
Phonon Scattering in Condensed Matter VII Eds.: M. Meissner and R.O. Pohl

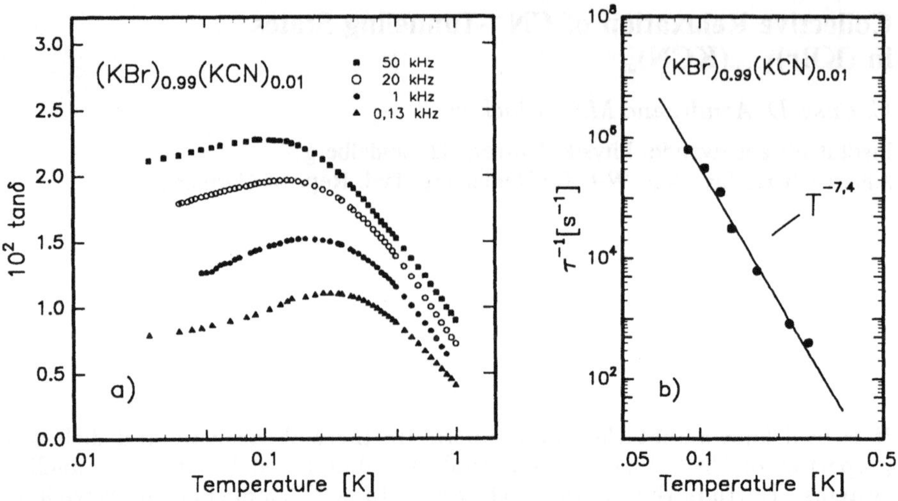

Fig.1: a) Dielectric loss angle of KBr doped with 1 % CN^- as a function of temperature for different frequencies. b) Relaxation rate of tunneling states in $(KBr)_{0.99}(KCN)_{0.01}$ vs temperature. The points are derived from the absorption peak temperatures and the associated frequencies. The solid line is a power law fit to the data.

by the usual relaxation of individual tunneling states via thermal phonons. On the other hand, a relaxation rate increasing on cooling has already been reported for other mixed crystals such as KCl:OH^- [1]. These results have been qualitatively explained by Janssen [2] by assuming an energy exchange among pairs or triples of tunneling states.

The temperature dependence of these collective relaxation processes can be very strong, depending on the distribution $P(\Delta, \Delta_0)$. Assuming, as mentioned above, a fixed tunnel splitting Δ_0/k_B of less than 20 mK [3] and a broadly distributed asymmetry energy, a calculation of the collective relaxation rate involving three tunneling systems leads to $\tau^{-1} \propto T^{-8}$. In fig. 1b the relaxation rate τ^{-1} is plotted as a function of temperature. The data were derived from the maxima of the $\tan(\delta)$ curves assuming that $\omega\tau = 1$ at this point. The best fit of the data by a power law gives an exponent of -7.4. We therefore conclude that in the system $(KBr)_{1-x}(KCN)_x$ the dominant relaxation process below approximately 300 mK is a collective relaxation of three tunneling states.

Indications for the same collective relaxation process were also found in samples with CN^- concentrations between $x = 0.001$ and 0.2. They are, however, most pronounced around $x = 0.01$.

[1] K. Knop, W. Känzig: Phys. Kondens. Materie **15**, 201 (1972).
[2] T. Janssen: Phys. Kondens. Materie **15**, 205 (1972).
[3] G. Baier, M.v. Schickfus, C. Enss: Europhys. Lett. **8**, 487 (1989)

Anomalous Low-Temperature Behavior in the Dipolar Glass $K_{1-x}(N(H,D)_4)_xI$

C. Bostoen[1], *J.-F. Berret*[2], *and M. Meissner*[3]

[1]D.S.M. Research, P.O. Box 18, NL-6160 MD Geleen, The Netherlands
[2]Groupe de Dynamique des Phases Condensées, U.A. CNRS 233, Université de Montpellier II, F-34095 Montpellier, France
[3]Hahn-Meitner-Institut Berlin, Neutronenstreuung 2, W-1000 Berlin, Fed. Rep. of Germany

Recently, $K_{1-x}(N(H,D)_4)_xI$ was introduced as a model system for orientationally disordered crystals showing a dipolar glass transition [1]. Despite its tetrahedral shape, the ammonium ion exhibits unexpected dipolar behavior. This property has been ascribed to the C_{3v}-symmetry of the ammonium in the alkali-halide lattice [2]. The (x,T)-phase diagram of the dipolar glass $K_{1-x}(NH_4)_xI$ was first investigated (see Fig. 1) by combining x-ray diffraction, Raman and neutron scattering measurements [1]. Three compositional ranges are observed at low temperatures. Above the critical ammonium concentration x_c, estimated between x = 0.72 and 0.82, concentration dependent phase transitions occur from NaCl to CsCl cubic structures. It should be noted that, as in pure ammonium iodide, large hysteresis effects were encountered in this range. For $0.3 \leq x < x_c$, an orientational glassy phase is characterized by dipolar short-range order of an antiferroelectric type. This was evidenced using elastic neutron scattering on deuterated compounds at x = 0.42 and x = 0.62 for which the onset of ND_4^+ freezing corresponds to the appearence of diffuse scattering patterns at the Brillouin zone boundary. Below x ~ 0.3, the transition is no longer collective, but replaced by a single-ion freezing.

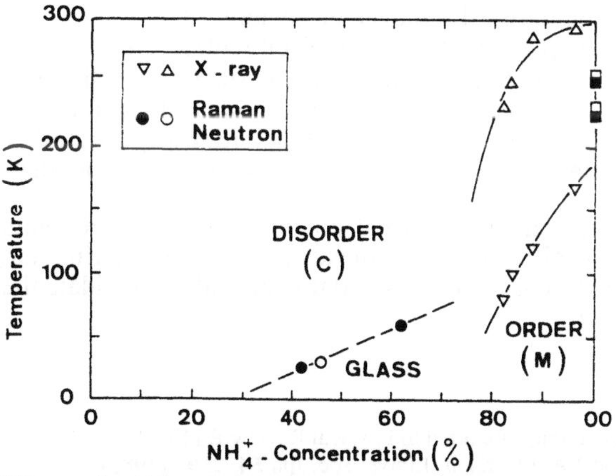

Figure 1 : Schematical (x,T)-phase diagram of $K_{1-x}(NH_4)_xI$ mixed crystals. Full symbols refer to deuterated compounds.

Springer Series in Solid-State Sciences, Volume 112
Phonon Scattering in Condensed Matter VII Eds.: M. Meissner and R.O. Pohl

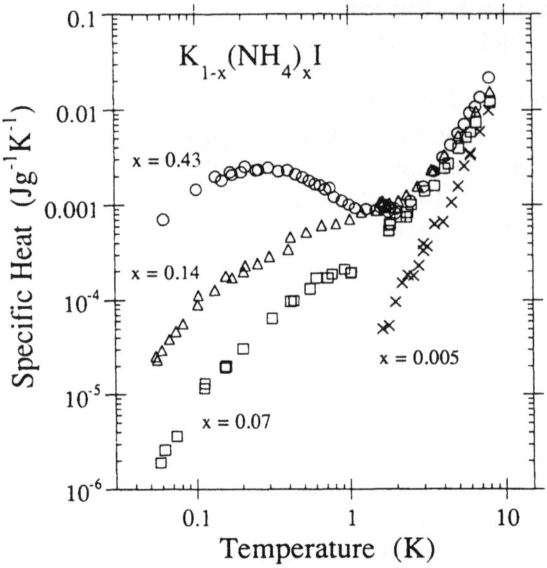

Figure 2 : Low-temperature specific heat of $K_{1-x}(NH_4)_xI$ mixed crystals for x = 0.005 (crosses), x = 0.07 (squares), x = 0.14 (triangles) and x = 0.43 (circles).

In view of this dipolar glass-like behavior, specific heat have been measured in $K_{1-x}(NH_4)_xI$ (with $x < x_c$) in order to test if this mixed crystal show the universal low-temperature properties of glasses. Time-dependent specific heat measurements were performed in the temperature range 0.05 K < T < 10 K for ammonium concentrations x = 0.005, 0.07, 0.14 and x = 0.43. The experimental set-up is described elsewhere [3]. Contrary to what would be expected from orientational glasses [3], a broad Schottky-like anomaly is measured (Fig. 2). One observes that the specific heat rises with increasing x and scales as : $C_p(x) \propto x^n$ with n = 2 - 3. This stands in contrast to KBr:KCN mixtures where the excess specific heat (T ~ 100 mK) passes through a maximum for a 10 % cyanide content.
The anomalous behavior in $K_{1-x}(NH_4)_xI$ is not still ellucidated. A possible explanation could be that at low temperatures a unique ground state level splitting is present for the $K_{1-x}(NH_4)_xI$ systems [4]. This would indicate that quantum fluctuations may destroy the frozen-in orientational disorder [5]. An other possibility would be to assume the existence of a transition to a low-temperature phase below 100 mK.
It would be very instructive to perform thermal conductivity measurements complementary to the present specific heat data. Inelastic neutron scattering and diffraction experiments at temperature below T = 100 mK may also ellucidate the nature of this unique ground state.

References

[1] J.-F. Berret, C. Bostoen and B. Hennion, submitted to Physical Review B (1992).
[2] Y. Ozaki, K. Maki, K. Okada and J.A. Morrison, J.Phys. Soc. Jpn **54**, 2595 (1985).
[3] J.-F. Berret, M. Meissner and B. Mertz, Z. Physik **87**, 213 (1992) and references therein.
[4] S. Grondey, M. Prager, W. Press and A Heidemeann, J. Chem. Phys. **85**, 2204 (1986).
[5] V. Dobrosavljevic and R.M. Stratt, Phys. Rev B **36**, 8484 (1987).

Influence of a Structural Instability on the Low-Temperature Behavior of Solids

P. Roth[1], *E. Gmelin*[1], *and E. Hegenbarth*[2]

[1]Max-Planck-Institut für Festkörperforschung,
Heisenbergstr. 1, W-7000 Stuttgart 80, Fed. Rep. of Germany
[2]Technical University Dresden, Low Temperature Physics Institute,
Mommsenstr. 13, O-8027 Dresden, Fed. Rep. of Germany

Glasses, amorphous, and other disordered solids are characterized by an additional entropy at low temperatures, which is given by a linear term, c_1T, in the specific heat generally observed below 1 K [1]. The linear term is attributed to low-energy excitations which can be described by tunneling of atoms in a double-well potential [2-3]. The tunneling states are mostly caused by various types of defects. However, it was also proposed that *structural phase transitions* should be important for the generation of such states [4]. To verify this, we have investigated the specific heat capacity C(T) of different ferroelectrics which exhibit a structural transition at T_c. Our results clearly indicate that some ferroelectrics show a pronounced linear term c_1T in the temperature dependence of specific heat at $T > 0.1$ K [5].

At low temperatures, the atoms responsible for the structural transition at the lowest T_c can occupy a set of nearly equivalent equilibrium positions. For instance, the Ti ion connected with the transition in $BaTiO_3$ can be displaced along six directions. The displacement correlates with an interaction energy which is in the range of k_BT_c. The movement of the ion can be described by a double-well potential with a barrier height of approximately k_BT_c. Because defects are incorporated always in solids, the positions of the atoms are not exactly equivalent; the potential has a small asymmetry. Around 1 K, the atoms cannot surmount the barrier by thermal activation, however, they can tunnel through the potential wall. The zero-point energy which scales with $k_B\Theta_D$ becomes important. At a certain ratio of $k_BT_c / k_B\Theta_D$, the tunneling of the atom is very probably. The value of c_1 is a measure for the density of tunneling states. To show a correlation, we have plotted the measured c_1 values of different ferroelectrics as a function of T_c/Θ_D in Fig. 1. The figure shows that c_1 has a maximum in the range of $1 < T_c/\Theta_D < 2$. Ferroelectrics with a critical ratio of about $T_c/\Theta_D = 1.5$ display glass-like properties, e. g., an anomalous thermal conductivity has been found at low temperatures. At small ratio $T_c/\Theta_D < 1$ and large ratio $T_c/\Theta_D > 2$, the linear term in C(T) and the other glass features gradually disappear [5].

Springer Series in Solid-State Sciences, Volume 112
Phonon Scattering in Condensed Matter VII   Eds.: M. Meissner and R.O. Pohl

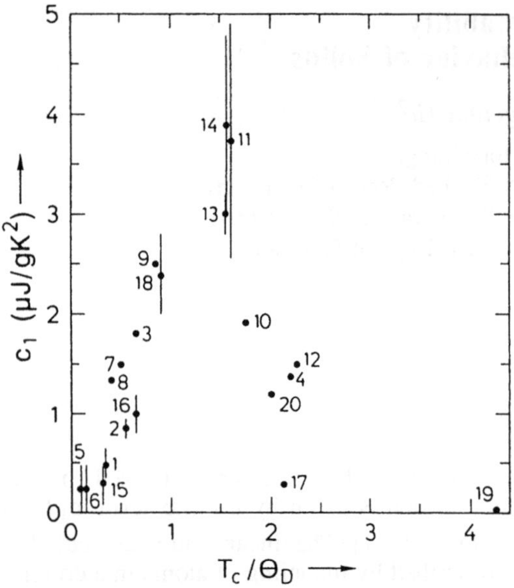

Fig. 1. The linear term in heat capacity, c_1, as a function of the ratio between phase transition temperature T_c and Debye temperature Θ_D [5].
1: $SrTiO_3$, 2: $BaTiO_3$, 3: $PbTiO_3$, 4: $CaTiO_3$, 5: $Ba_{0.03}Sr_{0.97}TiO_3$, 6: $Ba_{0.06}Sr_{0.94}TiO_3$, 7: $Ba_{0.20}Sr_{0.80}TiO_3$, 8: $Ba_{0.40}Sr_{0.60}TiO_3$, 9: $Pb_{0.30}Sr_{0.70}TiO_3$, 10: $PbZrO_3$, 11: $PbMg_{1/3}Nb_{2/3}O_3$, 12: $PbSc_{0.5}Nb_{0.5}O_3$, 13: $Pb_{0.915}La_{0.085}Zr_{0.65}Ti_{0.35}O_3$, 14: $Pb_{0.90}La_{0.10}Zr_{0.65}Ti_{0.35}O_3$, 15: KH_2PO_4, 16: K_2SeO_4, 17: $(NH_2CH_2COOH)_3 \cdot H_2SO_4$, 18: $Sr_{0.61}Ba_{0.39}Nb_2O_6$, 19: $LiNbO_3$, 20: $Pb_5Ge_3O_{11}$

In summary, we present an empirical relation which allows to estimate the value of $c_1(T_c/\Theta_D)$ in ferroelectrics and other solids with structural instability [5].

References

[1] R. C. Zeller, R. O. Pohl: Phys. Rev. B **4**, 2029 (1971)
[2] P. W. Anderson, B. I. Halperin, C. Varma: Phil. Mag. **25**, 1 (1972)
[3] W. A. Phillips: J. Low Temp. Phys. **7**, 351 (1972)
[4] K. K. Mon, N. W. Ashcroft: Solid State Commun. **27**, 609 (1978)
[5] I. Henning, M. Mertig, R. Plath, G. Pompe, E. Hegenbarth: Phys. Stat. Sol. A **73,** K 105 (1982), and to be published

Thermal Properties and Phase Transition of Ag_3AsS_3 (Proustite) at Low Temperatures

V. Efimov[1] *and E. Gmelin*[2]

[1]Institute of Solid State Physics, Russian Academy of Sciences, 142432 Chernogolovka, Moscow District, Russia
[2]Max-Planck-Institut für Festkörperforschung, Heisenbergstr. 1, W-7000 Stuttgart 80, Fed. Rep. of Germany

Abstract. Thermal conductivity and specific heat of single crystalline Ag_3AsS_3 are investigated in the temperature range from 2 K to 200 K where 3 phase transition (near 28 K, 49 K and 60 K, respectively) occur. The existence of the incommensurate phase (IC), 49K <T< 60K, is displayed by strong phonon scattering, which disappear in the commensurate (C) phase, below 49 K.

Proustite crystals show existence of (IC) phase in T between 49 and 60 K and near 28 K first order transition in ferroelectric state. In our experiment we wanted to find influence of these kind of transitions on thermal characteristics of monocrystal of proustite - thermal conductivity æ(T) (Fig.1) and specific heat $C_p(T)$ (Fig.2). The thermal history also indicated in these pictures.

First order phase transition at $T_{c1} \approx 28$ K show jump of æ(T) and hysteresis in 1 - 2 K at cooling and heating. $C_p(T)$ has the sharp, λ-like anomaly at 28 K found on virgin crystal, which is characteristic for a first order phase transition in ferroelectric phase with latent heat ($\Delta H \approx 32$ J/mole, $\Delta S \approx 1$J/mole*K). æ(T) at low temperature has a plateau what is typically for glass-like ferroelectrics.

Springer Series in Solid-State Sciences, Volume 112
Phonon Scattering in Condensed Matter VII Eds.: M. Meissner and R.O. Pohl

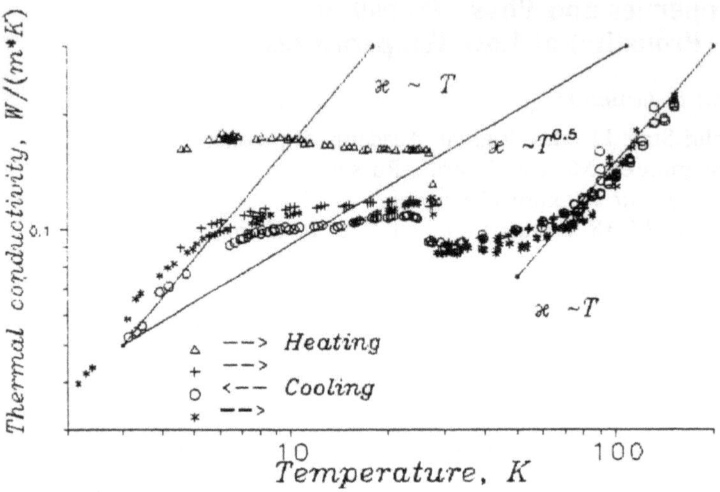

Fig.1

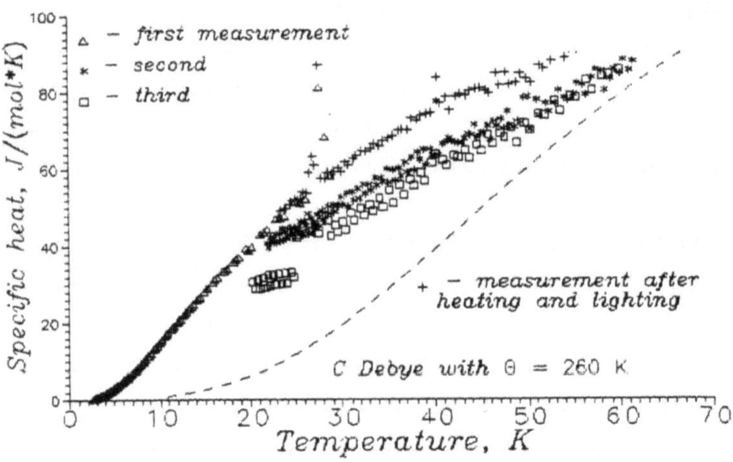

Fig.2

Second and later measurements after heating higher T_{c1} showed reduce absolute values both ϰ(T) and $C_p(T)$, what may show on existence in virgin and lighting crystal excite unequilibrium state, which disappear after heating up to 30-50 K - light sensitive hysteresis.

Second transition at $T_{c2}\approx 49$ K was registered only as change of temperature behavior C_p. For IC phase we offer model strong U-process for thermal phonons on 'freezing' incommensurate infringement of lattice period. Calculated from ϰ(T), $C_p(T)$ and sound velocity free length path of phonons λ_{ph} demonstrates correlation with IC-->C transition at T_{c2} as rising of $\lambda_{ph}(T)$.

Second order phase transition at $T_{c3} \approx 60$ K - didn't accompany change of symmetry of crystal - didn't register in both $C_p(T)$ and $\ae(T)$.

Influence of Electric Field and Hydrostatic Pressure on the Thermal Conductivity of Relaxor Ferroelectrics

E. Hegenbarth

Institut of Low Temperature Physics, Technical University Dresden, Mommsenstr. 13, O-8027 Dresden, Fed. Rep. of Germany

According our investigations at all ferroelectrics (FE) at low temperatures glasslike behaviour can be observed [1], [2], [3]. From the analysed data one can conclude, that for the generating of localized low energy excitations in these materials the structural phase transition (PT) is a necessary condition.

The close connection between PT and glasslike properties can be demonstrated by considering the behaviour of relaxor FE with a diffuse PT (DPT) and FE with a sharp PT (SPT). At FE with DPT like $Pb_{0.915}La_{0.085}(Zr_{0.65}Ti_{0.35})O_3$ (PLZT), $(Sr_{0.61})Ba_{0.39})Nb_2O_6$ (SBN), $Pb(Mg_{1/3}Nb_{2/3})O_6$ (PMN), $Pb(Sc_{0.5}Nb_{0.5})O_3$ (PSN) nearly the same behaviour as at pure a-solids is observed. This means the value of the linear temperature coefficient is nearly the same at all relaxor FE and the thermal conductivity (TC) shows T^2-behaviour and a plateau [2], [4]. At FE with SPT ($BaTiO_3$, KH_2PO_4, TGS and others) only deviations from Debye-behaviour was observed [5], [6].

Because FE are high polarizable materials it should be possible to influence the LS and therefore the low temperature properties by an electric field and by hydrostatic pressure. Fig. 1 shows as example measurments of TC at PMN [7] as function of temperature at different poling fieldes. Nearly the same behaviour was found at SBN [2], PLZT [1] and PSN [4]. The main influence of the poling field is found in the plateau region. Therefore one must look for an effective scattering process

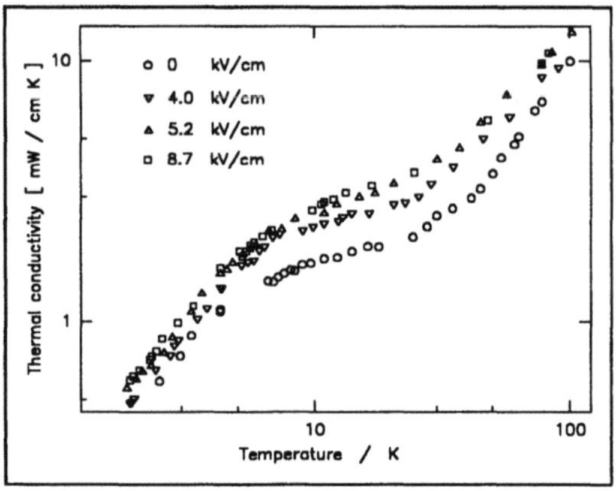

Fig. 1: Thermal conductivity of PMN-polycrystals at different electric fields

Springer Series in Solid-State Sciences, Volume 112
Phonon Scattering in Condensed Matter VII Eds.: M. Meissner and R.O. Pohl

in the plateau region, which can be influenced by the electric field, because specific heat measurments have shown that the density of LS is not changed by an electric field [5]. On the basis of the soft potential model [8] one could conclude, that the field changes mainly the coupling of resonance scattering at vibrational soft modes. But also a mechanism, which changes the phonon localization in polarized nano-regions by the field is possible.

At SBN single crystals the influence of hydrostatic pressure up to 960 MPa to TC was investigated [9]. From used method described in [9] one get the TC in the a-b-plane of these orthorhombic crystals, whereas by the usual method TC in c-direction is measured. A high anisotropy of TC was found. In c-direction in the plateau region the TC is about four times higher than in the a-b-plane. Also the influence of hydrostatic pressure is most effective in the plateau region. Fig. 2 shows the relative change of TC as function of pressure at different temperatures [9]. At about 640 MPa the change of TC has a maximum, which is most pronounced at temperatures in the plateau regions. Obviously by the pressure the density of LS and/or the coupling with the different scattering processes is decreased. To decide, which is the main process further investigations are necessary.

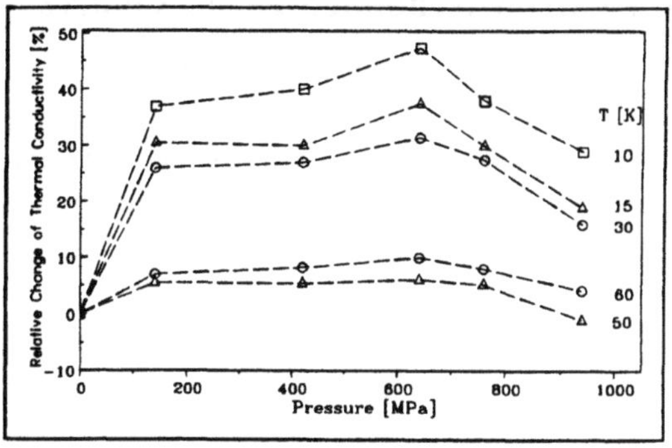

Fig. 2: Thermal conductivity of SBN as function of pressure at different temperatures

References

[1] E. Fischer, W. Hässler, E. Hegenbarth, V. J. Fritzberg; phys. stat. sol. (a) 66, K169 (1981)
[2] E. Fischer, W. Hässler, E. Hegenbarth; Phys. stat. sol (a) 72, K169 (1982)
[3] E. Hegenbarth; Ferroelctrics 78, 61(1988)
[4] H. E. Zimmer, J. Engert, P. Frach, I. Henning, D. Chat, R. Oszolinsk, E. Hegenbarth; Ferroelectrics Letters 7, 113 (1988)
[5] H. E. Zimmer, J. Engert, P. Frach, I. Henning, E. Hegenbarth; Ferroelctrics Letters 6, 33 (1986)
[6] H. E. Zimmer, J. Engert, E. Hegenbarth; Ferroelctrics Letters 6, 165 (1986)
[7] M. Fahland, E. Hegenbarth; to be published
[8] U. Buchenau, G. M. Galperin, V. L. Gurevich, H. R. Schober; Phys. Rev. B 43, 5039 (1991)
[9] K. Wagner, E. Hegenbarth; to be published

Evidence for the Anisotropy of the Tunneling States from Ultrasonic Attenuation Measurements in Neutron Irradiated Z-Cut Quartz

C. Laermans, V, Keppens, and M. Coeck

Katholieke Universiteit Leuven, Department of Physics, Celestijnenlaan 200D, B-3001 Leuven, Belgium

A z-cut quartz sample, labeled Z-N6, was irradiated up to a dose of 4.7×10^{19} n/cm^2 (E $\geq$ 0.3 MeV). On this specimen, measurements of the acoustic absorption have been performed as a function of temperature at three different frequencies: 70 MHz, 130 MHz and 280 MHz. Fig. 1 shows the results for 280 MHz in Z-N6 and, for comparison, the absorption at 330 MHz in an unirradiated z-quartz sample. For each sample, a temperature independent residual attenuation α_0, attributed to geometrical factors, was subtracted from the measured values. As can be seen, the irradiation causes a remarkable increase of the ultrasonic attenuation. In addition, a T^3-dependence is observed at the lowest temperatures, leveling off to a plateau at higher temperatures. This bevaviour is qualitatively the same as what has been found in irradiated x-quartz [4] and shows undoubtedly the presence of TS. Indeed, at very low temperatures, the tunneling model predicts a temperature dependent absorption [5]:

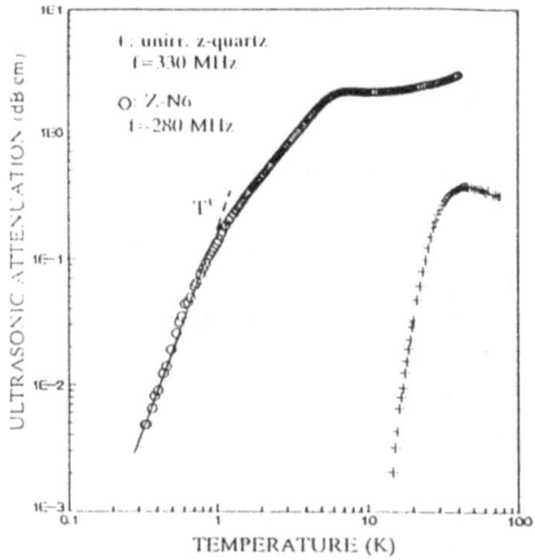

Figure 1: Ultrasonic attenuation as a function of temperature in unirradiated z-quartz and Z-N6.

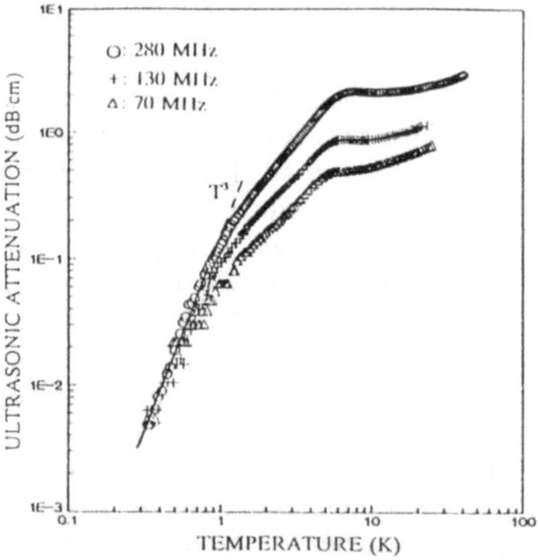

Figure 2: Ultrasonic attenuation as a function of temperature in Z-N6.

Springer Series in Solid-State Sciences, Volume 112
Phonon Scattering in Condensed Matter VII Eds.: M. Meissner and R.O. Pohl

$$\alpha = \frac{\pi^3 k^3 \bar{P}\gamma_l^2}{24\rho^2\hbar^4 v_l^3}\left(\frac{\gamma_l^2}{v_l^5}+\frac{2\gamma_t^2}{v_t^5}\right)T^3 \tag{1}$$

whereas at higher temperatures, the attenuation is given by [5]:

$$\alpha = \frac{\pi\omega\bar{P}\gamma_l^2}{2\rho v_l^2}. \tag{2}$$

In the equations above, $\bar{P}$ describes the density of states of the TS, γ_l and γ_t represent the coupling of the TS with respectively the longitudinal and transverse phonons.

In fig.2, the absorption in Z-N6 is given for three frequencies. It illustrates the frequency independent attenuation of the T^3-behaviour, which agrees with the predictions of the tunneling model (eq.1). It is also clear that the height of the temperature independent plateau increases linearly with the frequency, as described by eq.2: from 0.45 dB/cm at 70 MHz, the absorption increases to 2 dB/cm at 280 MHz.

In what follows, we will compare the absorption in Z-N6 with what has been found in N6, an x-cut quartz sample, irradiated with the same neutron dose of 4.7×10^{19} n/cm^2 [4]. Table 1 gives the TS-parameter $\bar{P}\gamma_l^2$ for both samples. These values are obtained from the absorption at the plateau (eq.2), taking into account the different longitudinal velocity v_l for the specimens. From this table, a striking difference between both samples appears: $\bar{P}\gamma_l^2$ is much smaller in the z-cut specimen compared to the x-cut N6. Since both samples were irradiated with the same dose, in the same reactor, we may assume that $\bar{P}_x = \bar{P}_z$. This implies that the observation $(\bar{P}\gamma_l^2)_z<(\bar{P}\gamma_l^2)_x$ can be reduced to a difference in γ_l. Thus, the coupling of longitudinal phonons will be smaller in z-quartz, compared to x-quartz. This gives a first evidence for an anisotropy of the TS.

Table 1: TS-parameter for N6 (x-cut) and Z-N6 (z-cut).

	Z-N6	N6 [4]
irradiation dose (n/cm^2)	4.7×10^{19}	4.7×10^{19}
longitudinal velocity (m/s)	6320	5750
$\bar{P}\gamma_l^2$ (10^6 g/cm s^2)	116	300

1. W.A.Phillips (ed.), Amorphous Solids: *Low Temperature Properties*, (Springer, Berlin, 1981).
2. P.W.Anderson, B.I.Halperin and C.M. Varma, *Philos.Mag.*25 (1972) 1.
3. W.A.Phillips, *J.Low.Temp.Phys.*7 (1972) 351.
4. A.Vanelstraete and C.Laermans, *Phys.Rev.*B42, (1990) 5842 and references herein.
5. P.Doussineau, C.Frénois, R.G.Leisure, A.Levelut and J.Y.Prieur, *J.Phys.* (Paris) 41 (1980) 1193.

The Slow Propagation Processes of Slightly Nonequilibrium Phonons in Mixed Crystals

S.N. Ivanov, A.G. Kozorezov, E.N. Khazanov, and A.V. Taranov

Institute of Radioengineering and Electronics, Russian Academy of Sciences, Mokhovaya Str. 11, 103907 Moscow, GSP-3, Russia

A strong anomaly in nonequilibrium phonon propagation in $Y_{3-x}Er_xAl_5O_{12}$ (YAG:Er) was observed and reported earlier [1]. In addition to the ordinary diffusion maximum which is common to all rare-earth-yttrium aluminium garnet mixed crystals and is due to strong point defect scattering, a second maximum delayed by 1-2 orders of magnitude with respect to the first was detected.

The dipole-dipole reservoir relaxation in YAG:Er was also studied. The main conclusion one reaches looking at these results is that there should exist a set of two-level systems (TLS) with energy splitting Δ of the order of a few kelvin.

We propose a new model of nonequilibrium phonon propagation in crystals containing TLS with proper energy splitting. For simplicity we consider the one dimensional problem. We also assume strong elastic scattering of phonons due to disorder in the mixed crystal. We are interested in the case when $\hbar\omega_c \approx \Delta$, where $\hbar\omega_c$ is the characteristic energy of the phonons in the heat pulse. We assume that the TLS specific heat is larger than that of the crystal lattice at low temperatures.

It is convenient to define the following three subsystems: TLS, nonresonant and resonant phonons. The interaction of the resonant phonons ($\hbar\omega = \Delta$) with TLS is the strongest. The interaction between the first two subsystems is either due to anharmonic three-phonon processes or to a coupling with TLS as an intermediate link. The direct interaction between the nonresonant phonons and TLS is a higher order process. The most significant processes for the interaction of the nonresonant phonons with TLS are presumably the Raman processes and the corresponding three-phonon processes. The equilibrium within the system of resonant phonons+TLS is restored much faster than that of the whole system.

Our main concern is to analyze the case when the energy exchange between the nonresonant phonons and resonant phonons+TLS followed by energy transfer and accumulation within the second system determines the transport processes.

If there are no TLS and we neglect the anharmonic scattering, then the heat pulse propagation in the mixed crystal is described as one dimensional diffusion for each group of phonons with the same energy $\hbar\omega$, the diffusion coefficient $D(\omega)$ being a function of ω. In crystals which contain the proper TLS the process of energy transfer from nonresonant phonons to TLS can become quite significant due to the pumping of the TLS by the heat pulse. Let $\tau_R(\omega)$ be the characteristic

Springer Series in Solid-State Sciences, Volume 112
Phonon Scattering in Condensed Matter VII Eds.: M. Meissner and R.O. Pohl

time for Raman scattering of phonons with the excitation or deexcitation of the TLS. In each Raman scattering event the energy of the phonons changes by an amount Δ. We assume $\tau_R(\omega) \gg \max\{\tau_0, \tau_{fs}\}$, where τ_0 is the the phonon elastic relaxation time and τ_{fs} is the resonant phonon lifetime with respect to the absorption by TLS. Each of the nonresonant phonons passes on average a distance $l_R(\omega) = \sqrt{D(\omega)\tau_R(\omega)}$ due to diffusion during a time $\tau_R(\omega) \gg \tau_0$. $l_R(\omega)$ is the characteristic length for the energy transfer from the nonresonant phonons to TLS. Whether this process is significant or not depends on the relation between $l_R(\omega_c)$ and the sample length L. If $B = (l_R/L)^2 < 1$ then the energy transfer is very significant since under the assumption $\hbar\omega_c \approx \Delta$ the energy accumulated within the system of resonant phonons+TLS is of the same order of magnitude as that of the heat pulse. The same condition can be rewritten as $\tau_R(\omega) < L^2/D(\omega)$. For $\omega = \omega_c$ this reads $\tau_R(\omega_c) < t_1 = L^2/2D(\omega_c)$, where t_1 is the time for the maximum response at the bolometer. The effect of energy accumulation at TLS can be important even if $\tau_R(\omega_c) > t_1$. This is related to the very slow decrease in the density of nonequilibrium phonons under one-dimensional diffusion. We must also take into account explicitly the TLS(or spin-)-lattice relaxation. Let τ_{sd} and τ_{sR} be the spin-lattice relaxation times with respect to the direct and Raman processes (the condition $\tau_{sd} \ll \tau_{sR}$ is supposed to be fulfilled). The slow processes of spin-lattice relaxation can nevertheless contribute to the transport of electronic excitations over the crystal since the diffusion coefficient due to the absorption and reemission of the resonant phonons by TLS is extremely small. The alternative approach is to take into account the spectral diffusion to provide efficient spatial diffusion at the wings of phonon line. If $\tau_{fs} \ll \tau_{sd}$, then the resonant phonons are captured by TLS and heat pulse energy is accumulated at TLS. If $\tau_{fs} \gg \tau_{sd}$, the TLS play the role of an intermediate link: first the nonresonant phonons pump TLS, then the latter swiftly emit the resonant phonons.

The theoretical analysis of the problem is based on the solution of the coupled kinetic equations for all the interacting subsystems [2]. It confirms the simple arguments we have used above.

We turn now to the discussion of the experimental results. A slow transport of slightly nonequilibrium phonons is not peculiar to the $Y_{3-x}Er_xAl_5O_{12}$ system. Such processes we observed in a long list of mixed crystals: $Y_3Al_5O_{12}$:Ho, $Gd_3Ga_5O_{12}$:Er, $MgAlO_4$:Mn, $Ca_3Nb_{1.69}Ga_{3.20}\square_{0.11}O_{12}$:Er, $YAlO_3$:Er, CaF_2:Er. We studied the solid solution $Y_{3-x}Er_xAl_5O_{12}$ in greatest detail. Fig.1 shows typical curves of the temperature dependence of the time of the maximum bolometer response in the slow process $t_2(T)$ for samples of various compositions and various lengths (from 0.065 to 1.3 cm). The ratio of the linear dimension of the generator to the sample length was chosen to be the same in all cases. The series of curves from a) to d) in Fig.1 corresponds to increasing Er concentration. Each of the series corresponds to a set of samples of various lengths; the length decreases with decreasing curve label. Particularly noteworthy is the experimental change in the sign of the derivative dt_2/dT for a series of samples having the same composition but differing in length (Fig.1c). The general picture of the transformation of the $t_2(T)$ curves is described completely by the theory [2].

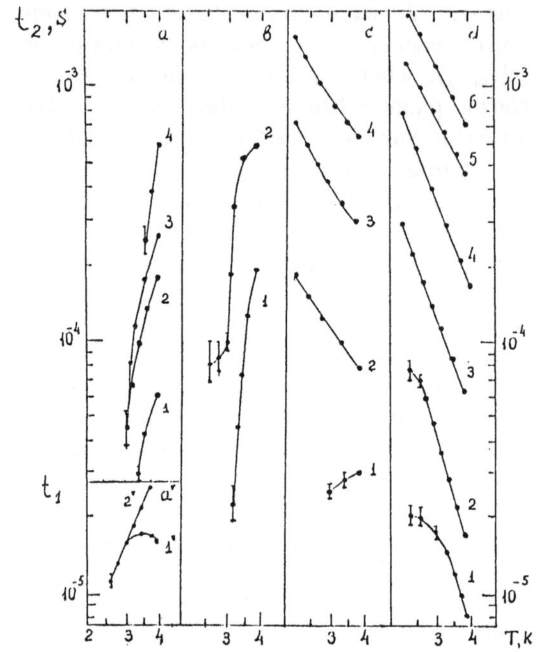

Fig. 1. $t_2(T)$ for $Y_{3-x}Er_xAl_5O_{12}$: a- x=0.2, L=0.4(1), 0.68(2), 0.95(3)cm; $Ga_{2.8}Er_{0.2}Gd_5O_{12}$ sample L=0.4cm (4); a′- the arrival time t_1 of the first diffusion maximum in $Y_{2.8}Er_{0.2}Al_5O_{12}$ (1′) and $Y_{2.8}Lu_{0.2}Al_5O_{12}$ (2′) L=0.68 cm; b-x=0.6, L=0.4 (1), 0.68 (2) cm; c-x=1, L=0.065 (1), 0.12 (2), 0.25 (3), 0.5 (4) cm; d- x=3, L=0.018 (1), x=0.03 (2), 0.07 (3), 0.14 (4), 0.22 (5) cm; sample of $Ho_3Al_5O_{12}$, L=0.03 cm (6).

The calculated dependence $t_2(T)$ for several value of the parameter B, related to the average temperature of the measurement interval, also shows remarkable change in sign of the curve's derivative. The theoretical and experimental results agree fairly well, revealing all the special features when $t_2(T)$ curves transform at about the same values of a characreristic parameter B. It is important to note that the latter was estimated from the independent experiments.

Analyzing the entire set of experimental data on slow propagation processes of slightly nonequilibrium phonons in mixed crystals [3] we can offer the following suggestion regarding the nature of the two-level centers.

In all the crystals which were used, except the calcium fluoride with rare earth ions (in which the presence of TLS has been established experimentally) and alumomagnesium spinel, there is a deviation of the composition from stoichiometry because of the presence of a certain number of oxygen vacancies. In the $Y_{3-x}R_xAl_5O_{12}$ (R-rare earth ion) crystals, for example, the number of oxygen vacancies is estimated to be 10^{20} cm^{-3}. Some of these vacancies are also associated with the incorporation of several ions in interstitial positions. Annealing $Y_{3-x}Er_xAl_5O_{12}$ crystals leads to a significant shift of t_2 toward a shorter arrival time, while the basic features of the slow transport of slightly nonequilibrium phonons are retained. It can thus be suggested that the formation of two-level systems is associated with the presence of vacancies on oxygen polyhedra which coordinate the a, c and d sites in the garnet lattice. It is natural to assume that these vacancies are distributed among the dodecahedra which coordinate the sites of yttrium Y and R atoms in proportion to composition. In other words, the vacancy concentration near R ions would be $(x/3)N_{vac}$. For the mixed crystals $Y_{2.8}Er_{0.2}Al_5O_{12}$, the value $N_{vac}=10^{20}$ cm^{-3} leads us to $6 \cdot 10^{18}$ cm^{-3}, which gives a fairly good numerical estimate of the τ_R. With increasing x, the number of R centers which contain oxygen vacancies in their nearest neighbourhood increases linearly. The appearance of TLS can be attributed to an elastic relaxation in the structure distorted by the presence of a vacancy. The different nature of the propagation of the slightly nonequilibrium phonons in $Y_{3-x}R_xAl_5O_{12}$ may stem from differences in the barrier height and in the size of the energy gap Δ of the TLS which arise. Slow transport processes are manifested in crystals containing Er and Ho. Interestingly, among the rare earths the erbium and holmium ions have ionic radii closest to that of the yttrium ion. The reason why there are no slow transport processes of slightly nonequilibrium phonons in other mixed $Y_{3-x}R_xAl_5O_{12}$ crystals may be a larger value of Δ due to a greater deformation of the structure. In a nonstoichiometric crystal matrix, there is the further possibility that Raman-pumping effects are not manifested because of the absence of strong elastic scattering.

In summary the results of our study of the propagation of slightly nonequilibrium phonons in mixed crystals has shown that it is possible to distinguish processes of a new type which are responsible for slow energy transport. The features observed in the signal propagation for various mixed crystals are evidence that slow transport processes play an important role. All the basic features of the propagation of slightly nonequilibrium phonons in these crystals can be explained satisfactorily by a model which assumes the existence of of TLS with a concentration of about 10^{18}-10^{20} cm^{-3} and energy splitting of several kelvins in these crystals. According to this model, the reason why the energy transport is slowed down in these crystals is a redistribution of energy between the heat pulse and the system of two-level centers, as a result of Raman pumping of centers by phonons. A study of the propagation of slightly nonequilibrium phonons in mixed crystals may thus prove to be a useful method for describing the transition from the crystalline to disordered state.

References

1.S.N.Ivanov, A.V.Taranov, E.N.Khazanov, V.A.Atsarkin and V.V.Demidov, Sov.Phys. JETP, **67**, 1021, (1988).
2.A.G.Kozorezov, Sov.Phys. JETP, **73**, 872, (1991).
3.S.N.Ivanov, A.G.Kozorezov, A.V.Taranov and E.N.Khazanov, Sov.Phys. JETP, **73**, 880, (1991).

Part VII

Two-Dimensional Electron Systems

Reduction in the Longitudinal-Optical Phonon Emission Rate in Polar-Semiconductor Quantum Wires and Quantum Wells

M.A. Stroscio[1], *G.J. Iafrate*[1], *K.W. Kim*[2], *A.R. Bhatt*[2], *M. Dutta*[3], *and H.L. Grubin*[4]

[1]U.S. Army Research Office, P.O. Box 12211, RTP, NC 27709, USA
[2]North Carolina State University, Raleigh, NC 27695, USA
[3]U.S. Army Electronics Technology and Devices Laboratory, Ft. Monmouth, NJ 07703, USA
[4]Scientific Research Associates, Inc., Glastonbury, CT 06033, USA

1. Introduction

This work addresses how confined and interface LO phonons in mesoscopic devices may be tailored through the judicious use of metal-semiconductor interfaces in such a way as to dramatically reduce unwanted emission of interface LO phonons.

2. Longitudinal-Optical Phonons in Metal-Semiconductor Systems

As originally proposed by Sakaki [1], the predicted high mobilities of quasi-one-dimensional wire-like regions of semiconducting material underlie many proposed quantum-wire system concepts [2,3], and have resulted in pioneering efforts leading to the actual fabrication of quantum wires [4,5]. Recently, however, theoretical studies of the interaction between longitudinal-optical (LO) phonons and carriers in a polar-semiconductor quantum wire [6] have revealed the presence of discrete LO phonon modes similar to those identified earlier for polar-semiconductor quantum wells [7-9]. As for the case of quantum wells [9,10], interface LO phonons are established at the semiconductor-semiconductor boundaries of quantum wires [11]. In addition, for carrier energies in excess of the interface LO-phonon energy, the inelastic scattering caused by carrier-interface-phonon interactions dominates over other scattering mechanisms when quantum-well [12] or quantum-wire [11] confinement occurs on a scale for about 50 Å or less. Herein, it is indicated that establishing a metal-semiconductor interface at the lateral boundaries of confined polar-semiconductor dramatically reduces or eliminates unwanted carrier energy loss caused by the interactions with interface LO-phonon modes.

Motivated by the recently demonstrated technology for the epitaxial growth of metals in intimate contact with polar-semiconductors [13], Stroscio et al. [14] have applied the dielectric continuum model of interface phonon modes [9,10] to determine the carrier-interface-phonon interaction Hamiltonian for carriers in a polar-semiconductor near a metal-semiconductor interface. It is demonstrated that [14] only those confined and interface modes of odd potential satisfy the correct boundary conditions at the metal-semiconductor interface. Hence, only those quantized LO phonon modes with vanishing potential at z=0 in a polar semiconductor slab bounded for $z < -a/2$ and for $z > a/2$ by polar semiconductors will be present in a polar semiconductor slab bounded for $z < 0$ by a perfect metal and for

Springer Series in Solid-State Sciences, Volume 112
Phonon Scattering in Condensed Matter VII Eds.: M. Meissner and R.O. Pohl

z > a/2 by a polar semiconductor. In particular, the symmetric interface mode, which has the largest electron-phonon coupling constant [9], is eliminated.

The assumption of a "perfect" metal with infinite conductivity at the metal-semiconductor interface is not critical to this result. Indeed, the plasmon frequency of the free electron gas in most metals is about two orders of the magnitude greater than the typical LO phonon frequency for a large variety of polar semiconductors. As demonstrated recently [15], the practical limit to this interfacial screening of interface LO phonons is caused by the Thomas-Fermi screening length in the metal; thus, the "free" electron gas in the metal will be capable of screening the LO phonon fields emanating in the semiconductor after they have penetrated only a few Angstroms into the metal. In practice it will be very desirable to form an epitaxial metal-semiconductor interface and to select metal-semiconductor systems such as Al-InAs [16] where Fermi level pinning does not result in the depletion of the semiconductor layer. The authors would like to thank Dr. J. Mink and Dr. M.A. Littlejohn for helpful discussions. This work is, in part, supported by the Office of Naval Research under grant N00014-90-J-1835 and the Army Research Office under grant DAAL03-89-0003-05.

3. References

[1] H. Sakaki, Jpn. J. Appl. Phys. **19**, L735 (1980).

[2] H. Sakaki, Jpn. J. Appl. Phys. **28**, L314 (1989).

[3] S. Luryi and F. Capasso, Appl. Phys. Lett. **47**, 1347 (1985).

[4] M. Watt, C.M. Sotomayor Torres, H.E.G. Amot, and S.P. Beaumont, Semicond. Sci. Technol. **5**, 285 (1990).

[5] M. Tsuchiya, J.M. Gains, R.H. Yan, R.J. Simes, P.O. Holtz, L.A. Coldren, and P.M. Petroff, Phys. Rev. Lett. **62**, 466 (1989).

[6] M.A. Stroscio, Phys. Rev. B **40**, 6428 (1989).

[7] K.L. Kliewer and R. Fuchs, Phys. Rev. **150**, 573 (1966).

[8] J.J. Licari and R. Evrard, Phys. Rev. B **15**, 2254 (1977).

[9] N. Mori and T. Ando, Phys. Rev. B **40**, 6175 (1989).

[10] K.W. Kim and M.A. Stroscio, J. Appl. Phys. **68**, 6289 (1990).

[11] K.W. Kim, M.A. Stroscio, A. Bhatt, R. Mickevicius, and V.V. Mitin, J. Appl. Phys. **70**, 319 (1991).

[12] M.A. Stroscio, G.J. Iafrate, K.W. Kim, M.A. Littlejohn, H. Goronkin and G.N. Maracas, Appl. Phys. Lett. **59**, 1093 (1991); K.T. Tsen, D.S. Smith, S.C.Y. Tsen, N.S. Kumar, and H. Morkoc, J. Appl. Phys. **70**, 418 (1991).

[13] J.P. Harbison, T. Sands, N. Tabatabaie, W.K. Chan, L.T. Florez, and V.G. Keramidas, J. Crystal Growth **95**, 425 (1989).

[14] M.A. Stroscio, K.W. Kim, G.J. Iafrate, M. Dutta, and H.L. Grubin, Phil. Mag. Lett. **65**, 173 (1992).

[15] A.R. Bhatt, K.W. Kim, M.A. Stroscio, G.J. Iafrate, M. Dutta, H.L. Grubin, R. Haque and X.T. Zhu (private communication).

[16] U. Mishra (private communication).

Theoretical Studies of Acoustic Phonon Absorption by a Two-Dimensional Electron Gas

K.A. Benedict

Department of Physics, The University of Nottingham,
University Park, Nottingham NG7 2RD, UK

There has been a good deal of recent interest in using the bulk phonons of a semiconductor material as a probe of a two dimensional electron gas embedded therein [1,2]. In this paper I would like to briefly describe a theoretical framework for the discussion of phonon absorption and to present results which indicate the possibility of a very rich structure in a strong magnetic field.

The simplified model used here assumes that the 2deg is embedded in an elastic medium, the absorption of phonons is via an interaction term in the Hamiltonian of the form

$$H_{int} = \sum_{s,\underline{Q}} M_s(\underline{Q})\rho_{\mathbf{q}}(a_s^\dagger(\underline{Q}) + a_s(-\underline{Q})) \quad ,$$

where $\rho_{\mathbf{q}}$ is the (2d) Fourier transform of the electron density operator and $a_s(\underline{Q})$ annihilates a bulk phonon of polarization s and (3d) wavevector $\underline{Q} = (\mathbf{q}, q_z)$.

The rate at which phonons are absorbed by a 2deg at $T_e = 0$ is given by

$$(\tau_s(\underline{Q}))^{-1} = |M_s(\underline{Q})|^2 Im\{\chi(\mathbf{q}, \omega_s(\underline{Q}))\} < n_s(\underline{Q}) > /\hbar \quad ,$$

where $\chi(\mathbf{q},\omega)$ is the polarizability of the 2deg, $\omega_s(\underline{Q})$ is the frequency and $< n_s(\underline{Q}) >$ the occupation of the phonon mode. Within the random phase approximation the polarizability can be written as

$$\chi(\mathbf{q},\omega) = \chi_0(\mathbf{q},\omega)/\{1 + U_{\mathbf{q}}\chi_0(\mathbf{q},\omega)\} \quad ,$$

where $\chi_0(\mathbf{q},\omega)$ is the polarizability of a non-interacting gas and U_q is the effective coulomb interaction between 2d electrons. The denominator modifies the non-interacting response function to include the effects of screening.

In a strong magnetic field, in the absence of interactions and disorder the energy spectrum of a 2deg would consist of discrete, macroscopically degenerate Landau levels, leading to some rather singular behaviour in the dielectric response, and only allowing the absorption of phonons at integer multiples of the cyclotron frequency $\omega_c = eB/m^*$. The fact there is inevitably

Springer Series in Solid-State Sciences, Volume 112
Phonon Scattering in Condensed Matter VII Eds.: M. Meissner and R.O. Pohl

some disorder in the electronic system, which broadens the Landau levels into bands allows some absorption at low frequencies ($\omega << \omega_c$) in which a quasiparticle and a quasihole are created in the same Landau level. If we assume that the eigenstates at the Fermi energy are characterized by a diffusion constant $D(\mu)$ then it can be shown that

$$\chi_0(\mathbf{q},\omega) \sim \frac{\rho(\mu)}{\pi}\frac{Dq^2}{\omega^2+D^2q^4}(Dq^2+i\omega) \quad , \quad l_c q, \omega/\omega_c < 1 \quad ,$$

where $\rho(\mu)$ is the density of states at the Fermi level. If we work at a finite low temperature such that the scaling of D is cut-off at an inelastic scattering length which is not too much longer than the cyclotron length then we can use the self-consistent Born approximation for $D(\mu)$, at lower temperatures it can be derived from the normal conductivity via the Einstein relation.

The results shown below are for the absorption of phonons at a fixed low frequency as a function of Fermi energy for a range of angles of incidence, clearly for near to normal incidence (such that q is not too large) the absorption follows ones naive expectation and is maximum when the Fermi energy lies at the centre of the Landau band, however at larger angles the fact that the screening is more effective when μ is near the band centre causes there to be a minimum in the absorption there.

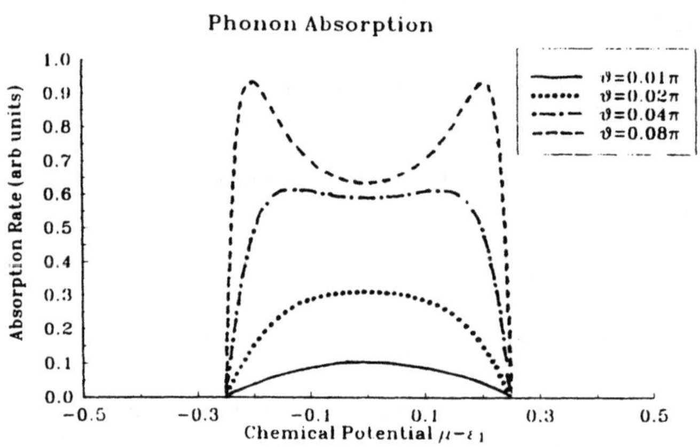

Hence one can see that the nature of the oscillation of absorption with Landau level occupation is very rich depending in detail on the competition between the effects of Fermi statistics on the occupation of quasiparticle states and the effects of screening.

References

[1] LJ Challis, AJ Kent and VW Rampton (1990), Semicon, Sci and Tech, **5** 1179

[2] KA Benedict (1991), J Phys Cond Matt, **3**, 1279

Phonon Absorption and Emission Experiments in Quantum-Hall-Effect Devices

W. Dietsche, F. Dietzel, U. Klaß, and R. Knott

Max-Planck-Institut für Festkörperforschung,
Heisenbergstr. 1, W-7000 Stuttgart 80, Fed. Rep. of Germany

1. Introduction

A two-dimensional electon gas (2DEG) can be formed at the interface between two different semiconductors. This electronic system interacts with the bulk phonons of the substrate material. The matrix element of this interaction contains an overlap integral of the respective wavefunctions of the electron and the phonons. In zero magnetic field this leads to the conservation of the in-plane momentum and of the energy. The interaction strength follows from the symmetry of the phonon strains and from parameters like the deformation potentials and the piezoelectric interactions.

The situation is less clear if magnetic field is applied which is strong enough to quantise the electron-energy spectrum in Landau-levels. The electron wave functions then become localized. Thus the in-plane momentum conservation rule no longer exists and the matrix element will take a more complicated form. A further complication arises because the current and electrostatic-potential distributions are no longer homogeneous over the sample surface.

In this contribution we report on our recent results about the phonon absorption and emission by a 2DEG in a strong magnetic field. The first ones were done by exposing a 2DEG sample to a flux of ballistic phonons. The phonon absorption strength was measured by observing the phonon-drag effect [1]. In the case of the phonon emission we concentrated on the question of the spatial distribution of the emission processes. This was done by utilizing the fountain-pressure of superfluid helium [2].

Springer Series in Solid-State Sciences, Volume 112
Phonon Scattering in Condensed Matter VII Eds.: M. Meissner and R.O. Pohl

Alternative approaches to study the elctron phonon interaction are being used by the group in Nottingham. Their work is being reviewed by A.J. Kent [3].

2. Phonon Absorption

The sample setup is shown as Fig. 1. The 2DEG was formed at the top of the 0.3 mm thick GaAs substrate at the interface to an MBE-grown [4] AlGaAs layer. The electron density was 6×10^{11} cm^{-2} and the mobility was 1.6×10^{6} $Vseccm^{-2}$. The 2DEG was etched into a shape resembling those used for Quantum-Hall-Effect measurements. There are two current contacts and two voltage probes which are connected to an amperemeter and a voltmeter, respectively. The narrow constriction (50x50 μm^2) forms the phonon-sensitive area. Phonons (heat pulses) are generated on the bottom side of the substrate by laser-heating an Al film. After travelling ballistically through the substrate the phonons are absorbed by the 2DEG and will accelerate the electrons leading to a phonon-drag voltage. The ballistic phonon transport is easily revealed by the phonon focusing pattern if the laser is scanned and an imaging experiment is performed [5]. Without a magnetic field there will be an induced current only but in a magnetic field there will be a phonon-drag Hall voltage in addition.

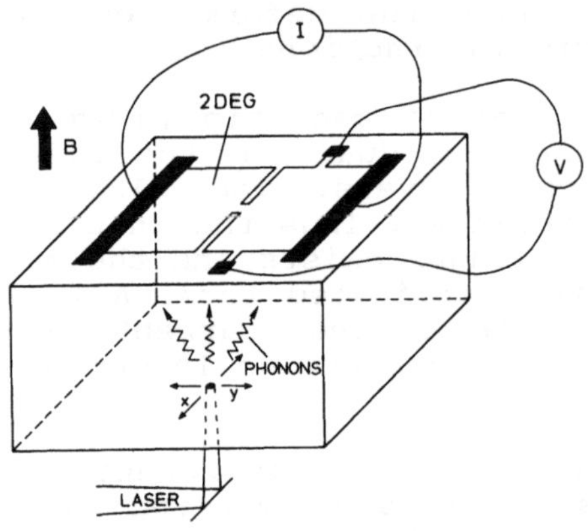

Fig.1. Experimental setup of the phonon-drag measurements in strong magnetic fields.

In zero magnetic field the results of the imaging experiment of Karl et al. were reproduced [1]. The phonon-drag image showed the focusing pattern folded with the piezoelectric interaction. There were several new effects, however, in strong magnetic fields. Firstly, there were two types of signal. One was due to ballistic phonons and allowed to take phonon-drag images. These had almost the same shape as without magnetic field and are therefore not reproduced in this publication. The normalized phonon-drag images do not depend on magnetic field indicating that there is no effect of the magnetic field on the phonon-strain dependence of the matrix element. The ballistic phonon signals were superimposed on a background due to thermal voltages of the Nernst-Ettinghausen type. This background was caused by temperature differences between the upper and lower sample edge in Fig. 1. Fortunately, this contribution could be subtracted out [6].

The overall strength of the electron-phonon interaction depended strongly on the magnetic field. This is obvious from the plot of both the phonon induced current and voltage as function of magnetic field in Fig. 2. The laser was kept in a fixed position and phonons with the fast transverse polarisation were detected. The signals oscillate with the field in a Shubnikov-de Haas like fashion. At fields where an integer number of Landau-levels are filled there is no signal since the typical phonon energies are much smaller than the cyclotron energies. The signal at intermediate filling factors must be due to transitions within the impurity broadened Landau-levels.

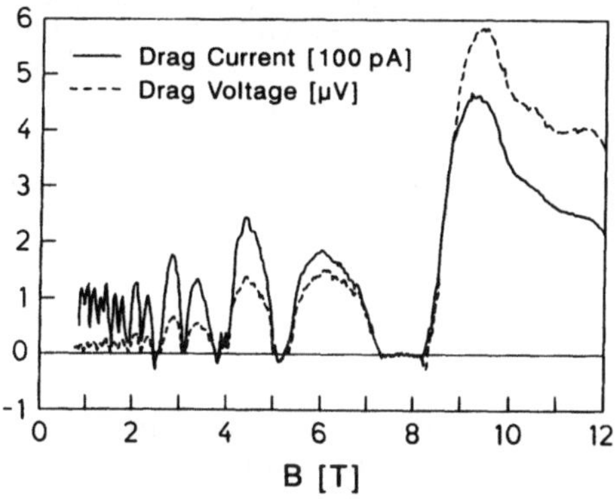

Fig.2. Phonon induced currents and Hall voltages as function of magnetic field.

In order to analyse these data we assume that a "phonon-wind force" $\mathbf{P}$ acts on the electrons in the same way as the electric-field force $e\mathbf{E}$ [7]. Thus the current density $\mathbf{j}$ of the 2DEG will be

$$\mathbf{j} = (\underline{\sigma}/e)\ (e\mathbf{E} + \mathbf{P}) \quad ,$$

where $\underline{\sigma}$ is the conductivity tensor of the 2DEG. The phonon-wind force is essentially the phonon flux times the interaction strength. Under the assumption that $\sigma_{xx} << \sigma_{xy}$ one finds in our geometry

$$\mathbf{E}_{Hall} = (1/e)(\sigma_{xy}/\sigma_{xx})\ \mathbf{P} \quad ,$$

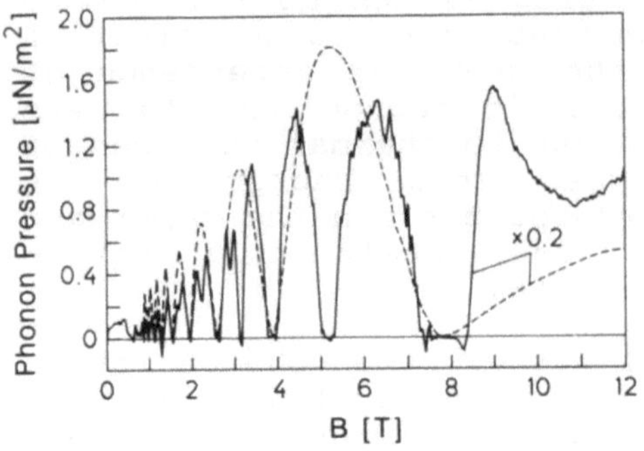

Fig. 3 Phonon-wind force as function of magnetic field.

which eventually allows us to make a comparison with theoretical models of the phonon absorption. A plot of the phonon-wind force as a function of the magnetic field is shown in Fig. 3. Also shown (dashed line) is a one parameter fit of a theory which was essentially based on the assumption that the phonon absorption strength is proportional to the density of electronic states at the Fermi edge. The main deviations are due to the fact that the spin splitting of the Landau levels and the evolving of the fractional quantum Hall effect (around 10 Tesla) was not included in the theory. Apart from these deficiencies the description is quite good. Thus, the conclusion is that the strain dependence of the electron phonon interaction is not affected by the magnetic field. The interaction strength is determined by the electronic density of states.

3. Phonon Emission

The electronic transitions which lead to the emission of phonons by a 2DEG in a strong magnetic field are either between states within a Landau level or between them. The latter would lead to phonon energies of the order of 10 meV at typical fields. One can argue that both types of transitions are only possible at low temperatures if a sufficiently strong electric field is applied. In this case the Landau levels become sufficiently tilted to allow both transitions between neighbouring levels by tunneling and within one level under phonon emission. The electric field distribution, however, is very inhomogeneous particularly near the current contacts. Only very recently there have been the first studies of its spatial distribution by electrooptical techniques [8].

In the experiments reported here we studied the spatial distribution of the phonon emission by the

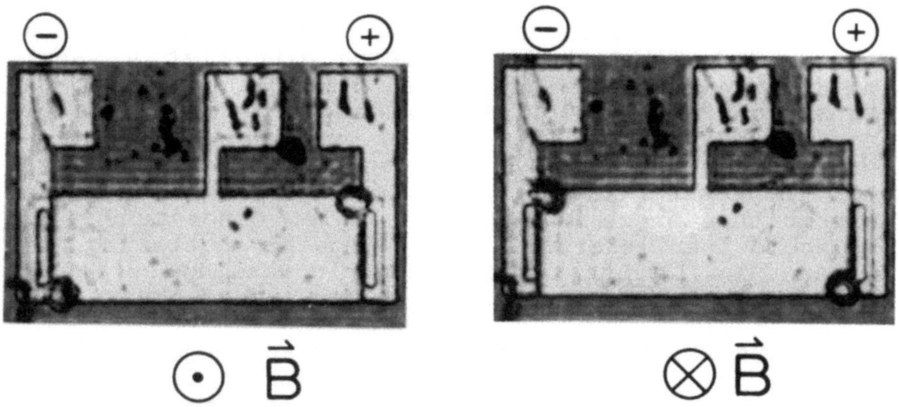

Fig.4 Fountain-pressure images of the phonon emission in a Si-MOSFET sample. The two images correspond to ± 6 Tesla. Current was 250 μA. Sample width was 1 mm.

fountain effect of the superfluid helium. The GaAs and the Si-MOSFET samples were partly immersed in superfluid helium in such a way that a thin helium film formed on the surface. The emission of phonons leads to an increase of the film temperature and consequently to an easily visible helium drop due to the fountain pressure. In the first experiments we studied the behavior near the current contacts. It has always been assumed that the electric field distribution in high electron mobility sample has two maxima in opposite corners of the current contacts. This leads indeed to strong phonon emission in those corners. In Fig.4 we show the example of a Si-MOSFET [9]. The two images were taken with magnetic fields of ±6 Tesla (filling factor 4), respectively. The helium drops in the two opposite corners are easily visible. With reversed field the drops were in the other two corners. There is no asymmetry between the current entry and exit points in this type of samples [10] in marked contrast to those based on GaAs heterostructures [2]. One can estimate the electric fields in the corners by assuming that the total Hall voltage is the maximum potential drop over a length equal to the drop diameter. This leads to an electric field of the order of 10 kV/m. This value is about the same as the one which is necessary to cause dissipation in the bulk of a QHE sample. It corresponds to drift velocities E/B of the order of 1000 m/sec. This is close to the sound velocities in Si or GaAs.

The spatial behavior of the phonon emission in the bulk of the samples has been studied in GaAs/AlGaAs heterostructures. In circular samples (Corbino geometry) it was found that the dissipation takes place along filaments [11]. Filaments were also observed in standard samples if the drift velocities exceeded the critical values. The filament first formed either at a sample imperfection or if the sample width changed at the transition from wide to narrow sample regions. The filaments usually formed a angles of about 30° with the sample edges. Interestingly these angles became smaller with increasing current. It is rather tempting to interpret this angle θ as a Hall angle. Since $\tan\theta = \omega_c\tau$ one would then interpret τ as being the scattering time between two Landau levels. With increasing current τ and θ would become smaller as it was indeed observed.

There is, however, no generally accepted theory to describe the onset of the phonon emission. In addition, the occurence of filaments has not been predicted. For further understanding it would be of great interest to perform phonon spectroscopic experiments. Unfortunately, there are no suitable techniques available in the magnetic fields. The inhomogeneity of the electric fields in the sample will also be a challenge in the interpretation of many possible experiments.

4. References

[1] H. Karl, W. Dietsche, A. Fischer, and K. Ploog. Phys. Rev. Lett. 61, 2360 (1988).
[2] U. Klaß, W. Dietsche, K. von Klitzing, and K. Ploog, Z. Phys. B **82**, 351 (1991).
[3] A.J. Kent, this volume.
[4] The sample was grown by R. Nötzel and K. Ploog.
[5] G.A. Northrop and J.P. Wolfe, in Nonequilibrium Phonon Dynamics ed. by W.E. Bron (Plenum, New York and London, 1985) p.165.
[6] F. Dietzel, Dissertation Universität Stuttgart, 1992.
[7] A.L. Efros and Yu.M. Galperin, Phys.Rev.Lett. **64**, 1959 (1990).
[8] F.P. Fontein, P. Hendriks, F.A.P. Blom, J. H. Wolter, L.J. Giling, and C.W.J. Beenakker, Surf. Sci. **263**, 91 (1992).
[9] This sample was provided by L.J. Challis, University of Nottingham.
[10] F.F. Ouali, L.J. Challis, and J. Cooper, Semicond. Sci. Technol., in press.
[11] U. Klaß, W. Dietsche, K. Von Klitzing and K. Ploog, Physica B **169**, 363 (1991).

Phonon Imaging in Two-Dimensional Electron Systems

A.J. Kent

Department of Physics, University of Nottingham, University Park,
Nottingham NG7 2RD, UK

1. Introduction

Phonon studies of two-dimensional electron systems are valuable, not only because they give directly information regarding the electron-phonon interaction, but also because they can give additional information regarding the physical nature of the electronic states. Over the past decade a number of phonon emission and absorption experiments have been carried out, for a review see [1], some of which have used the imaging technique. There is an intrinsic angular dependence of the phonon emission/absorption process which can be seen to arise partly from energy and momentum conservation considerations. The details of this angular dependence change upon the application of a strong magnetic field. In the quantum Hall regime phonon emission and absorption take place only at localised regions in the device. All of these effects can be studied in detail using phonon imaging. In this paper I will review some of the imaging experiments we have caried out at Nottingham.

2. Phonon Emission Imaging

When the 2DEG in the Si inversion layer or the GaAs-(AlGa)As heterojunction is heated above the lattice temperature, e.g. by passing an electric current through it, the electrons cool by emitting phonons. If the electron temperature is not too great, $T_{el} \leq 50K$, and the lattice is at liquid helium temperatures then acoustic phonons are emitted which travel ballistically through the substrate and can be detected on the opposite surface. By using pulsed currents, with a pulse duration shorter than the phonon time of flight, and a bolometer, the various phonon modes can be resolved owing to their different speeds. The angular dependence of the emission may be studied by using many small discrete bolometers or preferably an extended spatially sensitive bolometer.

A detector that we have developed for this purpose is based on cadmium sulphide (CdS). When vacuum evaporated under the right conditions, CdS forms a semiconducting film in which the conduction electrons freeze out upon cooling to liquid helium temperatures. Upon illumination with band-gap radiation, electrons are promoted to shallow traps a few meV below the conduction band and

Springer Series in Solid-State Sciences, Volume 112
Phonon Scattering in Condensed Matter VII Eds.: M. Meissner and R.O. Pohl

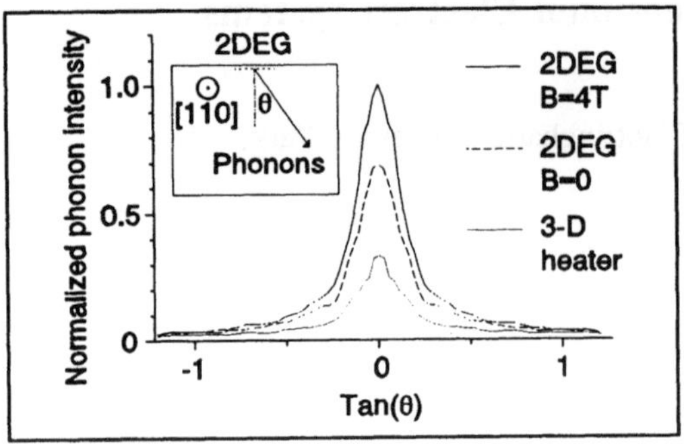

Figure 1. Linescans of the phonons emitted by a heated 1x1mm 2DEG in an (001) Si MOSFET at $B=0$ and 4T. Also shown is a linescan of the phonons emitted by a 3-D heater. In each case the electrical input power is the same, $P_{in}=40$mW.

the conductivity is very temperature sensitive. Large-area interdigital electrodes are formed using photolithography and a focused Ar^+ laser used to illuminate the exposed CdS between a pair of contact strips. In this way a selected spot on the bolometer is sensitised. This spot can be reset by heating the CdS with a current pulse. An image of the phonon flux falling on the bolometer can be made by repeating the sensitising and resetting operations at different positions.

Figure 1 shows linescans of the phonons emitted by a MOSFET on the (001) surface of a 5mm thick Si wafer. The linescans are for phonons propagating in the (110) plane. Also shown, for comparison, is a linescan of the emission by a conventional 3-D heater. The dominant feature in all of the scans is the focussing of the transverse (TA) mode phonons close to [001], however, it is clear that the intensity in the forward direction is greater in the case of emission by a 2DEG. There is a further increase in forward intensity when a quantising magnetic field is applied to the 2DEG.

Detailed theoretical treatment of the emission process can be found in a number of papers, see for example [2]. However, it is possible to give a semi-quantitative explanation of these results by applying some simple momentum conservation considerations to the emission process: The largest possible value of the in-plane component of the emitted phonon's momentum is $q_{\parallel max}=2k_F$, corresponding to the initial and final electronic states lying on opposite sides of the 2-D Fermi circle, this is the so-called $2k_F$ cutoff. Momentum is also conserved in a direction perpendicular to the plane of the 2DEG and $q_{\perp max}\approx 1/a_0$, where a_0 is the 'thickness' of the 2DEG sheet. This corresponds to the uncertainty in electron momentum associated with its confinement in the plane. In the experiment of figure 1, $k_F=(\pi N_s)^{1/2}=2\times10^8\text{m}^{-1}$, where $N_s=1.3\times10^{15}\text{m}^{-2}$ is the 2DEG sheet density and $a_0\approx 2$nm, therefore the most energetic (largest q-vector) phonons would be emitted at an angle of $\theta=\tan^{-1}(2k_Fa_0)=40°$ to the 2DEG normal. The phonons are directed still closer to the normal by focusing and hence the increase in forward flux relative to the 3-D heater.

Upon application of a quantising magnetic field the electronic density of states is broken up into a series of Landau sub-bands. At sufficiently high electron temperatures, $k_BT_e>\hbar\omega_c$, it is possible to have electronic transitions between a pair of Landau levels accompanied by cyclotron phonon emission. The in-plane confinement of the electrons is now on the scale of the magnetic length, $l_B=(h/eB)^{1/2}$ and so if the field is sufficiently strong the maximum angle at which a cyclotron phonon can be emitted, $\theta_{max} \approx \sin^{-1}(q_c l_B)=\sin^{-1}(\omega_c l_B/v_s)$, is less than θ_{max} at $B=0$ and so the forward intensity is increased still further.

3. Phonon Absorption Imaging

One of the problems with imaging the phonon emission is that heating smears out the electronic energy distribution particularly at small values of k_F. In absorption experiments the electron temperature remains close to that of the lattice, however, the physical details of the electron-phonon interaction remain essentially the same.

To image the phonon absorption/scattering in a Si MOSFET we used a 'phonoconductivity' technique in which a movable source of nonequilibrium phonons was generated by thermalising a pulsed Nd-YAG laser in a constantan film on the opposite side of the 5mm substrate and their effect on the device conductivity was measured as a function of laser position. Figure 2 shows an image of the phonon absorption/scattering recorded in this way, alongside it is the image recorded using an ordinary bolometer instead of the 2DEG, both detectors were 1x1mm in size. The first and most obvious difference is that the dominant features running diagonally in the bolometer image, due to fast transverse (FTA)

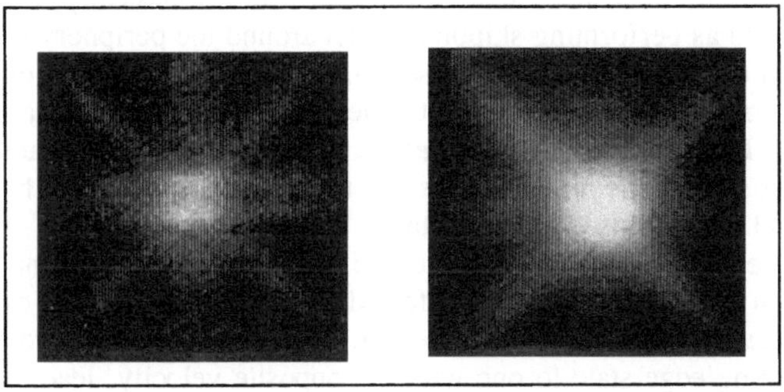

Figure 2. Phonon images recorded using a 2DEG detector (Left) and a CdS bolometer (Right). The images span 1cmx1cm on the 5mm thick Si substrate.

modes propagating in (100) and (010), are much weaker in the case of the 2DEG detector. This is because the FTA mode does not possess the necessary component of polarization perpendicular to the 2DEG plane to couple to the electrons. The residual signal is probably due to FTA modes that were mode converted by scattering at the Si-SiO_2 interface. In the case of the 2DEG detector the intensity of the slow transverse (STA) mode propagating in the (110) plane falls of more rapidly at large angles if N_s and hence k_F is reduced, see figure 3.

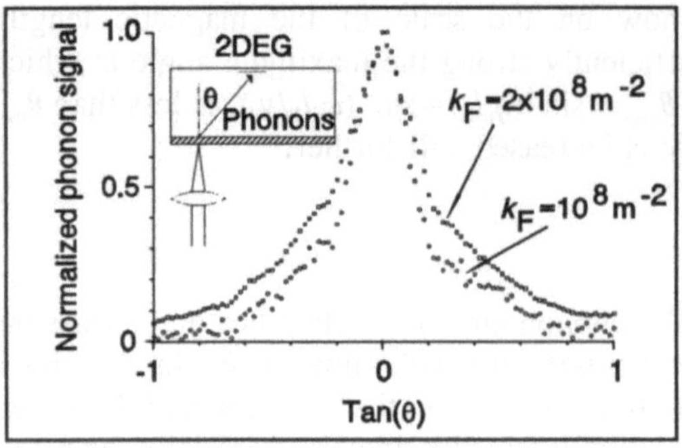

Figure 3. Linescans of STA phonons propagating in (110) and detected by an (001) 2DEG at two different values of k_F.

4. Phonon Imaging in the Quantum-Hall Regime

Phonon emission experiments to probe the location of dissipation in the QHR have been carried out showing that the dissipation in the bulk of a standard Hall bar drops to zero and moves to the corners at which current enters and leaves the device [3]. Another interesting question, on which phonon imaging techniques have recently shed some light, concerns the role of edge states in the QHE.

The condition that the electronic wavefunction must vanish at the edge of the 2DEG means that, within a few cyclotron radii of the sample edges, the energy levels are increased relative to the same Landau level in the bulk. Classically, the electrons may be visualised as performing skipping orbits around the periphery of the device. The direction of propagation of these so-called edge states depends on the magnetic field direction and on which side of the device they are on. At small currents and hence small Hall voltages, such that $eV_H << \hbar\omega_C$, the edge states are the only current carrying states. The net current is just the difference between the currents at each edge and the quantised Hall resistance depends on the number of edge states at the Fermi level (E_F) each of which acts as a pseudo-one-dimensional channel. The propagation of an edge state is unaffected by a moderate amount of disorder, however, processes that cause backscattering, i.e. transfer of a electron from a forward propagating edge state to one with the opposite velocity, lead to the loss of quantisation [4]. At large V_H there is also bulk current, however, loss of quantisation is still associated with the backscattering of electrons via those edge states of opposite velocity.

Experimentally we are able to stimulate backscattering using an intense beam of nonequilibrium phonons. The phonons are generated on the back surface of a 380μm GaAs wafer at 2K by thermalising a pulsed laser in a metal film. These traverse the substrate ballistically and fall on the device, a 3x1mm 2DEG in the GaAs/(AlGa)As heterojunction, the effect on its two terminal resistance is measured. Owing to the strong focusing of TA modes close to the [001] the largest phonon flux is in an area about 100μm across directly opposite the source. By moving the laser spot different areas of the device can be probed.

Figure 4 shows a pair of images corresponding to different positions of the E_F within a Landau gap (B=1.5T; filling factor=10; device current=100μA). With

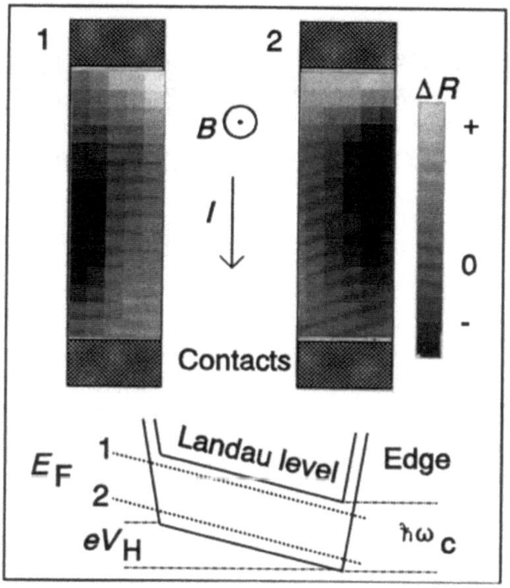

Figure 4. Images of the two-terminal phonoconductivity of a GaAs/(AlGa)As heterojunction 2DEG (N_s=7.8x10^{15} m^{-2}; mobility=70 $m^2V^{-1}s^{-1}$) in the quantum Hall regime. Dark areas represent an **increase** in the two terminal resistance. The electronic energy levels and the relative position of the Fermi level between the 10th. (filled) and 11th. (empty) Landau sub-bands are shown schematically below the images.

E_F just below the unfilled level the strongest signal is observed on one edge of the device. **This edge swaps sides upon reversal of current or magnetic field direction.** We believe this signal to be due to phonon assisted tunnelling of electrons from the edge states on the left of the device to the empty Landau level above, whereupon they can quickly move through the empty level to the right hand (counterpropagating) edge. It is necessary for the Fermi level to be close to the empty states because, owing to the condition $q_{\perp max} \approx 1/a_0$, only phonons of frequency $\leq$200GHz can interact with the electrons. If E_F is reduced to just above the bulk energy of the filled level, then it becomes possible for phonon assisted tunnelling of electrons from the bulk states within a few magnetic lengths of the right hand edge to the empty edge states near the Fermi level.

5. Conclusions and Acknowledgements

I have described some of the ways in which phonon imaging has been able to add to our knowledge of the behaviour of low-dimensional electronic systems. Phonon imaging is unrivalled in its ability to graphically illustrate the anisotropy of the electron-phonon interaction in 2-D, it also provides us with a tool to probe the spatial dependence of the properties of electronic states within a device.

I would like to take this chance thank all of those who worked on this project over the years, especially: G Hardy, P Hawker, D Hurley, D McKitterick, S Chapman, R George and E Varnell. I would also like to thank the SERC of the UK, The EEC and the Royal Society for financial support.

References

[1] Challis L J, Kent A J and Rampton V W 1990 *Semicond. Sci. Technol.* **5** 1179
[2] Challis L J, Toombs G A and Sheard F W 1987 *Physics of Phonons* ed. T Paszkiewicz (Berlin: Springer) 348
[3] Russell P A, Ouali F F, Hewett N P and Challis L J 1990 *Surf. Sci.* **229** 54
[4] Buttiker M 1988 *Phys. Rev.* **B38** 9375

Phonon Images of Crystalline GaAs Obtained by the Phonon-Drag Effect in Two- and One-Dimensional Electron Gases

D. Lehmann[1], *Cz. Jasiukiewicz*[2], *and T. Paszkiewicz*[2]

[1]Institut für Theoretische Physik, Technische Universität Dresden, Mommsenstr. 13, O-8027 Dresden, Fed. Rep. of Germany
[2]Institute of Theoretical Physics, University of Wroclaw, ul. Cybulskiego 36, PL-50-205 Wroclaw, Poland

Images of crystals provide a global information on their internal structures and dynamics. Dietsche and collaborators [1] contrived a new method of obtaining images of *GaAs*. They used the phonon-drag effect in the two-dimensional electron gas (2DEG) formed in *GaAs/AlGaAs* heterostructures.

We give a detailed kinetic picture of the nonstationary phonon-drag effect in low-dimensional electron systems. The electrons belonging to a 2DEG or 1DEG are coupled to 3D-bulk phonons by the deformation potential and the piezoelectric interaction. We describe the coupled subsystems by a pair of Boltzmann kinetic equations and derive general explicit expressions for the instantaneous (j) and the time integrated current density ($\bar{j}$) induced by a phonon pulse. Using the formulae we calculate the time integrated current density as a function of phonon propagation direction both for the case of a 2DEG lying in a (001) plane and for an 1DEG aligned along the [110] direction.

Results are displayed in form of patterns. We notice that their shape depends essentially on the frequency of phonon bursts and on the density of LDEG. Therefore we present separately the contributions of the deformation potential and the piezoelectric coupling for several frequencies and different electron densities (for figures and details see [2,3]).

The numerical results reflect the main features of experimental finding. They are a convolution of the phonon quasimomentum focusing (contains the information about the geometry of the slowness surface) [4] and the probability of phonon absorption by the LDEG. For the 2DEG and not too high phonon frequencies the patterns of time integrated current density provide useful information about the crystalline medium and relatively scarce information about the electron-phonon interaction. For higher frequencies or an 1DEG the situation is reversed - the electron-phonon interaction distorts deeply the quasimomentum focusing patterns and the information about the medium should be deconvoluted.

In order to compare our theoretical result for the 2DEG with the experimental pattern the contributions of both interacting mechanism are added. This leads to a ratio between piezoelectric and deformation potential coupling constant of $(2e\, e_{14}/\varepsilon_0\varepsilon_r)/\Xi_D \approx 9.7 \times 10^8\mathrm{m}^{-1}$. In agreement with experimental assumptions the frequencies of generated phonons seem to be close to 120GHz.

Springer Series in Solid-State Sciences, Volume 112
Phonon Scattering in Condensed Matter VII Eds.: M. Meissner and R.O. Pohl

In the case of a (quasi-)1DEG the dependency of $\overline{j_\zeta}$ on different confinement potentials is examined, but no evident modifications can be found in the patterns.

Finally, one should underline that the proposed microscopic theory depends on several parameters and some of them influence the results substantially. Therefore there is a need for further accurate phonon-drag experiments in the case of 2DEG and first systematic experiments in the case of 1DEG.

References

[1] For a review of the experiments, see W. Dietsche, in *Phonons'89*, edited by S. Hunklinger, W. Ludwig, and G. Weiss (World Scientific, Singapore, 1990), p. 767

[2] Cz. Jasiukiewicz, D. Lehmann, and T. Paszkiewicz, Z. Phys. B - Condensed Matter **86**, 225 (1992)

[3] D. Lehmann, Cz. Jasiukiewicz, and T. Paszkiewicz, *Phonon images of crystalline GaAs obtained by the phonon-drag effect in an 1-dimensional electron gas* (in preparation)

[4] Cz. Jasiukiewicz, D. Lehmann, and T. Paszkiewicz, Z. Phys. B - Condensed Matter **84**, 73 (1991)

Phonon Emission from a Hot Two-Dimensional Hole Gas

A.J. Kent, K.R. Strickland, and M. Henini

Department of Physics, University of Nottingham,
University Park, Nottingham NG7 2RD, UK

It has recently become possible to produce high mobility hole gases in the GaAs-(AlGa)As heterojunction by MBE growth on the 311A plane. Such layers have a strongly temperature dependent component of the conductivity even well below 4.2K. This would imply that the carrier-acoustic phonon scattering is stronger than in 2DEG devices of similar mobility in which there is typically little temperature dependence below 10K. Direct studies of phonon emission and absorption are preferred methods of obtaining information regarding the hole-phonon coupling since they do not involve the averaging over mode and q-vector implicit in electrical transport measurements. Experiments on the phonon emission by hot holes in multiple quantum wells have been reported in [1] and strong coupling to acoustic modes was evident from the nature of the observed heat pulse signals. Part of the reason for this is the increased effective mass of the holes, the 2D-carrier-acoustic phonon relaxation rate $\propto (m^*)^{3/2}$. In this paper we report a systematic study of the phonon emission by a hot 2DHG in a single heterojunction.

The device used was based on a Be-doped (311A) heterojunction grown on a 380μ thick chromium-free semi-insulating GaAs wafer, the hole areal density was $1.6\text{x}10^{15}\text{m}^{-2}$ and the 1.5K mobility about $10\text{m}^2\text{V}^{-1}\text{s}^{-1}$. Because of difficulties in obtaining a sufficiently low ohmic contact resistance for emission from the contact regions to be neglected, 1mm^2 capacitive contacts were used. The holes were heated by applying 20ns current pulses and the phonons detected on the opposite side of the wafer using two granular Al superconducting bolometers, one directly opposite the device and one at an angle of about 50° to the 2DHG normal. The device resistance and power were deduced from the relative sizes of the forward and reflected pulses on the 50Ω line from the pulse generator. The hole temperature was deduced following the calibration procedure described in [2].

Figure 1 shows a set of heat pulse signals recorded on the bolometer directly opposite the device, very little was detected on the bolometer at 50°. Also shown, for comparison, is the signal recorded using the same geometry but with a 3D heater in place of the hole gas. It is immediately noticeable that the holes couple more strongly to the LA mode than does the heater. The absence of a signal on the second bolometer is a result of the $2k_F$ cutoff [3]. At even the lowest hole

Springer Series in Solid-State Sciences, Volume 112
Phonon Scattering in Condensed Matter VII   Eds.: M. Meissner and R.O. Pohl

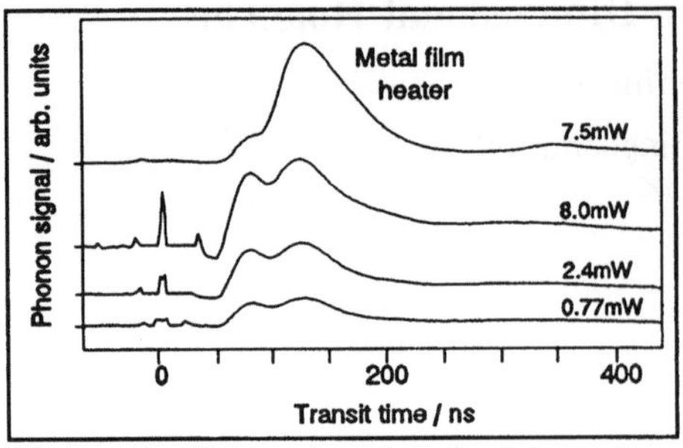

Figure 1. Heat pulse signals detected on bolometer opposite 2DHG

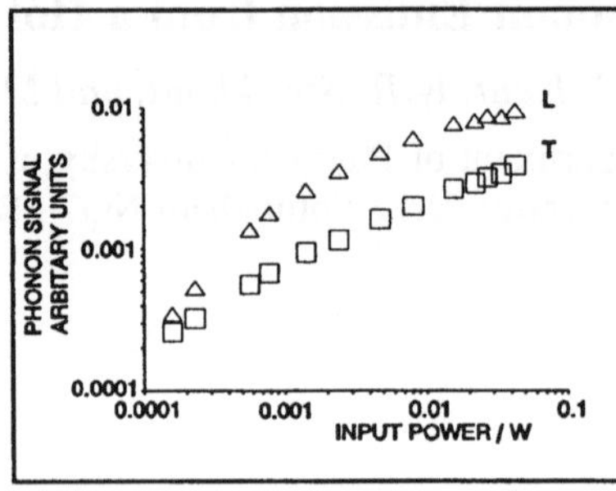

Figure 2. LA and TA mode signal heights as a function of power

temperatures hence smallest phonon q-vectors this would severely limit emission for angles greater than about 12°.

Figure 2 shows the size of the LA and TA mode signals as a function of power. The rate of increase of both modes tails off at high powers probably as a result of bulk scattering of the higher energy phonons, however, it is clear that the rate of increase of the LA mode is greater below about 3pW/hole ($T_h \approx 8$K). This is to be expected since the LA mode q-vector is less than that of a TA mode phonon of the same frequency, hence the LA mode is less severely affected by the cutoffs. In the near normal direction the condition $q_\perp < 1/a_0$ applies and from the hole temperature at which the increase of the LA mode starts to slow we can deduce that $a_0 \sim 2$nm which is considerably less than for an equivalent 2DEG. We find that the hot hole energy relaxation rate follows a T_h^5 dependence which is consistent with energy relaxation by acoustic phonon emission if the cutoffs are much higher than the peak in the phonon distribution corresponding to T_h [4]. This is somewhat surprising in view of the bolometer signals which suggest severe cutoffs. It should, however, be noted that the hole distribution is non-degenerate for hole temperatures above 5K ($kT_h > E_F$) which will relax the $2k_F$ cutoff. We observe no evidence of optic mode phonon emission at even the highest power we could apply (~ 120pW/hole) this is in contrast to the measurements on 2DEG's where optic mode emission was dominant above 5pW/electron.

This work is being supported by a grant from the SERC of the UK.

References

[1] Chin M.A, Narayanamurti V., Stormer H.L.and Gossard A.C, (1985) *Proc. 17th. Int. Conf. on Physics of Semiconductors*, Eds. J.D.Cahdi and W. A.Harrison (Berlin: Springer) p333

[2] Hawker P, Kent A.J, Hughes O. H.and Challis L.J. (1992) *Semicond. Sci. Technol.* 7 B29

[3] Kent A.J.in these proceedings

[4] Challis L.J, Toombs G.A.and Sheard F.W. (1987) *Physics of Phonons* ed.T. Paszkiewicz (Berlin: Springer) 348

Evidence for Virtual Phonon Exchange Between 2-D Electron Gases

T.J. Gramila[1], *J.P. Eisenstein*[1], *A.H. MacDonald*[2], *L.N. Pfeiffer*[1], and *K.W. West*[1]

[1]AT&T Bell Laboratories, Murray Hill, NJ 07974, USA
[2]Department of Physics, Indiana University, Bloomington, IN 47405, USA

This paper reviews our study of the interaction of phonons with two-dimensional electron gases (2DEG) using a direct probe of electron-electron (e-e) scattering rates. We find evidence for an e-e scattering process whose temperature and distance dependence strongly suggest a phonon-mediated interaction. The magnitude of the scattering rates, however, are much greater than can be accounted for by real phonon exchange. Instead, we argue that we have directly detected scattering due to the exchange of virtual phonons.

Our experimental technique [1] consists of measuring the voltage V, developed in one 2DEG when a current I, is driven through a second, parallel, 2DEG. Since the electron layers are electrically isolated, V is due entirely to interlayer e-e interactions and the ratio V/I directly determines the interlayer momentum relaxation rate τ_D^{-1}. The samples used are modulation-doped GaAs/AlGaAs double quantum wells grown by molecular beam epitaxy. By varying the barrier thickness between the wells, we change the electron layer spacing.

The dominant interlayer scattering mechanism for closely spaced layers is Coulomb scattering [1]. By separating the layers, however, this contribution to τ_D^{-1} is greatly reduced. Fig. A shows τ_D^{-1}divided by T^2 vs. temperature T, for barrier thicknesses of 175, 225, and 500Å. Two scattering processes are present. The uniform shifts in the data sets result from Coulomb scattering ($\propto T^2$), and the deviations from constant τ_D^{-1}/T^2 result from a second coupling mechanism. This mechanism clearly depends weakly on the layer spacing; one would expect little change from a phonon-mediated interaction, as acoustic phonon mean free paths are $\sim$ mm at these temperatures.

Because of its strong layer spacing dependence [1] ($\propto d^4$), Coulomb scattering is nearly negligible in the 500Å barrier sample, and $\tau_D^{-1}/T^2$is dominated by this second coupling mechanism. The temperature dependence of τ_D^{-1}/T^2 seen in this sample, is remarkably similar to the well-known acoustic phonon-limited mobility scattering rate, τ_{ph}^{-1}, for 2-D electrons in GaAs. The crossover [2] from a linear dependence for τ_{ph}^{-1} to T^5 or T^7 in the low temperature Bloch-Grüniesen regime occurs at a few degrees Kelvin. This provides additional support for a phonon-mediated interaction.

Springer Series in Solid-State Sciences, Volume 112
Phonon Scattering in Condensed Matter VII Eds.: M. Meissner and R.O. Pohl

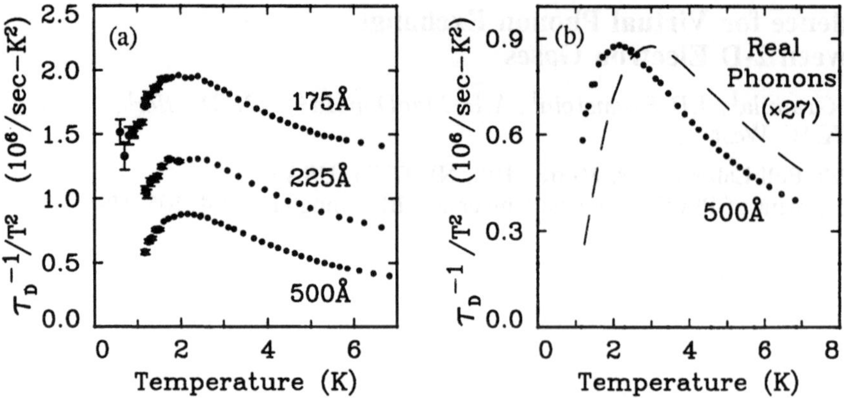

Fig. 1. A) Interlayer e-e scattering rates vs. temperature for three samples, with 175, 225, and 500Å barriers separating the electron layers. B) Data for 500Å barrier sample compared to 27 times the calculation of real phonon exchange.

The magnitude of τ_D^{-1} observed, however, is much greater than can be accounted for by real phonons. Recent calculations [3] based on real phonon exchange result in a scattering rate 20 to 100 times too small. Fig. B compares this calculation multiplied by 27 (dashed line) to τ_D^{-1}/T^2 for the 500Å barrier sample. It is clear, however, that a phonon mediated interaction captures characteristic features of the observed temperature dependence.

For these closely spaced layers, however, virtual phonon exchange must also be considered. Such a process is unusual because the electron systems are 2D while the phonon system is 3D. Consequently, momentum conservation normal to the 2D layers is relaxed and the phase space accessible to virtual phonons for a given electronic transition is considerably enhanced. Calculations [4] indicate the T dependence for a virtual phonon exchange interaction is similar to that for τ_{ph}^{-1} discussed earlier, with a magnitude in better agreement with our measurements.

Recent measurements [3] for very large layer spacings, however, provide experimental evidence for a virtual phonons. While exhibiting a T dependence identical to that of the 500Å barrier sample, we have found the range of this phonon-mediated interaction is less than 5000Å. This effectively rules out real phonon exchange, which would decrease significantly only when the layer spacing became comparable to the phonon mean free path, ~mm.

[1] T. J. Gramila J. P. Eisenstein, A. H. MacDonald, L. Pfeiffer, and K. West, Phys. Rev. Lett. 66 (1991) 1216, and Surf. Sci. 263, 446 (1992).

[2] P. J. Price, Solid State Commun. 51, 607 (1984).

[3] T. J. Gramila *et al.*, to be published.

[4] H. C. Tso, P. Vasilopoulos, and F. M. Peeters, Phys. Rev. Lett. 68 (1992) 2516, and A. H. MacDonald et al, (to be published).

The Use of Surface Acoustic Waves to Estimate the Conductivity of Small Areas of a 2-DEG

P. Hawker, K. McEnaney, P.J.A. Carter, M. Tonouchi, and V.W. Rampton

Department of Physics, University of Nottingham,
University Park, Nottingham NG7 2RD, UK

Introduction

The attenuation of surface acoustic waves (SAW) on a GaAs surface above a heterojunction with a two-dimensional electron gas (2-DEG), has been shown to be a function of the longitudinal conductivity, σ_{xx}, of the 2-DEG [1]. We have made experiments on the attenuation of SAW by an array of small areas of 2-DEG. We believe that this may be a useful method of estimating the value of σ_{xx} for an area which is too small to contact conventionally.

Experiments

We have used a GaAs/AlGaAs heterojunction grown by MBE to produce a 2-DEG. The 2-DEG had an electron sheet density of 3.53×10^{11} cm^{-2} and a mobility of 0.913×10^{6} $cm^2V^{-1}s^{-1}$ in the dark at 4.2K. Samples were prepared by etching away the heterojunction over most of the surface of the sample except for an array of 68,000 circular mesas of diameter about 700nm on a region of the wafer over which a SAW would pass. Interdigital SAW transducers with a fundamental frequency of 200MHz were formed on the sample surface, a transmitter on one side of the mesa array and the receiver on the other side.

Using r.f. pulses of 500ns the transmitted SAW signals were detected and averaged using a box-car averager. The SAW intensity was measured as a function of magnetic field up to 5T at a temperature of 4.2K.

Results and discussion

A magnetic field dependent attenuation is found which shows oscillations in the SAW transmission corresponding to the Shubnikov-de Haas oscillations in the longitudinal conductivity, σ_{xx}, figure 1. The oscillations decrease in intensity and eventually disappear as the incident SAW power is increased.

The following expression for the attenuation of a SAW has been given by Wixforth et al.[1]

Springer Series in Solid-State Sciences, Volume 112
Phonon Scattering in Condensed Matter VII Eds.: M. Meissner and R.O. Pohl

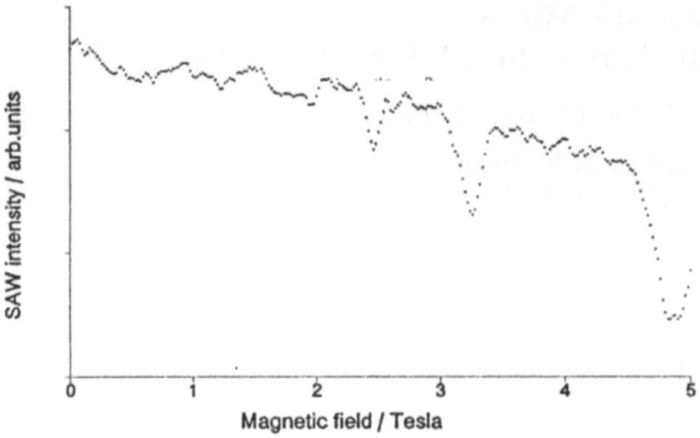

Figure 1 SAW transmission at 900MHz as a function of magnetic field.

$$\Gamma = k\frac{K_{eff}^2}{2}\frac{\sigma_{xx}}{\sigma_M}/\{1+(\frac{\sigma_{xx}}{\sigma_M})^2\}$$

where k is the SAW wave vector, K_{eff} is an effective piezo-electric coupling constant for SAW [2], σ_{xx} is the electron sheet conductivity and σ_M is the characteristic conductivity at which maximum attenuation occurs. In zero magnetic field we estimate the SAW attenuation to be $\Gamma = 4.0 \times 10^{-5}$ dB cm^{-1}, clearly too small to observe.

When the magnetic field is increased and we reach a quantum Hall plateau the conductivity σ_{xx} falls to a very low value. The maximum SAW attenuation occurs when σ_{xx} is equal to σ_M, and is about 3 dB cm^{-1} at 200MHz. We estimate that the fraction of the area through which the SAW propagates that is covered with mesas is approximately 0.018, so we expect to observe a maximum attenuation of about 0.06 dB cm^{-1}. Experimentally we observe a maximum attenuation of 0.14 dB cm^{-1} but as we have not seen any decrease in attenuation at the centre of the quantum Hall plateau when σ_{xx} becomes less than σ_M we cannot be sure that we have, in fact, reached the condition $\sigma_{xx} = \sigma_M$ in our sample.

The disappearance of the oscillations at high SAW intensity is possibly due to heating of the 2-DEG.

References

[1] Wixforth A., Schriba J., Wassermeier M., Kotthaus J.P., Weimann G. and Schlapp W., *Phys.Rev.* **40** 7874-7887 (1989).

[2] Deacon J.M. and Heighway J., *Elect.Lett.* **8** 6-7 (1971).

High-Frequency SAW Attenuation by a 2-DEG

V.W. Rampton[1], *K.B. McEnaney*[1], *P.J.A. Carter*[1], *A.G. Kozorezov*[2], *and C.D.W. Wilkinson*[3]

[1]Department of Physics, University of Nottingham, University Park, Nottingham NG7 2RD, UK
[2]Institute of Radioengineering and Electronics, Russian Academy of Sciences, Mokhovaya Str. 11, 10397 Moscow, GSP-3, Russia
[3]Department of Electrical and Electronic Engineering, University of Glasgow, Glasgow G12 8QQ, UK

1. Introduction

Surface acoustic wave (SAW) attenuation at frequencies in the range 70MHz to 500MHz, by a 2-DEG in a magnetic field, has been shown to be due to the acousto-electric interaction with the electrons in the extended states of the 2-DEG.[1]

We have made experiments on SAW attenuation by a 2-DEG in a magnetic field at higher frequencies, and find an extra contribution to the attenuation which we attribute to a deformation potential coupling to the localised electronic states. A theory of this attenuation which appears to account for our results has been given.[2]

2. Experiments

Experiments were made using a GaAs wafer on which a heterojunction had been grown by MBE. The specimens were prepared by photolithography and etched to leave a 2-DEG region between the positions for the SAW transmitter and receiver transducers. The transmitting and receiving SAW transducers, with a fundamental frequency of 1920MHz, were then made by electron-beam lithography at either end of the 2-DEG region with a SAW path length of 1.4mm in the (110) direction. A magnetic field up to 5T normal to the 2-DEG could be applied;the experiments were done at 4.2K. Rf pulses, 100ns long, were used to generate SAW. The detected signal was recorded as a function of magnetic field.

3. Results

A typical result is shown in Figure 1. The oscillations due to the acousto-electric coupling to the extended electron states is clearly seen, but in addition an attenuation which increases with frequency at first linearly and then as the square root of the magnetic field is apparent.

Springer Series in Solid-State Sciences, Volume 112
Phonon Scattering in Condensed Matter VII Eds.: M. Meissner and R.O. Pohl

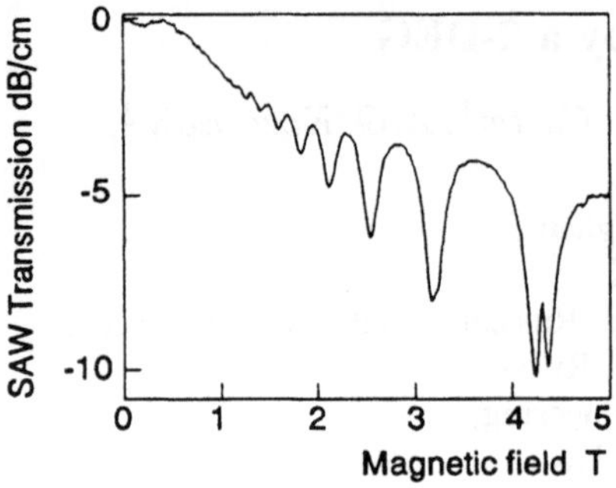

Figure 1 SAW attenuation at 1920MHz as a function of applied magnetic field.

4. Discussion

The SAW attenuation at half integer filling factor depends on the thermodynamic density of states, D_T, of the electrons. We assume a Gaussian form for the density of states of the electrons at a Landau level. At low magnetic fields when the Landau level broadening, due to disorder, is much less than the thermal broadening, then D_T is proportional to the magnetic field, B. At high magnetic fields when the thermal broadening is the smaller then D_T is proportional to $\sqrt{B}$.

It seems likely that a similar deformation potential coupling to localised states will also account for some of the attenuation of 9.3GHz bulk ultrasonic waves by a 2-DEG which we have reported [3].

5. References

[1] Wixforth A., Schriba J., Wassermeier M., Kotthaus J.P., Weimann G. and Schlapp W. *Phys.Rev.* **40** 7874 (1989)

[2] Rampton V.W., McEnaney K., Kozorezov A.G., Carter P.J.A., Wilkinson C.D.W., Henini M. and Hughes O.H. *Semiconductor Science and Technology* **7** 641 (1992)

[3] Carter P.J.A., Rampton V.W., Newton M.I., McEnaney K., Henini M. and Hughes O.H. *Phonon 89. Heidelberg 1989 ed. S.Hunklinger, W.Ludwig and G.Weiss, World Scientific, Singapore* p.998 (1990)

Interaction of Phonons with the 2-DEG in a MODFET

J.K. Wigmore[1], *M. Erol*[1], *M. Sahraoui-Tahar*[1], *M. Ari*[1], *J.H. Davies*[2], *M. Holland*[2], *C. Stanley*[2], *and C.D.W. Wilkinson*[2]

[1]School of Physics and Materials, University of Lancaster, Lancaster LA1 4YB, UK

[2]Department of Electrical and Electronic Engineering, University of Glasgow, Glasgow G12 8QQ, UK

Using nanosecond pulse techniques, we have studied the direct emission and absorption of phonons by the same two-dimensional electron gas (2DEG) in a GaAs/AlGaAs modulation-doped field effect transistor structure (MODFET) [1]. Pulsed emission of phonons from a heterojunction has been observed by Chin et al [2], and by Hawker et al [3], and absorption by Eisenstein et al [4]. In a MODFET, the additional feature of a Schottky gate allows the carrier concentration and hence the Fermi energy to be varied. It also enables the 2DEG contact resistance to be determined independently during a run at helium temperatures, essential for quantitative analysis.

The depletion MODFETs were fabricated on 480 μm thick wafers of (100) orientation. Mobility (μ) at 4 K was 45 or 200 $m^2\ V^{-1}\ s^{-1}$. On the opposite faces were patterned arrays of aluminium squares, defined at precise orientation. To detect phonons emitted by a 2DEG, these were used as superconducting bolometers, whilst for the absorption experiments they functioned as heat pulse generators. The phonons were detected via the change in mobility of a 2DEG. To complement the high spatial resolution (5 degrees), signals resulting from 10 ns exciting pulses were processed by a 3 GHz digital boxcar integrator.

Figure 1 shows some of the data obtained, a) to e) on the lower, and f) on the higher mobility samples. In emission, the high resolution allowed us to separate the phonons due to the 2DEG from those emitted by other regions of a device, such as the contact pads and the AlGaAs layer. Below a threshold of 2.3×10^{-12} W per electron the bolometer signal (1a) showed a sub-linear dependence on input power, which we interpret as a widening of the cone angle of acoustic phonon emission with increasing electron temperature, together with isotope scattering. Above the threshold, the power variation was strictly linear, implying that the phonon frequency distribution did not change. The pulse shape varied with power in the acoustic regime, but not in the optic regime. The figure for the crossover was considerably larger than a calculation based on deformation potential coupling, suggesting additional sources of inelastic scattering. Other evidence is the absence of longitudinal phonons in 1a), even allowing for phonon focussing in the (100) direction (TA/LA $\sim$ 12). The signal in 1b), due to the (100) quasi-diffusive transport of LO phonon decay products, was also totally transverse. However, LA phonons were seen in other directions (1c).

In absorption, data were taken for phonon propagation centred along the (100) axis in order to cancel out phonon drag, so that the response (1d) was due only to changes in electron temperature. In contrast to emission, both transverse and longitudinal phonon pulses were seen. The observed signal

Springer Series in Solid-State Sciences, Volume 112
Phonon Scattering in Condensed Matter VII Eds.: M. Meissner and R.O. Pohl

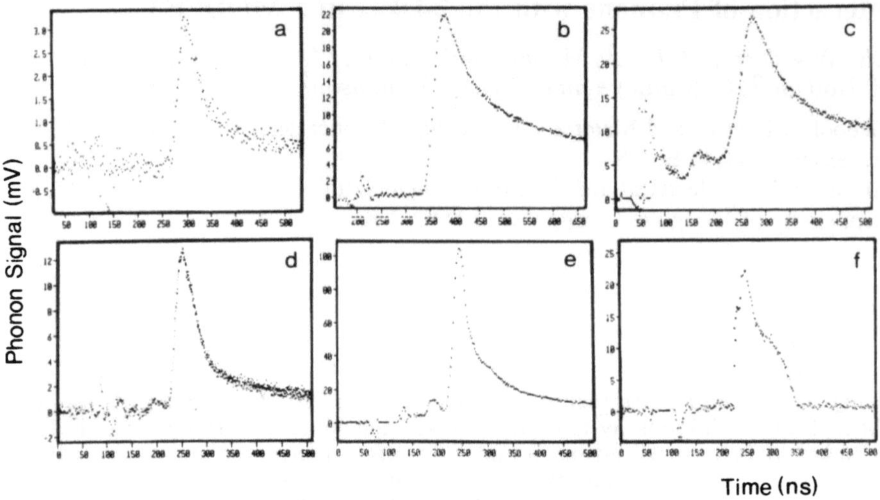

Figure 1 : a) Emission in (100) direction by $\mu = 45$ MODFET excited at 10 μW power level, b) as above at 80 μW power, c) emission at 30° to (100) at 200 μW excitation, d) absorption in (100) direction by same MODFET with 10 μa source-drain bias current, e) (100) absorption detected by channel-gate current, f) (100) absorption by $\mu = 200$ MODFET, at 20 dB higher input power level.

corresponded to an increase in sample mobility, confirming that the device was operating in the ionised impurity regime. From a comparison with the DC mobility-electric field and mobility-temperature curves we could determine absolutely the absorbed phonon flux and the resulting temperature increase of the carriers (several K). Figure 1e) shows the change in source-gate current, possibly phonon-enhanced tunnelling, due to phonon absorption in the (100) direction. Detailed comparisons between 1a), 1d), and 1e) should lead to clarification of the coupling mechanism. Figure 1f) is the response from the higher mobility MODFET, which has a faster response but much lower sensitivity. By gate control, the carrier density could be varied from zero to the equilibrium value of 2.65×10^{15} m^{-2}, and a phonon absorption coefficient increasing strongly with Fermi wavevector was found. In addition, by varying the heater power we could change the dominant phonon wavelength of the heat pulse distribution, and hence probe the dependence of the absorption function. The work is supported by the Science and Engineering Research Council.

[1] J.K. Wigmore, M. Erol, M. Sahraoui-Tahar and C.D.W. Wilkinson, J.H. Davies and C. Stanley, Semicond. Sci. Technol. 6 1 (1991).
[2] M.A. Chin, V. Narayanamurti, H.L. Stormer and J.C.M. Hwang, Phonon Scattering in Condensed Matter IV (Springer, Berlin) 328 (1985).
[3] P. Hawker, A.J. Kent, O.M. Hughes and L.J. Challis, Semiconduct. Sci. Technol. 7 B29 (1992).
[4] J.P. Eisenstein, V. Narayanmurti, H.L. Stormer, A.Y. Cho and J.C.M. Hwang, Phonon Scattering in Condensed Matter V (Springer, Berlin) 401 (1986).

Cyclotron Phonon and Photon Emission from Two-Dimensional Electron Gases (2-DEGs) in GaAs/(AlGa)As Heterostructures

L.J. Challis, N.N. Zinov'ev, R. Fletcher, B. Sujak-Cyrul, A.V. Akimov, and A.F. Jezierski

Physics Department, Nottingham University,
University Park, Nottingham NG7 2RD, UK

Abstract: Investigation has been made of the temperature and so phonon emission rate from a magnetically quantized 2DEG by measuring the intensity of the far infrared emission.

The processes of dissipation and energy loss in a magnetically quantized two-dimensional electron gas are still poorly understood. The Joule-heat $\geq I^2R_H$ produced by a current I is dissipated mainly in two small areas located at diagonally opposite corners of the bar as has been demonstrated by observing the phonon emission [1]. The frequency distribution of these phonons is not yet known but seems likely to include a component at the cyclotron frequency $\omega_c = eB/m^*$. However constraints of momentum conservation $q_c < l_B^{-1} = (\hbar/eB)^{-1/2}$, $q_c < a^{-1}$ ($q_c = \omega_c$/sound velocity, l_B=magnetic length, a=2DEG 'thickness') should greatly reduce the efficiency of this emission as the field increases suggesting that the temperatures of both the corners and bulk of the 2DEG might rise to quite high values even for relatively small power inputs.

To investigate this we are developing a technique to determine T_e by measuring the far-infrared intensity emitted by a 2DEG heated by current pulses of duration 10-120 μs with repetition rate < 1kHz. The intensity is measured by a broad-band hot electron InSb detector with constant responsivity up to 450GHz ($\sim \nu_c$ at 1T) and useful sensitivity up to 900GHz. The sample and detector are at 4K and mounted 18cm apart so that $B \sim 0$ at the detector. They are held in vacuo at either end of a light pipe. The back face of the 2DEG substrate is polished and in contact with a metal plate so acts as a good reflector. As a result the substrate of thickness d= 250μm acts like an etalon and the emission from the front face has minima when $2d = m\lambda$ (m=1,2..) as seen in fig 1(a) for a range of currents. The separation of the minima is $\Delta B = \pi m^* c/ed\sqrt{\epsilon_r}$ to within experimental error showing that the FIR emission is predominantly at ν_c as expected.

Fig 1(b) compares measured values of the detected FIR power E averaged over an oscillation with values computed using a theoretical expression for emissivity. E was measured as a function of power input P and for plots against T_e we use values of $T_e(P)$ in similar B-fields [2]. The satisfactory agreement below 15K

Springer Series in Solid-State Sciences, Volume 112
Phonon Scattering in Condensed Matter VII Eds.: M. Meissner and R.O. Pohl

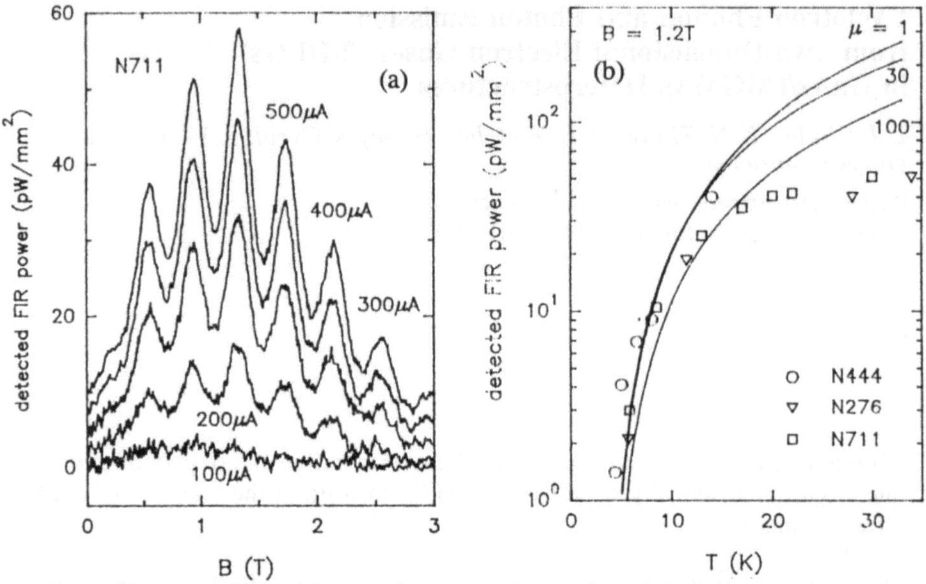

Fig 1. FIR power detected per unit area of 2DEG as a function of (a) B for various currents I and (b) T_e at 1.2T ; lines = theoretical values, the symbols=experimental points for 3 samples with sheet densities of $3x10^{11}$ m^{-2} and transport mobilities of 20, 150, and 80 $m^2V^{-1}s^{-1}$ for NU276, 444 and 711 respectively.

gives confidence in the FIR method and we plan to improve the technique by calibrating $E(T_e)$ using gated structures as we have for SI MOSFETS [3]. We note that the 2DEG is largely heated and cooled at the corners and suggest that T_e may rise to values where LO phonon emission becomes possible. The average 2DEG T_e is however lower and presumably set by electronic heat conduction from the corners and phonon emission from the 2DEG. If cyclotron emission becomes weak possiblities such as edge state emission [A.Y.Shik, private communication] or coupling to highly damped surface waves may become important.

We are very grateful to Dr M Henini for MBE growth of the samples and Mr B Hill for his help with the design and construction of the apparatus.

References

[1] P.A.Russell, F.F.Ouali, N.P.Hewett, and L.J.Challis, Surface Science **229**,54 (1990); U.Klass, W.Dietsche, K.von Klitzing, and K.Ploog, Z.Phys **B82**,351 (1991)

[2] K. Hirakawa and H.Sakaki, Appl Phys Letts **49**,889 (1986)

[3] A.V.Akimov, L.J.Challis, and C.J.Mellor: Physica B **169**,563 (1991)

Luminescence Detection of Phonons Emitted from the First and Second Sub-Bands of a 2-DEG in Silicon

C.J. Mellor[1], *J. Cooper*[1], *L.J. Challis*[1], *E.S. Moskalenko*[2], *and A.V. Akimov*[2]

[1]Physics Department, Nottingham University, University Park, Nottingham NG7 2RD, UK
[2]A.F. Ioffe Physico-Technical Institute, 194021 St. Petersburg, Russia

Abstract. An exciton cloud detector has been used to measure the frequency distribution of the phonons emitted by a heated 2DEG in a Si-MOSFET at power densities at which there is significant thermal population of an excited subband.

Studies of acoustic phonon emission from a heated two-dimensional electron gas (2DEG) have largely been carried out at sheet densities n_s and electron temperatures T_e for which $E_F<E_{ex}$ (E_{ex} is the energy of the first excited subband) and $k_BT_e<<(E_{ex}-E_F)$ so that the emission is very largely from transitions within the ground state subband. At higher values of E_F and T_e thermal population of an excited subband becomes significant and changes are expected in the nature of the emission. In the present work measurements have been made on (001) Si MOSFET samples for which it is known that E_F enters the first excited subband when n_s=4.9x10^{16}m^{-2} [1]. Furthermore, the values of power input P_m needed to produce significant population of E_{ex} had been determined as a function of n_s by observing the fall in resistivity for $P>P_m$ caused by the presence of higher mobility electrons in the excited subband [2].

The measurements used an exciton cloud as a phonon detector [3]. The cloud was formed at the substrate face opposite the 2DEG using a 0.3mm diameter cw argon laser beam. The substrate was immersed in liquid helium at 2K at which temperature all the excitons were bound (BE) by the acceptor impurities as could be seen from the luminescence arising from exciton annihilation which showed no trace of the free exciton (FE) spectrum. The 2DEG was then heated by 100ns pulses and the phonons incident on the exciton cloud created a pulse of FE luminescence and a decrease (negative-going pulse $I^B(t)$) of BE luminescence caused by the dissociation of excitons from acceptor impurities. Since the binding energy is known to be 3.8meV, $I^B(t)$ reflects the intensity of phonons of frequency >920GHz. The signal was integrated over a time of 500ns.

Fig.1 shows that for two values of input power $P>P_m$ the relative change S in BE luminescence fell significantly as n_s was increased (P was kept constant as n_s was varied). The area of the 2DEG was large compared with the exciton cloud so we discount the possibility that this decrease is due to changes in angular

Springer Series in Solid-State Sciences, Volume 112
Phonon Scattering in Condensed Matter VII Eds.: M. Meissner and R.O. Pohl

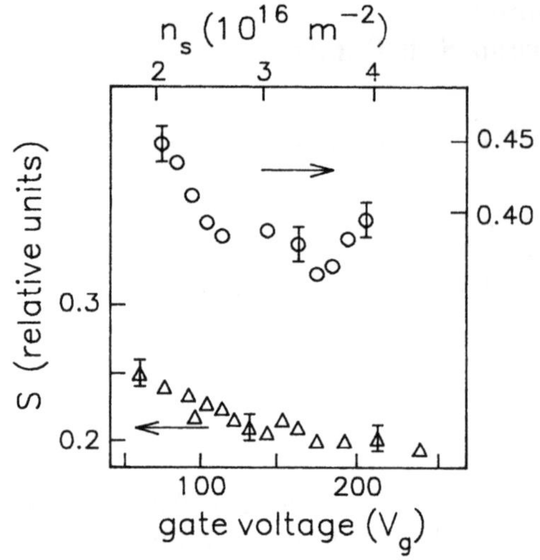

Fig.1. The relative decrease in BE luminescence S induced by a phonon pulse from the 2DEG as a function of n_s for two values of injected power density. Triangles - P=20 mW/mm^2; circles - P=60 mW/mm^2. $S=\Delta I^B/I_0^B$, where ΔI^B- the amplitude of the negative-going $I^B(t)$ pulse and I_0^B- the BE luminescence intensity before the phonon pulse is applied.

distribution and attribute it to a shift in spectrum to lower frequencies (<920GHz). This contrasts with the shift to higher frequencies expected from the emission from a single subband as a result of the increase in the cut-off values of both the in-plane ($2k_F$) and out-of-plane (a^{-1}) wavenumbers (k_F is the Fermi wavenumber and a the thickness of the 2DEG). However it would appear to be consistent with part of the emission arising from intrasubband transitions from an excited subband where k_F is lower and the electrons less strongly confined.

We are very grateful to Professor Kaplyanskii and other colleagues for very helpful discussions and to Dr C.M.K.Starbuck for the Si MOSFETs.

References

[1] N.P.Hewett, P.A.Russell, L.J.Challis, F.F.Ouali, V.W.Rampton, A.J.Kent, and A.G.Every, Semicond Sci and Technol **4**, 955 (1989)

[2] J.Cooper, F.F.Ouali, and L.J.Challis, Semicond Sci and Technol **7**, B570 (1992)

[3] A.V.Akimov, A.A.Kaplyanskii, and E.S.Moskalenko, Physica B **169**, 382 (1991)

Far-Infrared Response of $Al_xGa_{1-x}As/GaAs$ Superlattices

R. Fukasawa[1], *M. Wakaki*[2], *K. Shirawachi*[1], *S. Nishizawa*[1], *and K. Ohta*[3]

[1]JASCO Corp., 2967-5 Ishikawa-cho, Hachioji-shi, Tokyo 192, Japan
[2]Department of Electro-Photo-Optics Engineering, Tokai University, 1117 Kitakaname, Hiratsuka, Kanagawa 259-12, Japan
[3]Electrotechnical Laboratory, Tsukuba 305, Japan

The dielectric properties of semiconductor superlattices in the frequency range of phonon excitations have attracted much interest because they may show a novel dielectric response due to modulation of the vibrational properties such as the zone folding effect. Far-infrared spectroscopy is a powerful technique for studying novel dielectric properties related to the long-wavelength optical lattice vibrations of GaAs/AlAs binary superlattices[1].

We report the first experimental results on the aluminium (Al) compositional dependence of the far-infrared reflectance spectra of $Al_xGa_{1-x}As/GaAs$ alloy superlattices. On the basis of the standard theory[2] of thin-film optics, the shape of the infrared reflectance spectra is also discussed.

$(Al_xGa_{1-x}As)_{12}(GaAs)_{17}$ superlattices of alternating 17 GaAs monolayers (48 Å) and 12 AlGaAs monolayers (34 Å) were grown by Molecular Beam Epitaxy(MBE) on a semi-insulating GaAs substrate. The total superlattice thickness was 0.82 μm in all samples. The far-infrared reflectance measurements were performed at room temperature, using a Fourier transform infrared spectrometer(JASCO FT/IR-VM7) and a Triglycine Sulfate (TGS) pyroelectric detector. The reflectance spectra were obtained at near-normal incidence.

The solid lines in Fig.1 show the experimental reflectance spectra of the $(Al_xGa_{1-x}As)_{12}(GaAs)_{17}$ superlattices with the various Al mole fractions. The short horizontal bar in Fig.1 in each spectrum indicates a reflectance of 0.4. The enhanced spectral structure in the wave number region from 250 to 300 cm^{-1} is the GaAs-like reststrahlen (RS) band which may be formed by the superposition of the two RS bands originating from both GaAs and AlGaAs layers. On the other hand, the weaker RS band at around 360 cm^{-1} may come from the AlAs-like lattice vibration in the AlGaAs layer. With increasing Al mole fraction, the reflectance of the AlAs-like band becomes larger, as shown in Fig.1.

We discuss the experimental results of $(Al_xGa_{1-x}As)_{12}$ $(GaAs)_{17}$ on the basis of the theory of electromagnetic

Springer Series in Solid-State Sciences, Volume 112
Phonon Scattering in Condensed Matter VII Eds.: M. Meissner and R.O. Pohl

wave propagations in a multilayer structure[2]. In this theory, the propagation of electromagnetic waves through isotropic thin layers is treated by the characteristic-matrix method. The numerical calculations[3] of the reflectance spectra were carried out by modeling the finite $(Al_xGa_{1-x}As)_{12}(GaAs)_{17}$ superlattice with 100 periods corresponding to the specimens. The dielectric function of each layer is given for the long-wavelength optical vibrations by a classical oscillator model. The broken lines in Fig.1 show the calculated reflectance spectra. In the present calculations, we used the parameters of bulk crystal for x=0.14, 0.3 and 0.54 listed in Table II of ref.4. As a result, the simple theory of thin-film optics explains the essential features of the observed reflectance spectra of the ternary alloy superlattices. The overall profile of reflectance spectra in the ternary alloy superlattices is determined by the total superlattice thickness and the layer thickness ratio. However, the slight difference between the calculated and experimental spectra at around the RS bands could not be understood from the theory of thin-film optics with the dispersionless dielectric function. These spectra should be analyzed carefully by taking into account the contribution to the dielectric susceptibility of the superlattice vibrational modes[5].

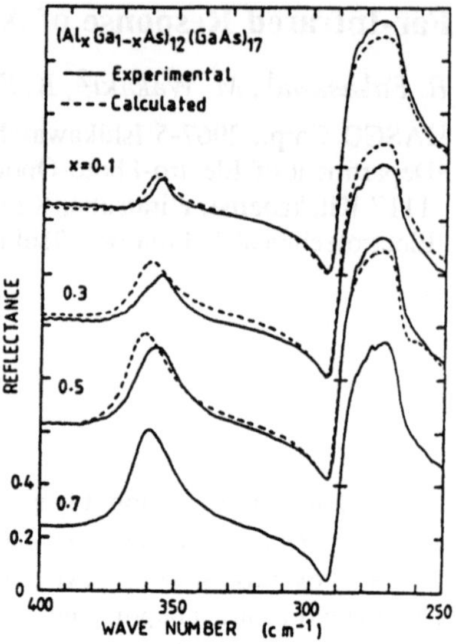

Fig.1.

In summary, we have reported an experimental study of the infrared reflectance spectra of $Al_xGa_{1-x}As/GaAs$ ternary alloy superlattices for various Al mole fractions. The observed infrared reflectance spectra are well reproduced by the calculation using the simple theory of thin film optics with the characteristic matrix in each layer.

References

[1]G.Scamarcio, L.Tapfer, W.Konig, A.Fischer, K.Ploog, E.Molinari, S.Baroni, P.Giannozzi and S.de Gironcoli: Phys. Rev. **B 43** (1991) 14754 and reference therein.

[2]F.Abeles: Ann. Phy. **5**, 596 (1950).

[3]R.Fukasawa, M.Wakaki, K.Shirawachi, S.Nishizawa and K.Ohta: Jpn. J. Appl. Phys. **31**, No.8 (1992) (to be published).

[4]O.K.Kim and W.G.Spitzer: J. Appl. Phys. **50**, 4362 (1979).

[5] H.Chu and Y.C.Chang: Phys. Rev. **B 38** (1988) 12369.

Studies of the Phonon Emission from the Current Entry and Exit Points in the Quantum-Hall Regime

F.F. Ouali, Z. Xin, L.J. Challis, J. Cooper, A.F. Jezierski, and M. Henini

Physics Department, Nottingham University,
University Park, Nottingham NG7 2RD, UK

Abstract: Studies have been made of the division of the total dissipation, I^2R_H between the current entry and exit points of a two-dimensional electron gas (2DEG) in the Quantum Hall regime. There is a striking difference between the behaviour of 2DEGs in Si and GaAs.

When a current is passed through a 2DEG and a strong perpendicular magnetic field is applied, the current enters and leaves the 2DEG at diagonally opposite corners. In the quantum Hall regime (QHR), $\rho_{xx}=0$ and the source-drain R_{SD} then equals the Hall resistance R_H to high precision. Dissipation in the centre of the device ($\propto \rho_{xx}$) vanishes and the total power I^2R_H is dissipated at the entry and exit corners of the 2DEG. This was first demonstrated directly by measuring the temperature opposite one of the corners [1]. Theoretical predictions differ as to how this dissipation should be distributed between the two corners and in the first experiments [1], which were on 2DEGs in Si, the dissipation was found to be equally shared to within experimental error. Similar results have also been obtained subsequently for Si using (i) an imaging bolometer technique [2], (ii) a similar but improved technique to that used in [1], [3] and (iii) a helium fountain prèssure technique [4]. The measurements reported in [3] indicate that the dissipation is shared in proportions $(50\pm x)\%$ (x is the experimental limit), where for example, at filling factor $i=8$, $x\leq 1.35$ for $P=8\mu W$.

Interestingly different behaviour was reported however in 2DEGs in GaAs using the fountain pressure technique [5]. For $i=2$, the dissipation was equally shared for $I<110\mu A$, but for higher currents it appeared to decrease substantially at the electron exit point relative to the entry point. We report here measurements on a 3×1 mm^2 GaAs/(AlGa)As heterojunction using the techniques of [1,3]. The phonon intensity was monitored by the temperature of a thermometer T_C, placed opposite one of the corners. This is measured relative to a reference thermometer situated some distance from the device towards the 1K heat sink. The measurements were made at constant power as a function of magnetic field so that changes in temperature are due to changes in phonon flux. Fig 1 shows the magnetic field dependence of T_C for both current directions, I^+ and I^-, for a power of 30 μW; also shown is the difference δT between the two signals. The

Springer Series in Solid-State Sciences, Volume 112
Phonon Scattering in Condensed Matter VII Eds.: M. Meissner and R.O. Pohl

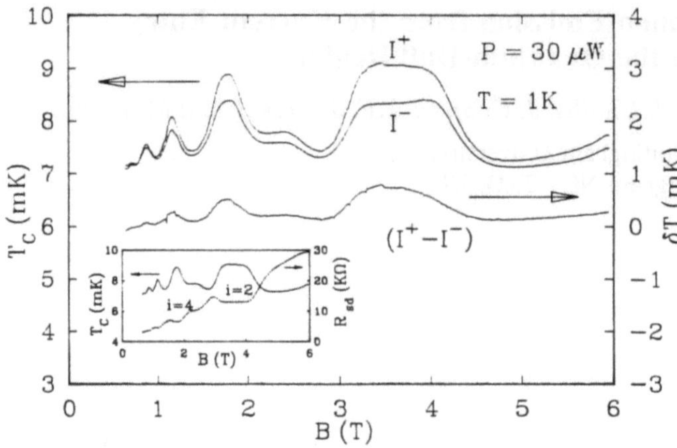

Fig 1: The variation of phonon intensity T_C with magnetic field opposite one corner

intensity peaks in the QHR in line with the fall in dissipation in the centre as shown in the inset of fig 1. The values of T_C are larger for current direction I^+ for which the corner is an electron entry point. Measurements have been made for 10 powers between 5 and 250 μW and in contrast with measurements in [5] the asymmetry (δT) persists to the lowest current values measured corresponding to I= 20 μA at i=2, although becomes weaker with decreasing current. We find that for i=2 and P=30 μW, about 55% of the total power is dissipated in the electron entry corner of the device with the remaining 45% in the exit corner.

We emphasise that these measurements cannot distinguish between asymmetry in magnitude of the dissipation and area over which it occurs since, if the area is larger at the electron exit corner, a smaller proportion of the emitted phonons strike the detector contact. Measurements are in progress to resolve this.

References

[1] P.A. Russell, F.F. Ouali, N. Hewett and L.J. Challis, Surf Sci **229** 54 (1990)
[2] A.J. Kent, Physica **B169**, 356 (1991)
[3] F.F. Ouali, L.J. Challis and J. Cooper, Semicond Sci Technol **7**, 608 (1992)
[4] R. Knott, U. Klaß and W. Dietsche, private communication.
[5] U. Klaß, W. Dietsche, K. von Klitzing and K. Ploog, Phys **B82**, 351 (1991)

Quantum Oscillations in the Cyclotron Phonon Emission from a Two-Dimensional Electron Gas

J. Cooper, L.J. Challis, F.F. Ouali, K.A. Benedict, and C.J. Mellor

Physics Department, Nottingham University,
University Park, Nottingham NG7 2RD, UK

Abstract: Oscillations have been observed in the phonon intensity I_N emitted normal to a heated 2DEG bridge in a Si MOSFET in a quantizing magnetic field as the sheet density is varied. The oscillations are attributed to variations in the intensity of cyclotron phonons (inter–Landau level transitions).

In a quantizing magnetic field, a heated two–dimensional electron gas (2DEG) can lose energy by both intra– and inter–Landau level transitions. Inter–Landau level transitions give rise to cyclotron phonon emission at frequencies $j\omega_c$, $j=1,2,3$.. For a constant electron temperature T_e, the $j=1$ emission should be greatest when $f_e(1-f_e)$ is a maximum, which occurs when the Fermi level is midway between two Landau levels and this is also the condition for minimum intra–Landau level emission. Oscillations with filling factor are therefore expected in the relative proportions of cyclotron to intra–Landau level emission.

To investigate this effect measurements were made of the phonon intensity I_N emitted normal to an intensely heated area of 2DEG formed by etching a $15 \times 180\mu m^2$ bridge at the centre of a $3 \times 1mm^2$ 2DEG in a (001) Si MOSFET. The magnetic fields used, $\leq 6T$, give rise to cyclotron phonons of frequencies $\lesssim 900GHz$ which travel ballistically across the $400\mu m$ substrate. Currents $\sim 100\mu A$ are estimated to produce local values of $T_e \sim 100K$ in the bridge leading to sizeable values of $f_e(1-f_e)$ for $B=6T$ ($\hbar\omega_c/k_BT_e \sim 0.5$).

The phonon intensity was monitored on a $0.25 \times 0.25mm^2$ contact connected to a sensitive thermometer. Figure 1 shows the variation of I_N as a function of sheet density, the power inputs indicated were maintained constant during the sweeps by computer control. At low powers e.g. $1\mu W$, the oscillations in I_N are in–phase with R_{sd}, with the minima occurring when E_F is midway between two Landau levels. These are similar to results obtained from 2DEGs devices without bridges [1] and are attributed to the fact that in the centre of the 2DEG, the dissipation varies as ρ_{xx}. However, figure 1 shows that as the power is

Springer Series in Solid-State Sciences, Volume 112
Phonon Scattering in Condensed Matter VII Eds.: M. Meissner and R.O. Pohl

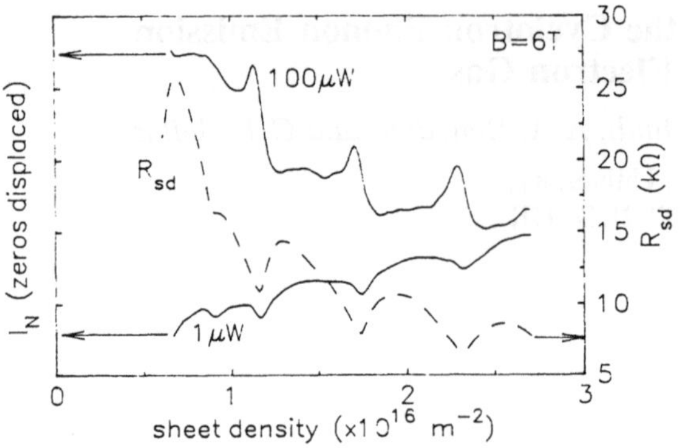

Figure 1: Variation with sheet density of I_N and R_{sd}

increased, I_N undergoes a phase change and now has maxima close to the points where E_F is midgap. Since the power is constant we attribute the changes in I_N to changes in the angular distribution of the emission. In-plane momentum conservation forces the large **q** cyclotron phonons to be more strongly emitted normal to the 2DEG than the small **q** intra-Landau level phonons and so the oscillations in I_N are in-line with the oscillations expected in relative intensity of these two emission processes. This explanation seems qualitatively consistent with measurements made over a wide range of fields [2] although it is not yet clear whether sufficient intra-Landau level emission can occur at these relatively high temperatures to account for the amplitude of the oscillations seen. It is of note that since the power in the device is kept constant the electron temperature may be expected to oscillate, thus an alternative explanation is that a small proportion of the dissipation is occurring by optical phonon emission from significantly higher Landau levels. The acoustic decay products of this emission will have a broader angular distribution and so this could in principle also account for the observed effects. Theoretical analysis is in progress in an attempt to distinguish these alternative explanations.

References

[1] P.A. Russell, F.F. Ouali, N.P. Hewett, and L.J. Challis, Surf. Sci. **229**, 54 (1990).

[2] J. Cooper, PhD thesis, University of Nottingham (1992).

Acoustoelectric Studies of the Integer and Fractional Quantum-Hall Effect

A. Wixforth, A. Esslinger, and J.P. Kotthaus

Sektion Physik, Universität München,
Geschwister Scholl Platz 1, W-8000 München 22, Fed. Rep. of Germany

Abstract. The interaction between surface acoustic waves and a two-dimensional electron system is used to study the quantum Hall effect in an alternative way. Momentum transfer from the wave to the electrons leads to strong quantum oscillations of both the acoustoelectric fields and currents. The off-diagonal voltage shows bipolar peaks at even filling factors and at filling factor 1/3. The acoustoelectric Hall resistance exhibits well developed plateaus at fractional filling factors not yet visible in transport experiments at the same temperature whereas well established DC Hall plateaus are replaced by asymmetric signals.

The interaction between the coherent phonon field of a surface acoustic wave (SAW) and a two-dimensional electron system (2DES) confined in a piezoelectric semiconductor has been proven before [1..3] to be very sensitive to the dynamic bulk-magnetoconductivity of the 2DES in a way quite distinct from standard resistivity measurements due to their integrating nature.

Here, we report on our recent observation of the acoustoelectric (AE) effect in a 2DES in the regime of the integer and fractional quantum Hall effect. We study the drag currents and electric fields which are induced in the 2DES. Whenever the diagonal magnetoconductivity of the 2DES drops to very low values of the order of $3 \cdot 10^{-7} \Omega^{-1}$ screening of the SAW field results in energy and momentum transfer to the 2DES. The former can be observed via SAW transmission experiments and described in terms of a relaxation model [1]. The latter gives rise to AE currents and voltages, which are subject of our present experiments. A recent theory predicts relatively strong acoustoelectric drag voltages in SAW propagation direction and a quantized AE current in the regime of the quantized Hall effect [4].

Our experimental setup has been described in detail before [1,2]. Here, we probe the AE fields and currents as a function of the magnetic field. A typical exam-

Springer Series in Solid-State Sciences, Volume 112
Phonon Scattering in Condensed Matter VII Eds.: M. Meissner and R.O. Pohl

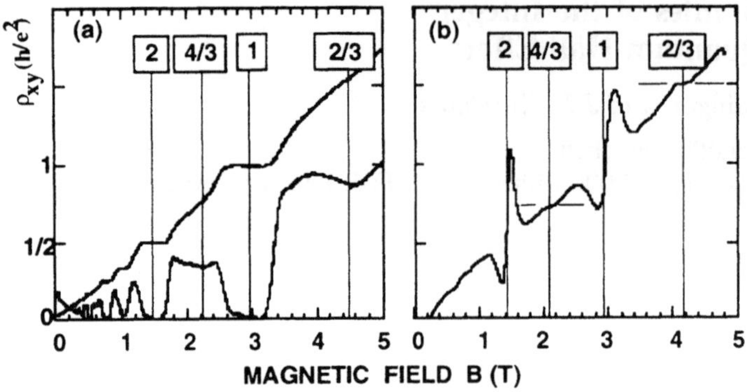

Fig. 1.: (a) Magnetotransport data, and (b) the corresponding AE Hall resistance, which shows asymmetric structure for in (a) well developed Hall plateaus and plateaus for fractional fillings.

ple of such a measurement is shown in Fig.1 together with DC transport data. In 'shorted' geometry, with source and drain connected the AE current j_x, and the Hall voltage V_y are accessible, and we can derive an AE Hall resistance $\rho_{xy} = V_y/j_x$. Well resolved (DC) quantum Hall states like $\nu = 1, 2, ..$ exhibit in the AE signals a characteristic asymmetric behavior, whereas for fractional filling factors like $\nu = 2/3$, 4/3 plateaus occur, which are not yet resolved in the DC data at this temperature. For $\nu = 1/3$, however, which has been omitted in the plot for clarity, again an asymmetric signal is observed. This is in clear contradiction to theory [4], predicting identical Hall plateaus for both acoustically and electrically generated currents.

In 'open' geometrie, however, all contacts are left open such that no macroscopic current flow is possible. Then V_x shows characteristic double peaks like the SAW attenuation [1] around integer filling factors as well as around $\nu = 1/3$, whereas V_y surprisingly exhibits bipolar peaks. This, too, is in complete contradiction to the theory of ref. [4], which predicts no Hall voltage at all in this geometry.
We thank the Deutsche Forschungsgemeinschaft for financial support.

References

[1] A. Wixforth, J. P. Kotthaus, and G. Weimann, Phys. Rev. Lett. **56**, 2104 (1986)

[2] A. Wixforth, J. Scriba, M. Wassermeier, J. P. Kotthaus, G. Weimann, and W. Schlapp, Phys. Rev. **B40** , 7874 (1989)

[3] R. L. Willett, M. A. Paalanen, R. R. Ruel, K. W. West, L. N. Pfeiffer, and D. J. Bishop, Phys. Rev. Lett. **65**, 112 (1990)

[4] A. L. Efros and Yu. M. Galperin, Phys. Rev. Lett. **64**, 1959 (1990)

Electron Mobility Enhancement in Modulated Phonon Structures

X.Th. Zhu[1], *H. Goronkin*[1], *G. Maracas*[2], *R. Droopad*[2], *and M. Stroscio*[3]

[1]Phoenix Corporate Research Laboratories, Motorola Inc., 2100 E. Elliot Rd., Tempe, AZ 85284, USA

[2]Electrical Engineering Department, Arizona State University, Tempe, AZ 85287, USA

[3]U.S. Army Research Office, Research Triangle Park, NC 27709, USA

We report results of a systematic study of the electron mobility in GaAs/AlAs quantum well samples with and without longitudinal-optical (LO) phonon confinement. This experiment demonstrates an enhanced electron mobility in structures with quantum well widths from 20Å to 70Å and phonon barrier widths of about 11Å. The experimental data are explained in terms of a transition from confined phonon scattering to interface phonon scattering.

The samples were grown by molecular beam epitaxy on GaAs substrates [1]. They differ only in the channel section. The conduction channel of one of the control samples, C8, consists of 44 monolayers (ML) of GaAs. PB8 is a multiple quantum well (MQW) structure with three equally spaced 4ML AlAs barriers in the host well and four 8ML GaAs miniwells. Thus, PB8 (or any one of the phonon barrier samples) is identical to C8 (or its corresponding control sample) except for the replacement of three layers of GaAs by AlAs in the channel.

The measured electron mobilities of C8 and PB8 are shown in Fig.1 as a function of temperature. The mobility of C8 (open circles) increases with decreasing T even for temperatures below 50K, characteristic of a sample with a good quality

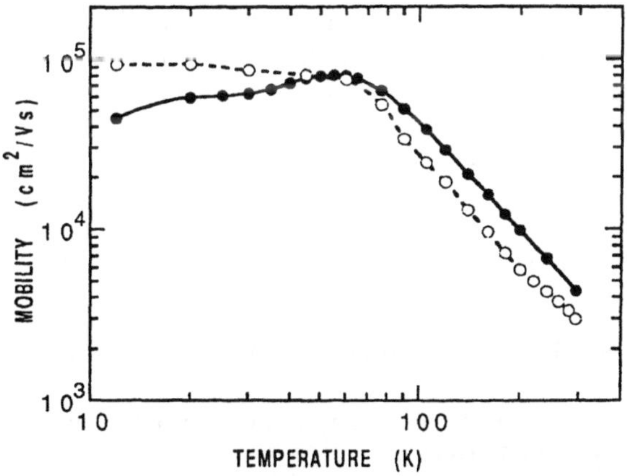

Fig.1. Temperature dependence of mobility for sample C8 in open circles and PB8 in solid circles.

Springer Series in Solid-State Sciences, Volume 112
Phonon Scattering in Condensed Matter VII Eds.: M. Meissner and R.O. Pohl

channel layer and a relatively large separation (90Å in our samples) between the doping layer and the channel. The mobility of PB8 (solid circles) decreases with decreasing T below 50K. For temperatures ranging from room temperature to 100K, the mobility follows the same exponential increase in both samples, as a result of dominant scattering by LO phonons.

Since the introduction of the low mobility AlAs layers into the conducting channel also causes additional scattering mechanisms, such as interface roughness scattering in the vicinity of the six new GaAs/AlAs interfaces, one would expect PB8 to have much lower mobility. In fact, these interface scatterings do affect the electron mobility as manifested by the decrease in μ with decreasing T in the low temperature region. However, the high temperature mobility of PB8 is significantly higher than that of C8; the enhancement is more than 55%. All the above observations equally apply for the other pairs of samples, e.g., PB16 and C16.

The apparent mobility enhancement in PB8 cannot be attributed to its modified conduction band structure. The ratio between the carrier sheet density of C8 and the sheet density of PB8 is found to vary by a factor of two for T > 100K. On the contrary, the mobility ratio between C8 and PB8 is essentially constant in the same temperature range. We have found [1] that, in this high temperature range, only the mechanism of reduced electron-LO phonon scattering due to phonon confinement can satisfactorily explain the large but constant mobility enhancement without any correlation to the electronic density variation.

We have also studied the dependence of the electron mobility enhancement on the miniwell width. The general behavior is the same for all temperatures T > 100K. As the miniwell thickness decreases from the bulk like samples, e.g., PB32, the mobility increases and reaches a maximum near the width of 16 MLs. As the well thickness further decreases the mobility decreases. In fact, the mobility of PB8 is about 7% lower than that of PB16.

It is well known that the electron-LO phonon scattering from interface modes increases with decreasing quantum well width. The observation of an enhanced mobility in sample PB8 indicates a upper transitional width at which the scattering from interface modes begins to dominate over that from confined modes. Using the scalar-potential model developed by Stroscio et al. [2], we obtain the following relation for structures in which the well width, a, is twice the barrier width, b, as for PB8: $qa_t \approx 3$. For GaAs LO phonon modes, $\omega_{LO} = 36meV$, the phonon wave vector corresponding to intrasubband back-scattering takes the value, $q \approx 0.15\text{Å}^{-1}$. It thus follows that electron-interface-phonon scattering will compete with electron-confined-phonon scattering for a PB8-like structure when the well width $a \leq 20\text{Å}$.

References

1. X. T. Zhu, H. Goronkin, G. N. Maracas, R. Droopad, and M. Stroscio, Appl. Phys. Lett. **60**, 2143 (1992).
2. M. A. Stroscio, G. J. Iafrate, K. W. Kim, M. A. Littlejohn, H. Goronkin, and G. N. Maracas, Appl. Phys. Lett. **59**, 1093 (1991).

Part VIII

Thermal Boundary Resistance

Interaction of Acoustic Waves with Metal Crystal Surfaces

K.N. Zinov'eva, D.A. Narmoneva, and A.S. Semenov*

Kapitza Institute for Physical Problems, ul. Kosygina 2,
117334 Moscow, Russia
*The invited lecture was given by K.N. Zinov'eva

1 Introduction

In this paper we present a brief survey of studies of acoustic energy transmission from liquid ^{4}He into metals. The angular spectra of transmission coefficient of acoustic phonons at surfaces of various metals were investigated experimentally in the temperature range 60 – 400 mK at frequencies 10 – 300 MHz.

It is well known, that there exists a considerable quantitative mismatch between experimentally observed values of Kapitza resistance R_K^{exp} and the value R_K^{th}, calculated with the Khalatnikov theory of acoustic mismatch [1]. One could achieve better agreement between theory and experiment using the advantages of generalized acoustic theory [2],[3],[4], which takes into account the acoustic wave dissipation in the solid. The generalized acoustic theory provides also an opportunity to calculate the acoustic energy transmission through the boundary of liquid ^{4}He with variety of media, either isotropic or anisotropic, for any crystal orientation, sagittal plane and angle of incidence. However the theory involves some phenomenological parameters, e.g. the value of dissipation of sound as a function of frequency and temperature $p(\omega, T)$. For this reason the direct experimental measurement of angular spectra of transmission coefficient $\alpha(\theta, \omega)$ in as wide a range of temperatures and frequencies as possible seems to be an important task.

Non-dissipative Khalatnikov model predicts that value of transmission coefficient for acoustic phonons in liquid ^{4}He is $\alpha(\theta) \simeq \frac{\rho_{He} c_{He}}{\rho_s c_s} \simeq .005$ for angles of incidence less than critical $\theta < \theta_c \simeq \frac{c_{He}}{c_s}$. At low values of acoustic wave dissipation $p \leq 0.01$ the principal distinctive feature of generalized acoustic theory is the existence of sharp peaks in transmission spectrum $\alpha(\theta)$ at angles of incidence $\theta_R > \theta_c$ caused by resonant excitation of surface Rayleigh waves and their dissipation. We have studied experimentally only the samples of different metals, since for other crystals there is no significant intrinsic dissipation of acoustic waves at low temperatures. In metals the phonons scatter conduction electrons, the transmission of phonons through their surface being

Springer Series in Solid-State Sciences, Volume 112
Phonon Scattering in Condensed Matter VII   Eds.: M. Meissner and R.O. Pohl

close to unit in the Rayleigh peaks [5]. The contribution of this *resonant transmission* of phonons into the total Kapitza resistance has a weak dependence on the value of dissipation parameter p, and is approximately equal to that of continuous spectrum, if this parameter p is reasonably large. At larger values $p >= 0.1$ the most important contribution becomes that of phonons, whose angles of incidence are far out of continuous spectrum $\theta > \theta_c$; this kind of behavior seems to be valid for the sample of polycrystalline gold.

2 Experimental technique

The measurements were carried out with four different samples, each being made of BCC-metal. They were: tungsten, copper, aluminum single crystals, and polycrystalline gold. All the samples had high chemical purity. The free surfaces of samples were mechanically and then chemically polished, their dimensions were $\sim 10\, mm$ in diameter and 1 – 2 mm in thickness.

All the studies were carried in the same experimental cell [6] filled with liquid ^{4}He at vapor pressure in temperature range 70 – 400 mK. The cell was attached to the mixing chamber of ^{4}He^{3}He dilution refrigerator. The operating temperature was restricted by strong dissipation of acoustic phonons in liquid ^{4}He and low Kapitza resistance values at higher temperatures, and by low cooling power and increasing times of thermal relaxation at the lower end of temperature range. The ultrasonic wave was generated by quartz transducer, whose basic frequency was 12.91 MHz. We used the same transducer for all measurements, the frequencies obtained being the odd multiples of fundamental frequency up to 21st mode (270 MHz). Higher frequencies are hardly available with this kind of transducer because operating at higher harmonics requires an extremely perfect quartz transducer. In addition the mean free path of acoustic phonons dramatically drops with the rise of frequency. The distance between sample and quartz was about 1 cm.

The angle of incidence θ was varied by rotation of sample in ultrasonic beam with a specially designed system. We used a capacity sensor, which was monitored by LC-generator to transduce angle of rotation into frequency. Typical speed of rotation was about 0.2 angular minutes per second.

Ultrasonic energy penetrating a sample caused its overheating with respect to the surrounding liquid. The resulting temperature rise of sample is given by

$$\alpha(\theta) = \frac{\Delta T\, S}{N\, R_K\, \sigma}$$

where N is the density of acoustic energy flow, S – total surface of sample and σ – area exposed to ultrasonic wave. The temperature rise of sample was registered with a germanium resistivity thermometer mounted on it, the identical one being immersed into surrounding liquid ^{4}He to stabilize its temperature.

3 Experimental results

3.1 Tungsten: acoustically isotropic media

The single crystal of tungsten is acoustically isotropic due to the occasional combination of its elasticity moduli. For this reason the crystallographic orientation of its free surface was arbitrary. Shown in fig. 1(a) is a typical experimental record of sound transmission into sample, measured at frequency 30 MHz and at three different temperatures. The energy penetrates into the sample only in the narrow cone restricted by $\pm 6^o$ (the critical cone). There are however well-resolved peaks with width of about 30 angular minutes at both edges of this continuous spectrum. They arise due to energy dissipation of resonantly exited Rayleigh surface waves. There are no minima between the regions of longitudinal and transverse modes nor at critical angles (they should appear according to the acoustic theory, at least at reasonably small values of p).

We have carried out a theoretical calculation of the value $\alpha(\theta)$ to compare with experimental results. As a result of this calculation we have found that for continuous part of spectrum where bulk phonons are excited the value of α is approximately 1.4×10^-3. There is also a sharp Rayleigh peak which parameters depend strongly on the value of p. Direct calculation of the Rayleigh waves contribution to the total Kapitza conductance is roughly equal to that of continuous spectrum.

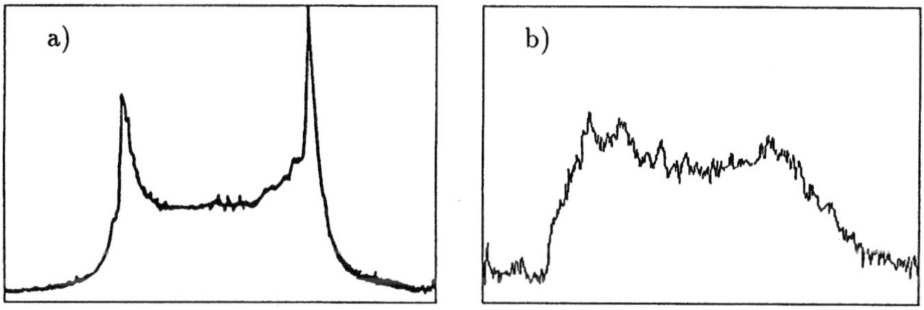

Figure 1: Experimental records of ultrasonic energy transmission into tungsten single crystal(a) and polycrystralline gold(b).

3.2 Polycrystalline gold: strong phonon scattering

A sample of polycrystalline gold was formed by pressing a gold plate between two polished quartz plates; the irregularity of its surface was found to be less than 0.5 μ m. An experimental record is shown in fig. 1(b). There are no explicit resonant peaks at the edges of spectrum, nor is there a sharp boundary between regions of transmission and reflection. Rather the spectrum has a smooth tail ranging to several degrees. All these features show that the internal dissipation of ultrasound in gold is large enough to raise the Kapitza conductance severalfold. According to this assumption the overheating of this gold sample had the least value under the same conditions among the four samples discussed in the paper.

3.3 Copper: strong acoustic anisotrophy

The factor of acoustic anisotrophy η of copper is very large. This is why there is a different kind of resonant surface excitation on the interface between copper and liquid ^{4}He , pseudo surface wave. Its role in evaluating the Kapitza conductance is essential since it gives a significant contribution even if there is no dissipation of acoustic energy in crystal: it has a natural radiative dissipation. This means that one should account for their contribution while calculating the Kapitza resistance of insulator or superconductor. We managed to observe both types of excitations on the surface of copper single crystal (see fig. 2). One could see that pseudo surface waves are much better described by the theory than Rayleigh waves. This is because their form is less sensitive to the assumptions about bulk scattering of phonons in crystal. The right and left sides of spectrum in fig. 2 are not identical, since the sample is tilted a bit in the rotating carrier and the opposite sides of spectrum correspond to different azimuth directions. The resonant peaks are much sharper than that of tungsten because of weaker electron-phonon interaction in copper.

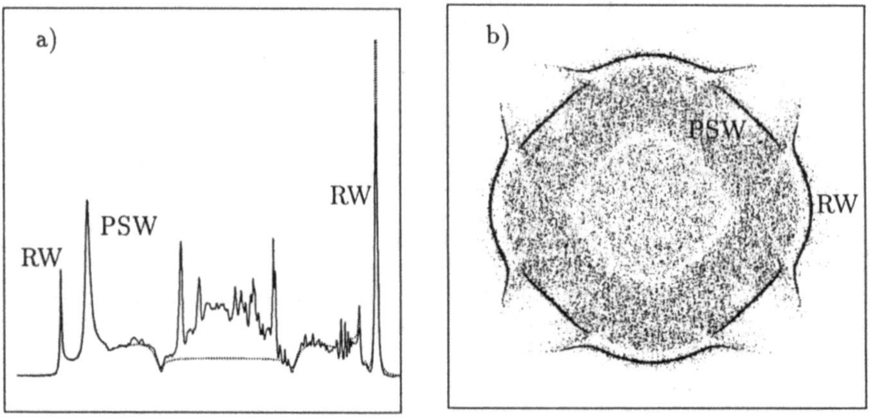

Figure 2: Experimental record (solid line) and theoretical curve (dotted line) of ultrasonic energy transmission into copper single crystal(a) and 2D theoretical spectrum for (001) copper plane(b).

3.4 Aluminum: superconductive-to-normal transition

We had also been interested in the nature of acoustic dissipation in metal single crystals which is responsible for scattering of surface wave. In order to see what is the contribution of conductivity electrons into this process we have studied angular spectra of ultrasound transmission into a superconductor in both normal and superconductive states. The corresponding spectra measured with aluminum sample are shown in fig. 3. As it was expected the acoustic dissipation in normal state turned to be much larger than in superconductive state. First, the Rayleigh peaks increased dramatically when a sample was turned into normal state with a magnetic field of 1 kOe. Second, a comb of multiple fine peaks disappeared. That comb is a result of interference of phonons, reflected from the back side of the sample. In the normal state these reflected phonons are strongly scattered in the bulk.

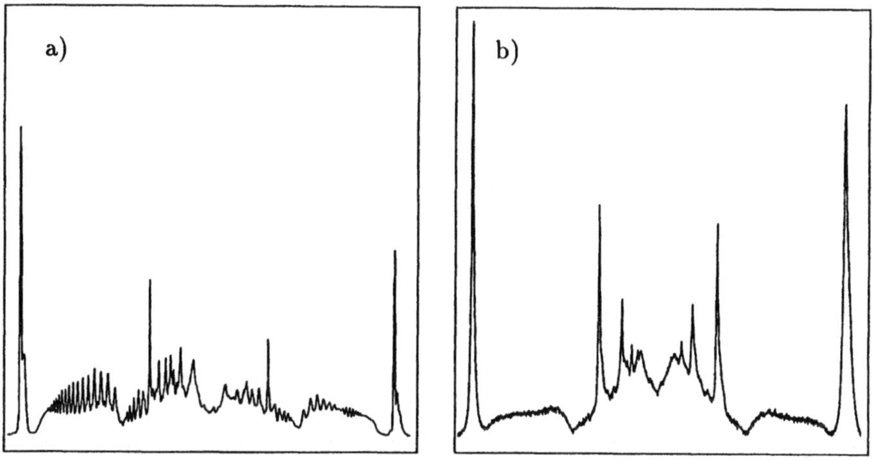

Figure 3: Experimental records of ultrasonic energy transmission into aluminum single crystal in superconductive(a) and normal(b) states.

4 Conclusion

The method we have developed for these measurements turned out to be sensitive and to have high angular resolution. It allows to identify different acoustic modes on the surface and in bulk crystal, to determine their contribution to the total Kapitza conductance, to estimate the parameter of dissipation p and to observe fine details of angular distribution of transient phonons. We have obtained a detailed picture of acoustic phonons transmission through the surface (001) of cubic-symmetry metal crystal. There were two different resonant surface modes observed experimentally – the Rayleigh surface wave and pseudosurface wave. It was shown that conductivity electrons are responsible for energy dissipation of surface wave in metal single crystals, and the grain boundaries are responsible for this process in polycrystalline samples.

In fact the developed technique is not only applicable to the problem of Kapitza resistance. It may be used also to study the properties of conductivity electrons since the peak region of spectrum is very sensitive to these properties. For example it may turn convenient to study *resonant scattering of surface phonons* on the parabolic belts of Fermi surface.

As a result of our investigations it was shown that experimental picture of phonon scattering at the interface between liquid ^{4}He and metal is qualitatively described by generalized acoustic theory in a wide range of energies and temperatures. The Kapitza conductance derived from our measurements by direct integrating of transmission coefficient however turned to be approximately two times smaller than experimentally observed value. For this reason either the generalized acoustic theory is not applicable to the nature of Kapitza resistance, or phonons with larger energy reveal some different kind of behavior. In fact the most "hot" phonons used in our measurements had an equilibrium temperature of about 5 mK, while the temperature of those responsible for boundary heat transfer was always higher then 100 mK. This means that it is very important to carry out some measurements at higher frequencies to ensure whether the former or the latter possibility is valid.

We have estimated theoretically the possibility to measure these angular spectra without a source of coherent phonons. It seems to be very easy to generate beams of thermal phonons using simply a hot single crystal plate. Most of phonons will be emitted into liquid ^{4}He resonantly at Rayleigh angles thus making it possible to create a narrow beam of thermal phonons. One could use this beam to see how it interacts with another surface. A particular case of this hypothetical arrangement consists of two identical single crystal plates with their axis aligned. In this case most of *resonantly emitted* by the hot source phonons will be *resonantly absorbed* at the opposite side by the cold receiver giving raise to the overheating of the receiver even with respect to the liquid intermediating in the heat transfer. A slight disorientation of the two samples will result in total absence of this effect since they are mutually acoustically "black" in very narrow region of solid angles. The 2D theoretical picture of this resonant heat exchange for Cu-Cu is given in fig. 4(b), and its cross-section along the [100] direction is given in fig. 4(a).

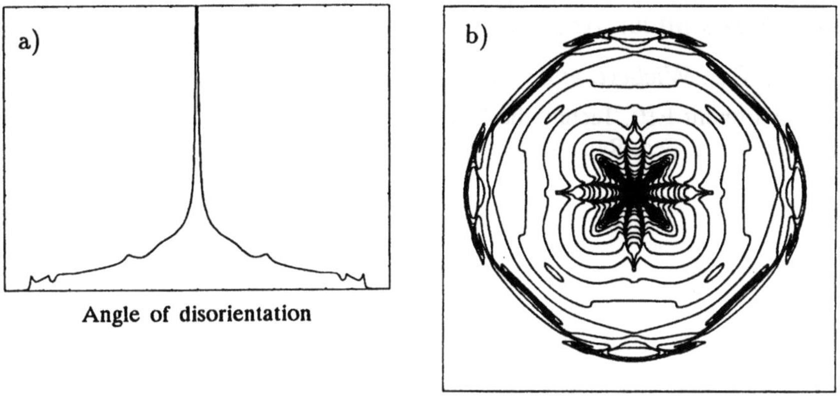

Figure 4: Theoretically calculated 2D picture of resonant heat exchange between two identical copper single crystals(b) and its cross-section along the [100] direction(a).

References

[1] I.M. Khalatnikov *Sov. Phys. JETP*, **22**, 687 (1952)

[2] R.E. Peterson, A.C. Anderson *J. Low Temp. Phys.*, **11**, 623 (1972)

[3] H. Haug, K. Weiss, *Phys. Lett.*, **A40**, 19 (1972)

[4] I.M. Khalatnikov, I.N. Adanenko *Sov. Phys. JETP*, **63**, 745 (1972)

[5] A.F. Andreev *JETP*, **43**, 358 (1962)

[6] K.N. Zinov'eva *JETP Lett.*, **28**, 269 (1978)

Kapitza Resistance: Angular Phonon Distribution in the Solid

H. Kinder, K.H. Wichert, and C. Höß

Physik Department E10, Technische Universität München,
W-8046 Garching, Fed. Rep. of Germany

1. Introduction

More than fifty years after its discovery, the Kapitza resistance is still under debate. New information can be obtained from experiments with angular resolution of the phonons emitted into the solid.

A model which allows predictions of angular phonon distributions was developed by Kinder and Weiss [1]. They assume a lossy layer of defects or adsorbed molecules on a flat surface of a lossless solid. The layer absorbs and emits phonons of the solid and of the helium and provides a very effective channel of heat transfer. The absorption angular distrubution is calculated by classical continuum acoustics [2] using for the layer complex elastic constants (modelling defects or force constant fluctuations) or a complex mass density (modelling mass defects). Due to detailed balance, the angular distribution of emitted phonons must be the same.

2. Experimental Technique

Fig. 1 shows the experimental setup we used to measure the angular phonon distribution on the solid side.

A PMMA support carrying a small heater film slides over the surface of a Si crystal which has a small Al junction detector on its far surface. Heat pulses are transmitted through the helium into the solid. The phonon flux

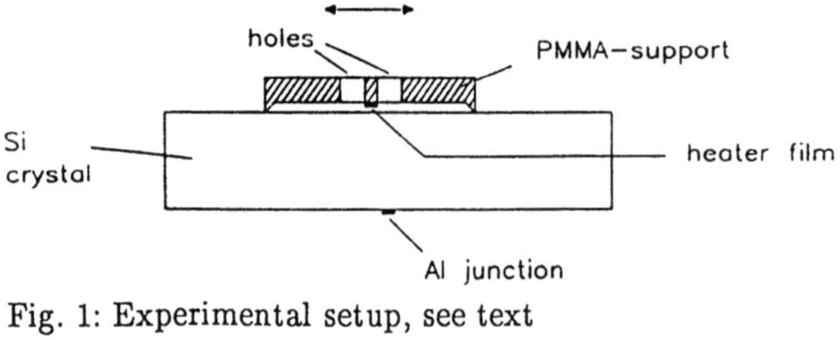

Fig. 1: Experimental setup, see text

Springer Series in Solid-State Sciences, Volume 112
Phonon Scattering in Condensed Matter VII Eds.: M. Meissner and R.O. Pohl

emanating on the solid side is detected under varying angles by scanning the heater position mechanically in x- and y-directions.

Both, heater and junction were 100 × 100 μm^2 wide, and the crystal thickness was 3 mm. The distance between the heater film and the crystal was 10 - 20 μm. Care was taken to avoid a lateral spreading of heat in this gap, by drilling two holes of 2 mm diameter, 100 μm apart, into the PMMA support, and placing the heater film between them. Usually, the gap was filled with bulk helium, although we could also work with helium films. The heater powers were around 5 mW, so that the phonon frequencies corresponded to the ambient temperature of 1.1 K.

3. Theory

Before we show the experimental results we wish to discuss the predictions of the lossy layer model for this experimental situation. Fig. 2(a) depicts, for comparison, the black body case where absorption and emissivity coefficient of the layer do not depend on the angle of the phonons

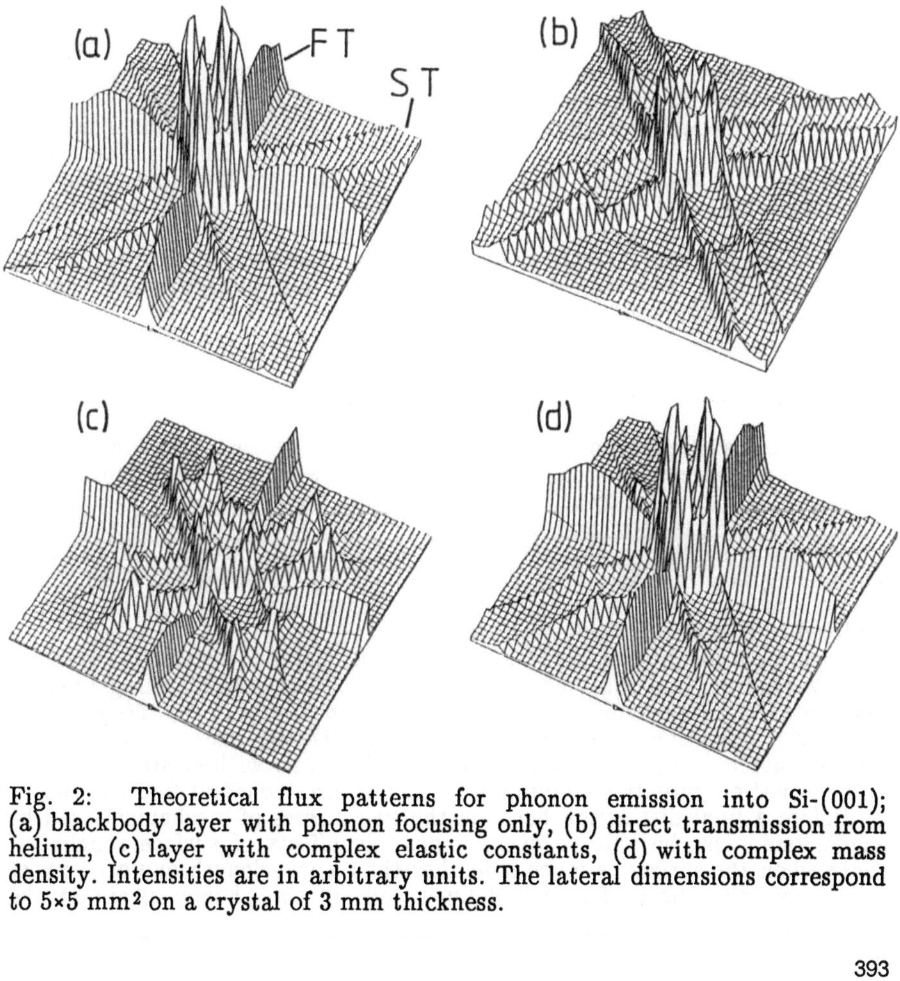

Fig. 2: Theoretical flux patterns for phonon emission into Si-(001); (a) blackbody layer with phonon focusing only, (b) direct transmission from helium, (c) layer with complex elastic constants, (d) with complex mass density. Intensities are in arbitrary units. The lateral dimensions correspond to 5×5 mm^2 on a crystal of 3 mm thickness.

in the solid. The flux is then just given by the focusing pattern which is well known for Si-(001). [3]

Fig. 2(b) is obtained without any layer, when only the direct acoustic transmission from a diffusive phonon field in the helium is remaining. The most striking change in comparison with Fig. 2(a) is the absence of the "FT-ridges" along the (100) and (010) directions. These transverse modes have polarizations nearly parallel to the surface and are therefore not excited by the longitudinal phonons of the helium. On the other hand, the ST-phonons along (110) and (1-10) are effectively transmitted because their polarizations have components perpendicular to the surface.

In Fig. 2(c) we have assumed a layer with complex elastic constants. We chose the same magnitudes as for Si, but a loss angle of 45°. The product of frequency and thickness is $\omega \cdot d = 1$ km/s, corresponding to, e.g., d = 0.8 nm at f = 200 GHz. The flux pattern has markedly changed in comparison with Fig. 2(a). Most characteristic for elastic losses is a decrease of the flux near normal incidence. The four maxima on the central peak have disappeared, and the intensity on the FT ridges decreases from outside to the center. At normal incidence, the elastic strain has a node at the free surface, so that the complex elastic constants cause little absorption.[4,1]

An additional new feature is a ring of intensity around the center which has pronounced maxima where it crosses the ST ridges. This marks the halo of "internal critical cone channeling" [5] where the FT and ST phonons couple via mode conversion to L phonons running parallel to the surface, and hence are very effectively absorbed by the layer.[6]

Fig. 2(d) shows the prediction for a layer with complex mass density, again assuming a loss angle of 45°, and $\omega \cdot d = 1$ km/s. This case resembles much more the black body case of Fig. 2(a), in that the four central maxima are remaining. But again there is the ring due to internal critical cone channeling. The appearance of this ring can therefore not discriminate between elastic and inertial losses. It does, however, indicate that the absorbing layer is thin, since it disappears for films thicker than a wavelength, according to the calculations.

4. Experimental results

A selection of experimental results is depicted in Fig. 3. In Fig. 3(a) we show the measurement of the total phonon flux (without time resolution) for a Si-(001)-surface which was Syton-polished and then cleaned by wiping with optical paper soaked in acetone. The FT ridges along the (100) and (010) directions are fully developed, and the halo due to internal critical cone channeling is nicely resolved. The former clearly confirms that the direct acoustic transmission is negligible, and the latter indicates that the heat is emitted from a thin surface layer indeed.

A closer inspection of Fig. 3(a) reveals that the four central maxima are clearly existing, and that the FT ridges do not decrease all along towards the center. This behaviour resembles much more closely to Fig. 2(d) than to Fig. 2(c), indicating that we deal here with inertial losses, due to mass defect scattering or density fluctuations, rather than with elastic losses due to deformation potential coupling of defects or force constant fluctuations.

Before a final distinction between the two classes of mechanisms can be made, one must be able to exclude possible effects of surface roughness. Only when the surface is much smoother than a wavelength one can expect to observe the effect of a well defined node of the phonon strain at the

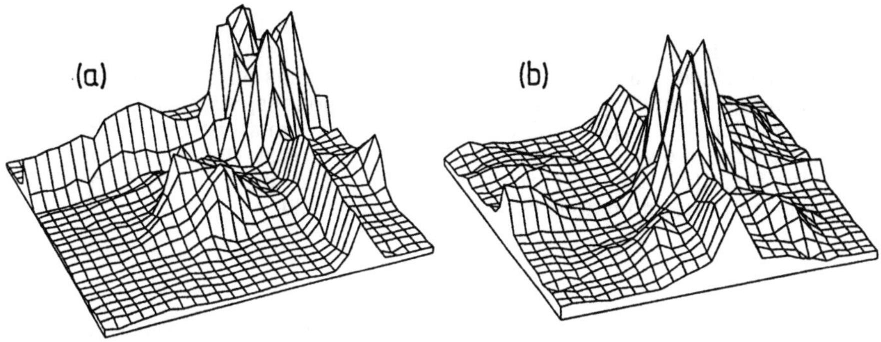

Fig. 3 Experimental results for the Si-(001) surface; (a) cleaned with acetone, (b) spin-etched with "clean spot". Lateral dimensions: (a) 2.7 mm²×2.7 mm², (b) 3×3 mm².

surface. Therefore we have made various attempts to make the surface as smooth as possible, although our technique of in situ-laser annealing [7] could not be used in the present experimental setup.

Best results were as yet obtained by "spin etching": After cleaning with acetone and methanol and depositing the Al junction on the back side, the Si crystal was mounted on a spinner and rotated with 5000 rpm. A mixture of alcohols (ARA), then some drops of etchant (1 % HF in ARA), and finally again ARA were dripped onto the surface. This should result in a rather smooth, H-terminated surface, which lasts at least 1 hour in air without reoxidation.[8] The samples were mounted and pumped on within 10 minutes.

A result for Si-(001) is shown in Fig. 3(b). The pattern is no longer symmetric, but reveals a "clean spot" on the left half of the surface, where one of the four central peaks is missing, and part of the FT ridge is depressed. We estimate a reduction of the anomalous Kapitza conduction at this spot by a factor of 5.

STM measurements [9] have revealed that large, atomically smooth terraces can be made ex-situ on air by a similar treatment of the Si-(111) surface. In Fig. 4 we show our most interesting result as yet. In Fig. 4(a), the pattern is extremely inhomogenous. The threefold symmetry of the (111) surface is hardly visible. Most strikingly, there are large areas with virtually

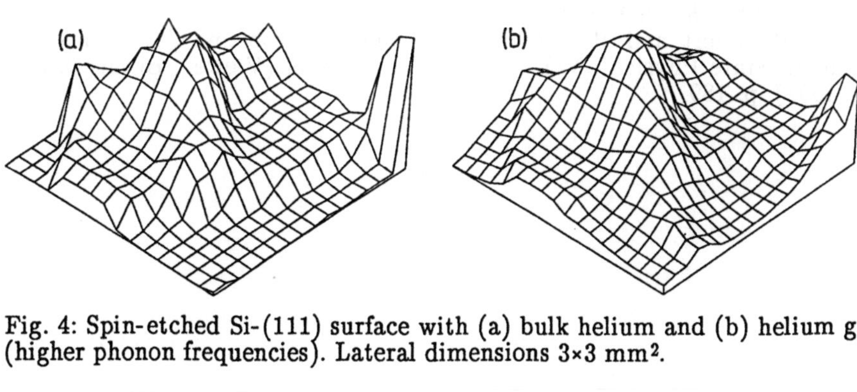

Fig. 4: Spin-etched Si-(111) surface with (a) bulk helium and (b) helium gas (higher phonon frequencies). Lateral dimensions 3×3 mm².

no signal at all. Fig. 4(b) was taken on the same surface on the same day, but with helium gas in the heater gap, and a nearly saturated helium film on the crystal surface. The condensation of helium atoms on the crystal surface leads to higher frequency, shorter wavelength phonons, which interact more effectively with the layer and thus have higher emissivities. Clearly, the clean area now shows a finite signal. So we can trust that the areas of no signal in Fig. 4(a) are exceptionally smooth and clean indeed, and the reduction factor of anomalous conduction is about 20.

5. Conclusion

Our experiments on the angular phonon distribution in the solid have shown that the anomalous Kapitza conduction is due to a thin lossy layer of contamination even on ex-situ prepared surfaces. The measurements are more likely in agreement with acoustic losses by mass defect scattering than by defects coupling via deformation potential (i. e. two-level-systems). The spatial resolution connected with the experiment allows to characterize the quality of surfaces by "Kapitza contrast". We hope to improve our ex-situ surface cleaning technique further to see the pattern of direct acoustic transmission.

References

1. H. Kinder and K. Weiss in "Phonon Scattering in Condensed Matter V", ed. by A.C. Anderson and J.P. Wolfe (Springer, Berlin 1986), p. 218, and to be published in J. Phys. C.
2. K. Aki and P.G. Richards, Quantitative Seismology (Freeman, San Francisco 1980)
3. F. Rösch and O. Weis, Z. Physik B25, 115 (1976).
4. H. Kinder, A. De Ninno, D. Goodstein, G. Papernò, F. Scaramuzzi, and S. Cunsolo, Phys. Rev. Lett. 55, 2441 (1985).
5. G.L. Koos, A.G. Every, G.A. Northop, and J.P. Wolfe, Phys. Rev. Lett. 51, 276 (1983).
6. H. Kinder in "Physics of Phonons" ed. by T. Paszkiewicz (Springer, Berlin 1987), p. 289.
7. H.C. Basso, W. Dietsche, H. Kinder, and P. Leiderer in "Phonon Scattering in Condensed Matter" ed. by W. Eisenmenger, K. Lassmann, and S. Döttinger (Springer, Berlin 1984), p. 212.
8. M. Morita, T. Ohmi, E. Hasegawa, M. Kawakami, K. Suma, Appl. Phys. Lett. 55, 562 (1989).
9. U. Memmert and R.J. Behm in "Festkörperprobleme / Advances in Solid State Physics", Vol. 31, ed. by U. Rössler (Vieweg, Braunschweig 1991), p. 189.

Dynamics of Nonequilibrium THz Phonons in Single-Crystal Ruby Fibers

S.A. Basun[1], *A.A. Kaplyanskii*[1], *S.P. Feofilov*[1], *and W.M. Yen*[2]

A.F. Ioffe Physico-Technical Institute, 194021 St. Petersburg, Russia
[2]University of Georgia, Athens, GA 30602, USA

In phonon dynamics in thin single-crystal fibers the role of phonon-interface interactions has to be accentuated. In experiments at LHT (see also [1]) the oriented single crystal fibers of 0.02 at.% ruby (diameter D =0.4 mm, length L=20 mm) grown by laser heated pedestal growth method [2] were used. Monochromatic 29 cm^{-1} phonons were detected using the R_2 fluorescence of Cr^{3+} [3] from small volume (d) excited by a cw Ar laser.

Broad-band phonon injection. 0.3 mm^2 constantan film "h" on the fiber surface was heated with 200 ns current pulses (Fig.1,inset). With the sample fully immersed in He II (level A,Fig.1a), $R_2(t)$ pulses were measured for different distances x=h-d. The arrival time of the steep leading edge of $R_2(t)$ indicates the ballistic phonon propagation with sound velocity. When the fiber is only partially immersed in He II and surrounded by He gas (level B, inset in Fig.1b), the shape of $R_2(t)$ pulses becomes typical for diffusive propagation.

The drastic dependence of $R_2(t,x)$ pulses on fiber surrounding proves the decisive role of diffuse scattering of phonons at the fiber surface in phonon transport along the fiber. In the case A diffusely scattered off-axial phonons can readily escape the fiber in contact with He II through the anomalous Kapitza transmission and only phonons ballistically propagating along fiber axis may reach the detector. In case B phonons cannot escape the fiber and their transport along fiber becomes diffusive like Knudsen flow as a result of multiple random-angle reflections at the surfaces while propagating ballistically between events.

Monoenergetic phonon generation. A pulsed (Δt=20 ns, f=10 kHz) Cu-vapour laser was used to excite 10 mm long part of fiber below (level A, Fig.2a) or above (level B, Fig.2b) He II level. This pumping results in formation of quasistationary $\bar{E}(^2E)$-excited Cr^{3+} population $N_{\bar{E}}$ and in pulsed generation of nonequilibrium phonon distribution with sharp peak at 29 cm^{-1}, originating from intralevel relaxation within the ^2E states [4].

Springer Series in Solid-State Sciences, Volume 112
Phonon Scattering in Condensed Matter VII Eds.: M. Meissner and R.O. Pohl

Fig. 1

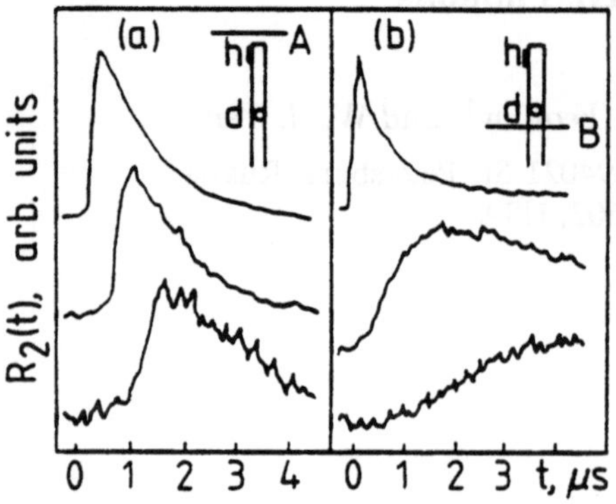

Fig. 2

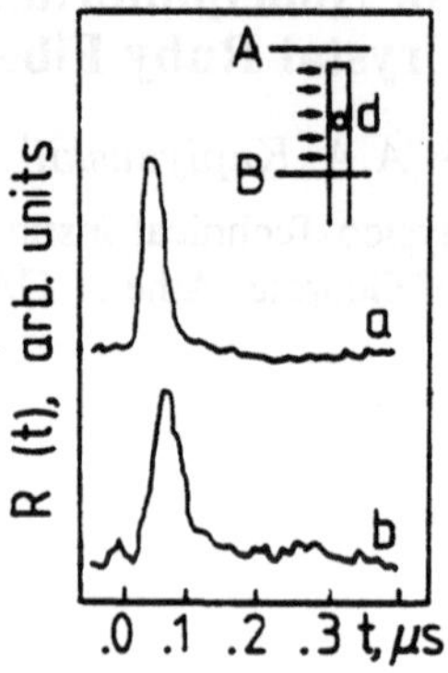

Fig.1. $R_2(t)$ pulses as a function of $x = h - d$. $x = 0, 4, 8$ mm (a) and $x = 0, 2, 4$ mm for pulses shown from above downwards, respectively.

Fig.2. $R_2(t)$ with monoenergetic phonon generation.

Fig.2 shows $R_2(t)$ pulses at low laser powers when the value of N_E provides a mean free path for resonant scattering of 29 cm^{-1} phonons from $Cr^{3+}(^2E)$ ions of $l \simeq$ 1 mm, comparable to D.Fast (t_1=150 ns) decay of 29 cm^{-1} phonons, independent on immersion conditions (A or B), may be due only to escape from the frequency mode in inelastic surface scattering events, independent on sample environment.The decay time is $t_1=D/v(1-p)^{-1}$ with p and 1-p the probabilities of elastic and inelastic 29 cm^{-1} phonon reflection, respectively. It follows from experimental value of t_1 that frequency conserving versus nonconserving events occur in a ratio of 3:2.

Thus, diffuse random-angle phonon scattering processes dominating in phonon-interface collisions are accompanied with 50% probability with the change of scattered phonon frequency. This must result in efficient frequency down-conversion of nonequilibrium phonon distribution during its transport along the fiber at LHT. The surface down-conversion scattering of high-frequency phonons preceeds also escape of the phonons into the He II when contact exists.

References

1. S.A.Basun, S.P.Feofilov, A.A.Kaplyanskii, W.M.Yen: Phys. Rev. Lett. 67, 3110 (1991); J. Lumin. (1992).
2. B.M.Tissue, Lizhu Lu, W.M.Yen: J.Lumin.45, 20(1990).
3. K.F.Renk, J.Deisenhofer: Phys.Rev.Lett.26,763(1971).
4. R.S.Meltzer, J.E.Rives: Phys.Rev.Lett.38, 421(1977).

Sensitive Measurements of the Phonon Scattering and Transmission at a Silicon–Helium Interface

J.R. Olson and R.O. Pohl

Laboratory of Atomic and Solid State Physics, Cornell University, Ithaca, NY 14853-2501, USA

1. Introduction

The transmission of phonons across interfaces has been of interest since Kapitza measured a thermal boundary resistance between solids and liquid helium[1]. Khalitnikov developed the acoustic mismatch model (AMM) to describe the transmission probability across interfaces[2]. No experiments to date have been able to test quantitatively the predictions of the AMM, because the predicted transmission rate for solids in contact with liquid helium is very small (generally less than 1%).

Through measurements of heat conduction in a silicon sample in contact with liquid helium, we have been able to perform such measurements. By varying the temperature from 0.05 to 2K, the dominant phonon wavelength is scanned from 13,000 to 300Å. These measurements are sensitive to transmission rates as low as 0.3%.

2. Experimental Details

High–purity silicon wafers, 5 cm diameter, 0.3 mm thick, are mounted to serve as the bottom plate of a sealed cell which can be evacuated or filled with liquid helium. The thermal conductivity is measured with and without helium in the cell, in a 2D geometry. Heat is generated in the center of the wafer and the temperature is measured at two different radii. The phonon mean free path ℓ is then calculated using $\Lambda = \frac{1}{3}Cv\ell$. High–purity silicon is used because there is little bulk scattering of the phonons below 5K, leaving only surface scattering. A thin wafer is used to allow many bounces at the silicon–helium interface to improve the sensitivity. Monte Carlo calculations are then used to analyze the mean free path with the sample cell either evacuated or filled with helium.

3. Results

The measured mean free path for polished silicon is shown in Fig. 1, measured in vacuum and in contact with liquid helium. At low temperatures, the sample

Springer Series in Solid-State Sciences, Volume 112
Phonon Scattering in Condensed Matter VII Eds.: M. Meissner and R.O. Pohl

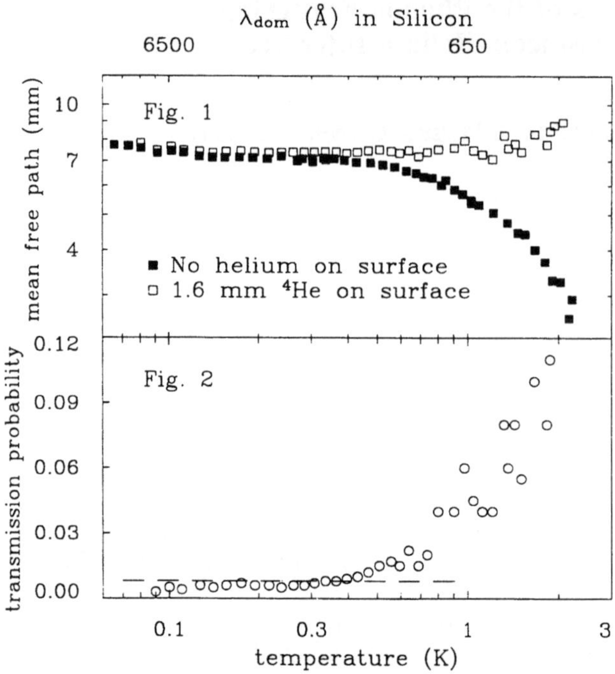

Figure 1. Phonon mean free path for polished silicon as a function of temperature, with and without helium in the sample cell. The sample was a 5 cm diameter, 0.3 mm thick silicon wafer, polished by the supplier.

Figure 2. Phonon transmission probability for polished silicon as a function of temperature, determined from Monte Carlo calculations. The AMM prediction is shown as a dashed line. The variation above 0.5K is due to residual surface roughness.

surface appears very smooth to the phonons (large mean free path), and there is very little effect upon adding helium to the cell. The results of Monte Carlo calculations using these data to determine the transmission probability are shown in Fig. 2.

This research was supported by the National Science Foundation, Grant No. DMR 91 15981.

[1] P.L. Kapitza, J. Phys. (Moscow) **4**, 181 (1941)
[2] I.M. Khalatnikov, Zh. Eksp. Teor. Fiz. **22**, 687 (1952)

Picosecond Optical Study of the Kapitza Conductance Between Metals and Dielectrics at High Temperature

R.J. Stoner and H.J. Maris

Department of Physics, Brown University, Providence, RI 02912, USA

We have used a picosecond optical technique [1] to measure the Kapitza conductance σ_K for interfaces between several dielectrics and metals. The Debye temperatures Θ_D of the dielectrics, BaF_2, sapphire and diamond ranged from 286 K to 2240 K, while those of the metals Pb, Au, Al and Ti ranged from 88 K to 426 K. Samples consisted of metal films of thickness $d_f < 1000$Å evaporated onto the substrates in high vacuum. In the experiment a 200 fsec laser pulse of wavelength 630 nm was focussed onto a spot on the film approximately 20 μm in diameter. Partial absorption of the pulse by the film gave rise to a sudden increase in its temperature of a few degrees, and the optical reflectivity of the film was changed by a small amount which we assume was proportional to the temperature rise. We obtained a record of the cooling of the film by measuring its reflectivity as a function of time with a second laser pulse delayed relative to the heating pulse by up to 2000 psec. A cooling curve for Pb on diamond is shown in Fig. 1.

Assuming that the temperature of the substrate beneath the heated spot remains constant as the film cools, the cooling curve takes the form

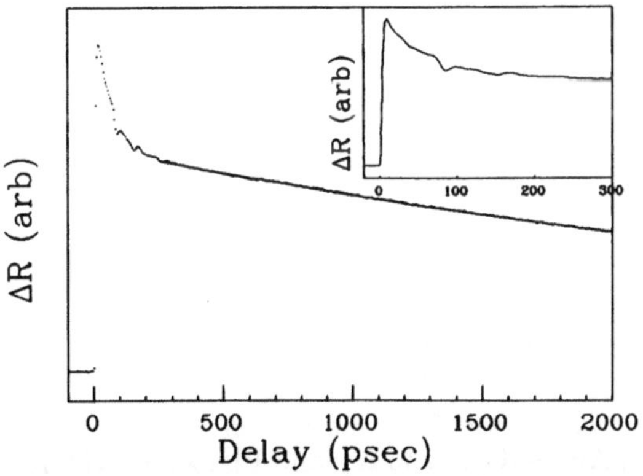

Fig.1 Cooling curve and exponential fit for 960Å Pb film on diamond at 300 K. Inset shows acoustic echoes generated by the heating pulse whose separation in time was used to deduce the film thickness accurately.

Springer Series in Solid-State Sciences, Volume 112
Phonon Scattering in Condensed Matter VII Eds.: M. Meissner and R.O. Pohl

$$T(t) = T(0)\exp(-\frac{\sigma_K t}{C_f d_f}) \tag{1}$$

where C_f is the specific heat of the metal. Cooling curves obtained for the diamond samples (the thermal conductivity at room temperature of our diamonds was nearly eight times that of copper) followed this exponential law for all four metals at all temperatures. The Kapitza conductance could therefore be obtained from the slope of $\ln(T(t))$ over the entire 2000 psec time range. For the lower thermal conductivity substrate materials exponential cooling in the first 500 to 1000 psec slowed noticeably at longer times due to substrate heating. Only the initial exponential region was used to obtain σ_K for these substrates.

Kapitza conductances obtained for the diamond samples at temperatures from 50 to 300 K are shown in Fig. 2. Also shown are theoretical curves based on a three dimensional lattice dynamical theory [2] for σ_K in which phonon dispersion, and the essential features of the densities of states of the film and substrate materials were accounted for in an approximate way. Qualitatively the results follow the expected tendency for σ_K to be greatest when the phonon spectra of the solids (which in the model are parametrized by their mass densities and Debye temperatures) are most similar. However, a comparison with the theoretical curves shows that the observed Kapitza conductances for the low Θ_D metals are much larger than expected, the most extreme example being Pb. A similar but less dramatic tendency was seen for the sapphire samples at high temperatures: for Ti and Al the measured conductances were in rough agreement with the lattice dynamical calculations; whereas, for Pb and Au the measured values were up to 10 times larger than predicted by the theory. A comparison of the BaF_2 results with theory is complicated by the existence of optical modes at low frequency. Therefore a discussion of these results will be given elsewhere [3].

Also shown in Fig. 2 are curves (dashed) showing the radiation limits [4] for Au and Pb on diamond. These curves were obtained using an elastic

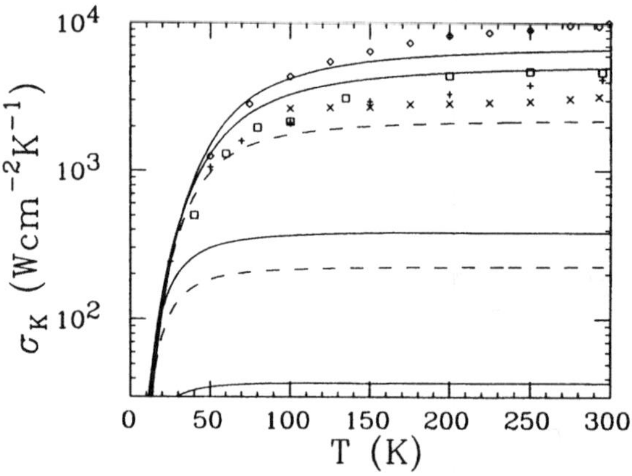

Fig.2 Measured Kapitza conductances for four metals on diamond: Ti (◇,Θ_D=426 K); Al(□,Θ_D=420 K); Au (+,Θ_D=165 K); and Pb (×,Θ_D=88 K). Solid lines correspond to lattice dynamical theory (from top to bottom Ti, Al, Au and Pb), and dashed lines give phonon radiation limits for Au (upper) and Pb (lower).

continuum theory in which we used cutoff frequencies for the metals taken from neutron diffraction data. The radiation limit gives a measure of the largest σ_K possible if the phonons emitted into the substrate from the metal all have frequencies at or below the metal cutoff. Since these curves fall below the experimental curves (in the case of Pb *far below*), we conclude that a large fraction of the energy entering the diamond is carried by phonons having frequencies exceeding the metal cutoff. Thus, for example, the results cannot be explained by elastic scattering of phonons in the Pb by interfacial defects –the enhancement must be due to an *inelastic* process. We have considered the possible effects on σ_K of electron-phonon and phonon-phonon interactions in the Pb-diamond case.

An interaction between electrons in the Pb film and phonons of all frequencies in the diamond near the interface could lead to accelerated cooling of the film. This possibility has been considered in the context of ^{4}He/metal interfaces [5], and found to give too small an enhancement to explain experimental observations. We tested the effect of electrons in our experiments by fabricating samples consisting of approximately 30Å of Bi sandwiched between 1000Å of Pb and the diamond substrate. Since Bi and Pb have very similar phonon spectra we assume that the presence of the Bi had little effect on the phonon flux reaching the interface. On the other hand, since the density of electrons in Bi is much less than in Pb it acted as a barrier between the diamond and Pb for electrons. Cooling curves for samples with and without the Bi interlayer were nearly indistinguishable. We take this as evidence that electrons do not significantly contribute to the cooling of the Pb.

One is left to consider how interactions amongst the phonons alone might lead to a large enhancement in σ_K. It is useful to consider a one-dimensional model in which the Pb atoms are viewed as classical oscillators joined together by a potential $V(r)_{Pb-Pb}$, and to the diamond by a potential $V(r)_{Pb-C}$. To excite a large flux of high frequency phonons in the diamond it is necessary for interfacial Pb atoms to exert forces on the diamond both of the correct frequencies and with sufficient strength. The energy transfer rate $\dot{Q}$ to the diamond depends strongly on the form of $V(r)_{Pb-C}$, about which little is known. Classically one can show that for a particle in a potential $V(r)$ $\dot{Q}$ is proportional to

$$\chi = \frac{1}{\tau}\int_0^{\tau}\left(\frac{dF}{dt}\right)^2 dt \tag{2}$$

where $F = -dV(r)/dr$ is the force on the particle, and τ is its period of oscillation. We evaluated χ assuming

$$V_{Pb-C}(r) = k_B\varepsilon\left[\left(\frac{a}{r}\right)^{12} - 2\left(\frac{a}{r}\right)^{6}\right] \tag{3}$$

for a range of ε (Fig. 3) at 300 K. For Pb-Pb bonds we estimate $\varepsilon \simeq 1,000$ K, while for C-C bonds $\varepsilon \simeq 55,000$ K. Also shown in Fig. 3 is a curve (dashed) which shows χ evaluated for the harmonic potential giving the same τ.

From Fig. 3 it is apparent that irrespective of the form of $V(r)$, the large enhancement of σ_K seen in the experiments cannot be accounted for if the Pb-diamond binding is taken to be Pb-Pb–like. For an enhancement factor of 100 it is necessary to have $\varepsilon \geq 15,000$ K for either the harmonic or anharmonic

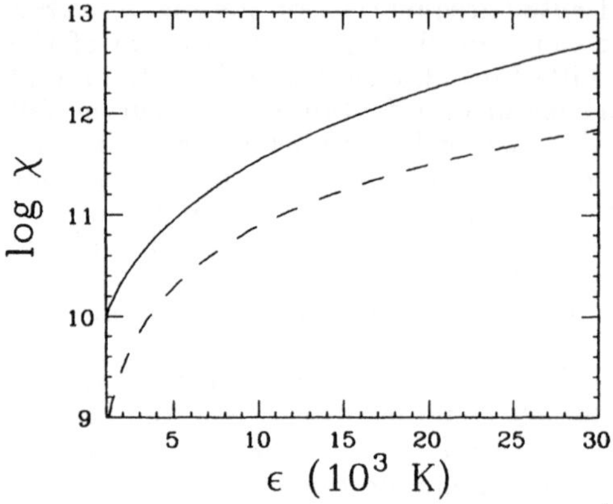

Fig.3 Energy transfer rate figure of merit χ for a Pb atom at 300 K as a function of the depth of the interfacial potential well ε. Solid curve gives the result for the anharmonic $V(r)$ of Eq.4, and dashed curve gives result when only the harmonic part of $V(r)$ is used.

potential [6]. Thus if the interfacial Pb atoms are bound strongly to the diamond, then an energy transfer rate large enough to sustain the observed anomalously large σ_K can be achieved. However, this does not guarantee that the cooling rate will be large, for it must also be true that the Pb atom can be *supplied* from the remainder of the Pb lattice with energy at least as fast. From the measured Kapita conductance we estimate the time required for a single layer of Pb atoms on diamond to cool to $1/e$ of its initial temperature (assuming a constant specific heat) to be of order 100 psec. The enhanced transfer rate can be sustained if the time required for the temperature of an interfacial row of Pb atoms initially at 0 K bound strongly to diamond (ε= 20,000) to come to thermal equilibrium with the remainder of the Pb lattice at 300 K is 100 psec or less. We have verified that this condition is satisfied by means of molecular dynamics simulations in which nearest neighbour interactions were assumed to occur via the *anharmonic* potentials discussed above [5]. It is not satisfied if harmonic potentials are assumed. Therefore we conclude that the anomalously large Kapitza conductances we have measured for Pb-diamond interfaces can result from the combination of anharmonicity in the Pb lattice and strong interfacial forces between the interfacial Pb layer and the diamond.

1. R. J. Stoner, H.J. Maris, T. R. Anthony and W. F. Banholzer, Phys. Rev. Lett. **68**, 1563 (1992)
2. D. A. Young and H. J. Maris, Phys. Rev. B **40**, 3685 (1989).
3. R. J. Stoner and H. J. Maris - to be published.
4. N. S. Snyder, Cryogenics **10**, 89 (1970).
5. W. A. Little, Phys. Rev. **123**, 435 (1961).
6. This potential is not well-understood to our knowledge; however, it is likely that it is at least partly covalent in nature.

Thermal Boundary Resistance at YBaCuO Film–Substrate Interface

A.V. Sergeev, E.E. Aksaev, I.G. Gogidze, G.N. Gol'tsman, A.D. Semenov, and E.M. Gershenzon

Moscow State Pedagogical University, M. Pirogovskaja str. 1, 119435 Moscow, Russia

Abstract. The nanosecond voltage response of YBaCuo films on Al_2O_3, MgO and ZrO_2 substrates to electromagnetic radiation of millimeter and visible ranges has been investigated. The analysis of experimental conditions for Al_2O_3 and MgO substrates shows that the resistance change is monitored by the Kapitza boundary shift of temperature during the temporal interval ~ 100 ns limited by the time of phonon return from a substrate into a film. The observed exponential voltage decay is described by the phonon escape time which is proportional to the film thickness and is weakly temperature dependent.

Films of YBaCuO were prepared by laser ablation and have the onset of superconductivity at about 90 K. We have measured the response of thin YBaCuO films to short pulse (20 ps) from Nd:YAG laser ($\lambda = 0.63$ μm) and also to modulated radiation from two backward-wave tubes ($\lambda = 0.3 \div 2$ μm). In visible light range the voltage signal has the sharp peak (nonequilibrium response) followed by more gradual nanosecond relaxation [1]. In millimeter range the relaxation time t is estimated from the modulation frequency dependence of signal. We have observed linear thickness dependence of t for samples with sapphire and magnesium substrates (• in Fig. 1). The value of τ rises weakly with temperature in the range $70 \div 90$ K and is independent on the film width. In Fig. 1 the same thickness dependence of from Ref. 2 is also demonstrated (⊙). It is interesting that width dependence is found $\tau \sim w^{\frac{1}{2}}$ for samples on zirconia substrates (⊕ in Fig 1).

In nonstationary measurements the new scale of time arises due to interface, namely the time of phonon return from substrate into the film $\tau_r = \tau_s/\alpha^2$ ($\tau_s \sim 0.1$ ns - phonon scattering time in substrate). Because of a small value of phonon transmission coefficient across interface $\alpha = 1/Rcu = 0.01 \div 0.05$ (R-thermal boundary resistance [2], c - acoustic phonon heat capacity, u - phonon group velocity) the value of τ_r is larger than the observed relaxation time. For $t > \tau_r$ the temperature continuity at the interface is valid and the measured transient response follows the $t^{-1/2}$ - dependence due to diffusion [3].

Springer Series in Solid-State Sciences, Volume 112
Phonon Scattering in Condensed Matter VII   Eds.: M. Meissner and R.O. Pohl

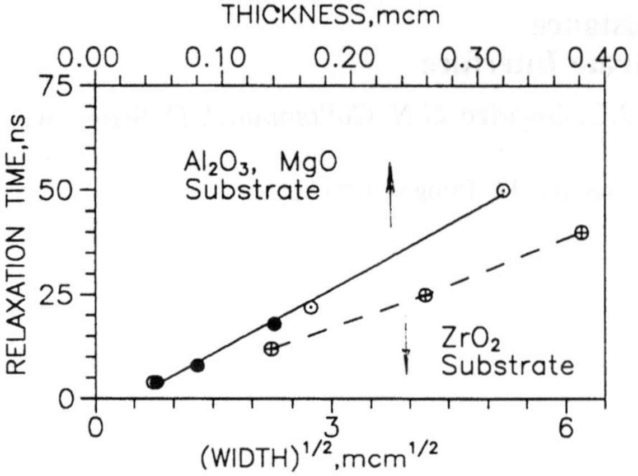

Figure 1: Thickness and width dependences of response time.

For $t < \tau_r$ the substrate phonons play the role of a heat sink. Under condition $l_f > \alpha \cdot d$ the temperature does not change across the film (phonon mean free path $l_f \sim 30 \div 100\text{Å}$). Therefore the voltage change is monitored by the transient Kapitza boundary shift of temperature under the radiation. The calculations show that the time of phonon escape from the film $\tau_{es} = d/\alpha u$ is proportional to the film thickness in spite of diffusion motion of phonons. All conditionsmentioned above are satisfied for samples with sapphire and themagnesium substrates and exponential decay with phonon escape time is observed. Our results of τ-measurements coincide with the data of Ref. 2, so we obtain the same value of Kapitza resistance R $= \tau_{es}/cd$.

The value of τ_r of samples with ZrO_2 substrate is smaller then phonon escape time due to very short $\tau_s \sim 1$ ps. For this reason there are no chances to observe the Kapitza resistance in such type of experiments.

References

[1] E.M. Gershenzon, G.N. Gol'tsman, I.G. Gogidze, A.D. Semenov and A.V. Sergeev. Physica C, 185-189, 1371 (1991).

[2] G.l. Caar, M. Quijada, D.B. Tanner et. al., Appl. Phys. Lett. 57, 2725 (1990).

[3] C.G. Levey, S. Etemad, A. Inam, Appl. Phys. Lett 60, 126 (1992).

Heat Propagation in High-T_c Films Investigated by Optical Response Measurements

H. Lengfellner, S. Zeuner, K.F. Renk, and W. Prettl

Institut für Angewandte Physik, Universität Regensburg, W-8400 Regensburg, Fed. Rep. of Germany

Thermal properties of granular Tl-Ba-Ca-Cu-O films were investigated by optical response measurements: An analysis of the response using a heat transfer model yields a thermal diffusivity $D = 10^{-3} cm^2/s$ at 150 K which rises to $6 \times 10^{-3} cm^2/s$ at a temperature of 30 K and allows for an estimation of the boundary resistance $R_{bd} \approx 10^{-3}$ Kcm^2/W between film and substrate. The dependence of the response time on film thickness obtained from the heat transfer model is compared with published data indicating that in many experiments the observed response is mainly bolometric in origin.

The optical response of high-T_c superconducting films is currently a subject of high interest. Response times between milliseconds and picoseconds are reported [1, 2]. We use a heat transfer model to analyse the bolometric response of polycristalline TlBaCaCuO films patterned into 200μm wide strips for response experiments. A laser pulse ($\lambda \simeq 10$ μm, pulse duration 100 ns) heats the film within the penetration depth. The time dependent temperature distribution within the film obtained from the heat transfer model together with the measured temperature dependent resistivity of the film allows the calculation of the bolometric resistance response [3].

For thick films, the thermal diffusivity can be obtained from a fit of the calculated to the measured response (Fig. 1), giving values ranging from $10^{-3} cm^2/s$ at 150 K to $6 \times 10^{-3} cm^2/s$ at 30 K. Comparison with the response of thin films, where the boundary resistance R_{bd} between film and substrate contributes significantly to the total thermal resistance, indicates $R_{bd} \approx 10^{-3}$ Kcm^2/W, in good agreement with values found recently for YBCO-films [1, 4]. The heat transfer model allows us to extract a characteristic bolometric response time, shown in Fig. 2. In addition, typical response times taken from published response experiments (see references in [3]) are displayed. We suggest that response times close to our calculated curve are mainly bolometric in origin, while large deviations indicate nonbolometric processes.

This work was supported by the Deutsche Forschungsgemeinschaft (H.L), the Bundesministerium für Forschung und Technologie, the Bayerische Forschungsstiftung through the Bayerischer Forschungsverbund Hochtemperatursupraleitung (FORSUPRA) (K.F.R.), and the European Communities (W.P.).

Springer Series in Solid-State Sciences, Volume 112
Phonon Scattering in Condensed Matter VII Eds.: M. Meissner and R.O. Pohl

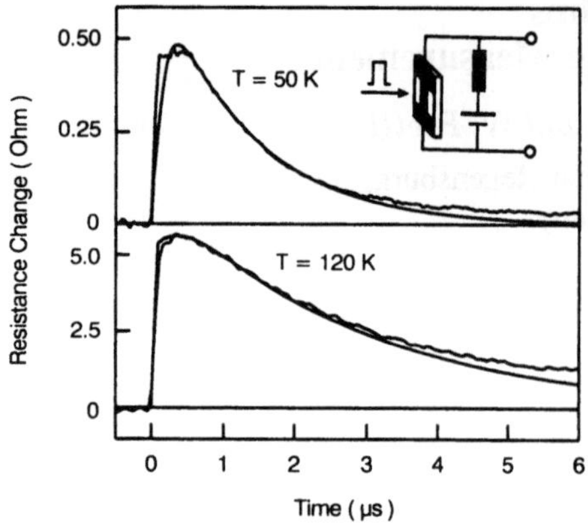

Fig. 1. Experimental arrangement, inset, and optical response at two temperatures, together with calculated curves from the heat transfer model.

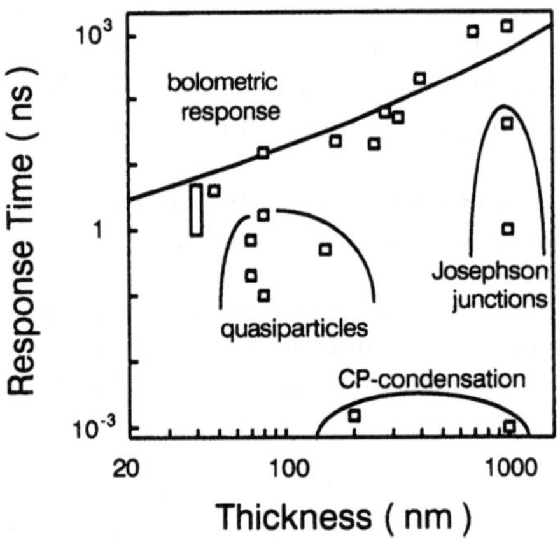

Fig. 2. Calculated response time (line) in comparison with published response time data (squares). The experiments are grouped for the eye according to the suggested response mechanisms. Processes in the picosecond time domain were attributed to Cooper pair (CP) condensation.

[1] G.L. Carr, M. Quijada, D.B. Tanner, C.J. Hirschmugl, G.P. Williams, S. Etemad, B. Dutta, F. DeRosa, A. Inam, T. Venketasan, and X. Xi, Appl. Phys. Lett. **57**, 2725 (1990).

[2] M. Johnson, Appl. Phys. Lett. **59**, 1371 (1991).

[3] S. Zeuner, H. Lengfellner, J. Betz, K.F. Renk, and W. Prettl, Appl. Phys. Lett., August 1992.

[4] M. Nahum, S. Verghese, R.L. Richards, and K. Char, Appl. Phys. Lett. **59**, 2034 (1991).

Giant Voltages Upon Surface Heating in Normal $YBa_2Cu_3O_{7-\delta}$ Films Explained by an Atomic-Layer Thermopile

H. Lengfellner, G. Kremb, J. Betz, K.F. Renk, and W. Prettl

Institut für Angewandte Physik, Universität Regensburg,
W-8400 Regensburg, Fed. Rep. of Germany

Large voltage signals betweeen lateral contacts on epitaxial $YBa_2Cu_3O_{7-\delta}$ films were observed upon heating the films by pulsed laser radiation or by heater wires. A model is proposed which explains the voltages by an atomic layer thermopile formed by the layered structure of the material.

Recently, unexplained transient voltaic signals have been observed at room temperature in $YBa_2Cu_3O_{7-\delta}$ films in response to pulsed laser irradiation [1]. We have performed experiments which show that the voltage signals are thermoelectric in origin. The magnitude of the voltages is attributed to an atomic layer thermopile formed by the layered structure of the material [2].

The experiment has been reported in detail elsewhere [2,3]. Irradiating an epitaxial $YBa_2Cu_3O_{7-\delta}$ film with a laser pulse ($\lambda = 1.06$ μm, pulse duration 100 ns, energy density of 10 mJ/cm^2) can lead, depending on sample, to a transient voltage signal of up to 1 V (Fig. 1a). From a heat propagation model [4] a maximum temperature difference of ≈ 50 K between film surface and film bottom is estimated. The signal decay (Fig. 1) is due to film cooling by heat diffusion. Irradiation of the film through the substrate leads to a reversed ΔT and a reversed signal polarity (Fig. 1b). The signal height is proportional to the pulse energy. Investigations of the spatial distribution of the electric potential

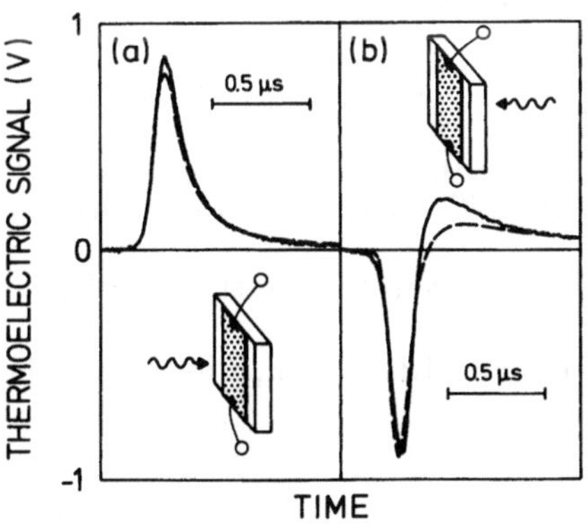

Fig. 1 Thermoelectric signal in response to pulsed laser heating for two beam geometries (a) and (b). Dashed line: Calculated temperature gradient, normalized to signal maximum.

Springer Series in Solid-State Sciences, Volume 112
Phonon Scattering in Condensed Matter VII Eds.: M. Meissner and R.O. Pohl

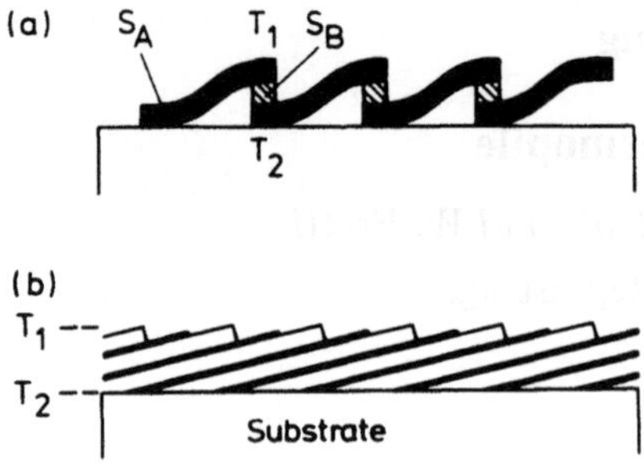

Fig. 2(a) Plane thermopile made from materials with Seebeck coefficients S_A and S_B, temperature T_1 of upper contacts, T_2 of lower contacts. (b) Atomic layer thermopile formed by CuO_2-planes and the layers between the metallic planes.

in the plane of the film [2] have shown that, within the heated spot, a dipolar source with „built-in" dipole axis – with respect to the sample – is generated. Experiments with thermal heat sources demonstrate that the signal voltage is purely due to heating [3]. In Fig. 2a, a plane thermopile is shown generating a voltage $U = N \cdot \Delta S_{AB} \cdot \Delta T$, where N is the number of thermoelements, $\Delta S_{AB} = S_A - S_B$ is the difference of Seebeck coefficients of the two materials involved, and $\Delta T = T_1 - T_2$ is the applied temperature difference. A similar thermopile may be formed by an $YBa_2Cu_3O_{7-\delta}$ film with a c-axis inclined by a small angle α with respect to the surface normal (Fig. 2b). For the $YBa_2Cu_3O_{7-\delta}$ atomic layer thermopile we find [2]

$$U = \frac{d}{D} \cdot \sin\alpha \cdot \Delta T \cdot \Delta S_{AB} \simeq 0.5 \text{ V} \quad ,$$

where d = 5 mm is the diameter of the heated spot, D = 5000Å is the film thickness, α is assumed to be $5°$, $\Delta T = 50$ K and $\Delta S_{AB} \simeq 10\mu V/K$ is the differential thermopower of the CuO_2 planes and the layers between the metallic planes, estimated from the anisotropy of S [5]. Small angles α in the range of several degrees occur due to miscut substrates. The dependence $U \sim \alpha$ for small angles has been proven in a very recent experiment [6].

This work was supported by the Deutsche Forschungsgemeinschaft (H.L.), the Bundesministerium für Forschung und Technologie, the Bayerische Forschungsstiftung through the Bayerischer Forschungsverbund Hochtemperatur-Supraleiter (FORSUPRA) (K.F.R.) and the European Community (W.P.).

[1] C.L. Chang, A. Kleinhammes, W.G. Moulton, and L.R. Testardi, Phys. Rev. B. **41**, 11564 (1990).
[2] H. Lengfellner, G. Kremb, A. Schnellbögl, J. Betz, K.F. Renk, and W. Prettl, Appl. Phys. Lett. **60**, 501 (1992).
[3] H. Lengfellner, G. Kremb, J. Betz, K.F. Renk, and W. Prettl, to be published in: Electronic Properties of High T_c Superconductors, Springer Verlag.
[4] S. Zeuner, H. Lengfellner, J. Betz, K.F. Renk, and W. Prettl, Appl. Phys. Lett., August 92.
[5] M.F. Crommie, A. Zettl, T.W. Barbee, and M.L. Cohen, Phys. Rev. **B 37**, 9734 (1988).
[6] H.S. Kwok and J.P. Zheng, preprint.

Evaluation of the YBaCuO/MgO Interface Thermal Boundary Resistance Through Measurement of the Photoresponse Frequency Dependence

M.M. Bonch-Osmolovskii, T.I. Galkina, A.Yu. Klokov, A.F. Plotnikov, and E.G. Romanov

P.N. Lebedev Physical Institute, Russian Academy of Sciences, Leninskii pr. 53, 117924 Moscow, Russia

1. Introduction

A novel interest has arisen to the nature of boundary resistance (R_{Bd}) of solid/solid interface in connection with the fact, for example, that the photoresponse of such systems as HTSC/dielectric, inevitably is determined by the heat transfer (phonon) processes via the interface after energy relaxation of excited particles. Up to now the role of R_{Bd} for explanation of photoresponse features is almost unknown.

2. Experimental

YBaCuO film ($d \approx 3500$ Å) was produced by magnetron sputtering without subsequent annealing. As a substrate the chemically polished MgO(100) was used ($d \approx 0.8$ mm; detailes in [1]). The He-Ne laser ($\lambda=0.63$ μm; P=1 mW) served as a radiation source. Nearly all the whole film was illuminated by the focused beam. Optical gate was applied for modulation of radiation. Bias current 0.3-1.0 mA was used in four-probe scheme. Fig.1 represents the experimental data (*).

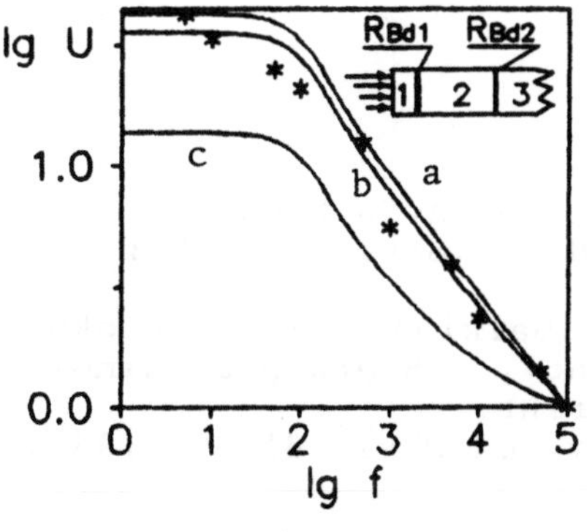

Fig.1
Experimental data (*), the least-square fitting of experimental data (b) and calculated dependences (a,c) of response amplitude vs frequency modulation. a-$R_{Bd1}=10^{-4}$; c-$R_{Bd1}=5\times10^{-3}$; $R_{Bd2}=0$ Kcm^2/Wt. Insert: 1-YBaCuO; 2-MgO; 3-bath.

Springer Series in Solid-State Sciences, Volume 112
Phonon Scattering in Condensed Matter VII Eds.: M. Meissner and R.O. Pohl

3. Model and discussion of results

We have calculated [2] the distribution of overheating in an onedimensional structure represented on the insert of Fig.1 the harmonic modulation of radiation being assumed. The boundary resistances were introduced at the interfaces YBaCuO/MgO and MgO/bath; at low values of overheating the film response is proportional to the average one. Fig.2 represents the calculated overheating for YBaCuO film (on MgO) vs frequency modulation of the radiation. While changing R_{Bd1} from 0 to 10 $K\times cm^2/Wt$ the dependence in the frequency range from 4×10^2 Hz to 10^8 Hz transforms from $f^{-1/2}$ to f^{-1}. The solid lines in the Fig.1 represent the calculated response frequency dependences for several magnitudes of R_{Bd1}. All the curves are normalized at the maximum frequency 10^5 Hz. The proper fit of the experimental points by the calculated ones allows to evaluate the boundary resistance of YBaCu/MgO interface at the transition temperature: $R_{Bd} \cong 6\times10^{-4}$ $K\times cm^2/Wt$. The acoustic mismatch theory evaluation [3] gives the magnitude by ≈50 times less than was found out in this work; therefore, the question about the nature and magnitude of the HTSC/dielectric boundary resistance in the vicinity of the superconducting transition requires more detailed consideration of the physics of phonon propagation via the interface HTSC/dielectric.

The authors are indebted to E.V.Pechen' for the excellent YBaCuO films.

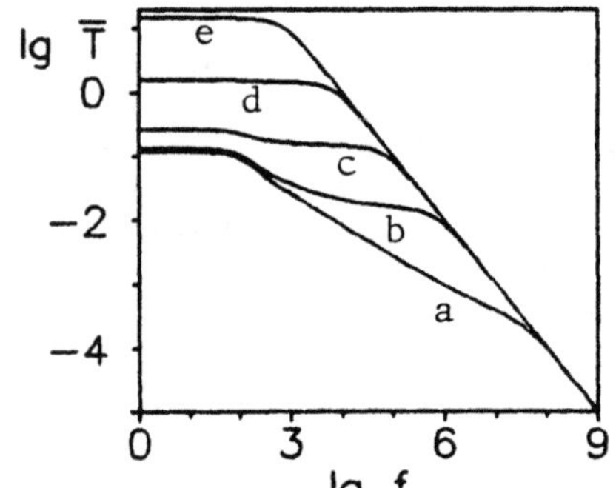

Fig.2
Calculated dependences of overheating average amplitude vs frequency. Parameter - boundary resistance of YBaCuO/MgO: a-R_{Bd1}=0, b-$R_{Bd1}=10^{-2}$, c-$R_{Bd1}=10^{-1}$, d-R_{Bd1}=1, e-R_{Bd1}=10. R_{Bd2}=0, $K\times cm^2/Wt$.

4. References

1. M.M.Bonch-Osmolovskii, T.I.Galkina, A.Yu.Klokov, E.V.Pechen', O.B.Pokrovskii and D.O.Saks, Fiz.Niz.Temp 17, 1470 (1991)
2. M.M.Bonch-Osmolovskii, T.I.Galkina, A.Yu.Klokov, A.F.Plotnikov and E.G.Romanov, Sverkhprovodimost' Fiz.Khim.Tekhn. 5, (1992) in print.
3. E.T.Shwartz and R.O.Pohl, Rev.Mod.Phys. 61,605(1989)

Anomalous Behaviour of the Phonon Heat Flux in NaCl–NaCl, KBr–KBr and KBr–Cu Point Contacts

A. Feher[1], P. Stefányi[1], R. Záboj[1], and A.G. Shkorbatov[2]

[1]Department of Experimental Physics, Šafárik University, Angelinum 9, 041 54 Košice, Czechoslovakia

[2]Institute for Low Temperature Physics and Engineering, Ukrainian Academy of Sciences, 310164 Kharkov, Ukraine

Point contacts (PCs) with dimensions smaller than the scattering lengths of charge and heat carriers have been intensively investigated in the recent years. First the conductive PCs with ballistic electric conductivity were studied [1,2]. The observed nonlinearity of the current-voltage characteristics of such contacts which is due to weak inelastic electron scattering is the basis of the method of PC spectroscopy of phonon state density in metals.

The PC is a unique physical device which enables to realize a strong nonequilibrium of electron and phonon systems. Ballistic transport in PCs is realized by practically noninteracting groups of carriers coming from the opposite banks of a PC. Near the junction no equilibrium temperature can be reached and the fluxes cannot be calculated in terms of the traditional thermodynamics of nonequilibrium processes.

Near the PC the ballistic acceleration of phonons occurs and therefore the heat flow through the single PC is not associated with appreciable change of temperature near the PC [3]. Therefore heat is transfered by groups of phonons whose distribution functions depend on the massive banks temperatures (T_1 and T_2). It is important to note that it is not necessary to place the thermometers in the immediate vicinity of PCs.

The low temperature dependence of the heat flux $\dot{Q}(T_1,T_2)$ is similar to that obtained by Little [4] who dealt with thermal resistance due to acoustic mismatch between the two media.

$$\dot{Q}(T_1,T_2) = K\,(T_2^4 - T_1^4).$$

If both banks of a PC are of the same material and phonons are not reflected by the PC boundary, the coefficient K is determined by the PC area A_o and the group velocities s^j (j is the phonon mode, k_B is the Boltzmann constant):

$$K = (\pi^2 A_o k_B^4/120\hbar^3)\sum_j (s^j)^{-2}.$$

As is shown by model calculations for two contacting fcc lattices [5], the simple law T^4 is valid for the heat flux even

Springer Series in Solid-State Sciences, Volume 112
Phonon Scattering in Condensed Matter VII Eds.: M. Meissner and R.O. Pohl

in the near-Debye temperature range. However, the thermal conductivity of real PCs can depend on temperature in a more complicated way. The review article published by Swartz and Pohl [6] leads to the conclusion that impurities or lattice distortions near the PC boundary can result in the T^{ε} law, where $\varepsilon<4$.

We realized the pressure-type contacts (a sharply pointed needle made of NaCl, KBr or Cu was pressed onto the surface of another NaCl or KBr crystal) with characteristic dimensions ranging from 50 to 500 nm. We studied the temperature dependence of the heat flux through NaCl-NaCl, KBr-KBr and KBr-Cu PCs in the temperature range from 2 K to 80 K. Most of PCs show ballistic heat transport with the heat flux proportional to T^4

In the $\dot{Q}(T_1, T_2)/(T_2^4 - T_1^4)$ vs. T_2 coordinates the heat flux demonstrates significant deviations from the simple model of ballistic transport. We found out that for every PC (NaCl-NaCl, KBr-KBr and KBr-Cu) the temperature dependence of the heat flux in these coordinates shows a maximum below 15 K (Fig. 1).

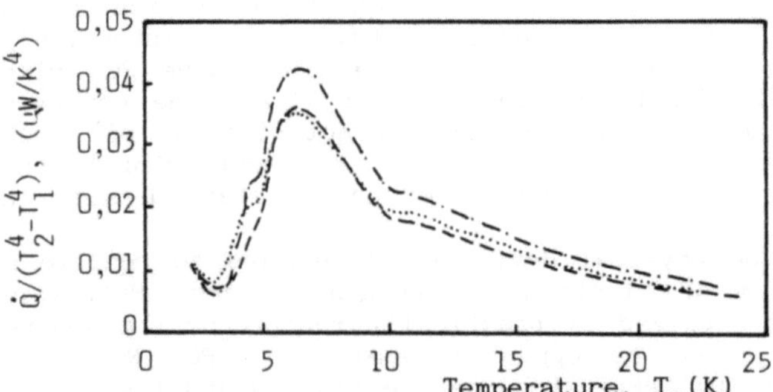

Fig. 1 Temperature dependence of the heat flux through PC for three KBr-Cu PCs

We propose a fairly acceptable explanation of this anomalous behaviour of the flux. We assume that on the contact surface, in addition to regions with strong coupling (perfect lattice), there are large regions with weak coupling (contact through impurity interlayers). We further assume that the presence of a layer of loosely-bound impurities at the interface (probably H_2O) can create such regions with weak coupling . Then, the heat flux through the PC may be divided into two parallel parts: dominant ballistic heat flux described by the T^4-law and an additional heat flux through the regions with weak coupling, this flux having resonance character. However, the interpretation of experiments below 10 K on such complex systems (a PC can consist of many smaller PCs) requires consideration of processes whose characteristic wave-length is comparable to the size of the particular PC.

References

[1] J.V.Sharvin, Zh. Exp. Theor. Fiz. 48, 984 (1965).
[2] I.K.Yanson, Zh. Exp. Theor. Fiz. 66, 1035 (1974).
[3] P.Stefányi, A.Feher, A.G.Shkorbatov, Fiz. Niz. Temp. 18, 154 (1992).
[4] W.A.Little, Canad. J. Phys. 37, 334 (1959).
[5] D.A.Young, H.J.Maris, Phys. Rev B 40, 3685 (1989).
[6] E.T.Swartz, R.O.Pohl, Rev. Mod. Phys. 61, 605 (1989).

Simple Method for Heat-Flux Investigation Through Point Contacts Between Two Solids

P. Stefányi and A. Feher

Department of Experimental Physics, Šafárik University,
Angelinum 9, 041 54 Košice, Czechoslovakia

1. Introduction

The tendency to miniaturization of electronic devices, which has stimulated progress in microelectronic, has led to the advent of very-large-scale integration devices whose elements have linear dimension approaching 100 nm. If such microcircuits representing various types of contacts operate at liquid nitrogen temperature, their geometric dimension can become comparable to or be even smaller than characteristic electron and phonon scattering lengths. Therefore the effects arising in such contacts, i.e. point-contacts, between two solids, as for example ballistic heat transport, are the objects of an intensive study.

In this paper we have proposed a new method of measuring the thermal flux in dielectric-dielectric and metal-dielectric PCs through which heat is transported by ballistic phonons. We found the inspiration in the relaxation method for measuring the heat capacity and also in theoretical works [1,2] discussing the phonon ballistic regime in PC.

2. Experimental set-up

The experimental set-up is shown in Fig. 1. The PC was realized between two single-crystals (or between single-crystal and copper needle in special cases): the lower single-crystal was kept in quasiadiabatic conditions, while the upper single-crystal was glued to a copper plate whose temperature was controlled between 2 K and 5 K via a copper thermal link with the liquid helium bath.

Using an external mechanism it was possible to control the pressure force (and thus the size of PC) between samples. To create quasiadiabatic conditions the samples were placed in a vacuum better than 10^{-4} Pa and surrounded by a thermal shield. The lower sample was suspended on four nylon fibers 0.08 mm thick. Electrical leads to the thermometer and heater were thermally anchored to the thermal shield.

Heat flux through the PC was measured by the following method. Using the electrical heater the lower bank of the PC was heated to

Springer Series in Solid-State Sciences, Volume 112
Phonon Scattering in Condensed Matter VII Eds.: M. Meissner and R.O. Pohl

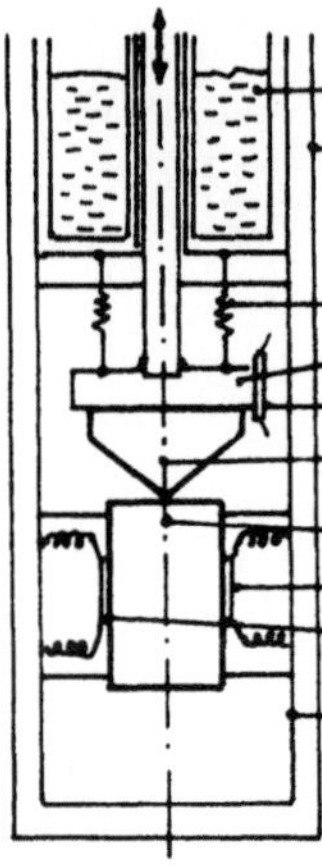

Fig. 1 Experimental set-up

the temperature of about 40 K to 80 K and the cooling curve $T_2(t)$ of this bank was measured down to the temperature of the upper bank. This "on-line" measurement took 2 hours to 30 hours, depending on the PC size. This experimental dependence was polynomically smoothed out to the second order through the choosen number of points (typically from 20 to 60) and numerically derivated. The temperature dependence dT_2/dt as a function of T_2 was calculated for the regimes with and without contact. The dependence for the regime without contact, which is determined by parasite thermal flows through the electrical leads, suspension, rest gas and due to radiation, was substracted from the dependence for the regime with contact. In our experimental arrangement the heat flux through the PC was

$$\dot{Q}(T_2,T_1)=C(T_2)dT_2/dt,$$

where C(T) is the heat capacity of the lower crystal.

3. Conclusion

To summarize, the proposed method enables to measure, in a simple way and with great sensitivity (<1.0%), the heat flux through a PC. Advantages of this method are:

- ability to study the contacts between different materials, such as dielectric-dielectric, dielectric -metal etc;
- ability to control the size of the contact without removing the sample from the cryostat;
- ability to obtain very high heat flux densities ~ 10^7 W/m^2K to 10^9 W/m^2K, at which the bulk sample would evaporate;
- ability to realize a strongly non-equilibrium system, possibly with non-linear effects, and systems with weak coupling;
- this method does not "record" the scattering processes taking place in the bulk material and enables to study the thermal boundary resistance at relatively high temperatures.

The main disadvantage of this method may be seen in the fact that the contact itself is not always well defined and its morphology is unknown.

4. References

[1] E.N.Bogachek, A.G.Shkorbatov, Sov.J.Low Temp.Phys.11,353 (1985)
[2] A.G.Shkorbatov, T.Z.Sarkisyants, Sov.J.Low Temp.Phys. 16,427 (1990).

Phonon Quenching in Splat Cooling

G. Barault, A. Khater, J.M. Greneche, Y. Labaye, and N. Randrianantoandro

Laboratoire de Physique des Matériaux, U.R.A. No. 807 CNRS, Université du Maine, F-72017 Le Mans Cédex, France

The interest in materials rapidly quenched from the melt to the solid state, is well known. In particular splat cooling [1] is one of the favorite techniques used to make wafers of amorphous metals destined to industrial use as well as for basic research [2]. The Miyazawa and Szekely paper, [3], raises the level of sophistication at which the quenching process has been considered, and like others makes the assumption that the heat transfer at the substrate-sample interface is Newtonian.

The purpose of this paper is to analyse the consequences of a heat transfer mechanism by exchange of phonons, [4], at the substrate-sample interface. Using some elements of the Miyazawa and Szekley model, we have developed a theoretical description of the splat that lends itself to the phonon exchange mechanism.

The problem of heat transfer at the free surface of a geometrically static semi-infinite solid has been treated previously, [5]. We adapt this model to splat cooling by introducing kinetic scaling, and the appropriate boundary conditions at the substrate-sample interface,

$$- k \frac{\partial T}{\partial z} \bigg|_{\text{interface}} = h(T^m - T^{*m}) \quad , \qquad (1)$$

where h is the heat transfer coefficient, and m = 1 and m = 4 correspond to Newtonian and phonon cooling, respectively. T* the substrate temperature, is taken constant.

Springer Series in Solid-State Sciences, Volume 112
Phonon Scattering in Condensed Matter VII Eds.: M. Meissner and R.O. Pohl

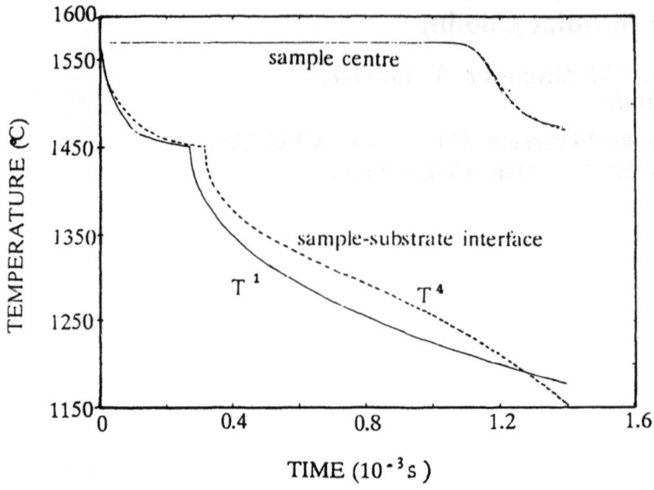

We present in Fig. 1 the variation of the temperature T as a function of time during the splat, at the centre of the sample and at the sample-substrate interface.

The model presented is relatively simple. However, it does give information as regards the differences between the Newtonian hypothesis, and the more realistic phonon description, for cooling at the substrate-sample interface.

References

1. **J. DUWEZ, R.H. WILLENS, and W. KLEMENT,** J. Appl. Phys. 31 (1960) 1136
2. **H. JONES,** Rep. Prog. Phys. 36 (1973) 1425
3. **K. MIYAZAWA and J. SZEKELY,** Metall. Trans. B 10B (1979) 349
4. **A. KHATER and J. SZEFTEL,** Phys. Rev. B 35 (1987) 6749
5. **J.C. JAEGER,** Proc. Camb. Phil. Soc. 46 (1950) 634

Anomalous Absorption of Bulk Phonons by a Two-Dimensional Transient Layer

Yu.A. Kosevich

Instituto de Ciencia de Materiales, E-28006 Madrid, Spain

It is known from the theory of waves in layered structures that a thin nonabsorbing layer on the free surface of a solid results only in a small change in the phase of a total reflection of bulk acoustic waves. The acoustic absorption in the layer leads to the variation (with the wave frequency) also of the reflection amplitude. When the layer posesses both the resonant and absorbing properties the phenomenon of anomalous absorption of bulk acoustic waves by a two-dimensional layer can take place. This phenomenon is characterized by an absence of the reflection and by an almost complete surface absorption of the incident wave. The origin of this effect is a strong enhancement of the resonant oscillations amplitude of a light and weakly bonded thin layer, which in the presence of the finite and relatively low absorption in the layer leads to an anomalously large "absolute" value of acoustic losses of the incident bulk wave. The width of the resonant absorption curve is proportional to the ratio between the effective elastic impedance of a two-dimensional layer and the acoustic impedance of a solid. This effect can be experimentally revealed on two-dimensional layers of helium (or other noble gases), adsorbed on a solid (see, e.g., [1]), or on two-dimensional polymer or liquid-crystal Langmuir-Blodgett adsorbed films (see, e.g., [2]). When a weakly bonded resonant two-dimensional absorbing layer is placed at the interface between two different crystals the phenomenon of anomalous surface absorption can take place for the incident bulk wave in the conditions of a total reflection (in the absence of the transient layer). In this case the phenomenon is characterized by an almost complete absence of both reflected and transmitted bulk waves. All vibrational energy of the incident acoustic wave is absorbed at the interface, and then is reemitted in the adjacent media via the thermal phonons. Such mechanism can contribute to the Kapitza thermal conductance across the interface, because it involves in the process the phonons, incident at the interface outside the critical cone. The anomalous interface absorption takes also place for the incidence of the first sound wave in superfluid helium at the high mobile interface with a solid helium at longitudinal critical angle at 1.3 K. Two-dimensional intermediate layer can also cause a total resonant transmission of phonons across

Springer Series in Solid-State Sciences, Volume 112
Phonon Scattering in Condensed Matter VII Eds.: M. Meissner and R.O. Pohl

the interface between two media in the case of their strong acoustic mismatch [3]. The origin of this phenomenon is the increase of the coupling between media due to the high amplitude of resonant oscillations of a weakly bonded transient layer.

At the interface between two crystals (a rigid and a soft) with very different elastic properties the existence is predicted of intrinsic resonant modes with frequencies inside the vibrational continuum of a soft crystal. The origin of these modes is a structure of several atomic layers of a soft crystal, adjacent to a rigid crystal. Since the strengths of the interlayer forces, holding a soft crystal to a rigid one, are in general case larger than the bulk interlayer forces of a soft crystal, the first several atomic layers of a soft crystal, adjacent to a rigid crystal, are strongly compressed with respect to the bulk layers of the crystal. Therefore these atomic layers form one "heavy interface layer", which has relatively low resonant frequencies, determined by the strengths of the bonds with a rigid crystal of the first atomic layer of a soft crystal, and by a total mass of a "rigid" interface layer. Surface resonant modes with flat dispersion exist at the interface between helium (liquid or solid) and ordinary solid [4], and can also exist at the interface between diamond and a soft crystal, because diamond has the highest sound velocity and Debye temperature of any material. Recent experiments [5] revealed that at the interface between diamond and such soft metals with very low Debye temperatures as *Au* and *Pb* the measured Kapitza conductances are as much as 100 times larger than expected from the conventional acoustic-mismatch phonon theory, as in the case of a helium-solid interface. From these observations follows, in particular, that phonons above the highest frequency in the metal must substantially contribute to the energy transfer [5]. The nonlinear interaction of the bulk phonons with two-dimensional intrinsic resonant layer leads to the subharmonic transmission across the interface of the phonons above the highest frequency in a soft crystal, and to the generation of phonons with frequencies, multiple to the resonant frequency. These inelastic processes can substancially contribute to the Kapitza conductance between solids with very different vibrational spectra (such as diamond and soft crystal), and also to the thermal conductance across a helium-solid interface.

1. C.H. Anderson and E.S. Sabisky, Phys. Rev. Lett. **24**, 1049 (1970).
2. F. Nizzoli, B. Hillebrands, S. Lee, G.I. Stegeman, G. Duda, G. Wegner and W. Knoll, Phys. Rev. B **40**, 3323 (1989).
3. Yu.A. Kosevich and E.S. Syrkin, Fiz. Tv. Tela (St. Petersburg) **33**, 2053 (1991) [Sov. Phys. Solid State **33**, 1156 (1991)].
4. H.J. Laute, H. Godfrin, C. Tiby, H. Wiechert and P.E. Obermayer, Surface Sci. **125**, 265 (1983).
5. R.J. Stoner, H.J. Maris, T.R. Anthony and W.F. Banholzer, Phys. Rev. Lett. **68**, 1563 (1992).

Resonance Heat Transfer Through an Impurity Monolayer

E.S. Syrkin[1], *A.G. Shkorbatov*[1], *and A. Feher*[2]

[1]Institute for Low Temperature Physics and Engineering, Ukrainian Academy of Sciences, 310164 Kharkov, Ukraine
[2]Department of Experimental Physics, Šafárik University, Angelinum 9, 041 54 Košice, Czechoslovakia

Abstract. The flux of phonon energy between two solids is calculated, taking account the impurity monolayer separating the crystals. It is shown that weak impurity-crystal interaction results in a resonance-type frequency dependence of the phonon energy transmission coefficient. The calculation enabled to describe the low temperature heat transfer anomalies recently observed in point contacts.

The problem of ballistic phonon transport in point contacts (PCs) is important because the inelastic phonon-phonon scattering length (l) rapidly increases with decreasing temperature T so that at T ~ 10K contacts with diameters less than 100 nm can be regarded as PCs whose size is small as compared to l [1-3].

An anomalous increase of the PC thermal conductivity at low temperatures has been revealed [1-3]. In the present paper, such anomalies are interpreted as resulting from the presence of a layer of weakly coupled impurities at the PC boundary.

For ballistic phonon transport, considering phonon reflection and refraction by the PC boundary, the heat flux from the contact bank having the temperature T to the bank kept at the constant temperature T_o can be written as

$$\dot{Q}_B = \frac{\hbar A}{2(2\pi)^3} \sum_j \int d\mathbf{k}\, \omega_j(\mathbf{k})\, |s_z^j|\, D^j(\mathbf{k}) \left[N(\omega_j/T) - N(\omega_j/T_o) \right].$$

Here D(**k**) is the coefficient of phonon energy transmission through the PC boundary and $s = d\omega/d\mathbf{k}$; $N(x) = 1/[\exp(\hbar x)-1]$; the z axis is normal to the PC boundary, A is the contact area, and j is the phonon mode.

In this paper the heat transport in PCs was mathematically modelled on the basis of the discrete [4] and "capillary" [5] models of phonon transmission through monoatomic impurity layers. Such a system of impurity atoms is characterized by weakly dispersed low frequency vibrations with $\omega_o \sim d^{-1}(2C_{ii}/\rho)^{1/2}$, where

Springer Series in Solid-State Sciences, Volume 112
Phonon Scattering in Condensed Matter VII Eds.: M. Meissner and R.O. Pohl

d is the thickness of the interlayer, ρ is its density and C_{ii} are the elastic constants of the interlayer. In the resonance range ($\omega \sim \omega_0$) for non-slipping angles of incidence the acoustic wave is fully transmitted through the plane defect ($D(\omega) \sim 1$). A similar effect appears for sufficiently long waves.

The figure shows the results of calculation of the reduced heat flux due to transmission of longitudinal waves. In the figure a maximum at T ~ 6K and the beginning of an increase at T ~ 2K are well seen. The latter is due to the fact that for phonons with maximum wavelengths D = 1. Note that such maxima can be induced by any of the phonon modes.

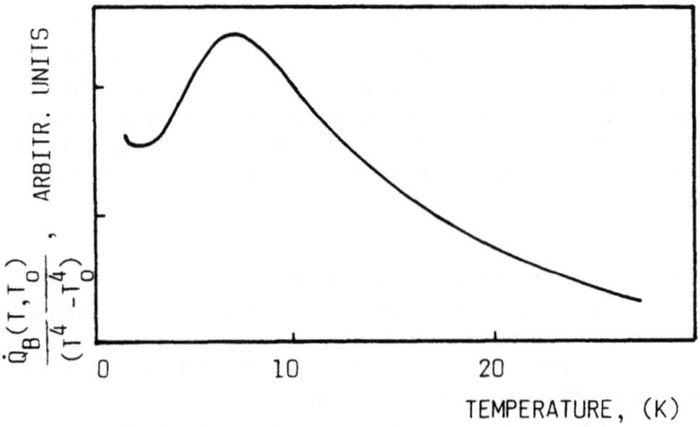

Fig. 1 The reduced heat flux of a KBr-KBr PC for the resonance model; $T_o = 2K$, $\omega_o = 3.0\ 10^{12}$ rad/s

References

[1] P. Stefányi, A. Feher, A. Orendáčová, Phys. Lett. A 143, 259 (1990); P. Stefányi, A. Feher, Physica B165-6, 911 (1990).
[2] P. Stefányi, A. Feher, A. G. Shkorbatov, Fiz. Niz. Temp. 18, 154 (1992).
[3] A. Feher, P. Stefányi, R. Záboj, A. G. Shkorbatov, this conference (1992).
[4] E. S. Syrkin, I. M. Gelfgat, Fiz. Niz. Temp. 12, 529 (1986); ibid. 4, 141 (1978).
[5] Yu. A. Kosevich, E. S. Syrkin, Fiz. Tv. Tela 33, 2053 (1991); Phys. Rev. B, to be publ. (1992).

Part IX

Films, Surfaces, and Interfaces

Phonon Localization in Mesoscopic Systems

N. Perrin

Laboratoire de Physique de la Matière Condensée de l'Ecole Normale Supérieure and Université Pierre et Marie Curie, 24 rue Lhomond, F-75231 Paris Cédex 05, France

An intense activity has been devoted to electron localization in thin wires and films, but the thermal analogue has received much less attention. Phonon states may be expected to be localized in the Anderson sense [1] in some particular cases such as one dimensional disordered solids [2], or two dimensional lattices with mass disorder [3].

The question we wish to address here is the possibility of observing phonon localization in two-dimensional dielectric films. We consider the phonon scattering mechanisms that can be at the origin of phonon localization and discuss the inelastic scattering processes in order to compare the inelastic mean free path ℓ_i to the localization length deduced from elastic scattering processes. We consider thin films of thickness d = 200 Å, lengh L and width ℓ; the phonon spectrum is spatially quantified in the thickness direction for free standing films and the phonon transport behave as in a two-dimensional medium. Moreover we consider films with real surfaces i.e. surfaces exhibiting defects and roughness on the usual scale of hundreds of angstroms. The scattering of phonons at the lateral surfaces may be either specular or diffuse. We introduce localized excitations in the form of two level systems (TLS) to represent the inelastic diffuse boundary scattering at these surfaces [4]. Following the extension of the work of Thouless [5] to the case of phonons by Jäckle [2,6] we define a localization length $\xi \leq L$ by

$$\hbar\left(\frac{\xi^2}{D_e}\right)^{-1} = \hbar\left(n_{2d}(\omega)\xi\ell\right)^{-1}.$$

With the phonon density of states per unit area $n_{2d} \propto \omega$, the localization time τ_ξ can be written

$$\tau_\xi = \frac{\xi^2}{2D_e} = \frac{3}{16\pi^2}\frac{\ell^2}{v^3}(\ell_e\omega^2)$$

where v is a mean phonon velocity and the diffusion constant D_e is determined by the elastic mean free path ℓ_e. Apart from the elastic boundary scattering that results in a mean free path $\ell_{eb} = \alpha\ell$ ($\alpha > 1$), Rayleigh scattering must be considered : in two dimensional systems, the isotopic scattering rate is $\tau_{2D}^{-1} = A\omega^3$. Including these two processes in τ_ξ yields :

Springer Series in Solid-State Sciences, Volume 112
Phonon Scattering in Condensed Matter VII Eds.: M. Meissner and R.O. Pohl

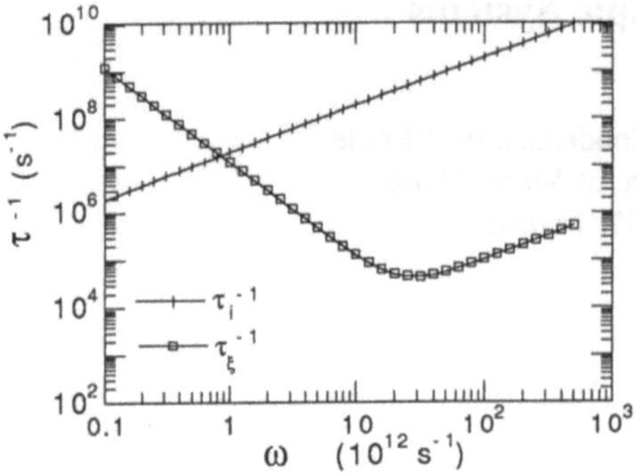

Fig. 1 : Inverse time τ_ξ^{-1} characterizing the localization length and inelastic scattering rate τ_i^{-1}.

$$\tau_\xi = \frac{3\,\ell^2}{16\pi^2 v^2}\,\omega^2\,\frac{1}{A\omega^3 + v/(\alpha\ell)} .$$

In order to get localization, we must have $\tau_\xi < \tau_i = \ell_i/v$.

Three phonon scattering processes are quite inefficient in a two-dimensional film; indeed, Lamb acoustic waves [7] have to be considered and they are expected to give decay phonon rates smaller than the decay rates obtained for Rayleigh waves by Tamura [8] i.e. macroscopic mean free paths. The resonant phonon absorption by the TLS yields :

$$\tau_{TLS}^{-1} = \frac{3\pi\gamma^2\bar{n}(E)}{\rho v^2}\,\omega$$

where ρ, γ and $\bar{n}$ ($E = \hbar\omega$) are respectively the mass density, the coupling constant and an effective density of states. At T = 1 K, we can estimate $\tau_{TLS}^{-1} = 6\pi\,10^{-6}\,\omega$ (with $\gamma^2\bar{n}(E)/\rho v^2 = 4\times10^5/2\times10^{11}$); with $A = 17{,}7\;10^{-32}\;s^3$ (the constants for Si have been used) and $\ell = 1\;\mu m$, $\alpha = 3$, we obtain the curves in Fig.1. It is seen that localization does exist for frequencies $\omega < \omega_0$ with $\omega_0 \sim 10^{12}\;s^{-1}$. The localization results from both the partially specular reflexion at the lateral surfaces and the surface defects : it is characteristic of the surface condition.

References

[1] P.W. Anderson, Phys. Rev. 109, 1492 (1958).
[2] K. Ishii, Prog. Theor. Phys. Suppl. 53, 77 (1973); J. Jäckle, Solid State Commun. 39, 1261 (1981).
[3] M.L. Williams and H. Maris, Phys. Rev. B 31, 4508 (1985).
[4] E.T. Swartz and R.O. Pohl, Rev. Mod. Phys. 61, (1989).
[5] Thouless, Phys. Rev. Lett. 39, 1167 (1977).
[6] We thank M.N. Wybourne for interesting and helpful discussions on the paper by Jäckle in [2].
[7] We thank A. Zarembovitch and B. Perrin for essential comments on Lamb waves.
[8] S. Tamura, Phys. Rev. B 30, 610 (1984).

Phonon Transmission and Localization in Random Superlattices

N. Nishiguchi and S. Tamura

Department of Engineering Science, Hokkaido University, Sapporo 060, Japan

Anderson localization is a general wave phenomenon due to the interference effect among multiply scattered waves and not unique to disordered materials at low temperatures where quantum effects are important. We study the Anderson localization of phonons in a synthetic random superlattice(SL) by analytically calculating the transmission rate and its fluctuations. We also compare the theoretical results with the numerical simulations based on the transfer matrix method[1,2]. The excellent coincidence is attained.

We consider the transverse acoustic wave which is normally incident on the SL from the substrate and detected in a layer on top of the SL. The constituent layers of the SL are AlAs (A) and GaAs (B) (thicknesses are 34Å for both layers). Those layers are stacked at random with the same occurrence probability. As the stiffness constant of AlAs is close to that of GaAs, we approximate the stiffness constants of the both layers by their averaged value. The phonon scattering is then attributed to the mass density fluctuation.

We formulate the phonon transmission in terms of the Green's function method[3]. As the system is essentially one-dimensional, the Green's function is calculated by taking account of the forward and backward scattering processes due to the mass density fluctuation. The contribution of all forward scattering processes can be summed up and the mass density fluctuation is renormalized into the phase-modulated complex field. The Green's function can now be expressed only by the backscattering due to the renormalized mass density fluctuation.

The transmission rate derived is a function of the ratio $t=L/\ell$ of the system size L to the mean free path given by $\ell=c\,\tau$. Here c is the sound velocity and $1/\tau$ is the backscattering rate in the Born approximation. We note that the scattering rate is proportional to the ensemble averaged squared structure factor, $\langle |S|^2 \rangle$, which can be derived analytically.

Figure 1 shows the frequency dependence of the theoretical averaged transmission rate $\langle T \rangle$ up to 1.3THz for the random SL consisting of 1500 basic layers together with the numerical data (open circles). The transmission rate exhibits the dips and peaks alternatively. The dips are attributed to the phonon localization in the randomly layered system. The transmission peaks are due to the resonances occurring when the wavelength becomes an integer-multiple of the thickness of the basic layers A or B.

The localized states of phonons are characterized by the Lyapunov exponent γ which is the inverse of the localization length ξ. The Lyapunov exponent is found to be $\gamma=1/2\ell$. Figure 2 shows the theoretical Lyapunov exponent γ together with the numerical data which are obtained for the random SL's with 1500 layers. The

Springer Series in Solid-State Sciences, Volume 112
Phonon Scattering in Condensed Matter VII Eds.: M. Meissner and R.O. Pohl

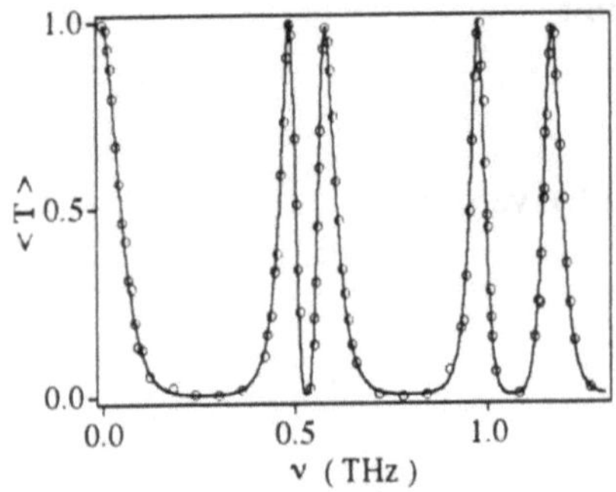

Figure 1. Frequency dependence of the average transmission rate of transverse phonons in the random superlattices with 1500 layers. The analytical result is plotted by the solid line and the numerical data are indicated by open circles. The numerical data are averaged over 150 different random configurations of layers.

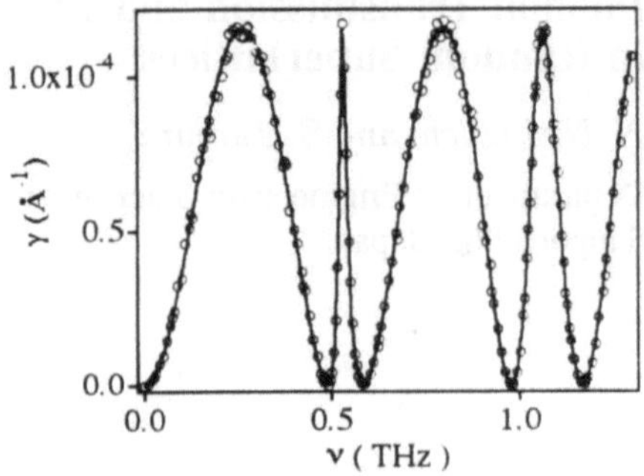

Figure 2. Frequency dependence of the Lyapunov exponent γ $(=\xi^{-1})$ in the random superlattices. The solid line is the theoretical result and open circles plot the numerical data. The numerical calculations are made under the same condition as Fig.1.

Lyapunov exponent γ oscillates regularly and has the maxima occurring periodically at the Bragg frequencies for phonons in the periodic SL consisting of A and B layers. The Lyapunov exponent γ vanishes at the resonance frequencies $\{\nu_{res}\}$ because the localization length diverges, and γ varies as $|\nu - \nu_{res}|^2$ in the vicinity of the resonance frequency ν_{res}, which is attributed to the Rayleigh scattering of phonons in one dimension.

For the random SL's with a finite size, phonons near the resonance frequency are considered to be extended. However, the transmission of phonons is still affected by the interference among the scattered waves and then fluctuates. The standard deviation of the transmission rates can also be calculated analytically. The resultant transmission fluctuation depends only on the scaling parameter t as well as the transmission rate. Therefore, the transmission fluctuation is regarded as a function of the averaged transmission. Figure 3 exhibits the theoretical standard deviation of transmission ΔT versus $\langle T\rangle$. Each dot denotes the numerical data obtained for 150 random superlattices. The transmission fluctuation has a maximum 0.28 at $\langle T\rangle \approx 0.40$ in the extended region. (The critical transmission rate T_{CR} is estimated from $L/\xi=1$ and is found to be $T_{CR}=0.27$ independent of the system size.)

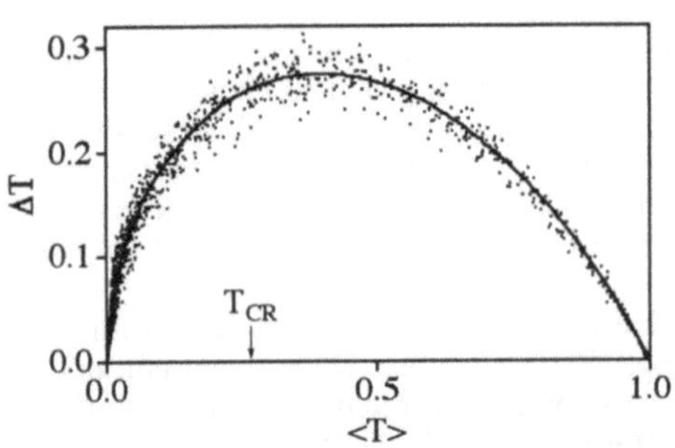

Figure 3. Fluctuations of phonon transmission versus average transmission. The solid line shows the analytical result and dots are the numerical data obtained from the transfer matrix method. Each dot is obtained by averaging over an ensemble of 150 superlattices.

It should be remarked that the excellent agreement of theoretical results with the numerical simulations is attained because in our calculations the phase shift of phonons associated with the forward scattering has exactly been taken into account.

[1] S. Tamura and F. Nori, Phys. Rev. B**41**, 7941 (1990).
[2] N. Nishiguchi and S. Tamura, (to be published).
[3] A. A. Abrikosov and I. A. Ryzhkin, Adv. in Phys.**27**,147(1978).

Third Sound Transmission Through Non-Periodic Lattices

K. Kono

Institute for Solid State Physics, University of Tokyo,
Roppongi, Tokyo 106, Japan

Recently, phonon propagation in modulated media has been receiving considerable attention. [1,2] Such structures can be realized recently in a semiconductor material such as GaAs/GaAlAs employing an modern MBE (Molecular Beam Epitaxy) technique. The most interesting aspects of these phenomena arise from the interference between multiply scattered partial waves. Even in random systems, the recent understanding of the wave localization is based on the interference, which had been believed to be completely absent. In order to observe these effects, it is important to preserve the phase coherency during the multiple scattering. In this sense the sound waves with little attenuation are favorable for the study of these phenomena.

Third sound is a wave propagating over the surface of superfluid helium films [3]. The attenuation is small. Since it propagates over the surface, the modulation of the media is easy and controllable. Recently we have developed an experimental method to measure the transmission spectra of third sound in one-dimensional lattices. The modulation of the surface was done with aluminum strips (width $d \sim 80\mu m$) which were periodically arrayed on a glass substrate. Periodic [4], Fibonacci [5], Thue-Morse [6], and random [7] lattices were investigated so far.

The Fibonacci sequence is generated by the following recurrence formula:

$$S_{n+1} = \{S_{n-1}, S_n\}, \text{ with } S_1 = \{B\} \text{ and } S_0 = \{A\}.$$

For example, $S_2 = \{AB\}$, $S_3 = \{BAB\}$, and $S_4 = \{ABBAB\}$. We have fabricated the Fibonacci lattice by mapping A to an aluminum strip on a glass substrate, whereas B to the bare surface of the same width. The Thue-Morse sequence is generated also by the similar formula as that for the Fibonacci sequence, which is neither periodic nor random.

Figure 1 shows the transmission spectra of third sound in these lattices, where k is the wave number of third sound. In the periodic lattice a transmission gap and bands are observed clearly. In the Fibonacci lattice the transmis-

Springer Series in Solid-State Sciences, Volume 112
Phonon Scattering in Condensed Matter VII Eds.: M. Meissner and R.O. Pohl

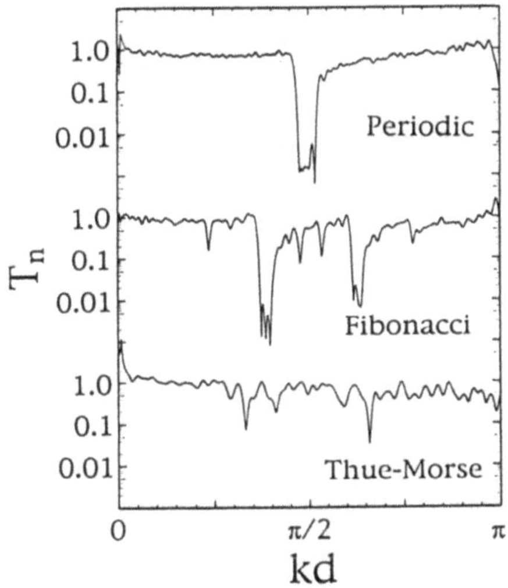

Figure 1: Transmission spectra of third sound in periodic, Fibonacci, and Thue-Morse lattices.

sion gap splits in two, whereas the bands are eroded away further by smaller gaps. The positional relation between these structures are characterized by golden ratio. Thue-Morse lattice shows weaker structures, implying the more homogeneous system than the former two.

The author is grateful to the Mitsubishi Foundation.

References

[1] R. Merlin, K. Bajema, J. Nagle, and K. Ploog, J. Phys. (Paris) **48**, C5-503 (1987).

[2] D.C. Hurley, S. Tamura, J.P. Wolfe, K. Ploog, and J. Nagle, Phys. Rev. B **37** 8829 (1988).

[3] K.R. Atkins and I. Rudnick, *Progress in Low Temp. Phys. VI ed.*, C.J. Gorter (North-Holland, Amsterdam, 1970) p. 37.

[4] K. Kono, S. Nakada, and Y. Narahara, J. Phys. soc. Jpn **60**, 364 (1991).

[5] K. Kono, S. Nakada Y. Narahara, and Y. Ootuka, J. Phys. soc. Jpn **60**, 368 (1991).

[6] K. Kono, S. Nakada, and Y. Narahara, J. Phys. soc. Jpn **61**, 173 (1991).

[7] K. Kono and S. Nakada, submitted to Phys. Rev. Lett.

Coherent Phonon Scattering by Submicron Structures

P.D. Vu and R.O. Pohl

Laboratory of Atomic and Solid State Physics, Cornell University,
Ithaca, NY 14853-2501, USA

1. Introduction

Using densely packed monolayers of Latex spheres as masks, we have evaporated two–dimensional regular arrays of metal dots onto polished silicon wafers[1]. Performing thermal conduction experiments in the boundary scattering regime[2], we observe peak phonon scattering when the dominant phonon wavelength in the silicon equals the spacing of such a diffraction grating on the silicon surface[3].

2. Experimental Matters

We use high–purity, n–type (111) cleaved silicon wafers as samples and measure a rough sandblasted sample, a smooth Syton polished sample, and samples with 300Å Au on 30Å Cr on polished silicon. The thermal conductance of these samples, in the temperature range from 5K down to 50mK, yields an inverse phonon mean free path which is proportional to the phonon scattering rate.

3. Results

Figure 1 shows the inverse phonon mean free path versus the average sample temperature. The top scale shows the dominant phonon wavelength in silicon. The solid triangles are scattering data for the rough sandblasted sample, in agreement with the Casimir prediction (solid line). The solid circles are data for the smooth Syton polished silicon sample. The solid squares and open circles are data for 300Å Au on 30Å Cr on polished silicon with the solid squares being the scattering of a continuous metal film covering 50% of the silicon surface and the open circles being the scattering of a metal grating of spacing 0.5μm covering 10% of the surface. Eventhough the grating covers less of the silicon surface than the film does, its scattering of phonons is greater than that of the film at temperatures well below 1K. In particular, when the dominant phonon wavelength is comparable to the grating spacing, the scattering reaches a peak. In the short wavelength limit, above 1K, the

Springer Series in Solid-State Sciences, Volume 112
Phonon Scattering in Condensed Matter VII Eds.: M. Meissner and R.O. Pohl

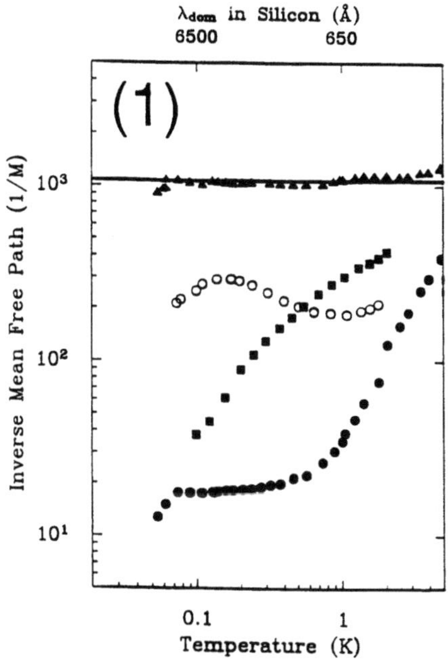

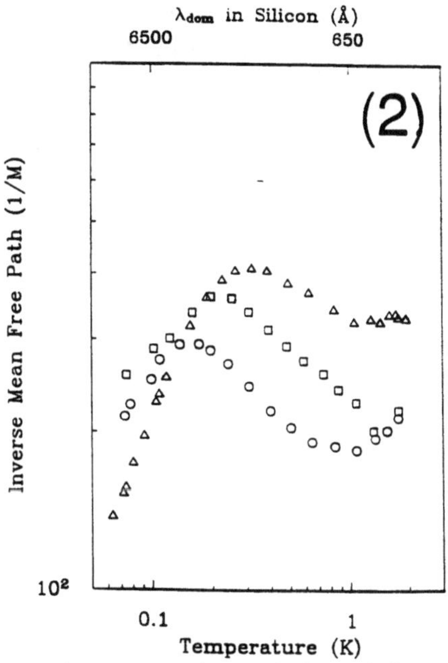

Figure 1 Phonon scattering data for four silicon samples with four different surface treatments. Symbols and results are discussed in the text.

Figure 2 The effect of changing grating spacing. See text for discussion.

scattering is geometric; and so, 10% coverage should produce less scattering than 50% coverage, as is the case. Figure 2 shows scattering profiles for 0.5μm (open circles), 0.25μm (open squares), and 0.125μm (open triangles) metal gratings, again with 300Å Au on 30Å Cr on polished silicon. Note that the peak in the scattering shifts accordingly. The 0.125μm data does not reach the same geometric limit as the other two for the following reason. While the 0.5μm and 0.25μm gratings each covers about 10% of the silicon surface, the 0.125μm grating has a high number of lattice defects of the kind that increases the surface coverage to much more than 10% but still less than 50%.

We have shown strong evidence for coherent scattering of thermal phonons from a two–dimensional diffraction grating. With further measurements, we hope to exploit the utility of this spectrometer in studying basic phonon physics.

This research was supported by the NSF, Grant No. DMR–91–15981 and by the U. S. Department of Education, Graduate Fellowship Program in Material Physics, Grant No. P200A10148.

[1] C.B. Roxlo et al., Science **235**, 1629 (1982).
[2] T. Klitsner, R.O. Pohl, Phys. Rev. B **36**, 6551 (1987).
[3] P.D. Vu et al., Bull. Am. Phys. Soc. **36**, 969 (1991).

Scattering Intensities Due to Acoustic Phonons Propagating Along the Axial Direction of Fiber Composites

R. Hotz and R. Siems

Theoretische Physik, Universität des Saarlandes,
W-6600 Saarbrücken, Fed. Rep. of Germany

We investigated an unbounded array of hard unidirectional infinite fibers in a soft matrix. As a basis for the interpretation of scattering experiments with $\underline{q} = (0,0,q)$ we calculated dispersion curves, normal modes and scattering intensities for various elastic constants, densities and fiber cross sections [1].

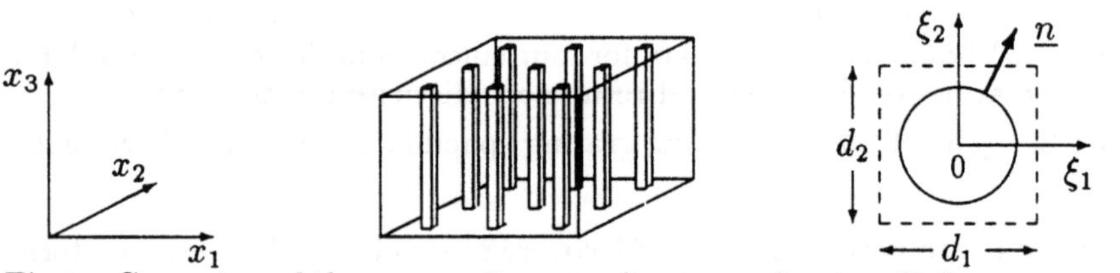

Fig.1. Geometry of the composite: coordinates and unit cell G.
$\xi_\nu = x_\nu - g_\nu d_\nu$: local coordinates with $|\xi_\nu| \leq d_\nu/2$; $g_\nu = 0, \pm 1, ...$; $\nu = 1,2$

Dynamical Structure Factor: For the scattering plane (010), phonon wave vector $\underline{q} = \underline{k}_I - \underline{k}_S$ and vertical-vertical polarized light the Brillouin intensity is proportional to the time Fourier transform of the autocorrelation function of $\delta\epsilon_{SI}(\underline{q},t) = \int_V \{P_{11}u_{2,2} + P_{12}(u_{1,1} + u_{3,3})\} exp(-i\underline{q}\underline{r}) d\underline{r}$. $P_{ik}(\underline{r})$ are Pockel's elasto-optical constants. The displacements may be expressed as superpositions of Floquet waves (with a simple plane wave dependence on x_3). $\underline{\tilde{u}}^j(\underline{k},\underline{r}) = e^{i(k_1 g_1 d_1 + k_2 g_2 d_2 + k_3 x_3)} \underline{e}^j(\underline{k},\xi_1,\xi_2)$ are the normal modes with frequencies $\Omega^j(\underline{k})$. The phonon polarization vectors $\underline{e}^j$ can be chosen such that $\int_G \rho(\underline{\xi})\, \underline{e}^j(\underline{k},\underline{\xi})\, \underline{e}^{j'*}(\underline{k},\underline{\xi}) d\underline{\xi} = d_1 d_2 \delta_{jj'}$ leading to the orthogonality relation for normal modes. The correlation function is obtained from the known thermal averages of the phonon occupation numbers. For large V one has

$$S(\underline{q},\omega) \sim \sum_j \frac{\hbar}{2\Omega_j(\underline{q})} \{(n_j(\underline{q})+1)\delta(\omega - \Omega_j(\underline{q})) + n_j(\underline{q})\delta(\omega - \Omega_j(\underline{q}))\} \mid J^j(\underline{q}) \mid^2,$$

$$J^j(\underline{q}) = \int_G e^{-i\underline{\xi}\underline{q}} \{P_{12}(\underline{\xi})\ (e^j_{1,1}(\underline{q},\underline{\xi}) + iqe^j_3(\underline{q},\underline{\xi})) + P_{11}(\underline{\xi}) e^j_{2,2}(\underline{q},\underline{\xi})\} d\underline{\xi} \qquad (1)$$

For $P_{ij} = const$ and $\underline{q} = (0,0,q)$, $J^j(\underline{q})$ reduces to the middle term.

Springer Series in Solid-State Sciences, Volume 112
Phonon Scattering in Condensed Matter VII Eds.: M. Meissner and R.O. Pohl

Dispersion Curves and Normal Modes: A variational procedure with independently varied displacements and stresses [2] is used for the eigenvalue problem in the unit cell, $\underline{r} = (\xi_1, \xi_2, x_3)$. With test functions $\varphi^{\underline{\alpha}} = e^{i(\underline{k}+2\pi\hat{\underline{\alpha}})\underline{r}}$, $\hat{\underline{\alpha}} = (\alpha_1/d_1, \alpha_2/d_2, 0)$, $\alpha_\nu = \{-M, .., M\}$ one obtains the approximations $\tilde{u}_l(\underline{r};\underline{k}) = \sum U_l^{\underline{\alpha}} \varphi^{\underline{\alpha}}$, $\tilde{\sigma}_{lm}(\underline{r};\underline{k}) = \sum S_{lm}^{\underline{\alpha}} \varphi^{\underline{\alpha}}$. The $3N = 9(2M+1)^2$ coefficients and the eigenvalue Ω^2 are solutions of $3N$ algebraic equations, the Galerkin equations, with k_3 as a parameter: $< \varphi^{\underline{\alpha}}, \tilde{\sigma}_{ij,j} + \rho\Omega^2 \tilde{u}_i >= 0$, $< \varphi^{\underline{\alpha}}, \tilde{u}_{i,j} - D_{ijkl}\tilde{\sigma}_{kl} >= 0$ (The brackets $<f, g>$ define the integral of $f \cdot g^*$ over the unit cell G and D_{ijkl} is the elastic compliance tensor). For symmetric fiber cross sections and $k_1 = k_2 = 0$ the system decomposes into two systems for modes with U_3 even, U_1, U_2 odd in $\underline{\alpha}$ and for modes with U_3 odd, U_1, U_2 even. We call the respective modes longitudinal and transverse in the mean.

Results: $S(\underline{q}, \omega)$ is nonzero only for the former. Except for the weight $P_{ij}(\underline{\xi})$ (it was taken constant) our dispersion diagrams could generally describe the one phonon contribution of inelastic coherent scattering experiments (cf. neutron results [3]). The low frequency region I where the composite behaves similar to a homogeneous elastic medium ends with Ω_0 (see the caption of Fig.2). In the adjacent range II the vibrations become nearly localized in the soft matrix between the rigid fibers. Since the scattering intensity is proportional to $(k_3/\Omega)^2$, it is larger in II than in I if the effective sound velocity in I is considerably larger than c_l. The scattering intensity depends also on

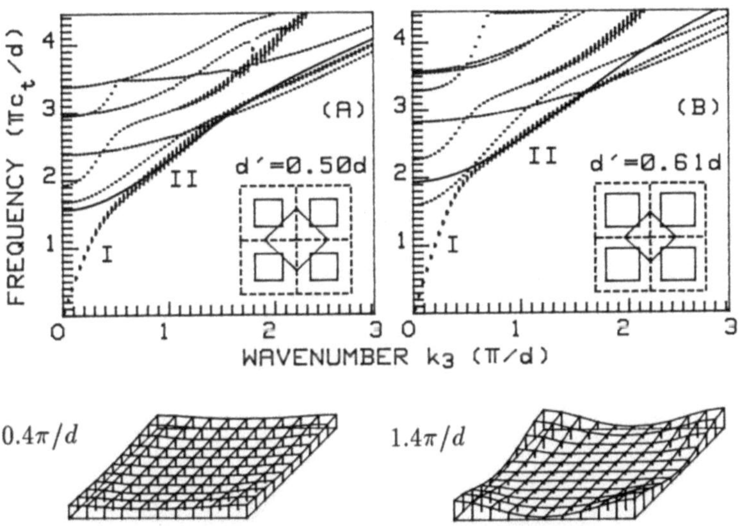

Fig.2. Dispersion relations and scattering intensities (hatching) for two ratios of fiber width d' and lattice constant d. Material constants: $\nu = \nu' = 0.3$, $\mu'/\mu = 60$, $\rho'/\rho = 3$. Matrix sound velocities: c_t, c_l. For comparison we show the lowest dispersion curve (full line) of a rod of matrix material clamped at its lateral surface with the cross section indicated in the insets. It starts as $\Omega^2 = c_l^2 k_3^2 + \Omega_0$, $\Omega_0 \approx 2.5 c_t/(d-d')$ (2).
Below: Lowest normal mode displacements $\underline{\tilde{u}}(\xi_1, \xi_2, x_3 = 0)$ of Fig.2B.

the active volume fraction. Although for very different elastic constants one may experimentally find only the hyperbola (2), not extrapolating to zero frequency.

1. R. Hotz, Thesis, Saarbrücken (1988)
2. S. Nemat-Nasser and M. Yamada, J. Appl. Mech., 48, 967 (1981)
3. H. Grimm, H. Stiller, C.F. Majkrzak, A. Rupprecht, V. Dahlborg, Phys. Rev. Lett. 59, 1780 (1987)

Relaxation and Propagation of Nonequilibrium High-Frequency Acoustic Phonons in Thin Crystalline Plates

A.A. Maksimov, D.V. Kazakovtsev, D.A. Pronin, and I.I. Tartakovskii

Institute of Solid State Physics, Russian Academy of Sciences,
142432 Chernogolovka, Moscow District, Russia

There is an important problem in the low temperature physics of solid state connected with the relaxation of nonequilibrium phonons. In the most practical situations the phonon system is not in the thermal equilibrium (it's usually so after interaction of the laser light with solid or after excitation by a current pulse). In this case space-frequency distribution function of the occupation numbers of phonons $n(t,\omega,r)$ may depends on r or even being spatially homogeneous, may deviate from the well-known Plankian function of ω. The way how the nonequilibrium phonons establish equilibrium temperature in the whole crystal space is one of the most interesting problem in phonon physics. Conditionally the study of this process may be reduced to the following two problems. The first is how does nonequilibrium frequency phonons distribution reach equilibrium one, this process is usually called relaxation. The second is that of propagation of the heat from excited region to cold parts of the crystal (heat pulses, heat conductivity).

To investigate phonon frequency relaxation we have measured the temporal evolution of the occupation numbers of high frequency nonequilibrium phonons in a thin anthracene crystal after intense laser pulse ($14cm^{-1} \leq \omega \leq 32cm^{-1}$, $0.5ns \leq t \leq 10ns$). We have used an optical method with a frequency selectivity and high temporal resolution [1] to detect acoustic phonons. We made also a quantitative comparison with the results of model-based calculations carried out in the "quasi continuous"-phonon-spectrum approximation. Good agreement was achieved, that allowed us to estimate accurately the lifetime of the acoustic phonons. This model describing the phonon system kinetics due to three-phonon anharmonic processes (decays and coalescences) was theoretically investigated also. By means of this model we investigated numerically the process of equilibrium formation after an essentially nonequilibrium initial phonon distribution being created. The results obtained were compared with the scaling solution used for such kind of non-linear equations. The validity range of the scaling approximation was determined [2].

Another interesting phenomenon, concerning phonon propagation, is so-called "hot spot", that is a small part of a sample

Springer Series in Solid-State Sciences, Volume 112
Phonon Scattering in Condensed Matter VII Eds.: M. Meissner and R.O. Pohl

where the temperature significantly exceeds that of the rest regions of the sample volume. We investigated the degradation of a hot spot obtained near one of the crystal surfaces of anthracene at helium temperatures. The experimental results were discussed in terms of the local thermal conductivity with both specific heat and thermal conductivity as well-known functions of temperature. The equation of thermal conductivity used for the numerical calculations is

$$C(T(x,t))\frac{\partial T(x,t)}{\partial t}=\frac{\partial}{\partial x}\left(K(T(x,t))\frac{\partial T(x,t)}{\partial x}\right),$$

where $C(T)=\alpha T^3$ is the specific heat capacity and the thermal conductivity coefficient $K(T)=K_0\cdot\exp(\theta/T)$, θ is Debye temperature.

A good agreement between experimental results and model calculations was achieved (Fig.1). Analyzing the results of calculations we draw a conclusion about strong sample temperature influence on the process of a hot spot formation immediately after the laser pumping. The role of the sample defectiveness is also discussed. The solution of non-linear thermal conductivity equation used to describe experimental results has been analyzed. It was found that initial conditions strongly affect the time evolution of a hot spot width (Fig.2).

Fig.1

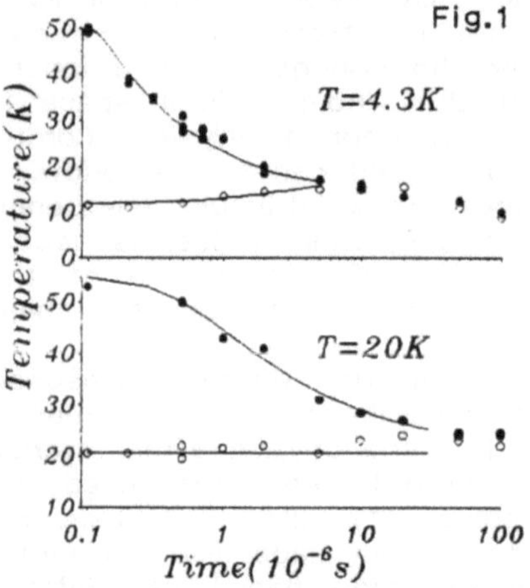

Fig.2

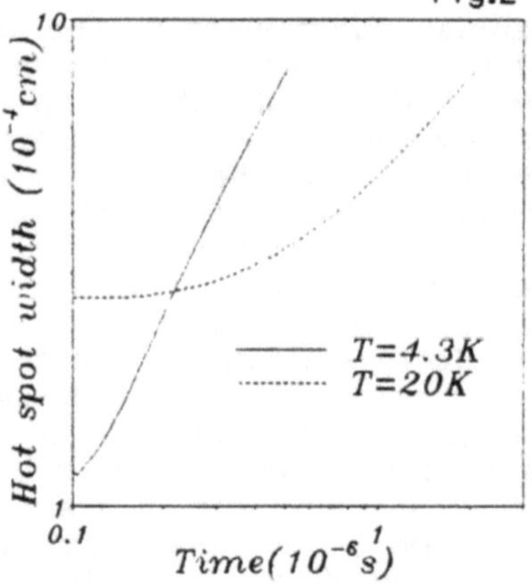

Fig.1 Temperature temporal dependences on the front (solid circles) and rear (open circles) surfaces after laser excitation at different initial temperatures T. The solid curves are the results of the calculations.

Fig.2 The calculated hot spot width.

References

1. D.V.Kazakovtsev, A.A.Maksimov, D.A.Pronin and I.I.Tartakovskii, Sov.Phys.JETP Lett., **49**, 61, 1989
2. D.V.Kazakovtsev, A.A.Maksimov, D.A.Pronin and I.I.Tartakovskii, Sov.Phys.JETP, 71, 819, 1990

Phonons and Resonances of Alloy Surfaces

H. Grimech[1] *and A. Khater*[1;2]

[1]Laboratoire de Physique des Matériaux, U.R.A. No. 807 CNRS, Université du Maine, F-72017 Le Mans Cédex, France
[2]SRSIM/CE Saclay, F-91191 Gif-sur-Yvette Cédex, France

It is now technically possible to prepare a surface alloy. In this paper we present theoretical results for the spectral densities of atomic vibrations of surface atoms in a surface alloy layer A_pB_{1-p}/A(001) of an fcc crystal, as a function of the concentration p, and for different wave vectors in the two-dimensional surface BZ.

These results are obtained using the matching technique [1], and Greens function thoery [2], in an effective field approach over the dynamics of two limiting pure systems, the ideal crystal surface, and the surface covered with a complete adsorbate monolayer. The configuration average restores the translational invariance. The calculation is in the harmonic approximation.

Numerical results are presented for the configuration averaged spectral densities $\rho_z(\ell = 1, \vec{k}_{11}, \Omega)$ of the z component displacements of the single surface layer alloy A_pB_{1-p}, in Fig.1. The numerical results are obtained using A-A and B-B nearest and next nearest neighbour force constants that correspond to a choice of A = Pt and B = Cu,and the arithmetic mean for the other A-B force constants. The masses are however hypothetical, and correspond to $m_B = 0.5\ m_A$. Figs. (1.a) and (1.b) give the spectral densities for randomly distributed concentrations p = 0.25 and 0.75, respectively. Note that the arbitrary unit scale is common. Ω is a dimensionless frequency.

Springer Series in Solid-State Sciences, Volume 112
Phonon Scattering in Condensed Matter VII Eds.: M. Meissner and R.O. Pohl

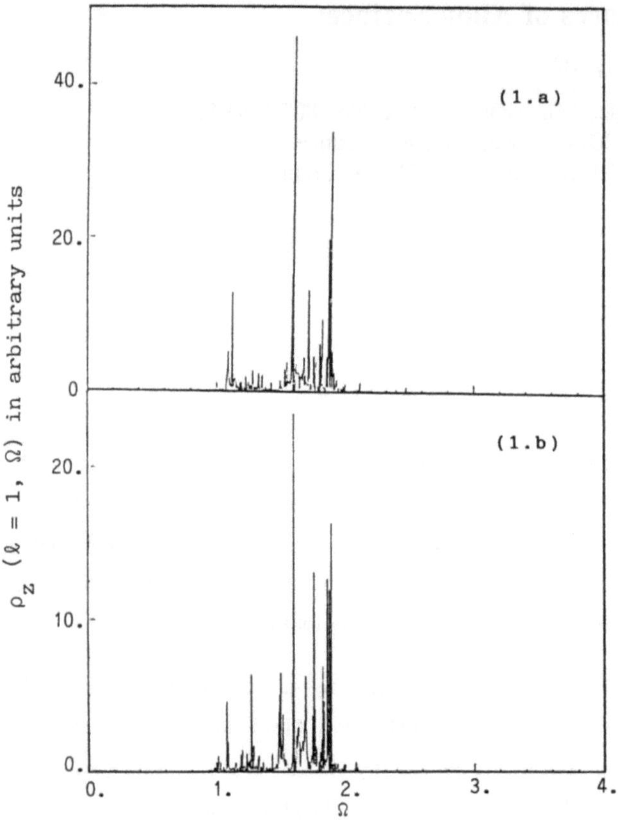

Fig. 1 The configuration averaged spectral density in the projected bulk phonon band, perpendicular to the surface; $\vec{k}_{11}$ corresponds to $\phi = \pi/4$ along the ΓM direction of the surface BZ.

References

1. **A. KHATER, N. AUBY, R.F. WALLIS,** Physica B 168 (1991) 273

2. **A.A. MARADUDIN and R.F. WALLIS,** Phys. Rev. 148 (1966) 945

Localized Vibrational States on Isolated Atomic Surface Defects Extending into the Bulk

O. Rafil[1] *and A. Khater*[1;2]

[1]Laboratoire de Physique des Matériaux, U.R.A. No. 807 CNRS, Université du Maine, F-72017 Le Mans Cédex, France
[2]SRSIM/CE Saclay, F-91191 Gif-sur-Yvette Cédex, France

Abstract : By an isolated atomic surface defect extending into the bulk is meant a surface defect which self-repeats gowing into the bulk. An example is an atomic plane A wedged between two semi-infinite crystals B. A two-dimensional model is presented of the localized surface states at the tip of a linear chain A wedged between two quarter-infinite squares B, to exemplify the problem.

The analysis of vibrational states in disordered surfaces differs in general from that in ordered surfaces, and is considerably more complicated, [1]. We present results for the localized vibrational states at the tip of a linear atomic chain A wedged between two quarter-infinite squares B. Nearest k_1 and next nearest k_2 neighbour interactions are considered in the harmonic approximation. The dimensionless frequency $\Omega^2 = \omega^2/\omega_o^2$ is defined with respect to $\omega_0^2 = k_1(B-B)/m(B)$. Note that $r_o = k_2(B-B)/k_1(B-B)$, $r_1 = k_1(A-B)/k_1(B-B)$, and $r_2 = k_2(A-B)/k_1(B-B)$, are ratios for defect host interactions. The defect mass is $m(A) = m(B)\,(1 + e)$ and $k_1(A-A) = k_1(A-B)$.

It is possible, using the matching method, [2], to define characteristic equations in the bulk, on the chain A, for evanescent vibrations parallel and perpendicular to the edge. The equations of atomic vibrations in the neighbourhood of the tip of the atomic chain A are obtained, and can be decoupled from the rest of the system by an appropriate use of the matching method in two dimensions.

Springer Series in Solid-State Sciences, Volume 112
Phonon Scattering in Condensed Matter VII Eds.: M. Meissner and R.O. Pohl

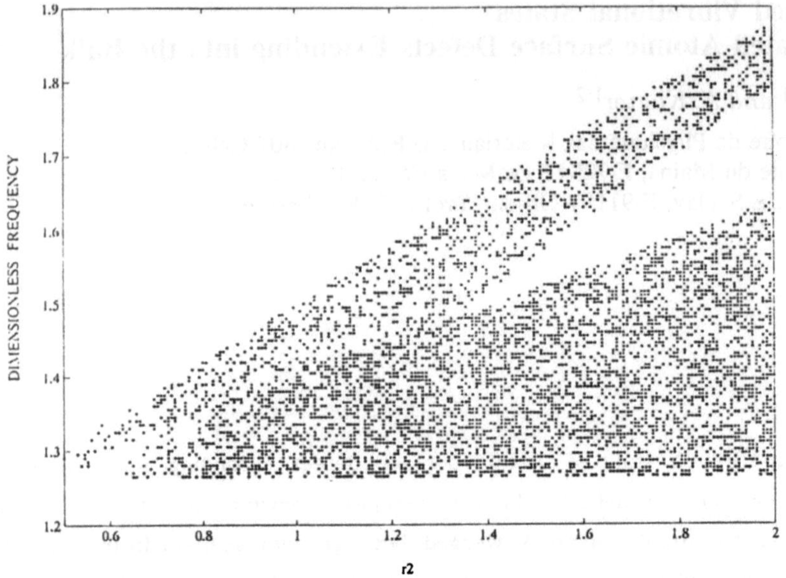

Fig. 1 The distribution of the energies Ω of the accessible localized states at the tip of an extended defect. See text for details.

Numerical results are presented in Fig. 1 for the energies Ω of the accessible localized vibrational states at the tip of the linear chain A, as a function of r_2, using model parameters r_o = 0.8, r_1 = 0.6, and e = 0.5.

The present method can be extended to realistic situations in three dimensions, such as the edge states of on atomic plane between semi-infinite crystals, and surface states of defects extending into the bulk created by rapid heavy ion bombardement.

References

1. **A.A. MARADUDIN and A.R. McGURN,** Phys. Rev. B 39 (1989) 8732.
2. **A. KHATER, N. AUBY, and R.F. WALLIS,** Surf. Sci. 217 (1989) 563

Noise in Scanning Tunneling Microscopy and Surface Phonons

R. Möller, B. Koslowski, C. Baur, and K. Dransfeld

Universität Konstanz, Fakultät für Physik,
Postfach 5560, W-7750 Konstanz, Fed. Rep. of Germany

Abstract. The spatial distribution of the fluctuations of the tunneling current occurring in a Scanning Tunneling Microscope are investigated. Experiments performed in ultra high vacuum on a cleaved (110) surface of GaAs reveal a noise pattern with a commensurate structure of atomic dimensions. The maxima in the noise distribution are located near the Ga-atoms while the maxima in the topography are located either near the Ga- or the As-atoms depending on the polarity of the applied bias. Due to the buckling relaxation of the surface the Ga-atoms have an almost planar coordination resulting in a large vibrational amplitude. The enhanced current noise at the position of the Ga-atoms appears to be associated to fluctuations of the occupation number of a specific librational surface phonon.

1. Introduction

Different Types of noise are expected to occur in Scanning Tunneling Microscopy. Since the tunneling current is composed of a large number of independent tunneling events of electrons the lowest possible limit of the fluctuation should be given by the shot noise. However, this has not yet been observed. Most experiments show that the power spectral density (PSD) of the observed current fluctuations largely exceeds the value for shot noise and it exhibits a 1/f frequency dependence [1] . If the voltage across the tunneling gap is small or equal to zero thermal noise becomes dominant [2]. For particular samples, e.g. a partially oxidized silicon surface, random telegraph noise has been observed due to the capture/release of charge carriers at single trapping sites [3].

Subject of the present paper is the local variation of the current noise amplitude which is found on the GaAs (110) surface. To explain the observation a model will be proposed relating the current noise to the fluctuations of a specific surface phonon [4].

Springer Series in Solid-State Sciences, Volume 112
Phonon Scattering in Condensed Matter VII Eds.: M. Meissner and R.O. Pohl

2. Experimental

Details of the STM are described elsewhere [5] . The coarse approach is achieved by a spring reduction. Vibrational damping is provided by a stack of metal plates spaced by pieces of viton. In addition the whole setup is mounted on an optical table with pneumatic damping. The instrument is operated in ultra high vacuum (UHV) at a pressure of $5 \cdot 10^{-11}$ mbar. The tunneling tips have been prepared by electrochemical etching of a tungsten wire of 0.5 mm diameter. Prior to the measurement they have been cleaned by field emission and desorption. The samples of n-GaAs have a dopant concentration of $2 \cdot 10^{18} cm^{-3}$ (Si). The (110) surface has been prepared by cleaving *in situ.*

The tunneling current is amplified by a low noise current-to-voltage converter with a gain of 10^9 V/A and an electronic noise of about 4.1 $fA/\sqrt{Hz}$ for the frequency range from zero to 30 kHz. The output is connected either to a dynamical signal analyzer (Hewlett Packard 35660) to analyze the spectrum of the current noise or to a "noisemeter" which allows to evaluate the PSD of the current noise at a frequency of 2 kHz. This is well above the cut-off-frequency of the feedback loop operated in the constant current mode. In addition the "noisemeter" allows to measure the magnitude of the current noise while the tunneling tip is scanning the sample at a slow scan rate of 0.1 Hz.

3. Results

A typical frequency spectrum $S_I(\omega)$ of the current noise observed on the GaAs (110) surface is shown in Fig. 1. The average tunneling current is 100 pA, the bias applied to the sample is - 1.8 V. The spectrum has a 1/f characteristic. At frequencies above 10 kHz thermal noise corresponding to the differential conductivity of the tunneling junction becomes dominant.

A simultaneous recording of the topography and the magnitude of the current noise is displayed in Fig. 2a and b, respectively. The scanned area is 40x40Å. The average tunneling current was 70 pA at a sample bias of -1.8 V, i.e. the electrons tunnel out of the As-derived states forming the valence band of the sample.

Due to the electronic structure of the surface the maxima in the topography correspond to the positions of the As-atoms [6]. The latter are spaced by 4Å along the $[1\bar{1}0]$ direction and by 5.6Å along the [001] direction. The corrugation is about 0.3Å. The lateral distribution of the current noise shows maxima along the $[1\bar{1}0]$ direction and a weak modulation along these rows. The ratio of the PSD in bright areas to the PSD in dark areas of Fig. 2b is about 8. However the cross correlation of noise and topography reveals that the maxima of the current noise do not coincide with the maxima in the topography, i.e. the positions of the As-atoms. Measurement at positive sample bias, i.e. tunneling into the Ga-derived states, reveal that the maxima of the noise are located near the Ga-atoms

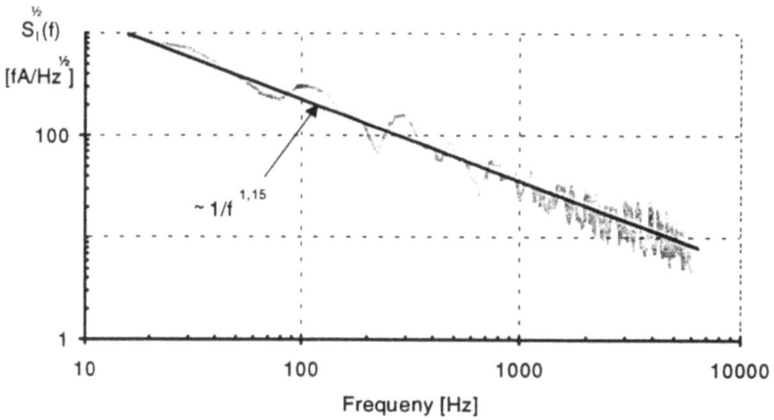

Fig. 1 Frequency spectrum of the current noise observed by STM on GaAs(110)

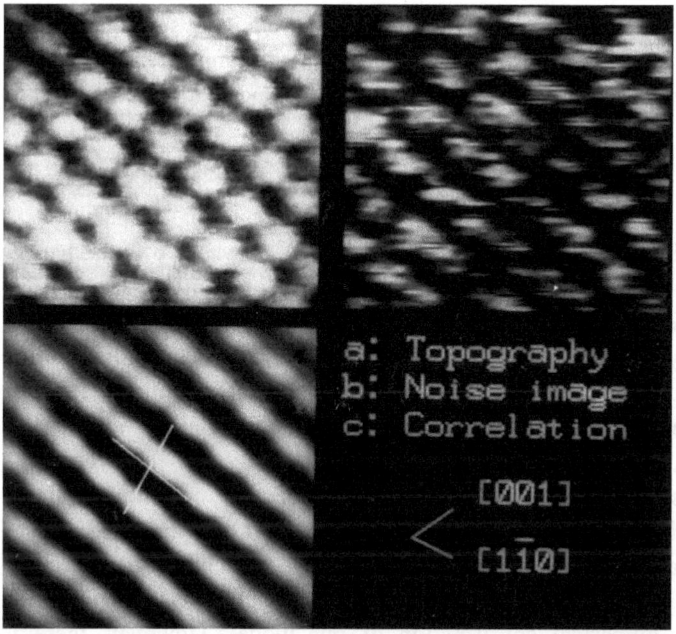

Fig. 2 Comparison between the topography (left top) and the lateral distribution of the tunneling current noise (right top); the calculated cross correlation of noise and topography is shown in Fig. 2c (left bottom), the cross indicates the origin.

independent of the bias polarity. Different tests have been performed to exclude artefacts, e.g. changing the scan direction or the scan speed. To verify that the lateral distribution shown in Fig. 2b is not simply related to other quantities but specific for the current noise the tunneling current I, dI/dz and dI/dV have been measured simultaneously.

4. Discussion

To explain the current noise in STM experiments different models have been proposed. However, they cannot explain our experimental observation. Our explanation is based on a specific property of the GaAs(110) surface. The buckling relaxation causes an important difference between As- and Ga-sites. The As-atom is pushed outward and the Ga-atoms are pulled inward. This leads to an almost planar coordination of the Ga-bonds giving rise to a large displacement of the Ga-atoms for a bond length conserving motion of the surface atoms. A surface phonon of this type has been observed experimentally [7] and calculated by *ab initio* methods [8] revealing a localized oscillation of a 'rotational surface phonon'.

To investigate the effect of a surface motion on the tunneling current it is useful to expand the tunneling current as function of distance z around an average position z_0:

$$I(z) = I_0 \, e^{-\alpha z} = I_0 \left[1 - \alpha (z-z_0) + \frac{1}{2} \alpha^2 (z-z_0)^2 \ldots \right]$$

with a mean tunneling current I_0 and $\alpha = 1.05 \sqrt{\phi}$ [Å^{-1}] where ϕ is the barrier height in [eV]. The phonon frequency is many orders of magnitude higher than the highest frequency detectable by the current-to-voltage converter. Therefore the latter is given by the time average over many periods. For sinusoidial motion of constant amplitude u_0 the linear term yields no average contribution. However the quadratic term results in an increase of the tunneling current of $\Delta I = \frac{1}{4} I_0 \alpha^2 u_0^2$, as discussed by Leavens and Aers [9]. If the vibrational amplitude fluctuates this leads to fluctuations of the tunneling current. The linear term could transfer fluctuations too, but it is not understood how the oscillation of the surface phonon at 2.5 THz could exhibit fluctuations at such low frequencies.

So far there is no theoretical model for 1/f type fluctuation of the phonon number allowing a numerical confirmation of our model. However, there is experimental evidence for such fluctuation by the investigation of Brillouin-scattering. Musha et al. [10] observed a weak variation of the intensity of Brillouin lines of a quartz sample with a 1/f spectral dependence. They concluded that this is due to fluctuation of the phonon number. Furthermore they could show that the noise scales with the inverse of the investigated volume. Hence if the latter is shrunken to a single unit cell strong effects are to be expected.

5. Conclusion

The analysis of the fluctuation of the tunneling current observed by STM on GaAs(110) reveals that the Ga-atoms are sites of enhanced noise. We propose that the observation is caused by the motion of the surface atoms, i.e. by the fluctuation of the phonon number of a rotational surface phonon which is connected to a large displacement of the Ga-atom. This would represent the first visualization of the displacement pattern of an optical phonon in real space.

6. References

[1] E. Stoll, O. Marti, Surf. Sci. **181**, 222 (1987)
S. Park, C.F. Quate, Appl, Phys. Lett. **48**, 112 (1986)
D.W. Abraham, C.C. Williams, H.K. Wickramasinghe, Appl. Phys. Lett. **53**, 1503 (1988)

[2] R. Möller, A. Esslinger, B. Koslowski, Appl. Phys. Lett. **55**, 2360 (1989)

[3] M.E. Welland, R.H. Koch, Appl. Phys. Lett. **48**, 724 (1986)

[4] A more detailed discussion will be given in:
B. Koslowski, C. Baur, R. Möller, K. Dransfeld, Proceeding of the 'Van der Ziel Symposium 1992', St. Louis
B. Koslowski, C. Baur, R. Möller, K. Dransfeld, submitted to Surface Science

[5] R. Möller, A. Esslinger, B. Koslowski, J.Vac. Sci. **A8**(1), 590 (1990)

[6] R.M. Feenstra, J.A. Stroscio, J. Tersoff, A.P. Fein, Phys. Rev. Lett. **58**, 1192 (1987)

[7] U. Harten, J.P. Toennies, Europhys. Lett. **4**(7), 833 (1987)

[8] Y.R. Wang, C.B. Duke, Surf. Sci. **205**, L755 (1988)

[9] C.R. Leavens, G. C. Aers, Surf. Sci. **181**, 250 (1987)

[10] Musha Toshimitsu, B. Gabor, S. Minoru, Phys. Rev. Lett. **64**, 2394 (1990)

Surface Acoustic Waves: A New Tool for Thin-Film Research and Sensor Technology

S. Hunklinger, M.v. Schickfus, and R. Hübner

Institut für angewandte Physik, Universität Heidelberg, Albert-Überle-Str. 3–5, W-6900 Heidelberg, Fed. Rep. of Germany

1. Introduction

For a long time the investigation of materials with ultrasonic waves has been a well established technique. The potential of this method resides in the fact that it is sensitive to small variations in mass density or of the elastic constants of the material to be investigated. With ultrasonic techniques structural inhomogeneities can be reliably detected. Another application of ultrasonic devices is that of a microbalance. With this device, generally a disk-shaped piezoelectric resonator, the thickness or mass of thin layers on the surface can be measured. This technique relies on the influence of the surface layer on the resonant frequency of the device. If the thickness of the layer is small with respect to the acoustic wavelength, this effect depends linearly on the mass and on the elastic properties of the layer.

Recently, the use of surface acoustic waves (SAW's) for such purposes has become increasingly attractive. Obviously, this wavetype is sensitive to properties of the surface of a solid and can therefore be utilized for the investigation of surface layers. One field of application is for chemical and biological sensors [1,2], another one is the non-destructive testing of surfaces.

Before presenting our recent results we shortly discuss some basic properties of surface acoustic waves. They are described by special solutions of the elastic wave equation for a semi-infinite solid [3]. The boundary conditions imply that at the surface the strain component normal to it must vanish. Additional boundary conditions occur if the solid is piezoelectric or if the surface is covered with a layer. For our work two of several possible SAW-type solutions are of importance: Rayleigh-type waves and the SH (shear horizontal) wave. Pure Rayleigh waves have two components of displacement: one in the direction of propagation and one perpendicular to the surface. Both components decay exponentially with distance from the surface within about one wavelength. Although the SH wave is in principle not confined to the surface, this can be achieved by the presence of surface layers or by the selection of suitable crystal cuts.

The confinement of the energy flux to a thin surface layer has two important consequences for applications: the SAW is extremely sensitive to variations of the surface properties and it is accessible for detection or modification at all points of the propagation path. One application making use of this property is for chemical or biological sensing devices: Small changes of the density or of the elastic properties of a sensing film on the device are detected by the

Springer Series in Solid-State Sciences, Volume 112
Phonon Scattering in Condensed Matter VII Eds.: M. Meissner and R.O. Pohl

associated variation of the SAW velocity. This technique will be discussed in section 2. In section 3 we shall present experiments utilizing SAW attenuation for an assessment of the adhesion of thin films.

2. Sensing with surface acoustic waves

In principle chemical and biological SAW sensors consist of a SAW device, either a delay line or a resonator, covered with the sensing film. The presence of this layer with a thickness $h \ll \lambda_{\mathbf{SAW}}$ will influence the velocity of the SAW like [4]

$$\frac{\Delta v}{v} = k\rho_{\mathrm{f}}\{4VA_1 v_{\mathrm{tf}}^2 - (A_1 + A_3)v^2\} \tag{1}$$

for a Rayleigh wave and like

$$\frac{\Delta v}{v} = k\rho_{\mathrm{f}} A_2\{v_{\mathrm{tf}}^2 - v^2\} \tag{2}$$

for the SH wave. In these equations ρ_{f} denotes the mass density of the film, $V = (1 - v_{\mathrm{tf}}^2/v_{\mathrm{lf}}^2)$ is determined by the longitudinal and the transverse sound velocity of the film, and v is the SAW velocity of the substrate. The A_i are given by $A_i = \omega\alpha_i^2/4P_1$, where α_i is the longitudinal ($i = 1$) or transverse ($i = 2,3$) component of displacement in the SAW, and P_1 the power flow density in the direction of the wave. Therefore the quantities A_i indicate the relative power density in the three modes.

The preceeding equations show that a variation of the density or of the sound velocity of the film upon incorporation of the substance to be detected will change the SAW velocity of the device. Since the elastic energy of the SAW is concentrated in a layer with a thickness of about one wavelength, the sensitivity of the device will increase with operating frequency. This is a definite advantage over bulkwave devices, where the elastic energy is distributed over the device volume. The SAW velocity variation can be detected with high sensitivity of better than 10^{-6} if the device is used as the frequency determining element in an oscillator circuit as shown in fig.1. Here the resonant frequency ω_{r} is directly related with the sound velocity v as $\Delta v/v = \Delta\omega_{\mathrm{r}}/\omega_{\mathrm{r}}$. For SAW devices a quadratic dependence of the sensitivity on the operating frequency is expected [5].
As can be seen in the figure, two SAW oscillators are used in our sensor, the output of which is fed into a mixer. The output of the mixer –the difference frequency of the two oscillators– is measured by a counter. One of the SAW tracks is used for sensing, the other serves as a reference for compensation of thermal drift. An additional advantage of this arrangement is that the

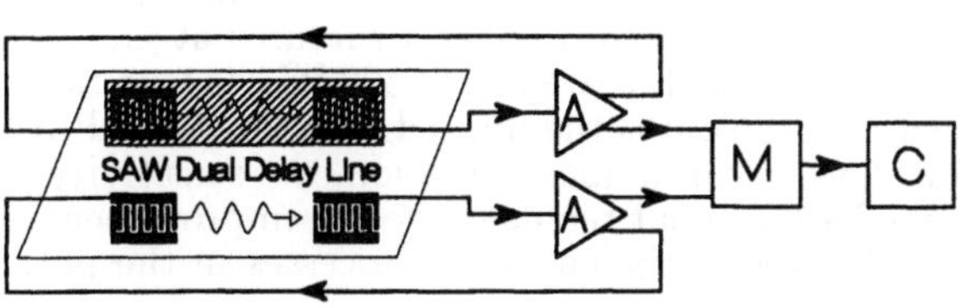

Fig. 1: Block diagram of a SAW sensor. A: Amplifier, M: Mixer, C: Counter.

mixing technique allows the use of a cheap low-frequency counter. Care must be taken for good thermal stabilization of the device, even if crystal cuts with low thermal expansion are chosen.

The most common use of SAW sensors is for gas sensing [1]. In our experiments we have used two sensor layers, namely a Pd metal layer for hydrogen sensing in a first experimental step [6] and phtalocyanine layers for NO_x sensing in a systematic study [7,8]. With a 300 nm layer of Pd our hydrogen sensor reached a sensitivity $\Delta\omega_r/\omega_r = 10^{-5}(p_{H_2}[\text{mbar}])^{0.5}$ for H_2 in nitrogen at an operating frequency of 100 MHz. The long response time of the order of 15 min was a problem, possibly caused by the diffusion limited access of the test gas to the Pd surface in the laminar gas flow.

For the NO_x sensor, based on phtalocyanine layers, we have developed a new measuring cell designed in such a way that high gas velocities are reached at the sensor surface. In this way laminar flow at the surface is suppressed and short response times are possible. A second condition for a rapid and reversible response is a high operating temperature, 130 °C in our case.

For the SAW device we have used two similar delay line designs, one working with Rayleigh waves, the other with SH waves. Both operate at a frequency of 600 MHz, about one order of magnitude higher than in usual designs, in order to achieve a high sensitivity. This technique allows us to work with sensor layers down to sub-monomolecular thickness. The sensitivity of both device types is similar. For reasons which we do not understand until now, frequency stability of devices covered with a sensor layer is considerably higher with the SH wave than with Rayleigh waves.

The interdigital transducers have a high number of finger pairs, so that the acoustic length under the finger pairs exceeds the delay. In this way operation in an oscillator circuit without disturbing mode jumps is possible. The phtalocyanine layers can be put on the device either by evaporation or by the Langmuir-Blodgett technique. Our devices were mainly produced by UHV sublimation of lead-phtalocyanine (PbPc). The layer thickness ranged between 1 and 15 nm. In this range the sensitivity depends linearly on thickness. The Langmuir-Blodgett technique promises more uniform, high-quality films, however, phtalocyanines suitable for this technique are difficult to obtain and an increased sensitivity of Langmuir-Blodgett devices has not been observed until now. In addition the Langmuir-Blodgett films available until now decompose at temperatures exceeding 60 °C, i.e. far below our normal operating temperature.

Fig.2 shows the response of a SAW NO_x sensor covered with 5 nm of PbPc after exposition to the exhaust gases of several motor cars. The response time of the sensor is sufficiently short, less than a minute, and the sensitivity is 500 Hz/ppm(NO_2). With a short-term stability of the oscillators of about 5 Hz, a detection limit of 10 ppb is reached by this sensor.

In this context we should also note some problems that must be overcome for a practical application of the SAW gas sensor. Cross-sensitivity is one that is common to almost all sensors and has not much to do with the SAW principle. Fortunately, this is not a serious problem for PbPc, however air humidity causes a frequency drift of the sensor. It has turned out that this drift is mainly caused by oxidation of the aluminum finger structures at elevated temperatures in humid air. Annealing the device at up to 300 °C has very much improved this behaviour. Another problem is the limited lifetime of the PbPc layer. In our tests, after a few days of continuous operation, the sensitivity of the device dropped, and vanished after about ten days. For an improvement of this situation a thorough investigation of the chemical processes in the phta-

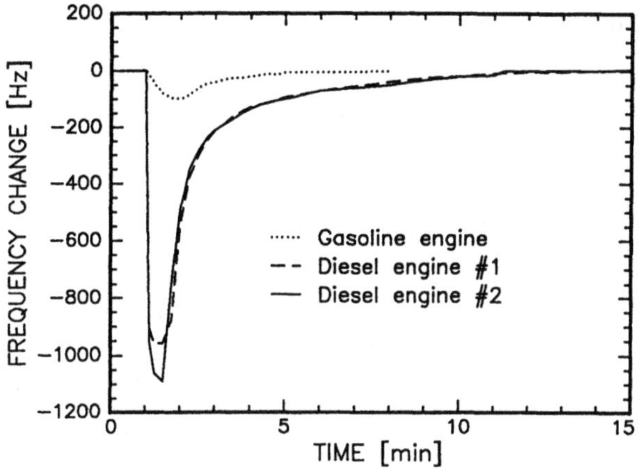

Fig. 2: Response of the SAW gas sensor to NO_2 in three samples taken from automobile exhausts. The sensitivity of this sensor is approximately 500 Hz/ppm.

locyanine layer is necessary. Finally, good thermal stabilization of the device to less than 0.1°C is indispensable. Immunity to temperature changes is also demanded for the amplifiers and the connection cables. In order to reduce the sensitivity of the SAW device to thermal stresses, new mounting techniques must be developed.

A vast field for future applications is sensing in liquids. Candidates are biological sensors, e.g. for medical diagnostic purposes, or sensors used in pollution control in sewage systems. Only shear wave type SAW devices can be used for this purpose, because Rayleigh-like waves couple to the longitudinal bulk mode in the liquid and thus get rapidly attenuated. Another problem with SAW liquid sensors is that in the conventional design the transducer fingers and the electrical connections are exposed to the liquid. This causes two problems: The transducer structures are chemically attacked by the liquid, e.g. ionic buffer solutions, and the high dielectric constant of water causes a high capacitive load in parallel to the device.

For the first problem we have investigated two solutions: Using noble metals for the transducers or covering them with an inert layer, sputtered SiO_2 in our case. Both techniques have led to satisfactory results. In order to increase the capacity of the substrate relative to the liquid, we have used $LiTaO_3$ with a dielectric constant of $\varepsilon \approx 40$ for the SAW device.

Most of our tests of the liquid sensor were performed with devices operating at a frequency of 108 MHz. They were originally designed as filters for communication systems. After proving that SAW propagation in the SH mode at a solid/liquid interface is possible with tolerable insertion loss in these devices, we have started to assess their sensitivity. For our tests we have used the same oscillator circuit as for the gas sensor. The system was first operated with pure water which, after a few minutes for thermal stabilization, was replaced by a solution of BSA (Bovine Serum Albumine) in water. BSA has the property of adhering preferrably to oxide surfaces. From the known thickness of a BSA layer and the frequency shift after adsorption we derive a mass sensitivity of about 10 ng/cm^2. This result proves that a SAW sensor operating in liquids can be made with a sensitivity sufficient for biological applications.

3. Study of film adhesion with acoustic surface waves.

Whereas a layer on the surface of a SAW device does influence velocity and attenuation of a SAW, it has also been found that the interface between a layer and the substrate can influence the attenuation significantly [9]. The crucial role of the interface for the attenuation behaviour of SAW devices became apparent when we investigated the influence of a very thin "adhesion layer" of a few nm of Al_2O_3 between substrate and metallization onto the attenuation of an Al-covered SAW device [10]. Without the adhesion layer, i.e. if the aluminum layer was evaporated directly onto the substrate, attenuation was strongly temperature dependent with a peak below room temperature (fig.3). This behaviour is typical for the thermally activated motion of dislocation lines. The effect of the adhesion layer seems to be that they either get pinned and thus immobile, or that the formation of dislocation lines is supressed by the adhesion layer because it reduces the effect of the lattice mismatch between layer and substrate.

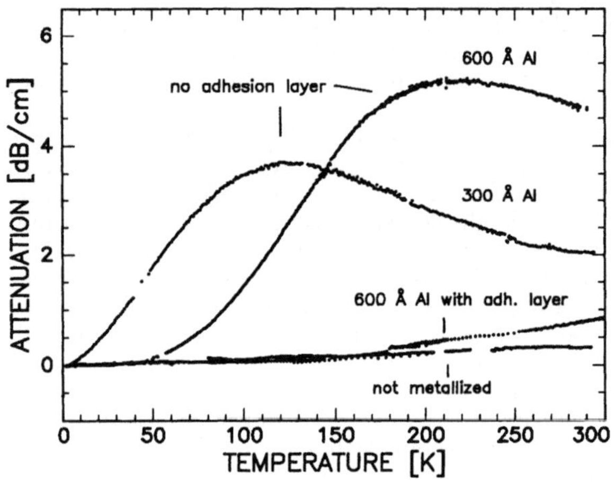

Fig. 3: Temperature dependence of the attenuation of a SAW delay line covered with aluminum films with and without adhesion layer. The measurements were carried out at 500 MHz.

This result has induced us to continue research on the adhesion of surface layers with SAW's. For such measurements it is indispensable to measure the value of attenuation with high absolute accuracy. This is difficult to achieve with the classical two- or three-transducer method because systematic errors caused by beam diffraction or by the electrical connection to the transducers may influence the result. We therefore installed a laser probe system utilizing the diffraction of a laser beam for a SAW intensity measurement [11] (fig.4). This system is capable of scanning the SAW profile in two directions and to deliver a locally resolved image of the SAW beam. In this way not only the attenuation but also beam diffraction and the influence of inhomogeneities like scratches can be detected.

As intensity is measured with a local resolution of about 0.5 mm, the influence of interference caused by diffraction and similar effects can be supressed in this system: Instead of integrating over the signal amplitude in the beam, which happens in an interdigital transducer, numerical integration over the intensity profile of the beam can be performed. Fluctuations of laser amplitude

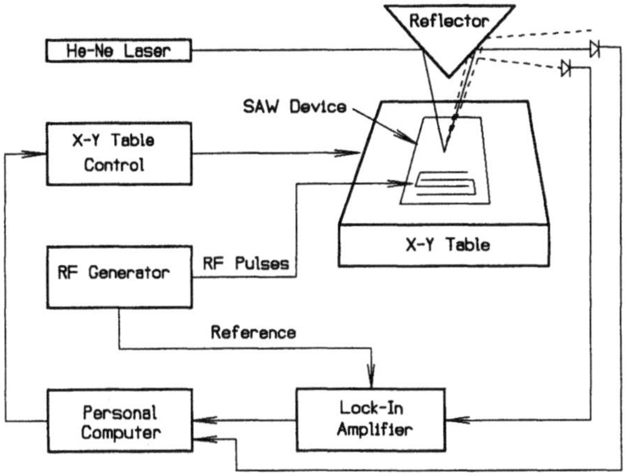

Fig. 4: The laser scanning system for SAW intensity measurements. Dashed line: Diffracted beam.

and polarization are taken into account by evaluating an undiffracted reference beam.

For a test of the capability of this system to indicate the adhesion of a film, we have prepared a SAW delay line covered with a sputtered 200 nm gold layer. For the first 10 mm, before evaporation of the gold, the device was covered with a 15 nm chromium layer which is known to improve the adhesion of additional metal films. The rest of the delay line was covered with the gold film alone. The attenuation behaviour of this device is shown in fig.5.

There is obviously a drastic difference in the attenuation between the two halves of the delay line. Attenuation of the Cr/Au film is 0.85 dB/cm, wheras the pure gold film attenuates with 1.5 dB/cm.

Of course we tend to correlate this attenuation difference with adhesion, although two important steps still have to be made: Complementary measurements will have to prove that the two films really do adhere differently, and the microscopic mechanism for the increased attenuation of the badly adhering film must be found. Further temperature dependent measurements will show whether the activation of dislocations discussed above for aluminum films can

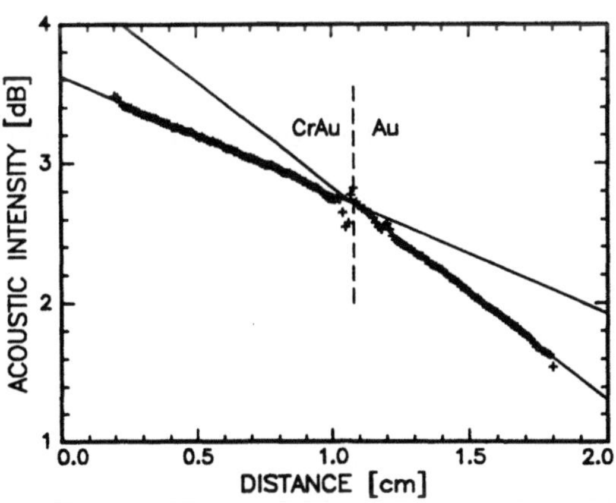

Fig. 5: Amplitude decay of a Rayleigh wave with distance. The $LiNbO_3$ substrate was covered with a 200 nm gold film sputtered onto a 15 nm chromium film (left half) and onto the bare substrate (right half). The measuring frequency was 270 MHz.

also explain our observations for the gold film. Finally, we will have to investigate in how far this result can be applied to other film/substrate combinations.

4. Conclusion

The localization of acoustic surface waves to a thin layer of the order of one wavelength makes them ideally suited for techniques where variations in the acoustic properties of thin films are to be detected. Chemical and biological sensors based on surface acoustic wave devices are sensitive to small concentrations of the substance to be detected and are useable for a wide range of applications. SAW sensors can be operated both in gas and in liquids.

Surface acoustic waves promise to be equally useful for studying the adherence of thin films to surfaces. The advantage of this technique is that it is non-destructive and reproducible. For a general application in materials testing, techniques for exciting acoustic surface waves in non-piezoelectrics must be developed. Simple optical methods should be a good candidate for this purpose.

References

[1] for a review see: C.G.Fox, J.F.Alder; Analyst **114**, 997 (1989)
[2] J.Andle, J.Vetelino, M.Lade, D.McAllister; 1990 IEEE Ultrasonics Symposium Proceedings, Institute of Electrical and Electronic Engineers, New York, p.291
[3] for a review see, for example: Acoustic Surface Waves, Topics in Applied Physics Vol.24, (A.A.Oliner ed.), Springer Verlag, Heidelberg (1978)
[4] G.W.Farnell; in [3], p.13
[5] G.Sauerbrey; Z.Physik **155**, 206 (1959)
[6] M.v.Schickfus, M.Rapp; acta phys.slov. **40**, 26 (1990)
[7] M.Rapp, D.Binz, I.Kabbe, M.v.Schickfus, S.Hunklinger, H.Fuchs, W.Schrepp, B.Fleischmann; Sensors and Actuators **4**, 103 (1991)
[8] M.Rapp, R.Stanzel, M.v.Schickfus, S.Hunklinger, H.Fuchs, W.Schrepp, H.Keller, B.Fleischmann; Thin Solid Films **210/211**, 474 (1992)
[9] D.W.Galipeau, J.F.Vetelino, R.Lec, C.Feger; Sensors and Actuators **B 5**, 59 (1991)
[10] U.Teubner, R.Dahint, M.v.Schickfus, S.Hunklinger; to be published in Appl.Phys.Lett. **61** (1992)
[11] A.J.Slobodnik; in [3], p.225

Contributions to this work by U. Teubner and R. Dahint are gratefully acknowledged.

Optical Generation of Coherent High-Frequency Acoustic Phonons in a Periodic Multilayer Structure

*P. Basséras, S.M. Gracewski, and R.J.D. Miller**

Mechanical Engineering Department, and
*Chemistry Department and Institute of Optics,
University of Rochester, Rochester, NY 14627, USA

Abstract A specifically-engineered [001]-oriented GaAs/AlGaAs quantum well structure, excited with nJ, .1 psec pulses from a Ti:Sapphire laser source is used to optically generate 50 GHz coherent acoustic waves. The acoustics are generated through photothermal expansion when the short light pulse incident on the structure is absorbed in the GaAs layers. The phonons are detected by monitoring the strain-induced change in reflectivity of the surface with a time-delayed probe pulse.

Theory

The dispersion relation in layered media was first calculated by Rytov using a continuum model.[1, 2] For waves propagating perpendicular to the interfaces

$$\cos(qd) = \cos\left(\omega\frac{d_1}{v_1}\right)\cos\left(\omega\frac{d_2}{v_2}\right) - \frac{1}{2}\left(\frac{\rho_2 v_2}{\rho_1 v_1} + \frac{\rho_1 v_1}{\rho_2 v_2}\right)\sin\left(\omega\frac{d_1}{v_1}\right)\sin\left(\omega\frac{d_2}{v_2}\right),$$

where ω is the frequency, q the wavevector, $d = d_1 + d_2$ the period of the structure; and d_1, d_2 are the thicknesses, ρ_1, ρ_2 the densities, and v_1, v_2 the sound velocities of layers 1 and 2, respectively.

The generation of surface and bulk modes in periodic multilayer structures has previously been demonstrated by Grahn et al. in a:Ge/a:Si multilayers.[3] The structure that we designed is a 6 × (500 Å, 500 Å), [001]-oriented GaAs/$Al_{.40}Ga_{.60}As$ quantum well structure, which has a much smaller impedance mismatch between layers. The reflectivity spectrum shows a strong rise in the region of the excitation wavelength. Because of the selective heating of the structure with the laser energy being deposited in the absorbing GaAs layers only, we expect a strain modulation with a period equal to the period of the multilayer. The calculated dispersion relation is shown in Fig. 1.

Experimental Results

The method for the generation and detection of the high frequency acoustic phonons is based on the pump-probe technique.[4] The output of the laser is split into two beams, the pump and the probe. The pump pulse excites carriers in the sample. The relaxation of the carriers generates thermal energy which results in

Springer Series in Solid-State Sciences, Volume 112
Phonon Scattering in Condensed Matter VII Eds.: M. Meissner and R.O. Pohl

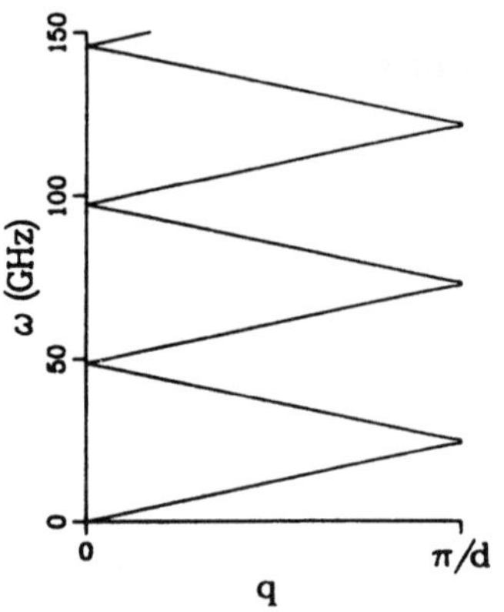

Fig. 1. Calculated dispersion relation.

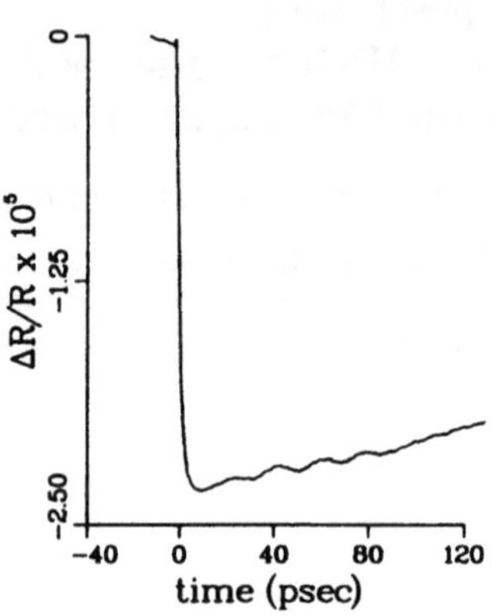

Fig. 2. Observed reflectivity changes.

material expansion and the excitation of acoustic waves. Both the excited carriers and the material strain cause a change in the optical constants and thus, a time-dependent change in reflectivity $\Delta R(t)$, which is monitored by the probe pulse.

The laser system consists of a Spectra Physics, Femtosecond Tsunami Ti:Sapphire Laser pumped by an Argon Ion Laser. The experiments are performed with the Tsunami laser wavelength tuned at 760 nm and pulses with duration ≤90 fsec. The average power is 1 Watt at 82 MHz, corresponding to an energy per pulse of 12 nJ. The pump and probe beams with intensity ratio 10 to 1 are focused to a 40μm diameter spotsize on the sample surface. Details about the experimental set-up will be given in a separate paper.

The experimental results are shown in Fig. 2. We observe a large change in reflectivity at t = 0 corresponding to the generation of electron-hole pairs. Superimposed on the nonexponential decay are oscillations with period $T \approx 20$ psec, corresponding to a frequency $\omega \approx 50$ GHz, that is approximately the lowest frequency of the zone center acoustic modes in the dispersion curve calculated for the structure.

Acknowledgments

The authors thank John Variano and Professor Gary Wicks from the Institute of Optics for the fabrication of the quantum well structure and for stimulating discussions. This work was supported by the National Science Foundation under Grant No. ECS-8913056.

References

1. S. M. Rytov, Sov. Phys. JETP 2, 466 (1956).
2. For example, B. Jusserand and M. Cardona: in Light Scattering in Solids, ed. by M. Cardona and G. Güntherodt (Springer, Berlin 1989).
3. H. T. Grahn, H. J. Maris, J. Tauc, and B. Abeles, Phys. Rev. B 38, 6066 (1988).
4. For example, H. J. Maris, C. Thomsen, and J. Tauc: in Phonon Scattering in Condensed Matter, ed. by A. C. Anderson and J. P. Wolfe (Springer, Berlin, 1986).

The Phonon Dimensionality in a Thin Metal Film

M.N. Wybourne and J.C. Nabity[1]

Department of Physics, University of Oregon, Eugene, OR 97403, USA

Nonequilibrium heating experiments on thin metal films have been used for sometime to obtain information about electron energy relaxation processes [2]. In an analysis of these experiments a knowledge of the phonon dimensionality in the film is important [3]. For perfect acoustic coupling between the film and a three dimensional substrate, the phonons will be three dimensional. In the other extreme of a self-supporting film in a vacuum, the acoustic phonons are expected to be spatially confined by the film thickness, d, and be of two dimensional character. In a real situation the acoustic coupling is between these extremes, thus the phonon dimensionality is difficult to evaluate.

We report an experiment to study the electron heating in a Au-Pd alloy film on silicon in which the acoustic coupling between the film and its surroundings could be changed in situ [4]. The films were 25 nm thick, 10 μm long and 1 μm wide and their resistance was measured by a four terminal ac technique to an accuracy of 1 part in 10^6. The electron heating was studied at a substrate temperature of 1.7 K by applying a dc electric field, E, to the sample and measuring the electric field dependence of the electron-electron resistance corrections found in Au-Pd films at low temperatures [5]. The acoustic coupling to the film was changed by having the free surface either in vacuum or liquid helium [4].

The equilibrium resistance change, $(R-R_0)/R_0 = \Delta R/R_0$, of the film as a function of temperature is shown in Fig. 1. Below the minimum equilibrium resistance ($R_0 \approx 100\ \Omega$) at 7 K, the resistance rises logarithmically, consistent with electron-electron interactions. The steady-state resistance change, $\Delta R(E)/R_0$, was dependent on the acoustic coupling, as shown in Fig. 2. When immersed in liquid helium, $\Delta R(E)/R_0$ falls over the whole range of applied fields. However, when the film is in vacuum, $\Delta R(E)/R_0$ reaches a minimum at about 8 V cm^{-1}. Under both surface conditions $\Delta R(E)/R_0$ falls below zero. These results show that at some fields R(E) becomes less than R_0. Furthermore, the extent to which R(E) falls below R_0 depends on the acoustic coupling between the film and its surroundings.

The equilibrium resistance can be described by the sum of three resistance terms, a Drude term originating from impurity scattering, a Drude term arising from electron-phonon scattering, and an interaction term, as shown by the solid line in Fig. 1. The Drude term arising from electron-phonon scattering depends on the electron and phonon temperatures, T_e and T_p, respectively, while the interaction term only depends on T_e. In the steady-state, $T_e > T_p$. Therefore, the contribution to the total resistance from each term is different than it is in equilibrium, which can cause $\Delta R(E)/R_0 < 0$ [4].

Springer Series in Solid-State Sciences, Volume 112
Phonon Scattering in Condensed Matter VII Eds.: M. Meissner and R.O. Pohl

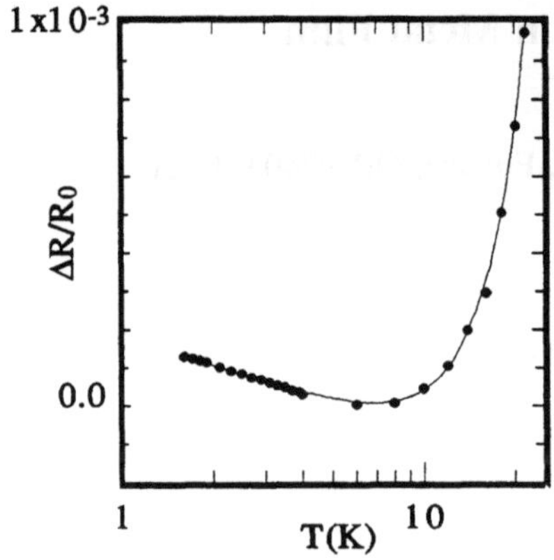

Fig. 1. The equilibrium temperature dependence of $\Delta R/R_0$.

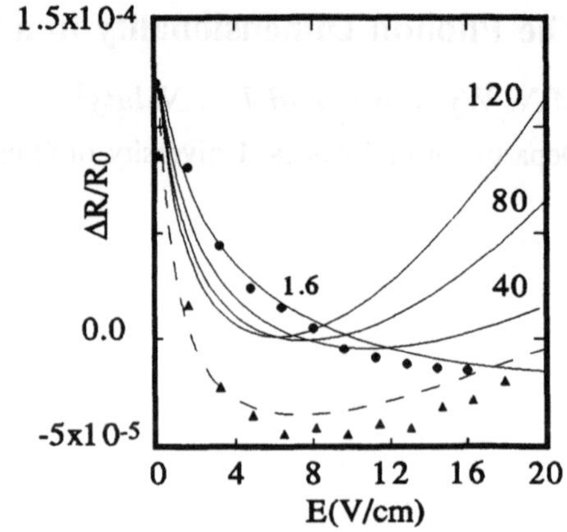

Fig. 2. The resistance change vs electric field for the film in vacuum (▴) and in helium (•). The numbers refer to the acoustic mismatch parameter η used to obtain the curves.

To explain the steady-state data shown in Fig. 2 we have used the Perrin and Budd model [6] to calculate the power per volume being transferred from the film to the substrate. We have used the Pippard [7] model for the electron-phonon relaxation time and have taken the phonon escape time from the film to be $t_{esc} = 4\eta d/V$, where V is the velocity of sound. For the film immersed in helium, the fit shown in Fig. 2 was obtained using a three dimensional phonon spectrum in the film, and η=1.6. For the film in vacuum, we attempted to obtain a fit by reducing the acoustic coupling, that is by increasing η, as shown in Fig. 2. No fit was possible by this procedure. To fit the vacuum data we used η = 3.2, a two dimensional phonon spectrum up to an energy where the anharmonic scattering length is equal to the film thickness, and an electron-phonon interaction modified to account for the two dimensional phonon spectrum [3]. This model gave the fit shown by the dashed curve in Fig. 2, and provides evidence of a reduced phonon dimensionality in thin Au-Pd films on silicon.

Supported by the National Science Foundation under Grant No. DMR-87-13884.

References

1. Present Address: J.C. Nabity Lithography Systems, P.O. Box 5354, Bozeman, MT 59717.
2. M.L. Roukes, M.R. Freeman, R.S. Germain, R.C. Richardson, and M.B. Ketchen, Phys. Rev. Lett. 55, 422 (1985).
3. D. Belitz, and S. Das Sarma, Phys. Rev. B36, 7701 (1987).
4. J.C. Nabity, and M.N. Wybourne, Phys. Rev. B44, 8990 (1991).
5. J.J. Lin, and N. Giordano, Phys. Rev. B35, 545 (1987).
6. N. Perrin, and H. Budd, J. Phys. (Paris) Colloq. 33, C4-33 (1972).
7. A. B. Pippard, Philos. Mag. 46, 1104 (1955).

Phonon Reflection from Buried Ultrathin Films Using Laser Picosecond Acoustics

O.B. Wright and K. Kawashima

Electronics Research Laboratories, Nippon Steel Corporation,
5-10-1 Fuchinobe, Sagamihara, Kanagawa 229, Japan

Abstract. We demonstrate using a time-resolved reflectance method the detection of optically excited longitudinal acoustic phonon pulses reflected from buried ultrathin films. Deviations from the predictions of continuum acoustics are observed for films of nm order in thickness.

High frequency phonons in the 0.1 THz range and above are scattered by atomic-scale inhomogeneities. For example, low temperature studies based on phonon pulse propagation or thermal transport have indicated anomalous phonon reflections from solid-solid interfaces.[1] Phonon reflection has often been found to be larger (but sometimes smaller) than the predictions of continuum acoustics (the acoustic mismatch model), especially for the higher phonon frequencies, but in many cases the microscopic mechanisms involved are still unclear. Recently room temperature heat transport experiments using a picosecond optical technique have indicated anomalously large phonon transmission at some interfaces[2]. We show in this paper how phonon pulse propagation measurements with this same time-resolved optical technique[3] can be used to probe such anomalies by the observation of the reflection of coherent phonons from buried ultrathin films.

Sputtered amorphous silicon films of thickness of nanometer order and above, sandwiched between chromium films of thickness 670 nm (upper) and 330 nm (lower) on a silica substrate, were investigated. Thicknesses were calculated from deposition times. Optical pulses of duration 2 ps (FWHM), photon energy 2 eV and pump pulse energy 0.4 nJ were used. The phonon frequencies excited, of order 0.1 THz, depend on the optical absorption depth of the samples. Figure 1(a) shows the reflectance change as a function of the delay time between the pump and probe pulses for an a-Si film of thickness $L = 16$ nm. Background signal components dependent on thermal changes have been subtracted to show the piezoreflectance variation. The echoes correspond to the

Springer Series in Solid-State Sciences, Volume 112
Phonon Scattering in Condensed Matter VII Eds.: M. Meissner and R.O. Pohl

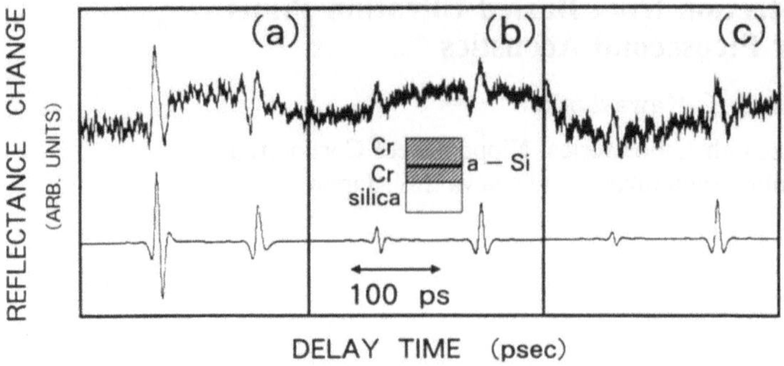

Fig. 1: Reflectance variation against delay time for three different thicknesses L of an a-Si film sandwiched between Cr films (see inset): a) $L=16$ nm, b) $L=3.2$ nm and c) $L=1.6$ nm. The zero of the delay time lies to the left, off the graphs. The lower, theoretical curves include the slight smoothing effect of the optical pulse durations.

phonon reflection from the a-Si film and from the substrate. We have modelled the pulse shapes (see lower curve) using the theory of Ref. 3 as a starting point, with the ratio of the photoelastic constants of Cr estimated from the substrate echo shape ($dn/d\eta:dk/d\eta \approx 1:1$, $n+ik$ the refractive index and η the longitudinal strain). Multiple phonon reflections inside the film distort the reflected pulse shape and also, to a lesser extent, the transmitted pulses. (These multiple reflections are accounted for up to order 7 in our model.) Figures 1(b) and 1(c) show results for $L=3.2$ nm and 1.6 nm, respectively, together with the theoretical fits. For the two thicker a-Si films the model is in reasonable agreement with the data, but for $L=1.6$ nm we observe an anomalously large reflection. For films of thickness less than 1 nm we could not resolve any echo from the a-Si film at the present noise level.

It is difficult to pinpoint the exact microscopic origin of the anomalous reflection. Various explanations, such as the presence of impurities, islands, or disorder, have been proposed to interpret similar low temperature data for interfaces or thin films[1]. However, studies with more characterized ultrathin films, for example with those grown epitaxially, should help elucidate these anomalies. The reflection of THz phonons is very sensitive to discontinuites in mass and bonding on an atomic scale, and laser picosecond acoustics should provide an important complementary source of data for such research.

[1]E. T. Swartz and R. O. Pohl, Rev. Mod. Phys., **61**, 605 (1989).

[2]R. J. Stoner, H. J. Maris, T. R. Anthony and W. F. Banholzer, Phys. Rev. Lett., **68**, 1563 (1992).

[3]C. Thomsen, H. T. Grahn, H. J. Maris and J. Tauc, Phys. Rev. B, **34**, 4129 (1986).

Phonon Spectroscopy by Time-Resolved Detection of Picosecond Surface Vibrations

O.B. Wright and K. Kawashima

Electronics Research Laboratories, Nippon Steel Corporation,
5-10-1 Fuchinobe, Sagamihara, Kanagawa 229, Japan

Abstract. We present a new optical technique for the spectroscopy of longitudinal acoustic phonon pulses in thin films by direct time-resolved detection of picosecond vibrations of the film surface. We apply it to thin films of amorphous silicon on chromium substrates at around 20 GHz.

The detection of optically excited coherent phonon strain pulses in thin opaque films with the laser picosecond acoustic method[1] is based on the measurement of time-resoved changes in reflectance [$\delta R(t)$] arising from the modulation of the film dielectric constants by the phonon strain. The shape of the acoustic echoes observed in this way can be written $\delta R(t) = \int f(z)\eta_{33}(z, t)\mathrm{d}z$, where η_{33} is the longitudinal strain, z the depth in the film, and $f(z)$ a damped oscillatory sensitivity function which depends on the photoelastic constants of the film. For materials in which the photoelastic constants are negligible it is not possible to observe phonon pulses. In addition, the function $f(z)$ distorts the phonon pulse shapes, and phonon spectroscopy with this method is limited to a comparison between successive echoes to determine, for example, the frequency-dependent ultrasonic attenuation[1].

For transparent films we recently observed phonon pulses through vibrations of the film surface by picosecond interferometry in the time domain[2]. As a natural extension of this research we describe a time-resolved method for picosecond phonon pulse detection in opaque films based on a technique first proposed by Rothenberg for transient thermal expansion measurements in bulk materials[3]: laser beam deflection combined with optical pumping and probing. Changes in the surface profile induced by optically excited phonon pulses bouncing back and forth inside the film are interrogated by the angular deflection ($\delta\theta$) of a probe laser beam. Since the beam deflection arises from changes in the surface slope, echo detection is still possible for materials in which the

Springer Series in Solid-State Sciences, Volume 112
Phonon Scattering in Condensed Matter VII Eds.: M. Meissner and R.O. Pohl

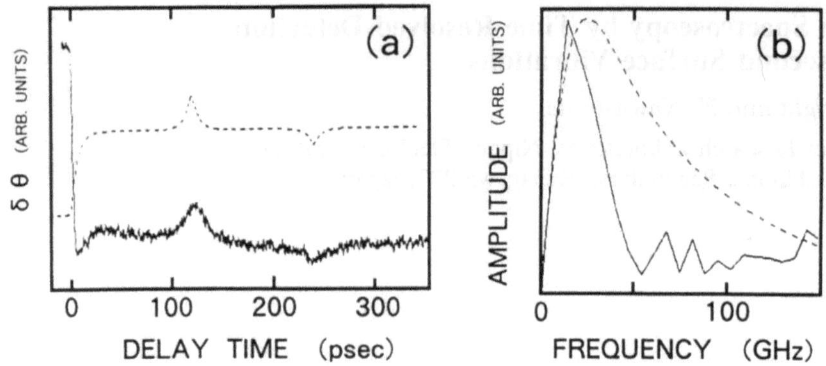

Fig. 1: a) angular beam deflection against delay time for a 500-nm a-Si film. b) phonon spectrum of the strain pulse corresponding to the first echo. The dashed theoretical curves include the slight smoothing effect of the optical pulse duration.

photoelastic constants are negligible at the probe photon energy used. We consider this case in the present paper.

Figure 1(a) shows $\delta\theta$ as a function of the delay time (t) between the pump and probe pulses for a 500-nm sputtered film of a-Si on a Cr substrate. Optical pulses of duration 2 ps (FWHM) with photon energy 2 eV and pump pulse energy 0.4 nJ are used. The negative step at $t=0$ is caused by thermal effects[3], not considered here. The echo pulse shape, proportional to the surface displacement $\int\eta_{33}(z, t)dz$, is calculated using a thermoelastic model (upper curve) using data for the optical absorption coefficient of a-Si from ellipsometry ($\zeta=53$ nm). The overall pulse shape is reproduced well. We found that the photoelastic constants for a-Si relevant to this experiment are negligible from time-resolved reflectance data for the same film: no echoes could be detected with this technique. These constants are also negligible for crystalline silicon[3].

For longitudinal strain pulses it can be shown from the elastic wave equation using a one-dimensional approximation that the surface velocity variation ($\propto d\theta/dt$) is directly proportional to the incident strain pulse shape. Therefore, by taking the Fourier transform (FT) of $d\theta/dt$, the frequency spectrum of the phonon strain pulse can be directly obtained without distortion. Figure 1(b) shows the Fourier transform of $d\theta/dt$ ($\propto$ FT of $\omega\delta\theta(t)$, ω the angular frequency) for the first echo, peaked at 14 GHz, together with the predicted curve. The deviations from theory should elucidate diffusion processes during pulse generation.

[1]C. Thomsen, H. T. Grahn, H. J. Maris and J. Tauc, Phys. Rev. B, 34, 4129 (1986).
[2]O. B. Wright, J. Appl. Phys., 71,1617 (1992).
[3]J. E. Rothenberg, Opt. Lett., 13, 713 (1988).

Laser Picosecond Interferometry in Double-Layer Transparent Films on Opaque Substrates

O.B. Wright, T. Hyoguchi, and K. Kawashima

Electronics Research Laboratories, Nippon Steel Corporation, 5-10-1 Fuchinobe, Sagamihara, Kanagawa 229, Japan

Abstract. Picosecond phonon pulses are optically excited and detected in transparent double-layer films of silica and amorphous silicon nitride on chromium substrates. By modelling the reflectance variation we derive the film thicknesses, sound velocities and photoelastic constants.

The laser picosecond acoustic technique shows great promise as a means for investigating coherent longitudinal acoustic phonon propagation and reflection in the 10 GHz to THz range in thin films[1]. There are also many technologically important applications such as the non-destructive testing of thickness, elastic properties and bonding for the nm-μm thickness range. With opaque samples, for example, the phonon pulse propagation time derived from the acoustic echo spacing allows the ratio of the thickness (d) to the sound velocity (v) to be determined.

For a transparent film on an opaque substrate the phonon pulses are generated at the interface between the film and substrate, and bounce back and forth inside the film[2]. The reflectance variation is more complicated than for opaque films, but permits a more quantitative analysis of the film properties. We present an extension of this analysis to a double-layer transparent film geometry[3] and compare it with data for sputtered films of a-Si_3N_4 on a-SiO_2 (labelled 2 and 1, respectively) of nominal thickness either both 500 nm or both 300 nm (on Cr substrates).

The upper curves in Fig. 1 show the measured reflectance variations obtained with 2-ps (FWHM) pump laser pulses of wavelength $\lambda = 630$ nm (photon energy 2 eV) and with normally incident and delayed laser probe pulses. The beat oscillations, proportional to the photoelastic constants of the films, arise from the changing phase of the light reflected by the moving phonon stress pulse in the transparent films[1]. These oscillations change in amplitude, phase and frequency at delay time $t = d_1/v_1$ when the stress pulse reaches the bottom of the upper a-Si_3N_4 film. The oscillation period τ for $t < d_1/v_1$ allows v_1 to be derived

Springer Series in Solid-State Sciences, Volume 112
Phonon Scattering in Condensed Matter VII Eds.: M. Meissner and R.O. Pohl

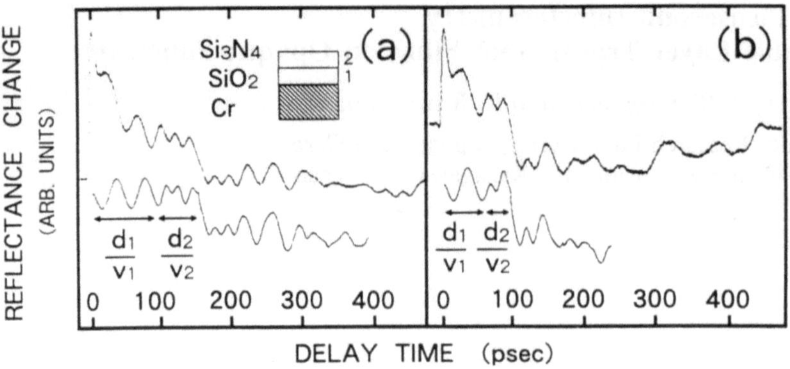

Fig. 1: Reflectance change against delay time for a-Si_3N_4/a-SiO_2 films with nominal thicknesses a) 500 nm and b) 300 nm. Upper curves: experiment. Lower curves: theory.

from $v_1 = \lambda/2n_1\tau$, where n_1 is the refractive index. At $t = d_1/v_1 + d_2/v_2$ the sharp change in signal level is caused by a combination of the ultrafast motion of the film surface when the stress pulse is reflected from it and the change in the transmission of light through the pulse at this time.

For $t > d_1/v_1$ the interpretation is complicated by multiple acoustic reflections, so we extend the theory for single-layer films (including the effect of multiple optical and acoustic reflections), as shown by the lower, fitted curves in Fig. 1. The beat oscillations are calculated by ignoring the optical reflection from the interface between the two films, but one such reflection (from either direction) is included in the calculation of the effect of the ultrafast motion and the light transmission through the stress pulse. The fitted constants are $v_1 \approx 5400$ ms^{-1}, $v_2 \approx 8300$ ms^{-1} (better estimates than in Ref. [3]—see Refs. [2] and [4]), $p_{12} = 0.28$ and 0.023 (Pockels coefficients for films 1 and 2), and $d_1 \approx d_2 \approx 480$ nm and 300 nm for Fig. 1(a) and 1(b), respectively. We use literature values $n_1 = 1.46$, $n_2 = 2.05$, and the known substrate refractive index $3.1 + 2.9i$ and υ (for stress pulse modelling). The good agreement with experiment (apart from the mainly thermal change near $t = 0$) suggests that this method can be used for the quantitative testing of two-layer films in practical situations, such as in the case of a transparent film deposited on an opaque surface covered with a thin, transparent oxide layer.

[1]H. T. Grahn, H. J. Maris and J. Tauc, IEEE J. Quantum. Electron., **25**, 2562 (1989).

[2]O. B. Wright, J. Appl. Phys., **71**, 1617 (1992).

[3]O. B. Wright, T. Hyoguchi and K. Kawashima, in *Photoacoustic and Photothermal Phenomena* III, D. Bicanic et al., eds. (Springer-Verlag New York, 1992) [in press].

[4]T. S. Hickernell, F. M. Fliegel and F. S. Hickernell, IEEE Ultrasonics Symposium, 445 (1990).

Picosecond Optics Measurements of the Acoustic Impedance of Water at Frequencies up to 280 GHz

C.J. Morath, R.J. Stoner, and H.J. Maris

Department of Physics, Brown University, Providence, RI 02912, USA

We report the first ultrasonic measurements of the acoustic impedance of water at frequencies as high as 280 GHz. We describe the picosecond optical technique and present preliminary results at 25 C for the reflection coefficient of bismuth/water interfaces as a function of frequency.

The basic idea of the experiment is shown in Fig. 1. A picosecond light pulse (the "pump") is absorbed in a thin metal (Bi) film, causing it to vibrate. The time-varying stress causes changes in the optical reflectivity of the sample $\Delta R(t)$ which is measured by a time-delayed "probe" light pulse. The metal film lies on a layer of photoresist which is in turn deposited on a silicon substrate. The large acoustical mismatch between the metal and photoresist causes the film vibrations to be slowly damped. When water is placed in contact with the free metal surface, acoustic waves are launched into the water resulting in additional damping. This extra damping gives the magnitude of the reflection coefficient of the water/metal interface. The frequency dependence of the reflection coefficient can then be measured by varying the thickness of the metal film between samples.

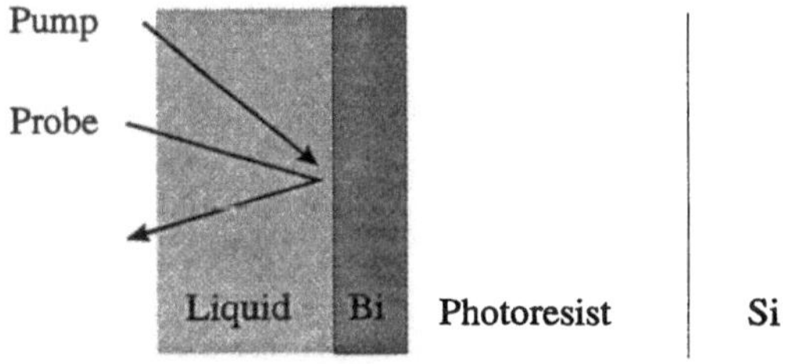

Fig.1 Schematic of the experiment. The pump pulse causes the Bi film to vibrate. The damping of the vibration is a measure of the reflection coefficient between the water and bismuth.

Typical data $\Delta R(t)$ are shown in Fig. 2a for a 40 Å bismuth film on photoresist (PR) on silicon. The slow damping rate reflects the large acoustic mismatch between Bi and PR. The damping of the oscillation gives the acoustic reflection coefficient of the Bi/PR interface. In Fig. 2b data are shown for the same sample in contact with water at 25 C. The additional damping is due to vibrations propagating into the water from the Bi film.

Springer Series in Solid-State Sciences, Volume 112
Phonon Scattering in Condensed Matter VII Eds.: M. Meissner and R.O. Pohl

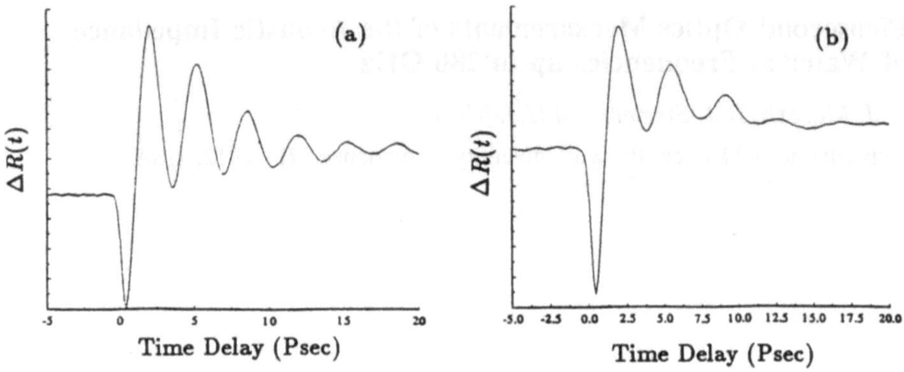

Fig.2 (a) Data $\Delta R(t)$ for a 40 Å Bi film on 2000Å photoresist on Si (b) Data for same sample immersed in water at 25 C.

The vibrating metal film satisfies the resonance condition

$$1 - r_s r_w e^{2ikd} = 0 \tag{1}$$

where r_s and r_w are the reflection coefficients for the substrate and water interfaces, respectively, $k = (\omega - i\Gamma)/v$, Γ is the damping rate at frequency ω, and v is the sound velocity in the metal. Solving the real part of Eq. 1 yields the relation $|r_w| = exp(-2\Gamma d/v)/|r_s|$.

Fig. 3 shows preliminary results for $|r_w|$ versus frequency at 25 C. A decrease in $|r_w|$ of about 10% is seen from 100 to 280 GHz. This effect may be due to the "stiffening" of water which occurs over a frequency range with $\omega\tau \approx 1$, where τ is the structural relaxation time of the water, thought to be approximately 1 psec at 25 C [1]. This stiffening would then be expected to occur over the frequencies measured.

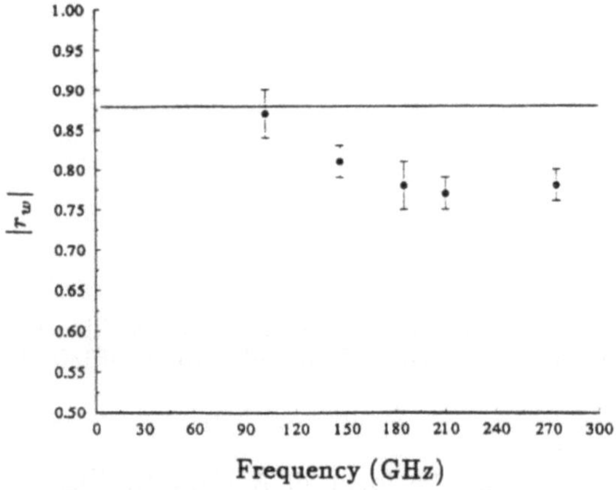

Fig.3 Results for the reflection coefficient of the Bismuth/Water interface $|r_w|$ as a function of frequency.

References

1. See for example C.M. Davis, Jr. and J. Jarzynski, in "Water: A Comprehensive Treatise" (F. Franks, ed.), Plenum Press, New York (1972).

Investigation of $Si_{1-x}Ge_x$ and $CoSi_2$ Films by Brillouin Scattering from Surface Acoustic Phonons

M. Mendik, M. Ospelt, C. Schwarz, H. von Känel, and P. Wachter

Laboratorium für Festkörperphysik, Eidgenössische Technische Hochschule Zürich, CH-8093 Zürich, Switzerland

The investigation of the dynamical properties of epitaxially grown thin films by inelastic light scattering techniques is an important task in materials science. Brillouin scattering (BS) detects thermally excited surface acoustic waves (SAW's) in the GHz range with a wavevector $\mathbf{k} \cong 0$ and a penetration depth of $\cong 3000$Å. The dynamical response of a perfect elastic medium characterized by its density ρ and the stiffness tensor C_{ijkl} (reduced to C_{11}, C_{44}, C_{12} for cubic systems) can be described by a wave equation for the particle displacements [1]. SAW's are described by plane waves propagating with a phase velocity v_{saw} and a wavevector $\mathbf{k_p}$ in the plane with a damped amplitude towards the interior of the bulk. For a single layer on a substrate boundary conditions for the substrate–layer and layer–vacuum interface have both to be met. In our case where layer and film exhibit cubic symmetry the velocity of the SAW's propagating on (100) oriented films can be computed iteratively and identified among the various bulk solutions that are present in layered media. Analyzing the angular dependent velocity of the SAW's, which depends on the bulk elastic constants and the densities of the substrate as well as the layer, allows to fit the elastic constants of the layer alone. This enables the characterization of individual layers or materials that can be grown to a limited thickness only.

For the experiments we used the TM polarized λ=514.5nm line of an Ar^+–ion laser as the excitation source. To avoid thermal induced modifications or phase transitions the power on the sample did not exceed 60mW. The backscattered light was analyzed by a (3+3) tandem Fabry–Pérot interferometer. During experiments the samples were in air and at room temperature. The incident angle of the laser beam relative to the sample normal was varied from 35° to 80° in order to change the length of $\mathbf{k_p}$ of the selected SAW, whereas rotating the sample around the normal allowed to select the direction of propagation of the SAW in the sample plane.

$Si_{1-x}Ge_x$ and $CoSi_2$ films of various thicknesses h ranging from 80Å to 5000Å were grown on n–doped Si(100) wafers by MBE and analyzed in situ by reflection high energy electron diffraction, X–ray photoelectron spectroscopy, scanning tunneling microscopy and ex situ by Rutherford backscattering spectrometry [2,3].

For the calculations, in contrast to other authors we avoided the use of the ultrasonic data for the substrate and utilized the elastic tensor obtained from the angular dispersion of SAW's on bare Si(100) wafers instead [4]: C_{11}=(161.4), C_{44}=(75.9), C_{12}=(61.3) GPa and ρ=(2300) kg/m³.

Springer Series in Solid-State Sciences, Volume 112
Phonon Scattering in Condensed Matter VII Eds.: M. Meissner and R.O. Pohl

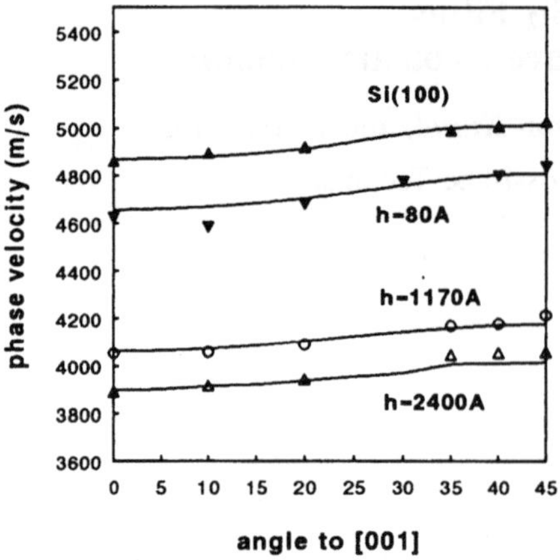

Fig.1) SAW's on $Si_{1-x}Ge_x$

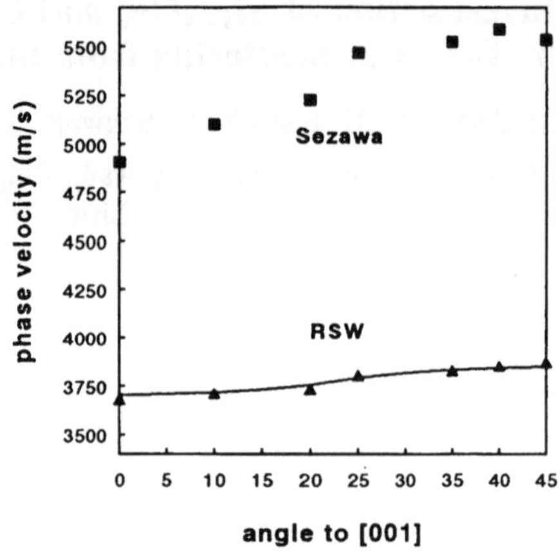

Fig.2) SAW's on $CoSi_2$

Fig.1 shows the angular dispersion of the SAW's on $Si_{1-x}Ge_x$ alloys. Thus we have fitted the following elastic constants: for $Si_{.50}Ge_{.50}$ (with h=80Å): C_{11}= (144.0), C_{44} = (67.3), C_{12}= (54.7) GPa with ρ=3800kg/m³; for $Si_{.65}Ge_{.35}$ (with h=1170Å, 2400Å): C_{11}=(149.5), C_{44}=(58.7), C_{12}=(48.1) GPa with ρ=3360kg/m³. These values agree with the linearly interpolated elastic constants of the constitutents Si and Ge.

Fig.2 demonstrates the angular dispersion of SAW's on $CoSi_2$(100) with h=660Å. In contrast to $Si_{1-x}Ge_x$ films, there exist higher order modes (Sezawa) that are illustrated in the same figure. Using the elastic constants for $CoSi_2$ obtained from ultrasonic measurements on large single crystals [5] as starting values for our fit, we computed the following elastic constants with ρ=4700 kg/m³: C_{11}=(200±5), C_{44}=(78±4), C_{12}= (138±8) GPa. Thus for the bulk modulus $B={}^1/_3\times(C_{11}+2\times C_{12})$ we obtained: (158 ± 6) GPa which is 15% lower than the theoretically calculated B for bulk $CoSi_2$ [6]. Comparing with the ultrasonic bulk data, C_{11} is lowered by 12% and C_{44} by 6%. Whether reductions can be explained by the presence of a thermic strain or the existence of dislocations will have to be figured out in the future.

In conclusion, we have shown that it is possible to obtain the elastic constants from the angular dispersion of SAW's on elastically anisotropic thin cubic films.

[1] G.W.Farnell, in *Physical Acoustics*, Ed.W.P.Mason, Vol.6 (Academic Press,New York, 1970) pp. 109–166.

[2] M.Ospelt,W.Bacsa,J.Henz,K.A.Mader and H.von Kanel, *Superlatt. Microstruct.***5** (1989)71.

[3] R.Stalder,C.Schwarz,H.Sirringhaus,H.v.Känel,*Surf.Sci.***271**(1992)355

[4] M.Mendik, M.Ospelt, H.von Kanel, P.Wachter, *Applied Surface science* 50 (1991) 303–307.

[5] G.Guenin,M.Ignat,O.Thomas,*J.Appl.Phys.*,**68**,(1990),6515.

[6] W.R.L.Lambrecht,N.E.Christensen,P.Blöchl,*Phys.Rev.B*,**36**,2493(1987)

Elastic Properties of YbX (X = As, P, N) Single Crystals Determined by Brillouin Scattering from Surface Acoustic Phonons

M. Mendik and P. Wachter

Laboratorium für Festkörperphysik, Eidgenössische Technische Hochschule Zürich, CH-8093 Zürich, Switzerland

Inelastic light scattering techniques probe the samples by a contactless way at a small laser beam spot and therefore experiments under various conditions are possible. Brillouin scattering (BS) detects thermally excited phonons with a wavevector $\mathbf{k} \cong 0$ in the GHz range. Recently performed Raman and Infrared absorption experiments [1] on stoichiometric single crystals spurred a complete determination of the elastic properties of the Yb monopnictides with respect to a complete identification of both optical and acoustic phonons.

For the experiments we used the TM polarized $\lambda = 514.5$ nm line of an Ar^{+}–ion laser as the excitation source. To avoid radiation damage the power on the sample has been held below 50 mW. The backscattered light was analyzed by a (3+3) pass tandem Fabry–Pérot interferometer. The incident angle of the laser beam relative to the sample normal has been kept fixed at 45° whereas rotating the sample around the normal has allowed to select the direction of propagation of the surface acoustic wave (SAW's) in the sample plane. The sample preparation is described in [3]. To avoid corrosion by air all samples have been prepared in a helium glove box and measured in He atmosphere. All (100) single crystals were finally oriented by Laue diffraction in order to fix the [001] directon in the sample plane. From the lattice constants determined by X–rays we obtained the densities for YbN, YbP and YbAs (all NaCl structure) : $\rho = 10880, 7720, 8520$ (kg/m^3), respectively.

The dynamical response of a perfect elastic medium characterized by its density ρ and the elastic stiffness tensor C_{ijkl} (cubic systems: C_{11}, C_{44}, C_{12}) can be described by a wave equation for the particle displacements. In particuliar, phonons localized at the surface are described by plane waves propagating with a speed v_{saw} and $\mathbf{k}_p$ in the plane and with a damped amplitude towards the interior of the bulk (penetration depth $\cong$ 3000Å). Hence, analyzing the angular dependent velocity of SAW's, which depends on the bulk elastic constants and the density alone, allows to fit the elastic constants C_{11}, C_{44} and C_{12} of the bulk as a whole.[4]

The angular dispersions of SAW's on (100) oriented planes of Yb pnictide single crystals are all shown in Fig.1 . On the (100) plane, the azimuthal angle α is measured from the [001] direction. From the experiment we determine systematically increasing SAW velocities going from YbAs to YbN, indicating increasing hardness. In addition, within the measurement error of 1–2% no changes of the SAW velocity as a function of the direction have been observed which is a sign for elastic "quasi" isotropy (i.e. the

Springer Series in Solid-State Sciences, Volume 112
Phonon Scattering in Condensed Matter VII Eds.: M. Meissner and R.O. Pohl

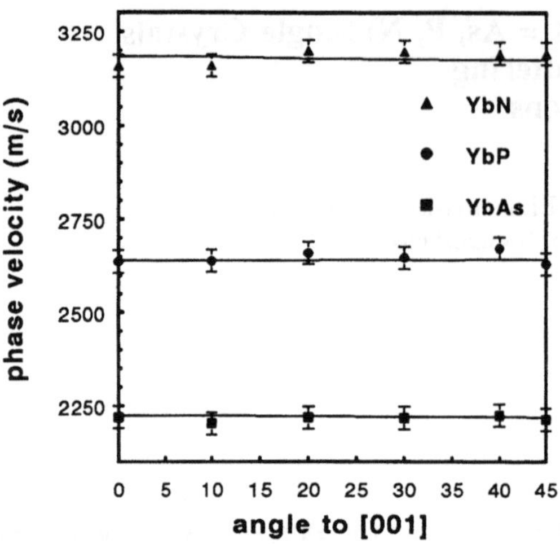

Fig.1) angular dispersion

anisotropy ratio $\eta = 2 \times C_{44}/(C_{11}-C_{12})$ is smaller than one which is in general found in most materials crystallizing in the fcc rocksalt structure[4]. From the angular dispersion we have fitted the elastic constants keeping the density fixed. The best fits have been obtained with: for <u>YbN</u>: C_{11} = (385 ±61) GPa, C_{44} = (129 ±9) GPa, C_{12} = (108.8 ±20) GPa, <u>YbP</u>: C_{11} = (270.3 ±49) GPa, C_{44} = (62 ±5) GPa, C_{12} = (150.4 ±30) GPa, <u>YbAs</u>: C_{11} = (250.5 ±45) GPa, C_{44} = (47 ±3) GPa, C_{12} = (147.8 ±29) GPa. The velocity of SAW's on quasi isotropic materials is closely related to the lowest bulk pure shear wave velocity $v_{shear}^2 = (C_{44}/\rho)$ which explains the mean errors of 17% for C_{11} and 20% for C_{12}. From the fitted elastic constants we have derived for the Poisson's ratio (along [001]) $\nu = C_{12}/(C_{11} + C_{12})$: ν = (0.2±0.1), (0.35±0.1), (0.37±0.1), for YbN, YbP, YbAs, respectively. Despite the error bar a clear destinction between the elastic properties of YbN and YbP or YbAs is possible. To obtain an expression for the longitudinal atomic force constant f_l along [001] we have combined the Phonon dispersion relation for a one dimensional, diatomic chain [5] with the equation for the velocity of a longitudinal phonon along [001]: $(\omega/\mathbf{k})^2 = v_{long}^2 = (C_{11}/\rho)$. Hence: $f_l = (2 \times C_{11} \times (M_1 + M_2))/(a^2 \times \rho)$, where a denotes the lattice constant, ρ the density and M_1,M_2 the atomic mass of the ions. Thus we obtained for f_l = (1.0±0.1), (0.8±0.1), (0.7±0.1) dyne/Å, for YbN, YbP, YbAs respectively. With the expression for the optical frequency ω_{opt} at $\mathbf{k} = 0$ given in [5] we obtained for ω_{opt} = (0.050± 0.01), (0.035±0.01), (0.025±0.01) eV which is in reasonable greement with the results obtained by infrared spectroscopy [1].
In conclusion, we have shown that it is possible to deduce some useful elastic properties of quasi isotropic single crystals from the angular dispersion of SAW's on (100) oriented planes.

[1] L.Degiorgi,G.Compagnini,S.Teraoka and P.Wachter,to be published.
[2] E.Kaldis, B.Steinmann, B.Fritzler, E.Jilek, A.Wisard, in *The Rare Earths in Modern Science and Technology*, ed. by J.McCarthy et al.
[3] M.W.Elmiger, J.Henz, H.v.Kanel, M.Ospelt, P.Wachter, *Surface and Interface Analysis* **14**,18(1989).
[4] R.E.Newnham, in *Structure–Property Relations*, Springer,1975,173.
[5] N.W.Ashcroft, N.D.Mermin,*Solid State Physics*,Holt–Saunders ed.,1981

Design of a Superlattice-Based Phonon Optics Device

S. Mizuno and S. Tamura

Department of Engineering Science, Hokkaido University, Sapporo 060, Japan

Phonon optics devices such as frequency-selective phonon filters and reflectors based on thin-film superlattice structures were suggested by Narayanamurti several years ago [1]. Here, we propose a new type of phonon optics device. The system we consider has a double barrier structure for phonons consisting of a bulk material sandwiched between periodic superlattices [2]. Schematic frequency-band diagram for this system is shown in Fig. 1. In this figure, the hatched regions indicate the frequency gaps of superlattices. The phonons within these frequency gaps cannot be transmitted through the system. In other words, the superlattices act as barriers for phonons with frequencies between ω_L and ω_H. Thus, we can regard the system shown in Fig. 1 as a double barrier system for phonons in the forbidden gap. This structure is similar to the double-barrier quantum-well structure for electrons [3].

We have derived the analytical expressions for the phonon transmission rate in this system. Here, we illustrate the frequency dependence of the transmission rate for a numerical example, calculated from the analytical expressions. As the numerical example, we choose (100)–AlAs/GaAs superlattices for the SL parts and AlAs for the BK part; the unit period of the SL parts consists of $(AlAs)_6(GaAs)_6$, and the number of period is $N = 12$, and the thickness of

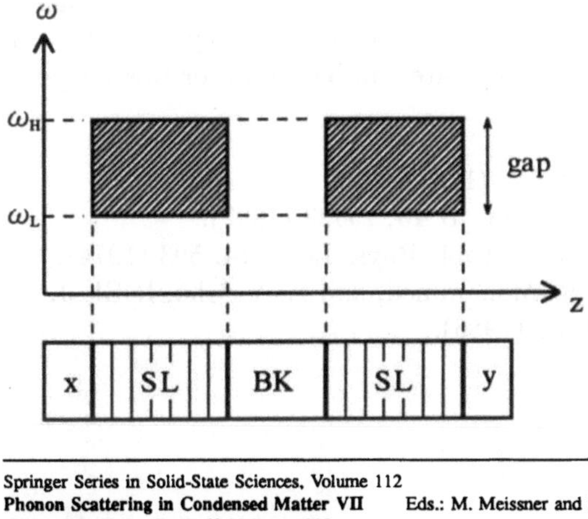

Fig. 1. Schematic frequency-band along a direction z, perpendicular to the layer interfaces of the double-barrier system. The parts SL, BK, x, and y mean the periodic superlattice, bulk material, substrate, and phonon detector, respectively.

Springer Series in Solid-State Sciences, Volume 112
Phonon Scattering in Condensed Matter VII Eds.: M. Meissner and R.O. Pohl

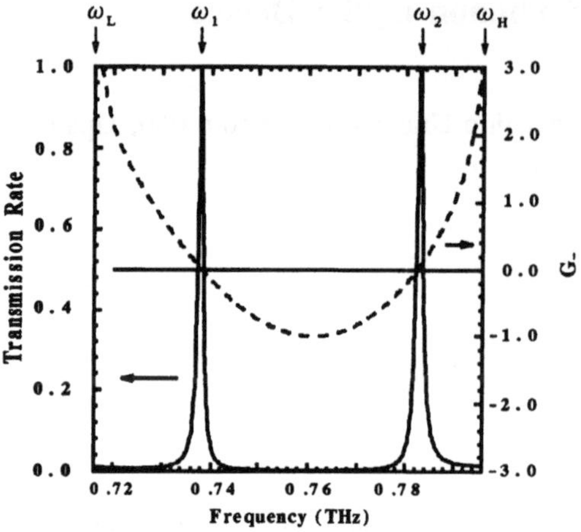

Fig. 2. The phonon transmission rate (solid line) and the function G_ (dotted line) within the lowest frequency gap in the double-barrier system.

the BK part is the same as the SL part. In Fig. 2, the transmission rate is shown in the lowest frequency gap (0.72–0.80 THz) of the SL part. We find two sharp enhancements within this frequency gap. These enhancements in transmission correspond to the resonance characteristic of the system. That is, we can regard the BK part as an impurity embedded in the perfect periodic superlattice. Thus, we can understand that the impurity levels are induced by this BK part. If the frequency of the incident phonons coincides with that of these impurity levels, the phonons in the gaps of the SL part are transmitted through both barrier without attenuation.

Based on the Green's function method, we derive the equation giving the frequencies of the impurity levels: $G_- = 0$. The dotted line in Fig. 2 shows this function G_-. We can find a good coincidence of the frequencies at the sharp enhancements in transmission with the frequencies satisfying $G_- = 0$.

Our results signify that the double-barrier system we consider is analogous to the double-barrier potential for electrons in which resonant tunneling occurs. This similarity suggests the potential for designing various phonon optics devices. In particular, the sharp resonant transmission in a wide frequency gap originating from the impurity levels can be used to design a phonon resonator for the detection or generation of monochromatic high-frequency phonons. Our analytical expression for the transmission rate will be useful for this purpose.

[1] V. Narayanamurti, Science **213**, 717 (1981).
[2] S. Mizuno and S. Tamura, Phys. Rev. B **45**, 13423 (1992).
[3] L. L. Chang, L. Esaki, and R. Tsu, Appl. Phys. Lett. **24**, 593 (1974). For a review, see F. Capasso, K. Mohammed, and A. Y. Cho, IEEE J. Quantum Electron. **QE-22**, 1853 (1986).

Low-Frequency Raman Scattering by Acoustic Vibrations Localized in Ultrathin Semiconductor Films

V.N. Novikov[1] *and N.N. Ovsyuk*[2]

[1]Institute of Automation and Electrometry, Russian Academy of Sciences, 630090 Novosibirsk, Russia

[2]Institute of Mineralogy and Petrography, Russian Academy of Sciences, 630090 Novosibirsk, Russia

Abstract. In this work thin strained and unstrained epitaxial films of Ge on Si and GaAs substrates, respectively have been studied by low-frequensy Raman scattering. We found out experimentally and established theoretically existance of the eigenmodes of layer acoustical vibrations in these systems. The frequency of such type of vibrations is of the order of v/L, where v is the longitudinal sound velocity and L is thickness of the layer.

1. Introduction

Currently great attention have been payed to theoretical investigation of the vibrational properties of composite layered structures [1]. Along with different kinds of surface vibrations at frequencies lower than the frequency of Brillouin peak the eigenmodes of layer vibrations must be observed in these systems. The frequency of such type of vibration is of the order of v/L, where v is longitudinal sound velocity, L - the thickness of the layer. In this work thin strained and unstrained epitaxial films of Ge on Si and GaAs substrates, respectively have been studied.

2. Results and discussion

The film was grown in the ultrahigh vacuum system at the remaining gas pressure lower than 10 Pa. Raman spectra were obtained in geometry of 90° scattering on Ramanor U-1000 spectrometer in (‖‖) polarization. Experimental Raman spectra are shown in Fig.1. The broad intensive peaks we connect with localized vibrational modes of the films. The eigenmodes of vibrations localized in a film and exponentially

Springer Series in Solid-State Sciences, Volume 112
Phonon Scattering in Condensed Matter VII Eds.: M. Meissner and R.O. Pohl

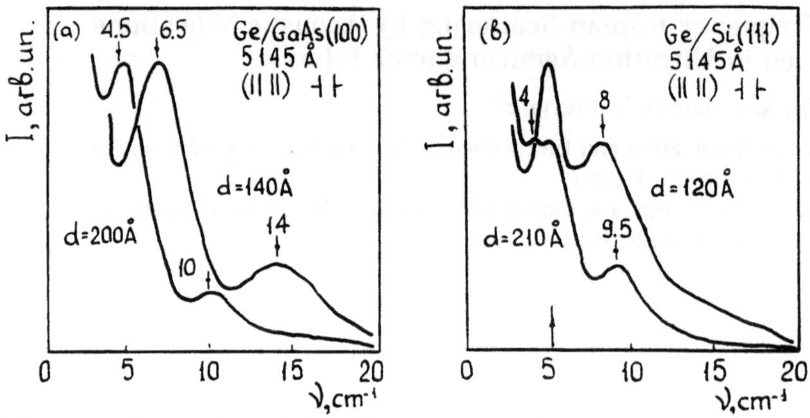

Fig.1. Experimental Raman spectra of Ge films on GaAs (a) and Si (b) substrates. The arrow indicate Brillouin peak at 5 cm^{-1}.

decaying in substrate can be found out by solving of respective equation from [1]. Let us denote by β_i force constant and by m_i average atomic mass, where i=1,2 for film and substrate respectively. Eigenfrequencies are equal to $\omega=2(\beta_1/m_1)^{1/2}sin(pL/2)$, where p is complex: $p=p_1+ip_2$. In our case the solution of the equation can be found out analytically due to smallness of the parameter $\beta_1 a/\beta L \approx 10^{-4}$, where a is interatomic distance and $1/\beta=1/\beta_1+1/\beta_2-2/\beta_I$, β_I is force constant between interface atoms. We are using approximation $\beta_I=(\beta_1+\beta_2)/2$. The value of eigenfrequency and its width are equal to:

$$\omega_n = \pi n(1+\beta_1 a/2\beta L)v/L \approx \pi n v/L \quad (1)$$

$$\Gamma = arcth(m_2\beta_2/m_1\beta_1)^{1/2}v/L \approx 0.9v/L \quad (2)$$

From (1) and (2) one can see that vibrational spectrum is nearly equidistant and the width of the modes is constant. The thickness of the film estimated from experimental peak energy with the help of eq.(1) is slightly lower (≈20 A) than one obtained by direct measuring. This is probably due to oxide layer which covers the film and results in the shift of the peaks to higher frequencies.

References

1. B.Sylla, L.Dobrzynski, H.Puszkarski, J. Phys. Cond. Matt.1(1989)1247.

Part X

Particle Detectors

Phonon-Mediated Detection of Particles

F. Pröbst for the Munich Cryogenic Detector Collaboration

Max-Planck-Institut für Physik, Föhringer Ring 6,
W-8000 München 40, Fed. Rep. of Germany

1 Introduction

At the Max-Planck-Institut für Physik and the Technische Universität München we are developing massive cryogenic detectors for scattering events with small deposited energy (< 1 keV). The type of detector I report on here consists of a dielectric absorber crystal, a strip of superconducting material evaporated onto it serving as thermometer, and a weak thermal coupling connecting the thermometer to a constant temperature bath. The detector is operated within the superconducting-to-normal transition of the thermometer, where a small temperature rise of the strip leads to relatively large rise of its resistance. The resistance of the strip (~0.1 Ω) is measured by passing a constant current through a circuit in which the strip is in parallel with a 0.05 Ω resistor and the pickup coil of a SQUID. A particle-induced rise in the strip resistance is measured via the current rise through the SQUID.

Here I report on tests of a detector with a 19 g Si crystal and an Ir/Au proximity-effect thermometer with a very wide transition (15 to 115 mK) of nearly constant slope, which enables systematic studies of the detector response in this temperature range. The energy resolution improves with decreasing operating temperatures. At 17 mK, single 60 keV photons have been detected with an energy resolution of 1 keV (FWHM).

The pulses have two exponential components: a fast nonthermal one with a decay time of ~1.5 ms at all temperatures, and a slow thermal one with a temperature-dependent decay time of 40 to 60 ms. At the higher operating temperatures the amplitude of the slow component is approximately as expected from the heat capacity of the absorber and the deposited energy. However, at lower temperatures it becomes smaller than expected from the heat capacity, with a factor of ~ 100 missing at 17 mK.

In the following section I present a thermal model of the detector which explains the time structure of the pulses and the suppression of the thermal component at low temperatures.

Springer Series in Solid-State Sciences, Volume 112
Phonon Scattering in Condensed Matter VII Eds.: M. Meissner and R.O. Pohl

2 The Model

The electrons and phonons of the metallic film are considered as subsystems weakly coupled by the electron-phonon coupling G_{e-p}. The phonon systems of the film and the absorber are coupled by the Kapitza coupling G_K across the interface, while the electrons are coupled to the thermal bath via G_{e-b}. The particle interaction creates high frequency phonons which do not thermalize in the absorber. These high frequency phonons are efficiently absorbed by the electrons in the metal film. Due to the strong interaction among the electrons, the energy transferred to a single electron is thermalized within the electronic subsystem. The absorption of high frequency phonons is thus a time-dependent power input P(t) into the electronic subsystem of the film, and is responsible for the fast signal component. Thermal phonons emitted from the electrons can either escape into the absorber or into the heat sink via G_{e-b}. The thermal signal is reduced by two mechanisms:

- Some of the phonons thermalized by the conduction electrons escape into the heat sink and do not heat the absorber.

- The thermometer does not "see" the full temperature rise of the absorber, because at low temperatures $G_{e-b} \gg G_{e-p}$.

2.1 Calculation of the thermometer response

The electrons are coupled to the heat bath at temperature T_b on the left side (x=0) of the strip by a Au-bonding wire of thermal conductance G_{Au}. Electrical connection is made by superconducting Al-bonding wires of negligible thermal conductance. A homogeneous heat input P(t) leads to a rise of the electron temperature $T_e(x,t)$. The resistance rise and hence the signal is assumed proportional to the average rise of electron temperature $\Delta\bar{T}_e = \bar{T}_e - T_b$. T_e and the absorber temperature T_a are calculated from:

$$C_e\frac{\partial T_e}{\partial t} - \kappa V_e\frac{\partial^2 T_e}{\partial x^2} + (T_e - T_a)\,G_{e-a} = P(t) \tag{1}$$

$$C_a\frac{dT_a}{dt} = (\bar{T}_e - T_a)G_{e-a} \tag{2}$$

where C_e, κ and V_e are the heat capacity, thermal conductivity and the volume of the electron system, C_a is the heat capacity of the absorber, and G_{e-a} the effective coupling of the electron system of the film to the phonon system of the absorber ($1/G_{e-a} = 1/G_{e-p} + 1/G_K$). Apart from a short period after "switching" on P(t) (the signal rise time), $C_e\partial T_e/\partial t$ in eq. 1 is negligible, and is therefore omitted.

With the boundary conditions $T_e = T_b + (P - C_a dT_a/dt)/G_{Au}$ at x=0 and $dT_e/dx = 0$ at x=L (L=length of the strip) equations 1 and 2 are solved by:

$$\Delta\bar{T}_e(t) = \frac{(1-\gamma)}{G_{e-a}} P(t) + (1-\gamma)\Delta T_a(t) \tag{3}$$

$$\Delta T_a(t) = \left(\frac{E}{C_a}\right)\left(\frac{1-\gamma}{1-\frac{\tau}{\tau_{eff}}}\right)\left(\exp\left(\frac{-t}{\tau_{eff}}\right) - \exp\left(\frac{-t}{\tau}\right)\right)\Theta(t) \tag{4}$$

with E the energy deposited in the crystal, $\Delta T_a(t) = T_a(t) - T_b$, and γ, λ, and τ_{eff} given by:

$$\gamma = \left(\frac{G_{e-a}}{G_{Au}} + \frac{\lambda L}{\tanh(\lambda L)}\right)^{-1}; \quad \lambda = \sqrt{\frac{G_{e-a}}{\kappa V_e}}; \quad \tau_{eff} = \frac{1}{\gamma}\frac{C_a}{G_{e-a}} \tag{5}$$

$\Delta\bar{T}_e(t)$ has two components, a fast and a slow one with the time structure of P(t) and $\Delta T_a(t)$, respectively. The thermometer "sees" $\Delta T_a(t)$ reduced with a factor $(1-\gamma)$. The absorber temperature rises with the decay time constant τ of $P(t)$ and decays with τ_{eff}; its amplitude is suppressed essentially by another factor of $(1-\gamma)$ compared to E/C_a .

$P(t)$ is calculated from:

$$P(t) = P_0 \exp\left(\frac{-t}{\tau}\right)\Theta(t) \tag{6}$$

with P_0 and τ given by:

$$P_0 = \frac{1}{2}\frac{E}{V_a} A < v_{g\perp}\alpha >; \qquad \tau = \frac{E}{P_0} \tag{7}$$

where A is the area of the absorber-thermometer interface, V_a the volume of the absorber, $v_{g\perp}$ the normal component of the phonon group velocity , α the transmission probability and $< .. >$ denotes an average over modes and angles of incidence of phonons impinging on the interface. Equation 7 assumes a homogeneous distribution of the high frequency phonons in the absorber, the absorption of all phonons transmitted into the thermometer, and exclusive thermalization in the thermometer.

3 Comparison with experiment

In this section the model is applied to the detector with the Ir/Au thermometer. The thermal conductivity κ of the Ir/Au film is estimated by inserting its normal conducting resistance of R_{nc}=0.4 Ω into Wiedeman-Franz law. This would yield a good approximation for κ within a narrow transition. With the wide transition of the Ir/Au thermometer (15 to 115mK), κ may be overestimated up to a factor of four at the lowest operating temperatures, where the resistance is 1/4 of R_{nc}.

A thermal conductance of $G_{Au} = 9.2\times10^{-7}\times T$ W K^{-2} of the Au-bonding wires, as calculated from Wiedemann-Franz law, and a Kapitza conductance of the thermometer-absorber interface of $G_K = 1.44\times10^{-3}\times T^3$ W K^{-4}, as calculated from acoustic mismatch, are used throughout the analysis.

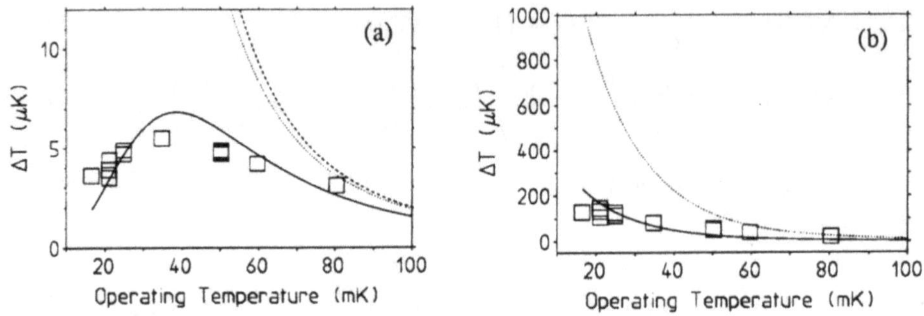

Figure 1: *Amplitudes of 60 keV pulses.*
a) Slow (thermal) component: The fully drawn curve is the best fit of β, the dotted curve is $\beta \to \infty$, and the dashed curve is $\Delta T = E/C_a$.
b)Fast component: The fully drawn curve assumes absorption of all high frequency phonons transmitted into the thermometer, the broken curve assumes that only those transmitted as longitudinal phonons are absorbed.

The electron-phonon coupling is parameterized as $G_{e-p} = \beta T_e^4$. The only free parameter of the model is β, which is determined by fitting the amplitude of the slow component, shown as the fully drawn curve in fig. 1a. This value of $\beta = 2.4 \times 10^{-3}$ W K^{-5} is then used to calculate the amplitude of the fast component and the signal rise time, shown in figures 1b and 2. With the volume $V_e = 2 \times 10^{-7}$ cm^3 of our Ir/Au film, this β translates into an electron phonon resistance of $R_{e-p} V T^4 = 0.83 \times 10^{-4}$ K^5 cm^3/W, which is close to $R_{e-p} V T^4 = 1.5 \times 10^{-4}$ K^5 cm^3/W measured for Rh:Fe [1] .

The calculated amplitude of the fast component, shown as the broken curve in fig. 1b, is considerably larger than the experimental values. Consequently, the thermometer does not absorb all high frequency phonons as assumed in eq. 7.

In the following I estimate the mean free path l for inelastic scattering of high frequency phonons in the thermometer from the value of β. The assumed temperature dependence of $G_{e-p} = \beta T^4$ corresponds to a frequency dependence of $l = \eta \nu^{-1}$ in the range of thermal phonon frequencies [2]. Anharmonic decay in Si [3] degrades the phonons to an average frequency of $\nu \sim 0.3$ THz within the ms time scale of P(t). Using η calculated from β by assuming the same scattering rate for all phonon modes, a naive extrapolation with $l = \eta \nu^{-1}$ to ν=0.3 THz yields l=700 nm. (l is an average of the mean free path in the Ir and Au films). With our film thickness of 100 nm this large l would lead to an incomplete absorption of high frequency phonons and thus to a reduction of P_0 compared to eq. 7.

At high frequencies $\nu > 10^{10}$ s^{-1} it is dominantly the longitudinal phonons which scatter with $l \propto \nu^{-1}$. At lower frequency transverse phonons also contribute, leading to a more complex frequency dependence of l [4]. Neglecting this contribution of transverse phonons and extrapolating to ν=0.3THz yields

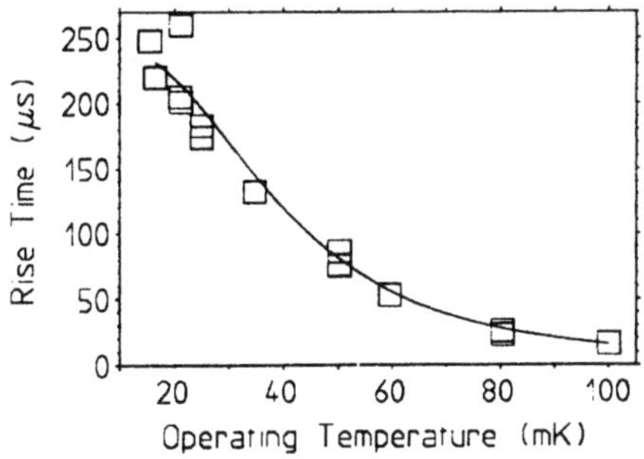

Figure 2: *Dependence of the signal rise time on the operating temperature.*

l=70 nm for longitudinal phonons. This value of l would lead to a complete absorption of all longitudinal phonons. Averaged over the Si-Au and Si-Ir interface areas 22% of all phonons in Si are transmitted into the film as longitudinal phonons. This reduces P_0 accordingly and brings the prediction, shown as the full curve in fig. 1b, closer to the data.

For the absorption of all high frequency phonons eq. 7 yields τ=3.4 ms. A reduction of P_0, which results from a smaller absorption probability, would increase τ accordingly, as long as all phonons are thermalized in the thermometer. This would lead to τ=15 ms in the case of exclusive absorption of longitudinal phonons. From the much shorter experimental value of τ=1.5 ms one has to conclude, that only ~10% of the phonons are thermalized in the thermometer.

During the rise of the pulse $\Delta Ta(t)$ can be neglected, so we obtain from eq. 3 a total thermal coupling of the electron system of $G = G_{e-a}/(1-\gamma)$. This gives a signal rise time of $\tau = C_e/G$ shown in fig. 2, which agrees nicely with the data.

References

[1] E.T. Swartz and R.O. Pohl in A.C. Anderson and J.P. Wolfe (eds.): Procceedings of "Phonon Scattering in Condensed Matter V" , Urbana, Illinois, June 2-6, 1986, Springer Series in Solid-State Sciences 68

[2] A.C. Anderson and R.E.Peterson, Phys. Letters 38A,(1972),519

[3] S. Tamura, Phys. Rev. B 31, (1985), 2574

[4] J.P. Harrison, J. Low. Temp. Phys. 37,(1979),467

Studies of Athermal Phonon Pulses in Silicon Crystals from Alpha, X-Ray and Neutron Bombardment Experiments

B. Cabrera[1], *B.L. Dougherty*[1], *A.T. Lee*[1], *M.J. Penn*[1], *J.G. Pronko*[2], *and B.A. Young*[1]

[1]Physics Department, Stanford University, Stanford, CA 94305, USA
[2]Lockheed Applied Research Laboratory, Palo Alto, CA 94304, USA

1. Introduction

We are developing a new class of elementary particle detectors capable of sensing weakly interacting particles such as neutrinos. Our detectors are based on the propagation of phonons in silicon crystals at low temperatures. During the past several years, we have made substantial progress towards the detailed understanding of phonon-mediated particle detection using silicon crystals with superconducting transition-edge sensors [B. A. Young, et al, *Nucl. Instr. Meth.* **A311**, 195 (1992)]. We call these devices silicon crystal acoustic detectors or SiCADs. The transition-edge sensors are biased in temperature below their superconducting transition and in current below the latching current, the boundary between self-extinguishing pulses and thermal runaway from I^2R heating. In this mode, the phonons from a particle interaction in the silicon crystal heat regions of the transition-edge sensors, driving those regions normal, and producing a voltage across the device that is proportional to the length of the normal region. In turn, the voltage defines a unique deposition energy once the depth of the event is determined from timing information.

2. Particle-Detection Experiments

Our group has built three generations of transition-edge devices. The first used superconducting aluminum films operated near 1.3 K using a ^{4}He cold-plate probe. With those devices we were able to detect alpha particle events through 300 μm thick silicon crystal [B. Neuhauser, et al, *Jap. J. of Appl. Phys.* **26**, 1671 (1987)]. In the second generation, we have used titanium films near 400 mK and developed a simple ^{3}He probe capable of a base temperature below 260 mK. With the second generation devices, we have been able to detect alpha particles through 1 mm crystals [B. A. Young, et al, *Phys. Rev. Lett.* **64**, 2795 (1990)] and 60 keV x-rays through 300 μm crystals. These studies have demonstrated that about 30% of the phonon energy transport is ballistic for alpha particle interactions which always occur near the surface and about 25% is ballistic for x-ray interactions which occur throughout the crystal. In addition, we have developed a neutron scattering facility, in collaboration with Lockheed in Palo Alto, to study the phonon processes present in nuclear-recoil events, the primary interaction mechanism of neutrinos and dark matter candidates [see paper by A. T. Lee]. These data on the phonon propagation processes within the silicon crystals are in good agreement with the Monte Carlo simulations that we have developed to understand the response of our detectors [also see paper by B. L. Dougherty].

Springer Series in Solid-State Sciences, Volume 112
Phonon Scattering in Condensed Matter VII Eds.: M. Meissner and R.O. Pohl

We characterize the performance of our titanium transition edge sensors by the phonon critical energy density needed to drive a portion of the film from the superconducting to the normal state. For our 40 nm thick films this critical energy density is about 0.5 eV/μm^2 and about 2 eV is needed to form a normal region across our 2 μm wide lines. Since the average phonon energy is about 5 meV, about 400 phonons must strike the same 2 μm × 2 μm patch for a resistance to develop across the pattern. Thus the detector is operating in the near phonon field of the events. For those events too far away from the crystal surface the phonons have become too dilute to be seen.

One way to increase the sensitive depth dramatically is to take advantage of phonon quantum mechanics. Since the average phonon quanta are about 5 meV, if a transition-edge device can be made that requires less than 5 meV to form a normal region across a line then we can enter the far field limit and the signals become roughly independent of distance from the crystal surface and roughly linearly proportional to the total energy. This quantum efficient operation does not allow us to observe single phonons because the signal from one normal region is below the amplifier noise level, but it does eliminate the detector threshold and makes the detector signal proportional to the deposition energy. This technique requires lower temperatures and has led us to our third generation experiments. This past year we have begun these third generation experiments with a dilution refrigerator operated at a temperature below 100 mK. We have already obtained transition temperatures below 60 mK with transition widths of 1 mK (10-90%) using tungsten films and a 90 mK sensor observed 6 keV x-ray events through 500 μm thick silicon [see paper by K. D. Irwin]. The tungsten film critical energy density in that experiment was about 16 meV/μm^2 of film, a factor of 25 times smaller than our titanium devices.

One very important aspect of this third generation is that we can enter the quantum efficient regime utilizing standard 1 μm wide line photolithography. We have already demonstrated sufficiently small critical energy density with the tungsten films. In order to observe the normal regions which are formed, we need a technique for confining the energy in this small region of the film. The time constant for excess heat within the tungsten films to thermalize with the silicon substrate is about 100 μs. However, the time constant for thermal transport along the film is nearly 1000 times greater. To prevent rapid cooling of quantum induced normal regions, we are studying the use of quasiparticle traps formed using an alternating chain of higher temperature superconducting links in series with tungsten. A higher temperature superconductor, such as niobium or aluminum, is well below its transition temperature and is thus a very poor thermal conductor, while remaining a perfect electrical conductor. We have already tested some of these ideas using aluminum/titanium structures and we will soon test devices in the quantum efficient region.

3. Conclusions and Acknowledgments

Imaging phonon-mediated detectors such as SiCADs are becoming a reality in the laboratory. The prospects look promising for obtaining the additional improvements necessary to achieve thresholds of a keV or better in large crystals. Such a detector would be of great interest for a number of experiments including dark matter searches for weakly interacting neutral particles and reactor neutrino mass experiments.

B. Neuhauser now at San Francisco State University has participated extensively. This work is funded in part by DOE Grant DE-FG03-90ER40569.

Charge-Carrier Collection in Superconducting Titanium Transition-Edge Sensors Deposited on High-Purity Silicon

B.L. Dougherty[1], *B. Cabrera*[1], *A.T. Lee*[1], *M.J. Penn*[1], *J.G. Pronko*[2], *and B.A. Young*[1;*]

[1]Physics Department, Stanford University, Stanford, CA 94305, USA
[2]Lockheed Research and Development Division, 3251 Hanover St., Palo Alto, CA 94304, USA
*Present address: Center for Particle Astrophysics, Physics Department, University of California, Berkeley, CA 94720, USA

Abstract. We have found that superconducting titanium transition-edge sensors (TES) deposited on high-purity, monocrystalline silicon operate in two distinct "modes", distinguished by different intrinsic gains. This difference may be due to changes in efficiency for prompt collection of the energy carried by electrons and holes; variously augmenting the dominant phonon-derived signals. Charge-carrier collection may be modulated by a space-charge region (Schottky-barrier) adjacent to the superconductor. We propose a model for this effect.

We have fabricated detectors composed of titanium TES deposited on high purity (resistivity ~ 2 kΩcm at room temperature) p-type silicon wafers [see B.A. Young et al., Nucl. Instr. and Meth. A311 (1992) 195]. Our experiments often involve measuring the spectroscopic response to low-energy ionizing radiation, in which both phonons and free charge carriers are liberated within the crystal bulk. "Calibration" spectra were taken using a 45 mCi ^{241}Am source mounted externally to our cryostat, providing event rates of roughly 10 sec^{-1}. Pulse heights for 60 keV events were noted just after cool-down ("mode I"). With the ^{241}Am source fixed in place (to expose our detector continually to ionizing radiation), for about 10 hours, similar pulse heights were returned. Over the next few hours however, gains appeared to increase until a new stable response was reached ("mode II"). Thereafter, 60 keV pulse heights had moved upward by 42%. So long as the detector was kept cold, this mode of operation continued indefintiely, with or without the ^{241}Am source being present.

Gain shifting may be caused by the following; in mode I little of the energy carried by electrons and holes is absorbed by our sensors, but in mode II nearly all of it is. The amount of increase depends on the ionization efficiency, or the fraction of total event energy deposited into charge carriers, which for photoabsorption in silicon, it is known to be constant at ≈31%. If charge-carrier collection efficiency becomes the same as that for phonons, the estimated mode II/I pulse-height increase would be around $0.31/(1-0.31) \approx 45\%$, almost exactly what was found. (In a separate series of elastic neutron-scattering experiments, in which recoiling silicon atoms were stopped in our detector, the mode II/I pulse-height ratio was lower, in correspondence with the reduced ionization efficiency expected for such events.)

The "extra" absorbed energy appearing in mode II probably does not originate as phonons in the silicon bulk. High kinetic-energy electrons and holes generated in any ionization event quickly shed phonons, to settle down in energy-level space near opposite band edges. For each electron-hole pair produced, there remains ≈1.2 eV worth of potential energy with which to create further phonons when capture and recombination occurs. At very low temperatures, and in silicon as

Springer Series in Solid-State Sciences, Volume 112
Phonon Scattering in Condensed Matter VII Eds.: M. Meissner and R.O. Pohl

lightly doped as our is, capture and recombination in the crystal takes a too much time to augment the prompt phonon flux substantially.

Charge-carrier propagation in the silicon bulk is dominated by neutral-impurity scattering, and the transit time for crossing our wafer is of order 1 μs, just short enough for many electrons and holes to reach the crystal surface in time with the prompt phonons. If charge carriers enter the sensor, their potential energy will be absorbed directly as heat, adding to the signal.

There exists an effective impediment to the absorption of charge-carrier energy, however: a space-charge region of electrons fixed at acceptor sites, induced by the Shottky-barrier active across the titanium-silicon boundary. A work function difference between these materials of almost 0.6 eV gives rise to a "depletion" volume nearly 10 μm deep extending from each Ti line. Since our TES lines are spaced 5 μm apart, these regions overlap. The entire crystal surface under a sensor is depleted, and contains roughly 10^{10} fixed charges per cm^2.

Due to charged-impurity scattering, the mean free path for electrons or holes propagating within this space-charge region is of order 0.1 μm. A large number of scatterings must then occur before charge carriers incident from the silicon interior could diffuse through to the crystal surface. This would take about as much time as was needed to cross the wafer, thus delaying any added delivery of energy into our sensors. It is more likely, however, that electrons and holes diffuse back out of the depleted volume. This may be especially true for holes, since there exists a repelling electric field inside this region. Charge carriers then re-enter the crystal bulk and are either lost elsewhere, are absorbed on the wafer's other side, or return roughly 1 μs later to diffuse again. Thus in mode I, or just after cool-down when this state obtains, little charge-carrier energy is absorbed quickly.

Due to small but finite capture cross sections, roughly 1% of the holes recombine on fixed electrons while diffusing through the depleted volume. If there is a continual source of holes (from constant exposure of our detector to ionizing radiation) then the entire space-charge region will gradually be neutralized. Since mean free paths are so long and temperatures are so low, these fixed charges are not likely to be replenished either by the trapping of free electrons or by thermally generated ionization. In time, the number of fixed charges will drop low enough to allow electrons and holes to impinge upon the crystal suface. Once mean free paths in this region become equal to those in the bulk, charge-carrier penetration is maximized. Recombination in the sensor (or at the crystal surface) rapidly releases the stored energy. Prompt TES pulse heights are then augmented a fixed amount.

The average penetration depth for holes is only a few diffusion lengths. Therefore the ≈10 μm wide space charge region will be neutralized not uniformly throughout, but more near its outer extremity, distant from the superconductor. As the last small fraction of this depleted volume is neutralized, charge carriers start to reach the crystal surface in larger numbers. Thus, mode shifting should occur relatively quickly compared with the time our device remains in mode I, as was observed. In fact, the absolute time required for mode shifting is about what one would expect given event rates, average ionization, recombination probabilities, and the number of fixed charges to be neutralized.

We thank Ira Chapman for help at Lockheed's Applied Physics Research Laboratory. This work was supported in part by DOE Grant #DE-FG03-90ER40569.

Quantum Efficient Detection of Phonons with Tungsten Thin Films

K.D. Irwin[1], *B. Cabrera*[1], *R. King*[2], *and B. Tigner*[1]

[1]Department of Physics, Stanford University, Stanford, CA 94305, USA

[2]Center for Integrated Systems, Stanford University, Stanford, CA 94305, USA

Abstract. We have developed superconducting tungsten transition-edge sensors (TES) to study athermal phonon processes induced by elementary particle scattering in silicon crystals. These sensors are roughly 140 times as sensitive as the titanium transition edge sensors we have been using. Several experiments using W TES have been performed with x rays and gamma rays. Due to the low transition temperature of the W films (57 mK) we have reached the film sensitivity necessary to construct a quantum efficient phonon detector. When a film becomes sensitive enough that single phonons can drive complete line segments normal, the detector becomes quantum efficient. Such operation dramatically reduces the detector threshold, allows us to instrument greater absorber thicknesses, and makes the detector more linear in energy. We are currently constructing a detector with line segments short enough to achieve quantum efficiency. This device can also be used as a phonon energy spectrometer, since by cooling the W films, we control the minimum phonon energy needed to drive a line segment normal.

We have developed W films to replace the Ti films used in our transition-edge sensors (TES) [see B.A. Young et. al., Nucl. Instr. and Meth. A311 (1992) 195]. Our Ti films have exhibited transition temperatures near 385 mK, with transition widths as narrow as 4 mK. Films with lower transition temperatures have better energy sensitivity as TES. Pure, bulk samples of tungsten go through a transition at 15.5 mK, the lowest of any elemental superconductor. Thus, as TES, W films potentially have dramatically better energy sensitivity.

Before this work, transitions in thin films of tungsten have always been reported between 0.4 and 4 K, much higher than bulk W transitions. One reason for this increase is the crystal phase. Bulk W is in the alpha phase (bcc). Thin films usually deposit in the beta phase (A15). We have explored various techniques to deposit W films with lower transition temperatures [see K.D. Irwin et. al., LTD-4 (IVth International Workshop on Low Temperature Detectors, Oxford, 1991), in press]. By sputtering our films with a negative bias voltage we have deposited alpha phase W films having transitions near 60 mK, with a width of 0.7 mK.

Several experiments using W TES have been performed with x rays and gamma rays. 6 keV x rays were observed through 500 μm of silicon. Since 60 keV gamma rays are barely detectable through 300 μm of silicon with Ti, this result demonstrates that our W films are at least 28 times more sensitive than Ti films. We have since deposited W films that we estimate to be 6 times more sensitive than the W films used in this experiment.

Springer Series in Solid-State Sciences, Volume 112
Phonon Scattering in Condensed Matter VII Eds.: M. Meissner and R.O. Pohl

When a film becomes sensitive enough that single phonons can drive complete line segments normal, the detector becomes quantum efficient. The character of the detector changes dramatically. When operating in a quantum efficient mode, the total length of film driven normal is longer, reducing the detector threshold. The detector can be sensitive to events throughout a much thicker crystal, limited only by phonon down-conversion and the detector area that must be covered. Thus, a greater absorber thickness may be used. When operated in a dilute regime, where incident phonon density drops to << 1 per line segment, the detector gains linearity in energy.

The threshold surface energy density (E_σ) is the energy required to drive a unit area of film normal. Near T_c, according to BCS theory,

$$E_\sigma \approx 5N(0)\Delta_0^2 \, d \, \Delta T/T_c$$

where N(0) is the density of states, $\Delta_0 = 1.76kT_c$ is the gap, d is the film thickness, and $\Delta T = T_c - T_{bias}$ is the transition width. Assuming the density of states to be the same as in bulk W, our best film to date has a calculated $E_\sigma = 2.2$ meV/μm^2, about 140 times less than our best titanium films. The mean phonon energy in our detectors is about 4 meV (1 THz). Thus, to achieve quantum efficiency with our W films, the superconducting segment area must be about 1 μm^2.

Superconducting W segments must be separated by thermally insulating segments to prevent the energy of the incident phonon from escaping down the W line before the signal can be measured. This insulation can be achieved by using superconducting segments with a much higher T_c than the W films. A potential problem is that the proximity effect may raise the T_c of the W segments. We do not expect this to be a difficulty in our case, but we will look further at this effect. The simplest insulated detector would be a tungsten-aluminum 2 layer device. Because of alignment constraints, the W segments must be several microns long. To keep their area small enough to be quantum efficient, they must be 0.25 μm wide. Note that we have successfully produced Ti meander patterns 0.25 μm wide [B.A. Young et. al., op. cit.].

We are also constructing a candidate quantum efficient device using the high-T_c, or beta crystal phase of W as the higher temperature superconductor. In this process, alpha phase (low-T_c) films are covered with an oxide layer, then patterned. Beta phase (high-T_c) films are then selectively deposited between the low-Tc segments by low pressure chemical vapor deposition (LPCVD). Since the process is selective, the beta W deposits only on the bare silicon and the exposed alpha tungsten, not on the portions of alpha tungsten covered by the oxide layer. Films with transitions as high as 2 K can be deposited this way [see C.A. van der Jeugd et. al., Appl. Phys. Lett. 57 (1990) 354-6] This process eliminates the alignment step, allows the alpha W segments to be much shorter, and greatly improves the detector area coverage.

The minimum phonon energy needed to drive a line segment normal is controllable. A quantum efficient device can therefore be used as a phonon energy spectrometer. The threshold energy density of the film, near T_c, is proportional to $T_c - T_{bias}$. Thus the phonon energy spectrum of a series of pulses can be measured by varying the bias temperature. The energy needed to drive a line segment normal is proportional to the area of the segment. A single detector could have line segments of different areas, allowing the measurement of the phonon energy spectrum of a single particle interaction.

This work was supported in part by DOE grant #DE-FG03-90ER40569 and DOE grant #DE-FG03-90ER45435.

Observation of Ballistic Phonons in Crystalline Silicon Induced by Gamma-Ray and Neutron Interactions

A.T. Lee[1], *B. Cabrera*[1], *B.L. Dougherty*[1], *M.J. Penn*[1], *J.G. Pronko*[2], and *B.A. Young*[1;*]

[1]Physics Department, Stanford University, Stanford, CA 94305, USA
[2]Lockheed Research and Development Division, 3251 Hanover St., Palo Alto, CA 94304, USA
**Present address: Center for Particle Astrophysics, University of California, Berkeley, CA 94720, USA

There has recently been much interest in the propagation of phonons in semiconductors at cryogenic temperatures, not only because of the solid state physics involved, but also because of its relevance to a new generation of semiconductor-based phonon-mediated particle detectors [1,2]. Phonons from particle interactions start out near the Debye frequency (~13.4 THz Si) where the scattering and decay rates are high. Phonons that are down-converted sufficiently for their mean free path to be comparable to the size of the crystal can reach the crystal surface with few or no additional scatterings and are termed ballistic. If the final down-conversion occurs sufficiently close to the original particle interaction, a ballistic phonon will contribute to a phonon focusing pattern on the surface of the crystal. It is necessary to understand the level at which particle interactions produce ballistic phonons to fully understand the response of a phonon-mediated detector. Furthermore, ballistic phonons can be exploited to improve both spatial and temporal resolution in a detector which leads to improved background rejection. One of the main goals of the group at Stanford is to build a detector capable of detecting non-baryonic dark matter. The improvement of background rejection is one of the most important considerations in developing a phonon-mediated detector for dark matter searches.

Recently, it was discovered that the fraction of phonon energy arrving at the sensor from a 59.57 keV gamma interaction in silicon is ~0.25 +.1/-.05 [3,4]. In the work presented here, the ballistic component from neutron interactions is compared to that from gamma interactions. Neutrons interact with silicon via nuclear recoil and there is speculation that the higher energy density of a such an event might lead to a higher ballistic phonon component.

The detector used in these experiments consists of a 300 μm thick monocrystal of silicon instrumented on both sides by titanium transition-edge sensors (TiTES) [5,6]. The transition-edge sensors consist of a meander pattern of 2 μm wide lines on a 5 μm pitch patterned from 400 A thick titanium. The entire detector is held just below the superconducting transition temperature of the titanium films (~380 mK) in a sealed ^{3}He cryostat. When phonons arrive at the surface of the detector sections of the line are driven into the normal state if the surface-energy-density is higher than the critical surface-energy-density E_σ. The detector is biased with a current and the signal is read out as a voltage which is proportional to the area driven normal. Since a ballistic phonon pattern has regions of high density it is expected that an enhanced ballistic component will result in detected interactions deeper in the crystal. In addition, it is expected that the maximum pulse height for the same energy deposition will be lower. The maximum pulse height for a number

Springer Series in Solid-State Sciences, Volume 112
Phonon Scattering in Condensed Matter VII Eds.: M. Meissner and R.O. Pohl

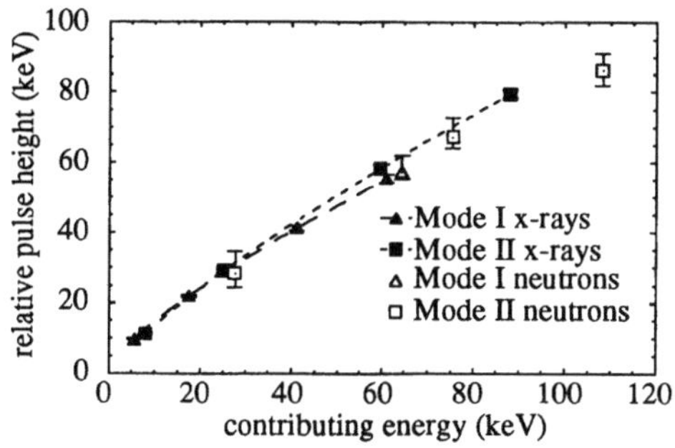

Fig.1. Maximum pulse height vs. total detected energy. Pulse height is normalized to a 59.57 keV gamma-ray detected in mode I.

of different gamma and x-ray energies is compared to the maximum pulse height for a number of neutron energies in Fig. 1. The energy of nuclear recoils and electron recoils is partitioned into phonons and electron-hole pairs, with 0.693 of the energy of an electron recoil generating phonons and ~0.85 of the energy of a nuclear recoil generating phonons (this fraction is energy dependent for nuclear recoils). The detector has been observed to operate in two "modes." In the mode I there is little or no electron-hole pair contribution to the detected energy, and in mode II most or all the electron-hole energy is collected (see paper by B.L. Dougherty in these proceedings, and [8]). The ionization-phonon partitioning and the ionization contribution to the collected energy is calculated to give the contributing energy in Fig.1. Within the 68.3% confidence limits the maximum pulse height for the nuclear recoils are consistent in magnitude with the electron recoil data. This indicates that the ballistic fraction for nuclear recoils and electron recoils is roughly the same [7].

This research was supported in part by DOE Grant No. DE-AT03-82ER40076 and DOE Grant No. DE-FG03-90ER40569.

[1] See, for example: Low Temperature Detectors for Neutrinos and Dark Matter III, eds. L. Brogiato, D. V. Camin, and E. Fiorini (editions Frontiere Gif sur-Yvette, 1990)

[2] H.J. Maris, *Phys. Rev.* **B41**, 9736 (1990)

[3] A.T. Lee, B. Cabrera, B.A. Young, N.I Maluf, LTD-4 (IVth International Workshop on Low Temperature Detectors, Oxford, 1991), eds. N.E. Booth and G.L. Salomon (editions Frontiere Gif-sur-Yvette, 1992) in press.

[4] A.T. Lee, B. Cabrera, B.A. Young, in preparation

[5] B.A. Young, B. Cabrera, A.T. Lee, *Nuc. Instrum. and Meth.*, **A311**, 195, (1992)

[6] B.A. Young, B. Cabrera, A.T. Lee, *Phys. Rev. Lett.*, Vol. **64**, No. 23, 2795-2798, (1990)

[7] A. T. Lee, B. Cabrera, B. L. Dougherty, M. J. Penn, J. G. Pronko in preparation.

[8] B.L. Dougherty, A.T. Lee, M.J. Penn, B.A. Young, and B. Cabrera in preparation.

A Massive Calorimeter for Single-Particle Detection

E. Umlauf, M. Bühler, and T. Fausch

Walther Meissner Institut, W-8046 Garching, Fed. Rep. of Germany

The compound detector, described here, consists of an absorber with 120 grams of silicon in thermal contact with a paramagnetic sample (about 150 mg) of $Tm_3Al_5O_{12}$ doped with 0.5% ^{168}Er (nuclear spin zero). With an applied magnetic field B=3 mT, the ground state Kramers doublet of Er exhibits a Zeeman splitting of the order of 10^{-6} eV. This is the excitation energy for spin-flip processes, which are induced on absorption of energy. The resulting change of the magnetic moment is measured with a SQUID, coupled directly to the magnetic sample [1].

The thermalized energy ΔE of a particle distributes itself in the compound detector according to its heat capacities. These are in an ideal case only the lattice heat capacity $C_l = a \cdot T^3$ and the heat capacity due to the Zeeman splitting of the Kramers doublets, described by $C_s = b \cdot B^2/T^2$. All additional heat capacities are included in $C_x(T)$. Therefore the energy is distributed as:

$$\Delta E = (C_s + C_l + C_x)\Delta T. \qquad (1)$$

It can be easily calculated that for our detector $C_l \ll C_s$ at $T \leq 50$ mK with B = 3 mT. The main problem is the additional heat capacity C_x arising mainly from magnetic interactions in the small magnetic sample. From the experiment we find actually: $C_l \ll C_x \approx C_s$.

Fig. 1a shows the signal height S for α-particles with 5 MeV as function of temperature with an applied field of 3 mT. The full line shows the calculations with $C_x = 0$. The experimental data are well below the calculated ones, due to $C_x(T)$. The noise level of our read-out signal has been found to be $3 \cdot 10^{-6}$ V/$\sqrt{\text{Hz}}$ [1]. From the signal to noise ratio we can therefore calculate a baseline energy resolution $\delta E = 8$ eV/$\sqrt{\text{Hz}}$, which would correspond to an energy resolution of 19 eV (FWHM) in a pulse height spectrum, if the signals are measured with a bandwidth of 1 Hz. Of course this energy resolution is not expected for α-particles due to additional effects, but with x-ray quanta we can expect an energy resolution of this order of magnitude.

The signal rise time τ_s is not only determined by the spin-lattice relaxation time τ_1 but in the ideal case we found [2]:

$$\tau_s^{-1} = \tau_1^{-1}(1 + C_s/C_l). \qquad (2)$$

The calculated values of $\tau_s(T)$ are plotted in Fig.1b by the full line. The experimental data show qualitatively the same behavior. The quantitative differences found at low temperatures can be explained by taking into account additional heat capacities.

Springer Series in Solid-State Sciences, Volume 112
Phonon Scattering in Condensed Matter VII Eds.: M. Meissner and R.O. Pohl

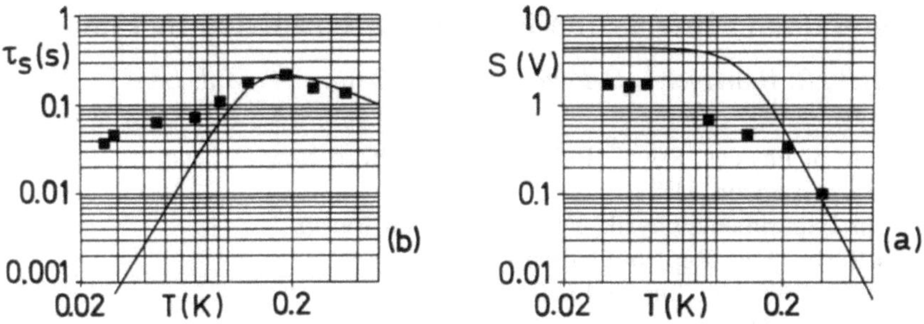

Fig1. (a) Signal height of 5 MeV α–particles on a compound detector of 120 g; full line: calculations for an ideal detector. (b) Rise time of the corresponding signals; full line calculated with equ. (2).

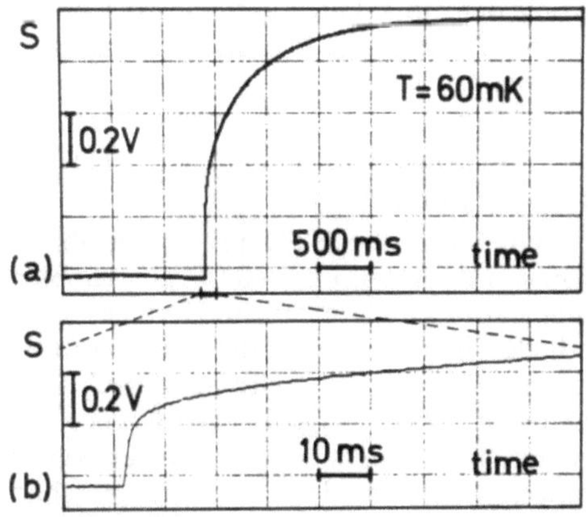

Fig.2. (a) Signal of a 5 MeV α–particle with a fast rise time. (b) a short interval of the same signal on an expanded time scale.

In addition we observe at very low temperatures (T< 100 mK) the superposition of a faster signal with a rise time $\tau_s^* \approx 1$ ms, as shown in Figs.2a and 2b. The proportion of the signal with τ_s^* increases with decreasing temperature, and with decreasing magnetic field. Furthermore, the fast signals are observed in any case when particles impinge on the magnetic sample, but when the particles impinge the absorber, the fast signals are not always observed.

We want to propose only one of several possible explanations. Assuming the ideal case, then τ_s^* would coincide with the calculated value. Therefore, one could assume that at first the energy-transfer from C_l to C_s occurs with this short time; but then, within a time scale of 1 ms, additional capacities C'_x become coupled to C_l and therefore the residual energy transfer is delayed.

Work supported by the Bundesministerium für Forschung und Technologie.

[1] E. Umlauf and M. Bühler, Proc. of the SPI-Conference on "EUV, X-Ray, and Gamma-Ray Instrumentation for Astronomy III", San Diego 1992, in print.
[2] M. Bühler, T. Fausch and E. Umlauf, to be published.

Electron–Phonon Interaction in NTD Ge Irradiated in 1980

P. Stefanyi[1], *R. Rentzsch*[2], *P. Fozooni*[1], *C.C. Zammit*[3], *M.J. Lea*[1], *and J. Saunders*[1]

[1]Department of Physics, Royal Holloway and Bedford New College, University of London, UK

[2]Institut für Experimentalphysik, Freie Universität Berlin, W-1000 Berlin, Fed. Rep. of Germany

[3]Department of Physics, Imperial College of Science, Technology and Medicine, University of London, UK

Neutron transmutation doped (NTD) Ge has been used by several groups as thermometric sensors [1] in cryogenic particle detectors. Pure Ge single crystals are irradiated with thermal neutrons in a reactor. Three of the five stable isotopes of Germanium are transmuted by nuclear reactions as follows:

$$\begin{array}{ll} {}^{70}\mathrm{Ge}(n,\gamma)\,{}^{71}\mathrm{Ge} \rightarrow {}^{71}\mathrm{Ga} & \text{shallow acceptor} \\ {}^{74}\mathrm{Ge}(n,\gamma)\,{}^{75}\mathrm{Ge} \rightarrow {}^{75}\mathrm{As} & \text{shallow donor} \\ {}^{76}\mathrm{Ge}(n,\gamma)\,{}^{77}\mathrm{Ge} \rightarrow {}^{77}\mathrm{As} \rightarrow {}^{77}\mathrm{Se} & \text{deep donor} \end{array} \qquad (1)$$

For germanium with natural isotopic abundance, the NTD process leads to semiconducting p-type material with a compensation ratio $\kappa = 40$ %. The samples studied here were irradiated in 1980 in nuclear research reactor in Germany with intensity ratio of thermal to fast neutrons better than 15:1. After decaying for more than six half life times, samples were annealed for 24 hours at 450°C. The irradiation doses for the two boules were: Ge 99: $2\mathrm{x}10^{18}$, Ge 3: $1\mathrm{x}10^{18}$ n/cm^2.
One of the advantages of neutron transmutation doping is that it produces boules with a homogeneous doping concentration which enables a series a matched sensors to be prepared. Samples cut from two boules, Ge 99 and Ge 3, were polished and contacted using In dots or evaporated gold pads on boron implants. The resistance-temperature characteristics $R(T)$ and the thermal conductance between the electrons and phonons were measured in a dilution refrigerator down to 20 mK. For the thermal conductance measurements a double sensor with Indium dots was used.
The measurements of $R(T)$ for samples from boules Ge99 and Ge3 have been analysed using the variable range hopping formula

$$\rho(T) = \rho_0 \exp(T_0/T)^x \qquad (2)$$

where the exponent $x = 1/4$ for a uniform density of states (Mott formula) or $x = 1/2$ for a Coulomb gap in the density of states (Shklovskii/Efros formula). A least squares fit of the data to eq (2) gave $x \approx 0.5$ for all sample. Figure 1 shows a plot of $\rho(T)$ versus $T^{-0.5}$ for these samples. The table shows the values of T_0 from a least squares fit to eq (2) assuming $x = 1/2$.
Figure 2 shows the measurements of thermal conductance between electrons and phonons for both of the double sensors. The data were analysed using "Hot electron model"

$$Q^o = g_{e\text{-}ph}\,(T_e{}^{\alpha+1} - T_{ph}{}^{\alpha+1}) \qquad (3)$$

and the fit with α=4 and $g_{e\text{-}ph} = 3.5\mathrm{x}10^{-4}$ W/K^5 is shown by a dashed line.

Springer Series in Solid-State Sciences, Volume 112
Phonon Scattering in Condensed Matter VII Eds.: M. Meissner and R.O. Pohl

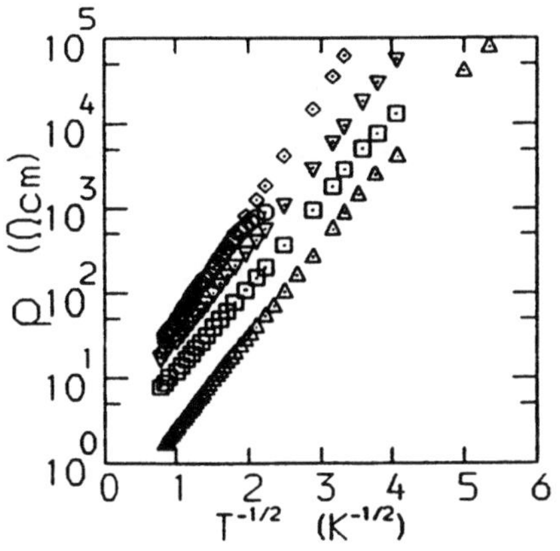

Figure 1. Plots of ρ(*T*) versus $T^{-0.5}$ for the NTD Ge sensors.

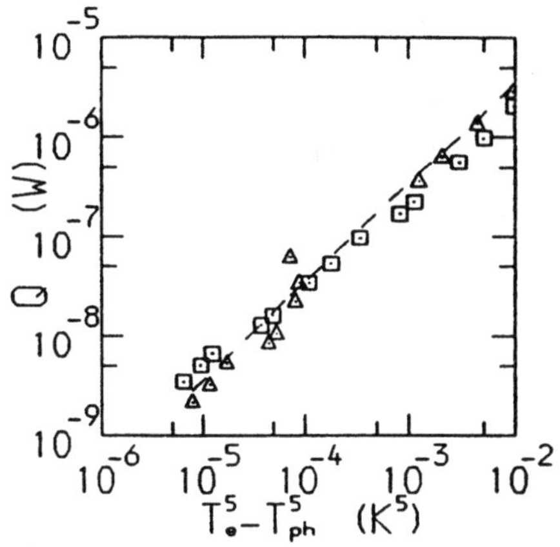

Figure 2. Heat flux from electrons to phonons versus $T_e{}^5 - T_{ph}{}^5$.

Characteristics of variable range hopping conduction in NTD Ge

Boule	Sample	T_0(K)
99	1	5.89
99	2	6.07
99	3	5.30
3	1	9.36
3	2	7.62

These samples are close to the metal-insulator transition and are comparable to thermometers cut from Boule 12 and Boule 19 of natural NTD Ge grown and irradiated by Professor E.E.Haller [2]. However one of the potential problems in using this material for thermometric sensors on targets used for cryogenic particle detection is radioactivity in the sensors themselves. This would be particularly serious in experiments where a very low background is required, such as a Dark Matter Detector [3]. The radioactive decays produced by the reactions in eq (1) all have relatively short half-lives (≤ 11 days) and would no longer be observed in a typical sensor several years after irradiation. However, a 30 g boule from Professor Haller was tested by Fiorini et al [4] for radioactivity soon after irradiation. As well as the radiation products expected from eq (1), radiation was detected from ^{68}Ge and ^{65}Zn (probably from fast neutrons in the source) and ^{95}Zr (probably from impurities), which have half-lives of 288, 244 and 64 days. At the contamination levels found, these would give a significant background count for many years in a sensor. The NTD Ge measured here has not been tested for radioactivity but the long time (12 years) since irradiation means that the residual radioactivity should be extremely low, and within the specifications for a Dark Matter Detector. Further experiments are in progress.

References

1. N.Wang, F.C.Wellstood, B.Sadoulet, E.E.Haller and J.Beeman, *Phys.Rev.* **B41** (1990) 3761.
2.E.E.Haller Workshop on Semiconductor Thermistors for Millikelvin Operations, Berkeley Calif. (1991).
3.P.F.Smith and J.D.Lewin, *Physics Report* **187** (1990) 205.
4.A.Alessandrello, C.Brofferio and D.V.Camin et al LTD-4 Oxford (1991)

Phonon Lifetime and Collection in Germanium Particle Detectors at 20 mK

T. Shutt[1], *P.D. Barnes Jr.*[1], *A. Da Silva*[5], *B. Ellman*[1], *J. Emes*[3], *E.E. Haller*[2;3], *A.E. Lange*[1], *R.R. Ross*[1;3], *B. Sadoulet*[1], *G. Smith*[3], *W. Stockwell*[1], *N. Wang*[4], *S. White*[1], *B.A. Young*[1], and *D. Yvon*[1]

Center for Particle Astrophysics, 301 Le Conte Hall, UC Berkeley, Berkeley, CA 94720, USA, and

[1]Department of Physics, University of California at Berkeley, Berkeley, CA 94720, USA

[2]Department of Materials Science and Mineral Engineering, University of California at Berkeley, Berkeley, CA 94720, USA

[3]Lawrence Berkeley Laboratory, 1 Cyclotron Rd., Berkeley, CA 94720, USA

[4]DAPNIA/SPP, CEN Saclay, F-91191 Gif-sur-Yvette Cédex, France

[5]Department of Physics, University of British Columbia, 6224 Agricultural Rd., Vancouver BC V6T 1Z1, Canada

We are developing low temperature particle detectors in which the ionization and the phonons created by a particle interaction are simultaneously measured. We describe in an accompanying paper[1] such a detector which has a mass of 60g and is operated typically at 21 mK.

We detect phonons with neutron transmutation doped germanium (NTD) thermistors[2]. Although designed primarily to measure the temperature rise of the crystal, these sensors are, intrinsically, even better detectors of high energy phonons. At the low temperatures at which we operate, thermal phonons tend to decouple from charge carriers in the thermistors[3]. On the countrary, phonons of energy greater than about 1K seem to readily couple, as suggested both by the rapid temperature dependence of the effective electron-phonon conductance and the fast rise time observed for particles impinging directly on one of the sensors.

The phonon sensors, which are typically 1.4x1.4x0.3 mm^3, are attached to the large Ge crystal using a thin gold-germanium eutectic[4]. For our 60g detector, approximately 1000Å of gold was deposited on each of the two polish-etched germanium surfaces to be bonded. The two pieces were pressed together and brought to the eutectic temperature (370°C) in an inert atmosphere, and then left to cool slowly. As shown in Figure 1a, this interfacing method provides signal rise times which are definitely faster than thin glue interfaces (≈25μ of silver epoxy), a fact that we interpret as evidence for higher phonon transparency. The phonon mismatch resistance between the NTD sensors and gold pads on top of them provide the weak link to the thermal sink needed for a calorimetric measurement.

Figure 1b shows a typical phonon signal from one NTD sensor on the 60g detector, where only 2.5% of the area is covered by sensors or metallic films. The time evolution of the pulse is well described by the superposition of three exponentials:

$$0.39 \exp(-t/64ms) + 0.61 \exp(-t/14ms) - \exp(-t/0.43ms).$$

The first time constant is readily identified as the thermal time constant, while the last one is compatible with the relaxation time of the sensor. We interpret the second component as the reverberation[5] for some 14ms of an out-of-equilibrium phonon population. Although phonons bounce ≈3500 times on the bare crystal surface, their energy presumably stays high enough for them to preferencially couple to the sensor rather than escaping. We model this phonon behavior as

Springer Series in Solid-State Sciences, Volume 112
Phonon Scattering in Condensed Matter VII Eds.: M. Meissner and R.O. Pohl

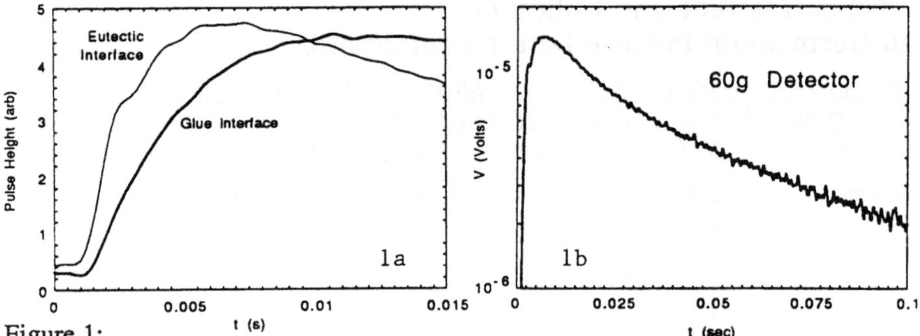

Figure 1:
a) Typical phonon pulses as observed through eutectic and thin glue interfaces.
b) A phonon pulse for ≈500keV γ interaction observed in a eutectically bound sensor in the 60g detector. Note the logarithmic vertical scale.

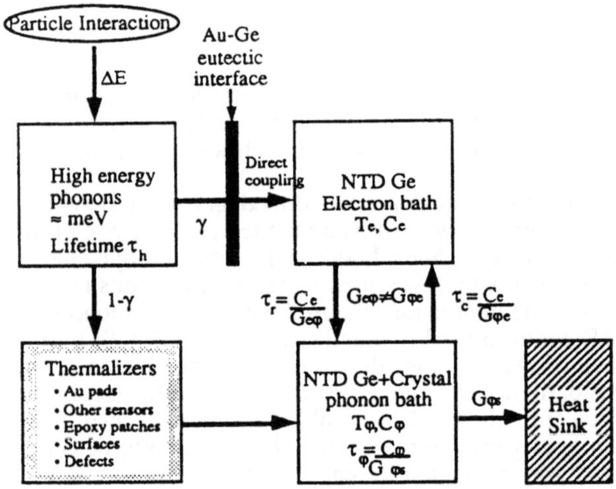

Figure 2: A schematic model of coupling between phonons and their sensors.

sketched in Figure 2. It is possible within this model to fit the three aforementioned exponentials in a way which is consistent with the known values of the measured thermal parameters of our device. We thus estimate that the transparency per bounce on the interface is between 5 and 10%. It is interesting to note that in the latter case, the six NTD thermistors appear to be the only efficient thermalizers in our device.

References

[1] Shutt, T. *et al.*, "Simultaneous Detection of Phonons and Ionization in a 60g Germanium Detector," this volume.

[2] Haller, E.E. *et al.*, *Infrared Physics*, *LBL-14649*, **25**, 257, (1985).

[3] Wang, N.; Wellstood, F.C.; Sadoulet, B.; Haller, E.E.; Beeman, J., *Phys. Rev.*, **41**, 3761, (1990).

[4] Wang, N. *et al.*, *Phonons 89*, World Scientific, Singapore, 1400, (1990).

[5] See e.g. Knaak, W. *et al.*, *Phonon Scattering in Condensed Matter V*, 174, (1986).

Simultaneous Detection of Phonons and Ionization in a 60 g Germanium Detector

T. Shutt[1], *N. Wang*[5], *B. Ellman*[1], *Y. Giraud-Heraud*[7], *C. Stubbs*[4], *P.D. Barnes Jr.*[1], *A. Cummings*[1], *A. Da Silva*[6], *J. Emes*[3], *E.E. Haller*[2;3], *A.E. Lange*[1], *R.R. Ross*[1;3], *B. Sadoulet*[1], *G. Smith*[3], *W. Stockwell*[1], *S. White*[1], *B.A. Young*[1], and *D. Yvon*[1]

Center for Particle Astrophysics, 301 Le Conte Hall, UC Berkeley, Berkeley, CA 94720, USA, and
[1]Department of Physics, University of California at Berkeley, Berkeley, CA 94720, USA
[2]Department of Materials Science and Mineral Engineering, University of California at Berkeley, Berkeley, CA 94720, USA
[3]Lawrence Berkeley Laboratory, 1 Cyclotron Rd., Berkeley, CA 94720, USA
[4]Department of Physics, University of California at Santa Barbara, Santa Barbara, CA 93106, USA
[5]DAPNIA/SPP, CEN Saclay, F-91191 Gif-sur-Yvette Cédex, France
[6]Department of Physics, University of British Columbia, 6224 Agricultural Rd., Vancouver BC V6T 1Z1, Canada
[7]Laboratoire de Physique Corpusculaire, College de France, F-75231 Paris, France

We have demonstrated simultaneous high resolution measurement of ionization and phonons created by particle interactions in a 60g Ge crystal operated at 21mK[1]. The detector is a 4cm diameter and 1cm thick disk of p-type germanium with net dopant concentration $\lesssim 2\ 10^{11}$ cm^{-3}. Phonons are detected with neutron transmutation doped (NTD) germanium sensors which are sensitive both to the temperature of the crystal and high energy phonons[2]. The ionization from the event is collected by means of a 0.5V bias applied across identical implanted (p+) ohmic contacts on both faces of the disk. The schematics of the electronics used are shown in Figure 1a. Signals are sensed using cold JFETs mounted in the cryostat. An AC biasing technique is used to avoid 1/f noise.

When irradiating the detector with 59.5keV γ–rays and 14 and 18keV X-rays from a ^{241}Am radioactive source, we observe the two dimensional distribution of ionization and phonon signals displayed in Figure 1c. The two energies have been renormalized so that the 59.5keV peak appears in both channels at this value. The r.m.s. energy resolution obtained at 59.5 keV is 700eV in ionization and 820eV in phonons. Figures 1b and 1c show the data projected onto the ionization and phonon E axes, together with random triggers which give a baseline noise of 500eV and 600eV respectively. An absolute normalization of the phonon signal shows that, contrary to commonly held opinion, the full recoil energy of particle interactions (not 2/3 for an electron recoil!) is recovered as phonons. This is true only when the ionization is fully collected. At an electric field below 250mV/cm this is no longer the case, and the study of the phonon signals shows that the electrons and holes are predominantly trapped on shallow levels.

As shown in Figure 2, one application of our simultaneous detection technique is to distinguish between nuclear and electron recoils in the crystal. For this measurement a ^{252}Cf neutron source shielded by 28cm of lead was used, and neutron interactions can be readily recognized as the accumulation of events with less ionization per unit phonon energy than the photon events on the diagonal. We

Springer Series in Solid-State Sciences, Volume 112
Phonon Scattering in Condensed Matter VII Eds.: M. Meissner and R.O. Pohl

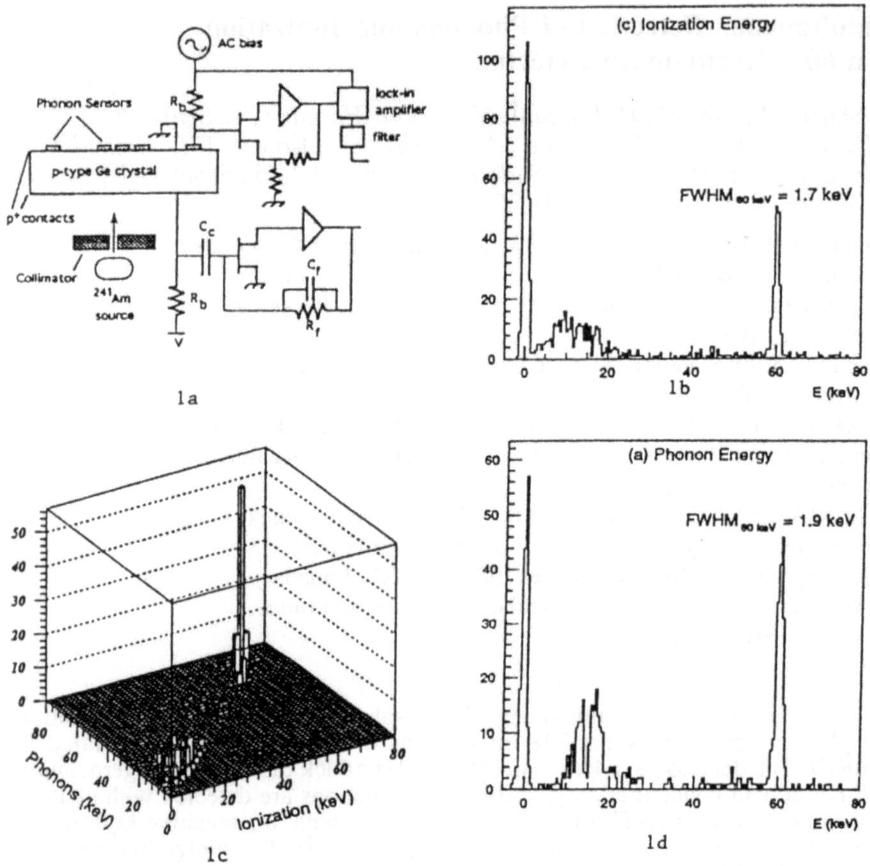

Figure 1: Measurement of ionization and phonons for γs from an ^{241}Am source.
a) Schematic of the detector, amplifiers, and radioactive source arrangement.
b) Projection onto the ionization energy axis.
c) Two dimensional distribution of energy in ionization and in phonons.
d) Projection onto the phonon energy axis.

find good agreement between our results and previous measurements of ionization yield from nuclear recoil in germanium. Such discrimination technique promises to greatly increase the sensitivity of searches for dark matter particles.

References

1 For more details and references to the literature, see Shutt, T. *et al.*, "Simultaneous High Resolution Measurement of Phonons and Ionization Created by Particle Interactions in a 60g Germanium Crystal at 25mK" and "Measurement of Ionization and Phonon Production by Nuclear Recoils in a 60g Crystal of Germanium at 25mK," submitted for publication.

2 Shutt, T. *et al.*, other contribution to this volume.

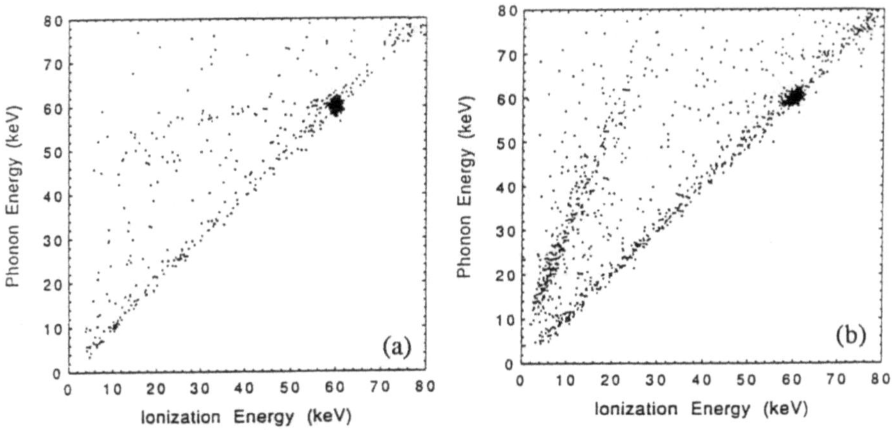

Figure 2: Phonon and ionization equivalent energies measured
a) γ-rays only. The 59.5keV peak from an ^{241}Am source provides an energy calibration.
b) In addition, neutrons and photons from a ^{252}Cf source.

SLAPS: Superconducting Large Area Phonon Sensors

S.M. Grannan[1], *S.E. Labov*[2], *A.E. Lange*[1], *B. Sadoulet*[1], *B.A. Young*[1], *and J. Emes*[3]

[1]Center for Particle Astrophysics and Physics Department, 301 Le Conte Hall, UC Berkeley, Berkeley, CA 94720, USA
[2]Laboratory for Experimental Astrophysics, Lawrence Livermore Natl. Lab., P.O. Box 808, L-401, Livermore, CA 94550, USA
[3]Lawrence Berkeley Laboratory, 1 Cyclotron Road, Berkeley, CA 94720, USA

We are investigating the use of thin film superconductors with NTD Ge thermistors for collecting phonons over large surface areas (i.e. ≈ few cm^2). The SLAPS (Superconducting Large Area Phonon Sensors) technique provides a simple, yet efficient means to increase detector sensitivity to athermal phonons. This increased sensitivity results from the use of novel quasiparticle trapping schemes. Our goal is to incorporate SLAPS technology into the design of cryogenic detectors currently under development for WIMP dark matter and neutrino experiments [1].

We report here on experiments performed using our first SLAPS detector, which consists of four symmetrically - located NTD Ge thermistors (1 mm x 1 mm x 300 μm) glued to a Ge substrate (Fig. 1). The thermistor contacts are evaporated with 4000 Å of Au. An 8000 Å Al film covers one surface of the Ge substrate and three of the four thermistors. These three thermistors are used to study the quasiparticle trapping mechanism, while the fourth is sensitive only to thermal phonons and is used for device calibration.

When an incident particle interacts in the Ge substrate, phonons are produced which propagate to the surface of the crystal and strike the Al film, creating quasiparticles. Our high-purity Al film allows relatively long quasiparticle lifetimes and mean free paths; as a result, the quasiparticles can travel large distances (several mm) to the NTD Ge thermistors on the ends of the device. The Al / Au / NTD region has a smaller gap (by about an order of magnitude) than the gap in the Al / Ge region due to the proximity effect. Quasiparticles which move into one of these low gap regions decay by phonon emission and become "trapped". This process heats the NTD Ge thermistor and produces a signal.

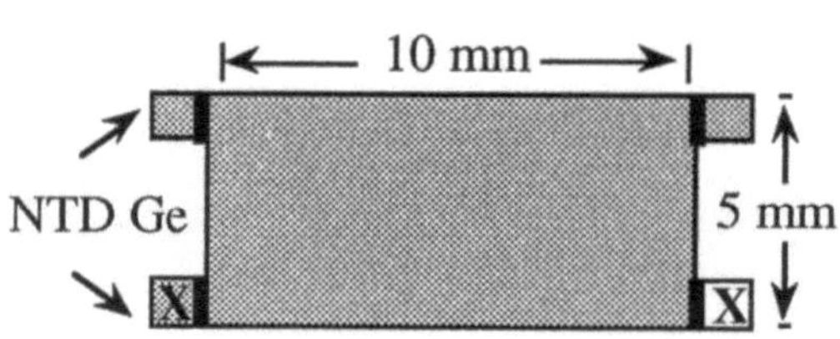

Figure 1. Top view of SLAPS I device with 4 glued NTD's. The X's denote sensors used in these experiments. The shaded region represents the Al film. The Ge substrate is ≈ 300 μm thick.

Springer Series in Solid-State Sciences, Volume 112
Phonon Scattering in Condensed Matter VII Eds.: M. Meissner and R.O. Pohl

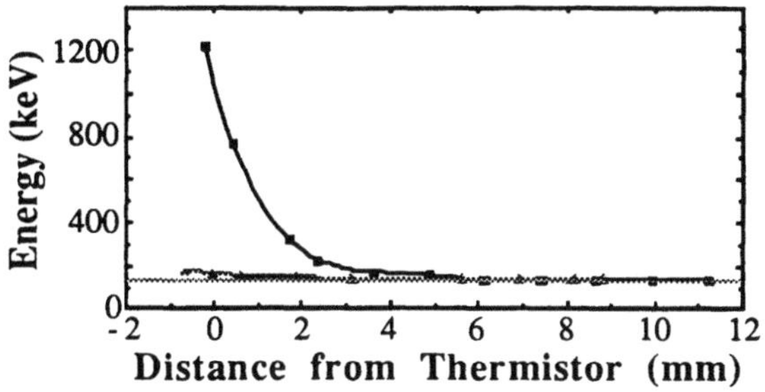

Figure 2. Detected energy of alpha event into Ge substrate vs. distance from thermistors (320 mK data) :
▫ NTD with Al film
▵ NTD no Al film

Our first tests of the SLAPS I device were made at 320 mK in a ^{3}He system. A collimated ^{241}Am alpha source was scanned between two NTD Ge thermistors, one with an Al / Au trap and one without. Three different types of signals were observed in the two thermistors: (1) 60 keV gammas directly striking the NTD; (2) 5.4 MeV alphas directly striking the NTD; (3) alphas striking the Ge substrate at various distances from the NTD's. The 60 keV pulses were used to calibrate the energy detected in each thermistor as a function of distance from the ^{241}Am source (Figure 2).

The quasiparticle trapping efficiency of SLAPS is affected by the ratio of the trap volume to the sensing film volume, impurities in the superconducting film, and the quasiparticle lifetime which increases at lower operating temperatures. The temperature dependent effects were studied by mounting the SLAPS detector and ^{241}Am alpha source in a dilution refrigerator. Three collimation slits were positioned at distances 0.5, 5.0, and 9.5 mm from the same two thermistors studied at 320 mK (trap / no trap). Spectra obtained at 60, 150, 225, and 300 mK indicate that the trapping efficiency improves significantly as the temperature is decreased. A full analysis of these data is underway.

We have demonstrated quasiparticle trapping into NTD Ge thermistors over distances of several mm at 60 mK $\leq$ T $\leq$ 320 mK. Our technique dramatically increases the sensitivity of bolometric detectors to athermal phonons. We have fabricated a second SLAPS device which will allow a careful comparison of six different trapping schemes. Data from SLAPS I and II will be combined to evaluate the feasibility and benefit of incorporating SLAPS technology into the design of a large scale dark matter detector [2].

This work was supported in part by a David and Lucille Packard Fellowship; the Center for Particle Astrophysics, an NSF Center operated under Coop. Agreement No. ADT-88909616; and under the auspices of the U.S. DOE by LLNL under Contract No. W-7405-ENG-48.

References

[1] *Low Temperature Detectors for Neutrinos and Dark Matter IV*, N.E. Booth and G.L. Salmon, eds., Editions Frontieres, Gif-Sur-Yvette, France (1992).
[2] T. Shutt, *et. al.*, (and contained references) in these proceedings.

Superconducting Thin Films as Coupling Media in a Si:P Detachable Bolometer: Experimental Results and Analysis

C. Boragno, Y. Shen, F. Buatier de Mongeot, and U. Valbusa

Centro di Fisica delle Superfici e delle Basse Temperature del CNR, Dipartimento di Fisica dell'Universita' di Genova and INFM, via Dodecaneso 33, I-16146 Genova, Italy

Si:P bolometers doped close to the metal insulator transition are suitable for detecting heat pulses owing to their fast response time, high responsivity [1] and possibility to work in magnetic field [2]. However, up to now these bolometers were realized directly on the surface of the same Si crystal in which the ballistic phonons propagated. This fact limited their application . To solve this problem a detachable Si:P bolometer has been realized using a superconducting thin film as coupling medium [3].

In this paper we report the study of the phonons transmission through a superconducting film (In, Sn and Pb) in presence of a magnetic field (parallel or perpendicular to the film).

The sample configuration and experimental set-up have been reported in ref.[3]. The phonons, generated by a thin gold film heater, propagate ballistically along <100> direction in a 10mm thick Si crystal and are detected by a detachable bolometer mechanically pressed on the surface of the crystal with thin film (In, Sn or Pb) as coupling medium.

The detachable bolometer is a Si:P chip doped by ion implantation of phosphorus with concentration $N=2\ 10^{18}\ cm^{-3}$ close to the metal insulator transition.

In Fig.1 the time of flight (TOF) spectra of ballistic phonons transmitted through a Pb film (thickness 4μm) are reported for various values of magnetic field **H** parallel to the film. The spectra show two peaks corresponding to longitudinal **L** and transverse **T** phonons. The signal intensity decreases monotonically for **H** greater than zero. This behavior has been observed for In, Pb and Sn films of 4 and 8 μm thickness..

The transmission of ballistic phonons through a superconducting film can be interpreted in the framework of the Bobetic model [4]. When $T/T_c << 1$, the film acts as a frequency filter: phonons having energy higher than the superconducting gap are strongly attenuated by Cooper pairs breaking mechanism. Instead, phonons with energy lower than the gap can be transmitted easily, because the quasi-particle scattering is negligible. By a detailed analysis of the signal intensity as function of film thickness and power dissipated W on the heater,the phonon mean free path in the normal state of the film can be derived [5] .

Springer Series in Solid-State Sciences, Volume 112
Phonon Scattering in Condensed Matter VII Eds.: M. Meissner and R.O. Pohl

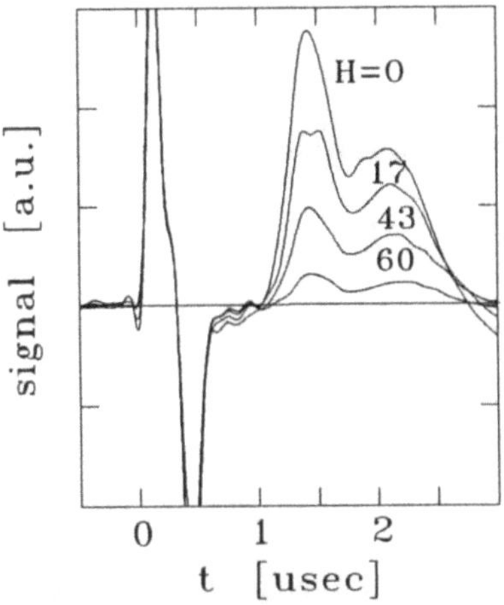

Fig. 1 - TOF spectra for Pb film, 4μm thickness, recorded for different **H** . W=5.8 W/mm^2 .

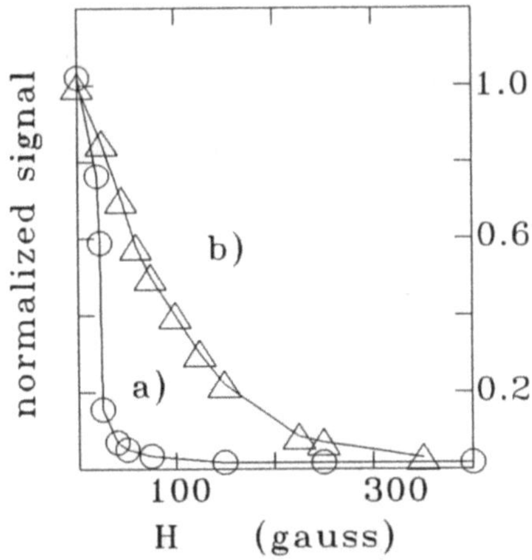

Fig. 2 - Normalized peak heigth as function of **H** (a) parallel or (b) perpendicular to a Pb 4μm film. W=5.8 W/mm^2

The **L** peak height, normalized to the value measured in absence of magnetic field, as function of **H** is shown in Fig.2. The curve (a) is the result for **H** parallel to the film and (b) for **H** perpendicular to it. In both cases, the signal intensity decreases strongly with increaseing the magnetic field, but surprisingly the effect is more pronounced in the parallel case. This behavior in perpendicular case can be qualitatevely explained by using the Landau model of the intermediate state [6]. Increasing **H**, the normal zones growth up and consequentely the phonon transmission through the superconducting zones decreases. However, according to the non-branching Landau model, the phonon intensity should decrease linearly in perpendicular case and it should be constant in the parallel one.
A more detailed analysis will be reported [5] .

References

[1] C.Boragno,U.Valbusa and G.Pignatel, Appl. Phys. Lett. **50** , 583 (1987)
[2] C.Boragno,C.Ferdeghini and U.Valbusa, J.Phys. E:Sci.Instr. **22**, 50(1989)
[3] Y.Shen,C.Boragno,F.Buatier de Mongeot and U.Valbusa Rev. Sci. Instr. (accepted for pubblication)
[4] V.M.Bobetic, Phys. Rev. **136** , 1535 (1964)
[5] Y.Shen,C.Boragno,F.Buatier de Mongeot and U.Valbusa, to be published
[6] Magnetic flux structures in Superconductors, ed. by R.P.Huebener, Springer-Verlag (Berlin) ,1979

Calorimetric Detection of 5.5 MeV α-Particles in Superfluid ^{4}He

F.S. Porter, S.R. Bandler, C. Enss, R.E. Lanou, H.J. Maris, T. More, and G.M. Seidel

Department of Physics, Brown University, Providence, RI 02912, USA

Many groups are currently developing low temperature techniques for the detection of low energy particles. The primary goal is to detect small amounts of deposited energy ($\sim 10^{-13}$-10^{-16}J) in a large target volume. Our work focuses on the properties of ^{4}He at low temperatures which allow the interaction energy of a low energy particle to be extracted from the target material into a low mass calorimeter with a reasonably high efficiency. In this case the sensitivity of the detector depends only weakly on the volume and in principle very large detectors can be built. We have constructed a prototype of this device using two liters of superfluid ^{4}He as the target material, and a thin silicon or sapphire wafer as the calorimeter.

The experimental arrangement is shown in Fig. 1(a) and consists of a 1.5 liter volume of isotopically pure ^{4}He maintained at a temperature of 70-100 mK. Two thin wafer calorimeters are suspended in a film free region above the helium bath using a baffled film removal device [1]. The wafers are anchored to the mixing chamber of a dilution refrigerator and are maintained at a base temperature of 30 mK. An NTD Ge thermistor is attached to each wafer, and is the temperature

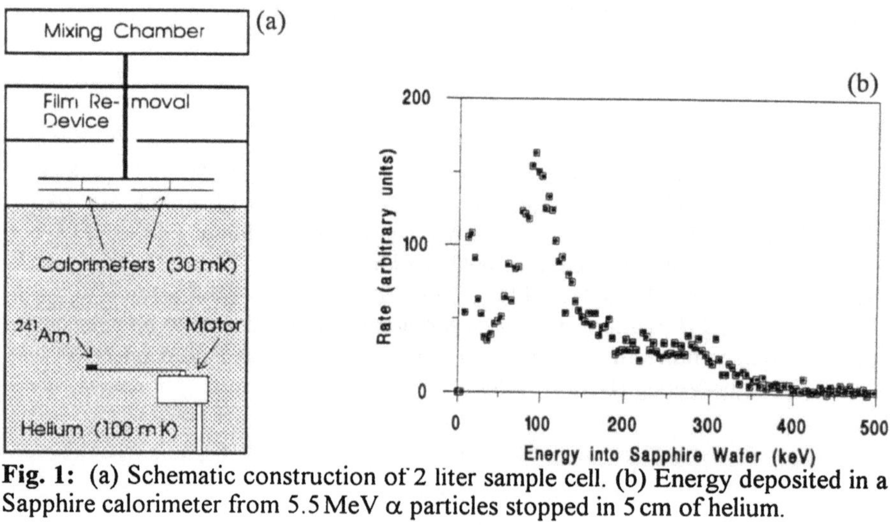

Fig. 1: (a) Schematic construction of 2 liter sample cell. (b) Energy deposited in a Sapphire calorimeter from 5.5 MeV α particles stopped in 5 cm of helium.

Springer Series in Solid-State Sciences, Volume 112
Phonon Scattering in Condensed Matter VII Eds.: M. Meissner and R.O. Pohl

sensing device for the calorimeter. A 5.5 MeV ^{241}Am α source is used to calibrate the detector and to investigate the overall efficiency of the detection model.

A low energy particle interacting with the liquid produces an impulse of excitations which at sufficiently low temperatures travel ballistically to the walls and to the free surface of the helium in the container. The excitations reaching the free surface with sufficient energy ($E > 7.16K$) will evaporate a helium atom with a finite probability [2]. The evaporated atoms bind to the silicon wafer with an energy estimated to be 100K providing a heat pulse to the calorimeter. The additional binding energy gained by adsorbing onto bare silicon amplifies the signal at the calorimeter. The amplification which is a product of the additional binding energy and the sticking coefficient to silicon was measured to be about 10 [3].

The calorimeters have been calibrated using an α source directly incident on the calorimeter. In this configuration the sensitivity of the 0.9 g silicon wafer has been measured to be 4 keV at 17 mK, and about 15 keV at 30 mK where the helium experiments have been conducted. The heat capacity of this wafer has been measured between 17 mK and 120 mK showing both a T^3 bulk silicon component and a linear factor most likely due to two levels systems in the surface oxide layer and the addenda associated with the thermistor. A similar investigation of a sapphire wafer in the range between 50-150 mK yielded a T^3 dependence in agreement with Θ_D.

A typical spectrum of 5.5 MeV α particles stopped in bulk liquid ^4He is shown in Fig. 1(b). Both the peak and the high energy shoulder are due to the alpha source although the specific cause of the broadening and the extended structure has not been determined. Background events due to environmental radiation have been reduced by surrounding the apparatus with 5 cm of lead shielding. Any remaining background events directly into the calorimeters have been eliminated by using two independent wafer systems in coincidence. An event originating in the helium target will provide a simultaneous signal on the two calorimeters.

A superconducting stepper motor has been used to move the α source into a shielded enclosure during the experiment. The resulting background spectrum showed a very low rate and no overall structure. Additional investigations using the movable source have shown a very large angular dependence of the evaporation signal on the source position.

The fraction of energy deposited in the calorimeter during the detection process, estimated from both the ^{241}Am α spectrum and the γ's produced by an external 5 mCi ^{137}Cs source, is approximately 5%[3]. This represents, however, a convolution of little known physical parameters including the efficiency and angular dependence of evaporation by phonons and rotons, the adsorption probability of helium onto silicon, and the reflection mechanism for excitations at the walls. Detailed experiments investigating these parameters are accessible in the current apparatus and are in progress. Increased sensitivity in the apparatus is achieved by lowering the base temperature of the calorimeters and by increasing the resolution of the thermistor. Development work in these areas is proceeding rapidly.

[1] R. Torii, S. R. Bandler, T. More, F. S. Porter, R. E. Lanou, H. J. Maris, and G. M. Seidel. Rev. Sci Instrum. **63**, 230 (1992).
[2] F. R. Hope, M. J. Baird, and A. F. G. Wyatt. Phys. Rev. Lett. **52**, 1528 (1984).
[3] S. R. Bandler, R. E. Lanou, H. J. Maris, T. More, F. S. Porter, G. M. Seidel, and R. H. Torii. Phys. Rev. Lett. **68** 2429 (1992).

Generation of Solitons by Recoiling Nuclei in 2D Lattices

T. More and H.J. Maris
Department of Physics, Brown University, Providence, RI 02912, USA

Recent interest in using low temperature phonon–based detectors to search for dark matter has emphasized the need to understand the spatial distribution of the phonon flux produced by a recoiling nucleus. The phonon distribution may reflect the recoil's direction, which in turn can be used in identifying events caused by scattering of dark matter. It has been suggested [1] that there is an anisotropy in the primæval phonon flux (the phonons created immediately after the recoil) due to the creation of solitons whose range and amplitude depend on the recoil direction and energy. The solitons decay, emitting phonons. One can then look for traces of this wake in the measured distribution of phonons at the surface.

A soliton is an excitation which transfers energy, but not mass, along one of the crystalline directions and whose shape does not change as it propagates. Calculations using a Lennard-Jones potential between neighboring atoms [1] indicate that such excitations are indeed formed in HCP crystals, though in lattices of more than one dimension their amplitude decreases as they propagate, unlike the case for an ideal soliton. We present results for calculations in two dimensions for cubic crystals, such as the alkali-halides, that have long range Coulomb forces. Such crystals are of interest in the search for dark matter since both practical and theoretical considerations may dictate their use in future detector designs [2].

The potential we use in our simulation is the Coulomb potential plus a core repulsion:

$$u(r_{ij}) = (q_i q_j / r_{ij} + A/r_{ij}^{12}) \, .$$

Although energy losses due to ionization were not considered, they are not expected to significantly affect our results with respect to the existence of the solitons and their dependence on recoil energy and direction. We study recoils with energies in the range expected for dark matter scattering events, up to about 1keV.

We have found that solitons can propagate along both the close packed $< 100 >$ directions and, at sufficient recoil energy, along the $< 110 >$, the second closest packed set of directions. The number of solitons that appear depends on the recoil direction and on the recoil energy. The velocities of atoms associated with a soliton propagating along the [100] direction are shown in Fig. 1. In Fig. 2a we show the distance of the soliton from the recoil atom as a function of time. The soliton travels away from the recoil, until it encounters the edge of the lattice and reflects. The amplitude, i.e. the velocity of the atom at the center of the soliton, decreases, showing the decay of the soliton (Fig. 2b). We find that the range and relative amplitude

Springer Series in Solid-State Sciences, Volume 112
Phonon Scattering in Condensed Matter VII Eds.: M. Meissner and R.O. Pohl

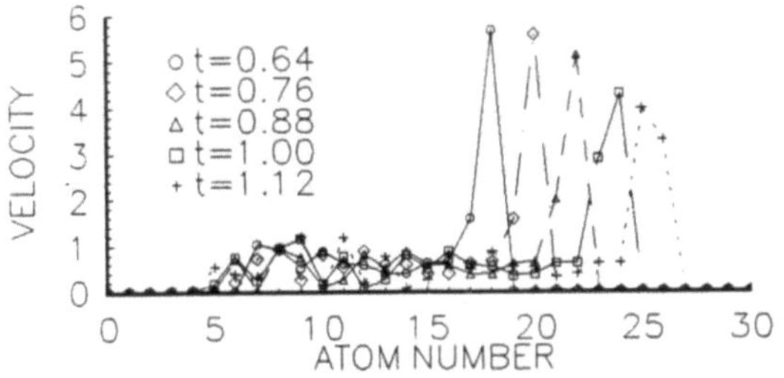

Figure 1. Velocities of atoms along the [100] direction for a series of times. At time 0 atom 0 was given an initial velocity at an angle of 20° to the x axis.

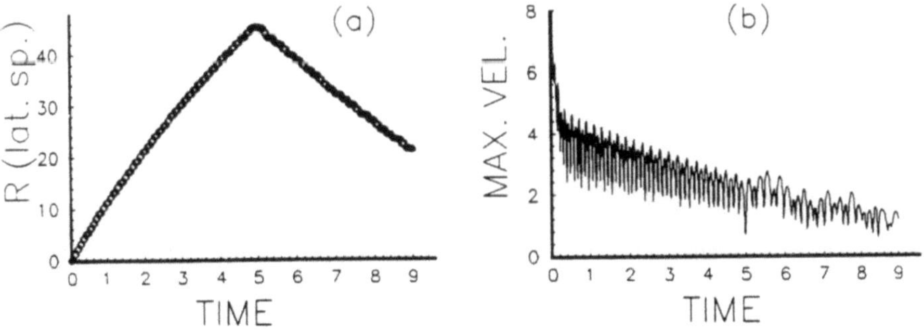

Figure 2. (a) Distance from recoil point to fastest atom as function of time. (b) Highest atomic velocity in lattice as function of time. Mass ratio was 1 :1.2.

of the different soliton branches depend on both the initial recoil energy and the recoil direction. We have also investigated the dependence on the model used by varying the exponent in the hard core potential. Our results show an interesting trend toward more chaotic behavior with softer potentials. Likewise, the solitons do not form as clearly when the masses of the anion and cation are mismatched.

Using an array of detectors sensitive to ballistic phonons, it may be possible to measure the resulting phonon anisotropy at the surface. Such measurements may then be used to deduce the direction of the recoil. Because of the motion of the sun through the galaxy, many of the scattering events due to dark matter should produce nuclear recoils in the direction opposite to the sun's motion. Thus, information about the recoil direction could help distinguish genuine signals due to dark matter interacions from those due to background processes.

[1] H. J. Maris and S. Tamura submitted to Phys. Rev. B.
[2] See, for example, P. F. Smith and J. D. Lewin, Phys. Rep. 187 (1990) 203, especially p. 230.

Part XI

Lattice Dynamics

Ab Initio Calculation of Phonons, Effective Charges, and Dielectric Tensor in α-Quartz

X. Gonze[1;3], *D.C. Allan*[2], *and M.P. Teter*[1;2]

[1]LASSP, Cornell University, Ithaca, NY 14853, USA
[2]Applied Process Research, SP-PR-22, Corning Incorporated, Corning, NY 14831, USA
[3]Unité P.C.P.M., Université Catholique de Louvain, B-1348 Louvain-la-Neuve, Belgium

1. Introduction

Understanding the lattice dynamics of α-quartz has important consequences, not only because α-quartz is of technological interest, but also because it is the simplest tetrahedrally bonded form of silica, and can help to understand silica glasses. Although experimental data are numerous, the generation of models is difficult due to the complexity of the interaction between silicon and oxygen ions. In this context, a parameter-free ab initio study proves valuable.

We invite the reader to consult reference [1], an account of this work in more detailed form. In the present paper, we briefly summarize that account, and refine the analysis.

2. Method and results

A new variational expression for the computation of response to small perturbations within density-functional theory (DFT) is used : the accuracy of DFT is retained, but without the computational drawbacks of some previous methods, like the supercell method or the dielectric matrix method.

The theoretical equilibrium lattice parameters and atomic positions are first computed (within about 2% of the experimental values). The components of the dielectric tensor, when calculated using the local-density-approximation, are about 7% off the experimental values, but the inclusion of the scissors-operator correction brings the agreement to 0.5%. Such a correction is not needed for phonon frequencies. The dynamical matrix at Γ, including non-analytical Coulomb contributions due to the Born effective charge tensors, is obtained ab initio.

The effective charge tensor for Oxygen atoms is represented in Fig.1, in the coordinate system defined by the Si-O-Si plane. In that coordinate system, it is nearly diagonal. Nevetheless, it is *not* isotropic : the xx component is much larger in magnitude than the yy or the zz component. It is even larger than the nominal charge of the oxygen ion (-2).

The comparison of theoretical phonon frequencies and LO-TO splittings with experiments are shown in Fig.2. Most phonon frequencies are obtained within 3%

Springer Series in Solid-State Sciences, Volume 112
Phonon Scattering in Condensed Matter VII Eds.: M. Meissner and R.O. Pohl

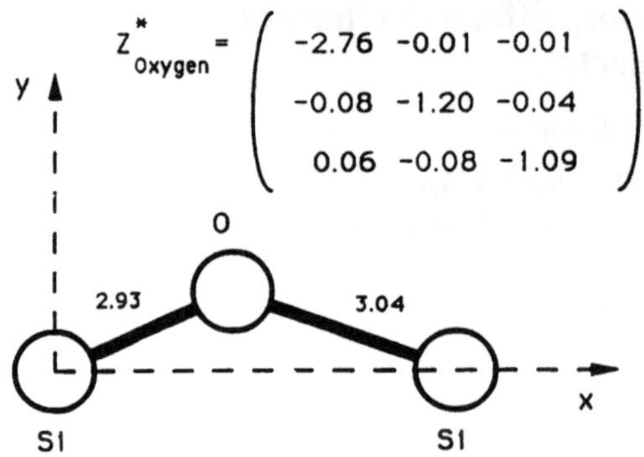

Fig. 1. The effective charge tensor for oxygen, defined with respect to the coordinate system of the bonds (x,y in the Si-O-Si bonding plane, and z out of plane) : a unit displacement along x (y and z respectively), creates a dipole whose direction is described by the first line (the second and the third respectively) of the matrix. Bond lengths are given in atomic units (bohrs).

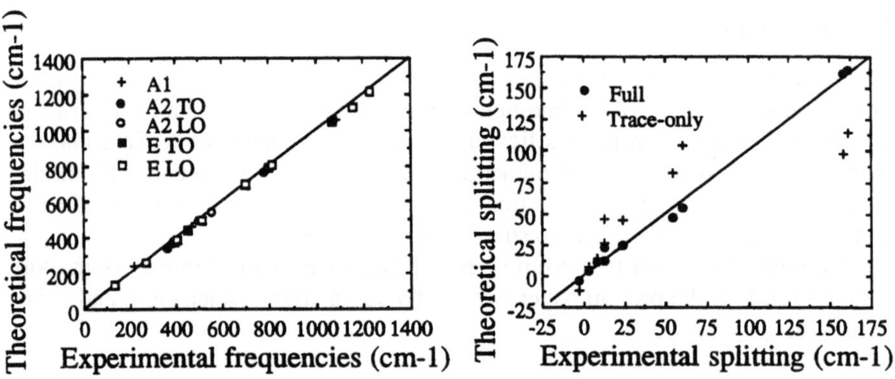

Fig. 2. (left) Comparison between experimental and theoretical phonon frequencies at Γ. The symmetry representation, as well as LO or TO character, are indicated. (right) Comparison between experimental and theoretical LO-TO splittings. Dots : the full theoretical effective charge tensor is used; crosses : only the isotropic part is used.

of the experimental values, and most LO-TO splittings are also well-reproduced. To emphasize the importance of the non-isotropic behaviour of the effective charge tensor, we have also computed the LO-TO splittings with an hypothetical isotropic effective charge tensor, using the trace of the ab initio tensor. The good agreement that was obtained with the full tensor is destroyed.

Reference

[1] X. Gonze, D.C. Allan, and M.P. Teter, Phys.Rev.Lett. **68**, 3603 (1992).

Resonant Raman Study of Phonon States in Ge Small Particles

Y. Sasaki and C. Horie

Department of Basic Science, Ishinomaki Senshu University, Ishinomaki 986, Japan

We have studied phonon states of Ge small particles with the diameters ranging from about 3 to 20 nm by Raman scattering resonating with the E_1~$E_1+\Delta_1$ gaps. Since the electronic transition energy depends on the particle size as a result of confinement of electron and hole in the particle, we can selectively excite the particle with a specific diameter by tuning the incident laser wavelength to match the transition energy. The small particles have been prepared by evaporation of Ge in Ar gas with the pressure of 0.5 to 2 Torr. We operate a non-evaporation getter pump in the gas-evaporation process to keep purity of Ar gas in the chamber. Small particles are collected on a sapphire substrate.

The Raman spectra have been obtained with a standard Raman system. Figure 1 shows an example of the resonant Raman

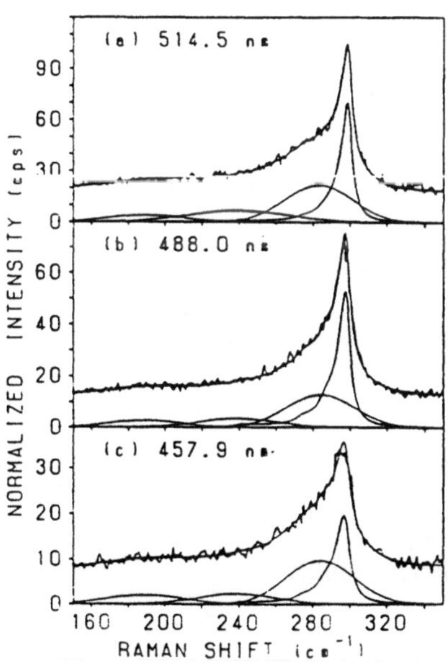

Fig. 1 Raman spectra of gas-evaporated Ge small particles.

Springer Series in Solid-State Sciences, Volume 112
Phonon Scattering in Condensed Matter VII Eds.: M. Meissner and R.O. Pohl

spectra obtained for a sample with different laser wavelengths. The sharp line located just below 300 cm^{-1} is due to the $k \sim 0$ optical phonon confined in the small particle. The position of this line is shifted toward lower frequency for an excitation with shorter laser wavelengths. It should be noted here that the confinement of the phonon in small particles yields the discrete values for **k** and relaxation of the **k**-conservation in the Raman scattering process. As a result the Raman peak shifts toward lower frequency and the line shape becomes asymmetric as the size of Ge particles becomes smaller than about 20 nm. On the other hand, the confinement of the electron and hole in smaller particles increases the electronic excitation energy. Consequently, the size of particles responsible to the resonant Raman scattering is smaller for an excitation with a shorter laser wavelength. A simple calculation indicates that the particle with its size of 10 nm (6 nm) resonates with $E_1 \sim E_1+\Delta_1$ gaps for the incident laser wavelength of 514.5 nm (457.9 nm), whereas the resonance peak is located at 556 nm in the bulk crystal [1]. The solid lines in Fig. 1 show the result of curve fitting by means of the confined optical phonon (the asymmetric line around 298 cm^{-1}) and other Gaussian lines corresponding to amorphous Ge, or very small (~ 1 nm) particles. The latter components are located at 190, 240 and 283 cm^{-1}. Details of the analysis are published elsewhere [2].

The particle size may be estimated from the experimental result of the resonant Raman scattering by two methods; (1) particle size dependence of the electronic excitation energy in the resonant scattering and (2) size dependence of the Raman spectrum of the confined optical phonon. It is found from the analysis of our data that the particle sizes estimated from the second method are smaller than those obtained by the first method by a factor of about 2. This discrepancy can be explained in terms of the frequency shift and the enhancement of decay rate of the optical phonon in smaller particles due to the additional anharmonic interaction with surface or interface vibrational modes. In fact, the frequency of the optical phonon shifts toward lower frequency than that in the bulk crystal by 0.5 cm^{-1} for the 10 nm particle and 2.5 cm^{-1} for the 6 nm particle, respectively. The damping increases by 1 cm^{-1} for the 10 nm particle and by 3 cm^{-1} for the 6 nm particle, respectively, compared with the damping in c-Ge. The similar effect of the anharmonicity has been reported for CdSe particles smaller than about 10 nm [3].

This work is supported in part by Grant-in-Aid for Scientific Research on Priority Areas and by Grant-in-Aid for New Program both from Ministry of Education, Science and Culture in Japan.

References

1] F. Cerdeira et al., Solid State Commun. **10**, 591, (1972).
2] Y. Sasaki and C. Horie, to be published.
3] A. Tanaka et al., Phys. Rev. **B45**, 6587 (1992).

Spectral Hole-Burning Study of Neutron-Irradiated Diamond

I. Sildos[1], *U. Bogner*[2], *and A. Osvet*[1]

[1]Institute of Physics, Estonian Academy of Sciences,
Riia 142, EE2400 Tartu, Estonia

[2]Institute of Physics III, University of Regensburg,
W-8400 Regensburg, Fed. Rep. of Germany

A spectroscopic investigation of a neutron-irradiated IaB-type natural diamond crystal was carried out. Persistent spectral hole-burning in the absorption lines of n-irradiation induced defect aggregates at 649.5, 655 and 681 nm was performed at 5÷200 K. By virtue of the high Debye temperature of the diamond the holes burnt in 655 nm line were observable up to ∿200 K. A weak temperature dependence of the holewidth indicates to the interaction with the localized structural instabilities, created by heavy doze irradiation.

1. Introduction

Diamond crystal is unique in many aspects. Thus, by virtue of the high Debye temperature (∿1750 K) it is possible, even at room temperature, to find out "low-temperature" features in spectra of defects [1]. The purpose of this work was to find out the possibilities for spectral hole burning (SHB) in electron spectra of n-irradiation-induced defects of diamond. Further, by means of the burnt-in holes to investigate low-energy excitations in such a crystal.

2. Experimental

After thermal annealing at 950^{0} C neutron-irradiated sample (neutron flux 10^{19} neutrons/cm^{2}) became partially transparent in the red part of the visible. At HeT comparatively weak lines at 644.1, 649.5, 651.4, 655 and 681 nm can been seen, where all the lines are placed on the top of a quite strong continuous background (optical density ∿3 at 600 nm).

In case of exciting the sample into the 649.5 nm line (by using pulse or cw dye-laser) a selective SHB occurs, where the photoproduct reveals itself in the 644.1 and 651.4 nm lines (Fig. 1). The difference spectrum (Fig. 1,c) indicates the correlation between the educt and the photoproduct, where photoproduct shifts the 644 nm line to the red side of spectrum. For 655 nm line SHB was performable by using only an excimer-pumped pulse dye-laser, while for 681 nm line both type of lasers were suitable.

At 5 K the HB with narrow-line ring laser ($\delta \sim 2$ MHz) gave homogeneous widths of 649.5 and 681 nm lines 0.7 and 2 GHz, respectively. The holewidth for the 655 nm line is quite large, about 3 cm^{-1}.

Springer Series in Solid-State Sciences, Volume 112
Phonon Scattering in Condensed Matter VII Eds.: M. Meissner and R.O. Pohl

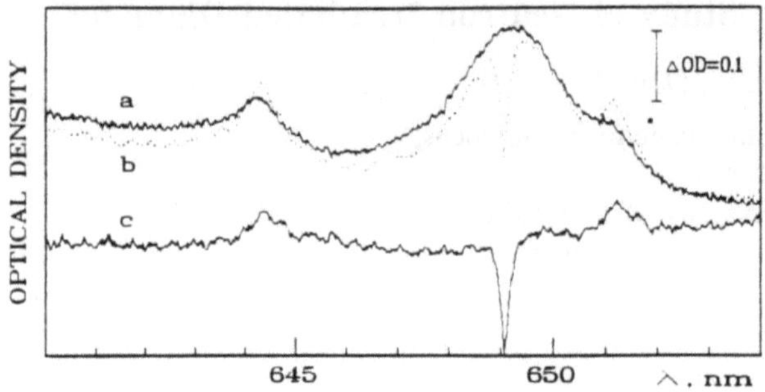

Fig. 1. Absorption spectrum of n-irradiated diamond. (a) The initial spectrum. (b) The spectrum after 5 min burning with cw dye-laser. (c) The difference spectrum. Burning conditions: power 2 W/cm^2, spectral width of laser 0.3 cm^{-1}. All spectra measured at 5 K.

The holes in 649.5 nm line are quite temperature sensitive and irreversible refilling of holes starts at ∿10 K. The broadening of holes versus temperature is $\sim T^2$. The spectral holes in the 681 nm and the 655 nm lines are more stable and are observable on heating the sample up to 100 K and 200 K, respectively. Measured broadening rates versus temperature are extremely slow, following $T^{0.5}$ law up to 75 K and T^2 towards higher temperature.

3. Discussion

We should note that very little is known about defects, which are responsible for spectral lines, used in this investigation [2].

In case of SHB in 649.5 nm line we may propose, that the burning mechanism could be light induced defect site distortion, which occurs during optical excitation (by assistance of phonons, created during relaxation of excitation energy?). Such kind of HB mechanism has been observed in case of $Cm^{4+}:CeF_4$ and $Pr^{3+}:D^-:CaF_2$ crystals [3,4].

A slow broadening of holes in 655 nm line versus temperature indicates to the weak coupling of this defect with crystal modes. Observed broadening is determined by interaction with localized structural instabilities, caused by the heavy doze of neutron irradiation. The results could be analyzed by introducing asymmetric double-well-potentials, which are characteristic for all disordered systems including crystalls with local disorder [5].

1. P.A.Growther and P.J.Dean, J. Phys. Chem. Solids (GB) 28, 1115 (1967).
2. Y.Nisida, Y.Mita, K.Mori, S.Okuda, S.Sato, S.Yazu, M.Nakagawa and M.OKada, Materials Science Forum 38-41, 561 (1989).
3. G.K.Liu, J.V.Beitz, Chem. Phys. Letters 171, 335 (1990).
4. R.M.Macfarlane, R.J.Reeves, and G.D.Jones, Opt. Lett. 12, 660 (1987).
5. T.Attenberger, I.Sildos, G.Jones, U.Bogner and Max Maier, Radiation Effects and Defects in Solids 119-121, 325 (1991).

Digital Storage by Phonon Hole Burning

F. Tsuruoka

Department of Physics, Kurume University, Kurume, Fukuoka 830, Japan

Spectral hole burning is observed in a condition that individual absorption centers have narrow width and slightly different absorption frequencies from each other due to, for example, the effect of each surrounding circumstance and the whole absorption line of the centers is broad and moderate. When narrow band energy source with some frequency is applied to excite and deform only the centers with the frequency, it produces a decrease in the absorption line, that is, a hole at the frequency.[1]

An array of present and absent holes in the absorption line at certain frequency intervals has been used to encode binary data. Optical source has been applied to storage information.[1]

In phonon hole burning, rf electric field energy is absorbed by piezoelectric particles. They have acoustic resonance frequencies which were determined by their dimension, shape and crystallographic angles. Powders obtained through destruction with a mortar and pestle and sieving with standard mesh screens have inhomogeneous resonance frequencies.[2]

We used $KBrO_3$ powders with diameter of 105 - 125 μm. Their resonance frequencies ranged from 10 to 20 MHz. The Powders were loosely packed in a parallel plate capacitor. The number of particles contained is about 10^6. Holes were written by applying of large amplitude rf electric field pulses. The shape and depth of holes were controlled by the pulse amplitude of up to 300 V/mm, the width of 5 - 100 μs and the total number of pulses. After applied writing pulses, particles were taken out of the capacitor and thoroughly stirred and replaced into the capacitor. Reading out of holes was performed by measuring electric impedance of the capacitor containing sample powder. The depth of holes was detected as reduction of the impedance by less than a few percent. We measured the change of the real part of the difference between the impedances of the sample capacitor and a reference variable capacitor using a differential transformer to increase detection sensitivity. Before writing a hole, we tuned the variable capacitor to show null output. These procedures, including the stirring treatment, were performed at pressures around 10^{-2} mmHg.

An example of digital storage observed is shown in Figure 1. A burned hole encodes a digital "1" and no hole "0". The array of

Springer Series in Solid-State Sciences, Volume 112
Phonon Scattering in Condensed Matter VII Eds.: M. Meissner and R.O. Pohl

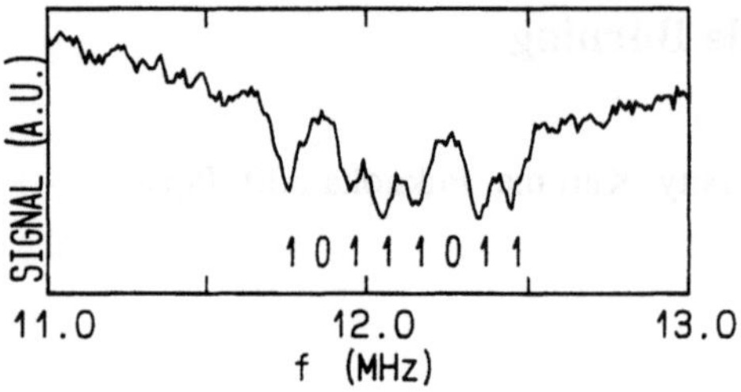

Fig. 1. Example of digital storage by phonon hole burning in piezoelectric powder $KBrO_3$

holes shows the 8 bit information of "10111011". Each hole was written by rf pulses with the field amplitude of 250 V/mm, the pulse width of 25 μs and the total pulse number of several thousands. The left hole was written by the pulses with the frequency of 11.70 MHz. The frequency intervals between bits were 100 kHz.

The storage is stable but the stored information density in the frequency domain is not so high. The ratio of the minimum frequency interval and the resonance frequency is 0.1 MHz / 10 MHz = 100. There are two possible ways to increase this number. The first one is to use particles with smaller absorption width or smaller damping constant. The second one is to use smaller particles with higher resonance frequencies. Apart from the possibility of hole burning, acoustic resonat oscillation of particles has been observed at microwave frequency range.[3]

The mechanism of hole burning is considered to be plastic deformation of particles and changing of the characteristics of resonant oscillation. Particles obtained by destruction contain many crystal dislocations, which change the oscillation feature, namely, particles resonance frequency and damping constant through dislocation multiplication introduced by large amplitude strain and plastic deformation.

As a conclusion, we showed the possibility of digital storage by phonon hole burning in piezoelectric particles.

References

1. W.E.Moerner: Persistent Spectral Hole Burning: Science and Applications (Springer Verlag, Berlin, Heidelberg, New York 1988) Chap. 1, p. 1.
2. F.Tsuruoka: Jpn. J. Appl. Phys. **28**, 43 (1990).
3. K.Fossheim, K.Kajimura, T.G.Kazyaka, R.L.Melcher, and N.S.Shiren: Phys. Rev. **B17**, 964 (1978).

Anharmonicity of a Thermally Unstable Lattice Defect

A. Rosenberg[1], *B.P. Clayman*[1,2], *A.J. Sievers*[1], *K.W. Sandusky*[3], and *J.B. Page*[3]

[1]Laboratory of Atomic and Solid State Physics and Materials Science Center, Cornell University, Ithaca, NY 14853-2501, USA

[2]Simon Fraser University, Burnaby British Columbia V5A 1S6, Canada

[3]Department of Physics and Astronomy, Arizona State University, Tempe, AZ 85287-1504, USA

1. Introduction

The low-temperature dynamics of the KI:Ag^+ defect system are well-described by a harmonic breathing-shell model which includes substantial force-constant weakening [1]. The force-constant changes are determined by the measured frequencies of the far-IR gap and resonant modes. Recent far-IR measurements have confirmed the model's striking prediction of three nearly-degenerate gap modes of different symmetries having unusual "pocket-like" displacement patterns, strongly localized on the defect's fourth neighbors [2]. In spite of the model's success, this defect system's anomalous thermal instability continues to be of interest: the low-temperature on-center configuration changes to an off-center one by ~20 K, each one having an associated set of far-IR modes. The pronounced anharmonicity suggested by this temperature dependence has been probed by uniaxial stress and static electric field measurements.

2. Uniaxial-Stress Measurements

Whereas "standard" gap modes (*e.g.*, KI:Cl^-, KI:F-center, KI:Cs^+ and KI:Rb^+) couple strongly to both fully-symmetric and tetragonal strains, as reflected by the comparable values of their A and B stress coupling coefficients, the KI:Ag^+ pocket gap mode has a similar A but a nearly vanishing B; thus, this mode couples almost exclusively to fully-symmetric strains. The ratio of the coupling coefficients B/A of the previously-measured KI:Ag^+ low-frequency resonant mode, which is ***not*** a pocket mode, is more similar to the B/A values of the standard gap modes than to those of the KI:Ag^+ pocket gap mode [3]. This suggests that a small B/A ratio may be a stress-behavior signature of a mode's pocket nature [4].

In trying to explain the unique pocket mode stress behavior qualitatively, we have extended our harmonic model for the low-temperature dynamics of KI:Ag^+ to describe the IR-active gap mode frequency shifts under uniaxial stress. We were able to fit the cubic anharmonic terms between the fourth-neighbor (200) site and its nearest neighbors to the experimental stress data. However, our fit anharmonicity between the (100) and (200) sites suggests that the interaction between these sites is described by an anomalous attractive-dominated potential, in contrast to the usual rapidly-varying repulsive potential with a weaker

Springer Series in Solid-State Sciences, Volume 112
Phonon Scattering in Condensed Matter VII Eds.: M. Meissner and R.O. Pohl

Coulomb part. The computed stress-induced displacement patterns point toward another possible explanation of the unusual stress coefficients observed for the T_{1u} pocket gap modes which does not require anomalous anharmonicity. If the (100)-(200) longitudinal force constant were weakened beyond the value we obtained by fitting the T_{1u} gap and resonant modes, the (200)-(300) separation would actually increase under an E_g stress, in contrast to the decrease which occurs in the pure crystal; if the (200)-(300) potential were dominated by a standard strong repulsive term, this would weaken the (200)-(300) longitudinal force constant under stress, thus partially compensating for the stress-induced stiffening of other force constants. This could produce a small B value, such as that observed experimentally, without the presence of anomalous anharmonicity.

3. Static-Electric-Field Measurements

Application of a static electric field should induce a mixing of the three nearly-degenerate KI:Ag^+ pocket gap modes, similar to that produced by the natural occurrence of $^{41}K^+$ isotopes in the host lattice [2]. However, our measurements indicate only extremely small shifts of the T_{1u} pocket gap mode, ~0.01cm^{-1} at ~100 kV/cm. Compared to the case of gap modes due to other monatomic substitutional impurities, these shifts have comparable magnitude but different polarization dependence. Previous measurements on the low-frequency resonant mode of KI:Ag^+ had also uncovered frequency shifts comparable to those of resonant modes due to other lattice/defect systems [5]. The gap-mode static-electric-field results are *not* consistent with the stress-fit anharmonicities in our two-parameter harmonic-force-constant model; however, a sensible one-parameter extension of the model leads to a consistent explanation of the data within essentially the same pocket-mode framework.

4. Conclusions

Both the uniaxial-stress and static-electric-field results on KI:Ag^+ imply that the anharmonicity of the potential near the defect site is small and similar in size, if not distribution, to that present near other monatomic substitutional defects. This result is consistent with the success of the harmonic perturbed-phonon model in describing this defect system's low-temperature configuration. Therefore, the cause of this defect system's thermal instability remains an open question, although we have now ruled out the presence of unusually-large anharmonic terms in the potential.

5. References

1. J. B. Page, J. T. McWhirter, A. J. Sievers, H. Fleurent, A. Bouwen, and D. Schoemaker, Phys. Rev. Lett. **63**, 1837 (1989).
2. K. W. Sandusky, J. B. Page, A. Rosenberg, C. E. Mungan, and A. J. Sievers, Phys. Rev. Lett. **67**, 871 (1991).
3. I. G. Nolt and A. J. Sievers, Phys. Rev. **174**, 1004 (1968).
4. A. Rosenberg, C. E. Mungan, A. J. Sievers, K. W. Sandusky, and J. B. Page, to be published in Phys. Rev. B.
5. R. D. Kirby, Phys. Rev. B **4**, 3557 (1971).

Internal Rotation–Phonon Coupling in Lattice Dynamics of p-Xylene

J. Kalus[1], *J. Wolfrum*[1], *F. Wörlen* [1], *K. Holderna-Natkaniec*[2], *I. Natkaniec*[2], *M. Monkenbusch*[3], *and M. Prager*[3]

[1]Physikalisches Institut der Universität Bayreuth, W-8580 Bayreuth, Fed. Rep. of Germany
[2]Frank Laboratory of Neutron Physics, JINR, 141980 Dubna, Russia
[3]Institut für Festkörperforschung der KFA Jülich, W-5170 Jülich, Fed. Rep. of Germany

Abstract. Solid p-$C_6H_4(CH_3)_2$ and p-$C_6H_4(CD_3)_2$ have been studied by Raman-(R) and Inelastic Incoherent Neutron Scattering-(IINS) at temperatures from 15K up to melting at 287K. Decreasing frequencies and strong broadening of optical phonons and methyl torsions have been observed in the R and IINS spectra with increasing temperature. A lattice dynamics model based on the atom-atom potential parameters with free internal rotation of methyl groups satisfactorily explains the Raman modes and phonon frequency spectra measured at 15K. In the framework of this model mixing of different modes can be studied in details.

Solid p-xylene provides a relatively simple example of methyl group rotation in a weak potential caused by intermolecular interactions, which also determine the lattice dynamics [1]. The crystal structure is know from X-ray data [2] at 180K and high resolution neutron powder diffraction [3] at 5K. Detailed studies of Raman and IINS spectra in a solid phase supply necessary data to submit a realistic model of molecular dynamics in crystalline p-xylene [4]. The IINS spectra have been studied with a different resolution at the FRJ2 reactor of the KFA Jülich and at the IBR-2 high flux pulsed reactor [5] of the JINR at Dubna.

Temperature dependence of the generalized phonon density of states - G(ω) measured with the KDSOG spectrometer at Dubna is shown in Fig. 1. The first broad maximum at an average energy of about 65 cm^{-1} does not depend significantly on methyl deuteration and temperature. So it should correspond mainly to translational lattice vibrations and librations around the direction perpendicular to the benzene plane. It seems that these modes are not strongly coupled with the methyl rotation. Peak 2 should correspond to the intermolecular libration modes which are coupled with the methyl rotation and it seems that this coupling is increasing with increasing temperature. Its energy corresponds well with a strong Raman mode seen at about 90 cm^{-1}. This band shifts only slightly with increasing temperature. Peak 3 is very intense and shows a doublet structure at 15K. With increasing temperature the width of this peak becomes significantly broader. The decrease of the energy as a function of temperature of this peak is much stronger as it was found for all the other maxima in the *G(ω)*. The energy of

Springer Series in Solid-State Sciences, Volume 112
Phonon Scattering in Condensed Matter VII Eds.: M. Meissner and R.O. Pohl

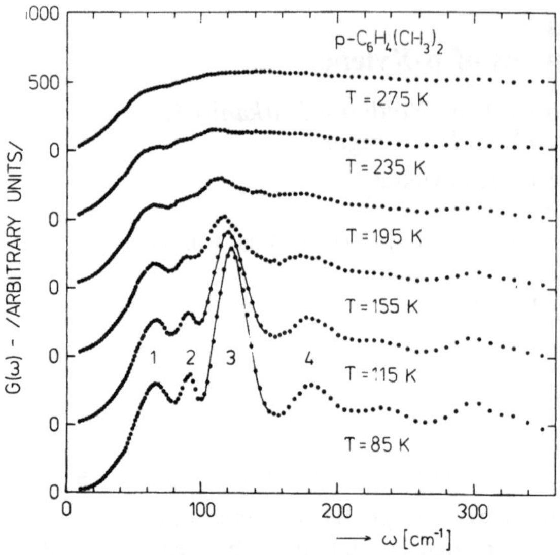

Figure 1. Temperature dependence of the amplitude weighted phonon density of states G(ω) of protonated p-xylene. The numbers indicate prominent peaks in the spectra.

this peak corresponds to Raman lines at 121, 136.5 and 142 cm^{-1}. As one can see in Table 1 two infrared active modes have energies in this region too. All these modes show large librational amplitudes in the calculations. Broadening and smearing of the *G(ω)* singularities with increasing temperature is caused by phonon scattering on reorientational jumps of methyl groups. The frequency of this jumps increase with temperature and seems to be comparable with the optical phonon frequencies at above 250K.

The lattice dynamics calculations performed with the Williams IVb set of atom-atom potential parameters [6], which describes well the dynamics of a benzene crystal, cannot explain satisfactorily the dynamics of the p-xylene crystal. Therefore we looked for a better suited set of potential parameters. We altered the magnitude of Williams potentials but not of the positions of the potential minima. Besides we had to differentiate between interaction of hydrogen belonging to the methyl groups and to the benzene ring, respectively. The modified set of parameters describe both, the IINS- and R-spectra with a reasonable accuracy and allow one to study in more details the coupling of the internal methyl rotation with the lattice modes.

In Table 1 the calculated optical phonon frequencies are compared with the measured Raman modes at 14K. The modes with R-character are Raman active, whereas the modes with T-character are sensitive to IR-spectroscopy. An orthogonal *x, y, z* - coordinate system is fixed to the molecules at their center of gravity, *x* and *y* lie in the benzene plane and *z* is normal to this plane. The *x*-axis goes through the C-atoms of the methyl groups. With respect to the symmetry operation connecting two molecules in the unit cell the phase of the displacement vectors is defined as in phase (+) and out of phase (-). The results presented in Table 1 will be discussed in more details in the paper of Kalus et al. [4].

Table 1. The calculated optical phonon frequencies (ħω in cm^{-1}) for protonated and partially deuterated p-xylene. The experimental data are due to Raman-experiments at 14 K. In the column "assignment" the character of these modes is indicated. (R=rotational, T=translational character).

$C_6H_4(CH_3)_2$			$C_6H_4(CD_3)_2$		
exp	calc	assignment	exp	calc	assignment
	33.7	T_a (+)		32.6	T_a (+)
58.0	58.8	R_z (+)	54.0	53.6	R_z (+)
	65.0	T_b (-)		61.3	T_b (-)
71.5	70.0	R (-)	67.0	61.8	R_x (-)
87.0	87.2	R_x (+)	73.0	75.1	R_x (+)
	87.6	T_c (+)	77.0	80.6	R (-)
90.0	96.6	R (-)		85.1	T_c (+)
111.0	110.2	R_y (+)		87.5	T_{CD_3} (-)
	119.8	T_{CH_3} (-)		89.9	T_{CD_3} (+)
121.0	122.0	R_y (-)	98.0	96.9	R (+)
	126.0	T_{CH_3} (+)	110.0	111.0	R (-)
136.5	138.2	R_x (-)		124.3	R_x (-)
142.0	143.0	R_x (-)	122.5	125.4	R_x (+)

Here we mention, that the concept of a universal set of parameters for a large variety of molecules has to be abandoned. It seems, that this concept at least is not useful for molecular crystals where methyl groups are present.

References

[1] Prager, M., Hempelmann R., Langen H., Müller-Warmuth W.; J. Phys.: Condens. Matter 2 (1990) 8625.

[2] Van Koningsveld H., Van den Berg A.J., Lansen J.C., De Goede R.; Acta Cryst. B 42 (1986) 491.

[3] Prager M., David. W. I. F., Ibbberson R. M.; J. Chem. Phys. 95 (1991) 2473.

[4] Kalus J., Monkenbusch M., Natkaniec I., Prager M., Wolfrum J., Wörlen F. (to be published)

[5] User Guide, IBR-2 Experimental Facilities, JINR, Dubna 1991.

[6] Taddei S., Bonadeo H., Marzocchi M.P., Califano S.; J.Chem. Phys. 58 (1973) 966.

On the Rotation of Isolated Methyl Groups in a Phonon Bath

T. Kranjc[1] *and J. Peternelj*[2]

[1]Department of Physics, University of Ljubljana, Jadranska 19, 61000 Ljubljana, Slovenia

[2]J. Stefan Institute, University of Ljubljana, 61000 Ljubljana, Slovenia

Molecular rotational tunneling is a quantum-mechanical phenomenon occurring when there exist, for a symmetrical molecular group, equivalent orientations due to the environment. In this contribution the rotational motion of an isolated methyl group (CH_3) in a molecular or crystalline solid will be considered. The methyl group may be regarded as a rigid rotor with moment of inertia, J, and its motion can be described by a single angular coordinate φ in a hindering potential $V(\varphi)$ due to the environment.

The system - methyl group plus environment - will, as usually, be described by the Lagrangian of the form [1,2]

$$L = \frac{1}{2}J\dot{\varphi}^2 + b\dot{\varphi} - V(\varphi) + \frac{1}{2}\sum m_j(\dot{x}_j^2 - \omega_j^2 x_j^2) - \sum f_j(\varphi)x_j \ . \tag{1}$$

The term $b\dot{\varphi}$ describes the interaction of the methyl group with an external magnetic field, B, $b = 3\,|e|\,\phi/h$, and $\phi = \pi r^2 B\cos\beta$ is the magnetic flux through the circle of radius r traced out by the protons of the CH_3 group. $V(\varphi) = V_0\,(1 - \cos 3\varphi)$, $\{x_j\}$ are the coordinates of the normal modes of the lattice, $f_j(\varphi) = V_{int}(\varphi)g_j$ are coupling parameters, and we take $V_{int}(\varphi) = V\sin 3\varphi$.

Studying the dynamics of the methyl group, we are specifically interested in the transition probability that the methyl group being in the state $|\psi_i\rangle$ at the initial time $t = 0$ is in the final state $|\psi_f\rangle$ at a later time t. To this end, we calculate the density matrix describing the methyl group, using the Feynman-Vernon formalism [3,4].

The density matrix in the coordinate representation is $\rho(\varphi x, \varphi' x'; t) = \langle\varphi x|\rho(t)|\varphi' x'\rangle$ and $\rho(t) = U(t,0)\,\rho(0)\,U^\dagger(t,0)$. The path integral representation of the time displacement operator $U(t,0) = e^{-iHt/\hbar}$ is

$$\langle\varphi x|U(t,0)|\varphi'' x''\rangle = \sum_{n=-\infty}^{\infty} \int_{\tilde{\varphi}(0)=\varphi''}^{\tilde{\varphi}(t)=\varphi+2\pi n} D\tilde{\varphi} \int_{x(0)=x''}^{x(t)=x} Dx\; e^{\frac{i}{\hbar}S[\tilde{\varphi}(t),x(t)]} \tag{2}$$

$$S[\tilde{\varphi}(t), x(t)] = \int_0^t dt'\, L(\tilde{\varphi}, x; \dot{\tilde{\varphi}}, \dot{x}) \ .$$

We have introduced $\tilde{\varphi} = \varphi + 2\pi n$, where $\varphi \in [0, 2\pi)$, $\tilde{\varphi} \in [-\infty, \infty)$, i.e. $\varphi = \tilde{\varphi}(\mathrm{mod}\, 2\pi)$, and the (integer) winding number n characterizes the homotopy class.

Springer Series in Solid-State Sciences, Volume 112
Phonon Scattering in Condensed Matter VII Eds.: M. Meissner and R.O. Pohl

Being interested in the dynamics of the methyl group only, we introduce the reduced density matrix as

$$\rho_R(\varphi,\varphi';t) = \int dx\, \rho(\varphi x,\varphi' x;t)\,. \qquad (3)$$

We assume the initial density matrix $\rho_R(0)$ describing the state of the methyl group at $t = 0$ to be $\rho_R(0) = |\psi_i\rangle\langle\psi_i|$, where $\langle\varphi|\psi_i\rangle \equiv \psi_i(\varphi - \varphi_i)$ is a gaussian-like wave function peaked sharply around φ_i and normalized to unity. Then $\rho_R(\varphi'',\varphi''';0) = \langle\varphi''|\rho_R(0)|\varphi'''\rangle = \psi_i(\varphi''-\varphi_i)\psi_i^*(\varphi'''-\varphi_i)$. Similarly, we also take the final wave function $\psi_f(\varphi-\varphi_f)$ to be a gaussian sharply peaked around φ_f. We then arrive at the following expression for the probability for the methyl group to go from the state $|\psi_i\rangle$ at the time $t = 0$ to the state $|\psi_f\rangle$ at a later time t

$$P_{i\to f}(t) = |C_i|^2|C_f|^2 \sum_{n,m} e^{i\frac{b}{\hbar}2\pi(n-m)} K_{nm}(\varphi_f,t;\varphi_i,0) \qquad (4)$$

with $C_{i,f} = \int_{-\pi}^{\pi} d\varphi' e^{-i\frac{b}{\hbar}\varphi'}\psi_{i,f}(\varphi'-\varphi_{i,f})$. The propagator K_{nm} is given as a path integral over two independent orbits $\tilde{\varphi}_1(t)$ and $\tilde{\varphi}_2(t)$ (forward-backward path integral)

$$K_{nm}(\varphi_f,t;\varphi_i,0) = \int_{\varphi_i}^{\varphi_f+2\pi n} d\tilde{\varphi}_1 \int_{\varphi_i}^{\varphi_f+2\pi n} d\tilde{\varphi}_2\, e^{\frac{i}{\hbar}(S_o[\tilde{\varphi}_1]-S_o[\tilde{\varphi}_2])}\, e^{\frac{i}{\hbar}\Phi[\tilde{\varphi}_1,\tilde{\varphi}_2]}\,, \qquad (5)$$

$e^{\frac{i}{\hbar}\Phi[\tilde{\varphi}_1,\tilde{\varphi}_2]}$ being the influence functional [3], with the influence phase

$$i\Phi[\varphi_1(t),\varphi_2(t)] = -\sum_j \int_0^t du \int_0^u dv\, \{\, [f_j(\varphi_1(u)) - f_j(\varphi_2(u))]\times$$
$$G_j^{(R)}(u-v)[f_j(\varphi_1(v)) - f_j(\varphi_2(v))] + i[f_j(\varphi_1(u)) - f_j(\varphi_2(u))]\times$$
$$G_j^{(I)}(u-v)[f_j(\varphi_1(v)) + f_j(\varphi_2(v))]\,\}\,. \qquad (6)$$

The correlation functions are $G_j^{(R)}(u-v) = (\cos(\omega_j(u-v))/2m\omega_j)\cdot\coth(\beta\hbar\omega_j)$ and $G_j^{(I)}(u-v) = -\sin(\omega_j(u-v))/2m\omega_j$.

The effect of the gauge potential representing the coupling with the magnetic field is to produce, in the path integral formulation, interference between paths with different winding numbers. This is a consequence of the nontrivial topology of the configuration space $\varphi \in [0,2\pi)$; if φ had been treated as belonging to R from the beginning, only the term $n = m = 0$ would persist in the sum over winding numbers.

The details of the calculation will be published elsewhere.

Acknowledgment. One of the authors (T.K.) would like to express gratitude for financial support by SMELT-Ljubljana and Gospodarska zbornica Slovenije.

References

1. J. Peternelj, I. Jenčič, B. Cvikl and M.M. Pintar, Phys. Rev. **B39**, 6374 (1989)
2. M.W.G. Whittall and G.A. Gehring, J. Phys. C: Solid State Phys. **20**, 1619 (1987)
3. R.P. Feynman and F.L. Vernon, Ann. Phys. (N.Y.) **24**, 118 (1963)
4. R.P. Feynman and A.R. Hibbs, Quantum Mechanics and Path Integrals, McGraw-Hill, New York (1965)

Vibrational Relaxation and Line Shift at Surfaces Due to Substrate Anharmonicity

Z. Ye and P. Piercy

Department of Physics, University of Ottawa, Ottawa, Ontario K1N 6N5, Canada

1. Introduction

The vibrational line shape of an atom or molecule adsorbed at a surface is caused, in part, by phonon-induced dephasing and decay processes. The effectiveness of an anharmonic adsorption potential in mediating these processes was shown by Persson and Ryberg[1] to be severely reduced by adiabaticity.

We show here how the anharmonicity of the substrate itself directly affects the vibrational line shape of an adsorbed molecule. This result is based on (i) the interpolation of the molecular vibration and the phonon modes, derived in an exact treatment of the harmonic Hamiltonian which couples the admolecule and substrate, and (ii) a perturbative treatment of the anharmonic potential between substrate atoms.[2]

2. Simple theoretical model

We study the perpendicular vibration of an adsorbed atom or molecule, corresponding to stretching of its bond to the surface. The case in which this localized vibrational mode has frequency above the phonon band is considered here. The vibrational displacement u of the adatom is coupled to the surface displacement v of the surface atom beneath it by an adsorption potential $V(u-v)$. In a harmonic approximation, the bilinear term $\propto uv$ causes a mixing of the vibrational coordinates of the adsorbed atom and the surface atom. An exact solution[3] of the coupled equations of motion for u and v gives the interpolation relations $u = au^0 + bv^0$ and $v = cv^0 + du^0$, in terms of the free harmonic amplitudes u^0 and v^0 derived without the bilinear term. Not only is the molecular vibration delocalized by the bilinear coupling, but the phonon displacement v has a local component. It is via this latter effect that the substrate anharmonicity couples to the molecular vibration.

The substrate atom beneath the adatom is assumed to interact with its neighbours as well, through the anharmonic potential $\phi(v-v_i)$. Taking the interpolation relation for the surface atom displacement v into $\phi(v-v_1)$ couples the free adatom coordinate u^0 anharmonically to the modified phonon coordi-

Springer Series in Solid-State Sciences, Volume 112
Phonon Scattering in Condensed Matter VII Eds.: M. Meissner and R.O. Pohl

nate $q^0 = cv^0 - v_1^0$. In this initial study, only the perpendicular displacement v_1 of neighbouring atoms in the second layer are considered in a long wave length limit, a bulk Debye model is used, and a Morse potential is used for ϕ.

The vibrational line shift and width are contained in the self energy of the Green's function for the molecular coordinate u, which is calculated by a perturbation expansion in ϕ, using a finite temperature field theory. Keeping third and fourth derivatives of ϕ, lowest order terms differ in sign but give a net red shift. A line width term describing vibrational damping due to two-phonon emission and absorption is also calculated.[3]

3. Results and summary

Applications are made to the adsorption of CO at on-top sites on the Pt(111) and Ni(100) surfaces. All interaction parameters are fixed by elastic properties of the solids. On Pt, the moelcule-surface vibrational frequency is known from experiments[4] to be 470cm^{-1}. The interpolation of the phonon field with the molecular displacement is given by $v = 0.925v^0 - 0.144u^0$. The calculated frequency shift over the temperature range from 100 K to 290 K is –1.8cm^{-1}, in close agreement with the experimentally observed shift of –2cm^{-1}.[5] The vibrational frequency of this system is about 2.8 times the Debye frequency of the solid, so the two-phonon damping term mentioned above is not applicable.

For the CO/Ni(100) case, the vibrational frequency is 476cm^{-1}, compared to a Debye frequency of 271cm^{-1}, and the two-phonon damping process gives a line width of 12cm^{-1}at 300 K, in reasonable agreement with the experimental obserserved width of 15cm^{-1}by Chiang et al.[5] A red shift is predicted theoretically, but we are not aware of experimental data to compare with.

In summary, it has been shown that the anharmonicity of the substrate plays a major role in defining the vibrational line shapes in some adsorption systems. This occurs because the adsorbate local mode interpolates with the continuum phonon band of the solid. Without fitting parameters, the theory agrees surprisingly well with experimental data on the CO/Ni(100) and CO/Pt(111) systems, considering the simplicity of the model used.

This work is supported by the Network of Centres of Excellence program and the Natural Sciences and Engineering Research Council of Canada.

References

1. B.N.J. Persson and R. Ryberg, Phys. Rev. B40, 10273 (1989).
2. A.A. Maradudin and P.A. Flinn, Phys. Rev. 129, 2529 (1963).
3. Zhen Ye and P. Piercy, submitted 1992.
4. R. Ryberg, Phys. Rev. B40, 8567 (1989).
5. S.Chiang, R.G. Tobin, P.L. Richard and P.A. Thiel, Phys. Rev. Lett. 52, 648 (1984).

Heat Capacities of Hydrogenated Nickel–Zirconium Alloys from 100 K to 400 K

M.E. Patt, D.M. Brennan, I.T. Sacolick, B.E. White, Jr., U. Thomanschefsky[a], *and E.J. Cotts*

Department of Physics and Materials Research Center, Binghamton University, P.O. Box 6000, Binghamton, NY 13902-6000, USA

1. Introduction

The heat capacities of crystalline ZrNi transition metal hydrides and deuterides have been measured over the temperature range from 100 K to 400 K by means of differential scanning calorimetry. We find that the heat capacity, c_p, of our hydrogenated and deuterated ZrNi samples is increased relative to the as-prepared alloys. We believe that this increase in c_p is due to the contribution from optical phonon modes of interstitial H or D atoms. Using an Einstein oscillator approximation [1], we estimate the hydrogen vibration frequency, ω, and find it to be consistent with previous neutron scattering data [2].

2. Results

Figure 1(a) shows the increase in the heat capacity of crystalline ZrNi upon hydrogenation. Deuterated samples show a similar, but larger, increase in c_p. Figure 1(b) is a plot of the difference, Δc_p, between the heat capacities of $ZrNiD_y$ and $ZrNiH_y$ for $y = 1.0$ and $y = 2.7$, normalized to the number of hydrogen atoms. We assume that Δc_p is mostly due to the difference in the masses of the H and D atoms, and not due to differences in the lattice heat capacity. Modeling Δc_p with an Einstein oscillator approximation [1], we estimate a hydrogen vibration frequency of 138 meV for the samples with $y = 2.7$, in good agreement with previous neutron scattering measurements [2]. The model does not produce good agreement with the data for $y = 1.0$, although it is clear that the hydrogen vibration frequency is lower.

Springer Series in Solid-State Sciences, Volume 112
Phonon Scattering in Condensed Matter VII Eds.: M. Meissner and R.O. Pohl

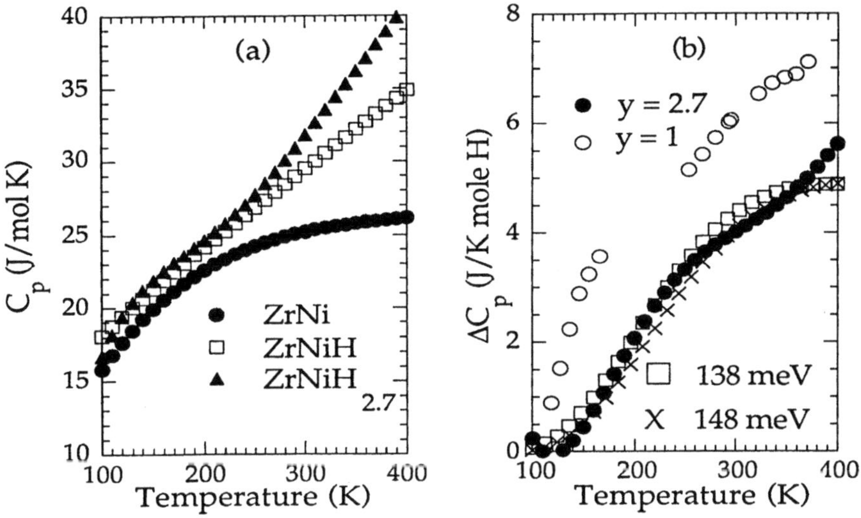

FIG. 1. (a) Heat capacities of crystalline ZrNi, $ZrNiH_{1.0}$, and $ZrNiH_{2.7}$. (b) Difference in heat capacities of $ZrNiD_y$ and $ZrNiH_y$ for y=2.7 and y=1.0, along with the predictions of the Einstein oscillator model at two different phonon frequencies.

3. Conclusions

We have measured the heat capacity of $ZrNiH(D)_y$ alloys from 100-400 K. The difference in heat capacity between the hydrogenated and deuterated samples has been analyzed using the Einstein oscillator model. The best agreement between the model and the heat capacity data for y = 2.7 is obtained for a H vibration frequency of 138 meV. This value of the frequency is in good agreement with previous measurements.

This work was supported by Aerojet ElectroSystems and the Petroleum Research Fund of the American Chemical Society.

4. References

[(a)] Permanent Address: Department of Physics, Pennsylvania State University at Erie, The Behrend College, Erie, PA 16563-0203.

[1] H. E. Flotow and D. W. Osborne, J. Chem. Phys. **34**, 1418 (1961).
[2] K. Suzuki, J. Less-Common Met. **89**, 183 (1983).

Index of Contributors

Springer Series in Solid-State Sciences

Editors: M. Cardona P. Fulde K. von Klitzing H.-J. Queisser

1 **Principles of Magnetic Resonance**
3rd Edition By C. P. Slichter

2 **Introduction to Solid-State Theory**
2nd Printing By O. Madelung

3 **Dynamical Scattering of X-Rays in Crystals** By Z. G. Pinsker

4 **Inelastic Electron Tunneling Spectroscopy**
Editor: T. Wolfram

5 **Fundamentals of Crystal Growth I**
Macroscopic Equilibrium and Transport Concepts. 2nd Printing
By F. Rosenberger

6 **Magnetic Flux Structures in Superconductors** By R. P. Huebener

7 **Green's Functions in Quantum Physics**
2nd Edition By E. N. Economou

8 **Solitons and Condensed Matter Physics**
2nd Printing
Editors: A. R. Bishop and T. Schneider

9 **Photoferroelectrics** By V. M. Fridkin

10 **Phonon Dispersion Relations in Insulators** By H. Bilz and W. Kress

11 **Electron Transport in Compound Semiconductors** By B. R. Nag

12 **The Physics of Elementary Excitations**
By S. Nakajima, Y. Toyozawa, and R. Abe

13 **The Physics of Selenium and Tellurium**
Editors: E. Gerlach and P. Grosse

14 **Magnetic Bubble Technology** 2nd Edition
By A. H. Eschenfelder

15 **Modern Crystallography I**
Symmetry of Crystals.
Methods of Structural Crystallography
By B. K. Vainshtein

16 **Organic Molecular Crystals**
Their Electronic States. By E. A. Silinsh

17 **The Theory of Magnetism I**
Statics and Dynamics. 2nd Printing
By D. C. Mattis

18 **Relaxation of Elementary Excitations**
Editors: R. Kubo and E. Hanamura

19 **Solitons.** Mathematical Methods for Physicists. 2nd Printing
By G. Eilenberger

20 **Theory of Nonlinear Lattices**
2nd Edition By M. Toda

21 **Modern Crystallography II**
Structure of Crystals
By B. K. Vainshtein, V. M. Fridkin, and V. L. Indenbom

22 **Point Defects in Semiconductors I**
Theoretical Aspects
By M. Lannoo and J. Bourgoin

23 **Physics in One Dimension**
Editors: J. Bernasconi, T. Schneider

24 **Physics in High Magnetic Fields**
Editors: S. Chikazumi and N. Miura

25 **Fundamental Physics of Amorphous Semiconductors** Editor: F. Yonezawa

26 **Elastic Media with Microstructure I**
One-Dimensional Models. By I. A. Kunin

27 **Superconductivity of Transition Metals**
Their Alloys and Compounds
By S. V. Vonsovsky, Yu. A. Izyumov, and E. Z. Kurmaev

28 **The Structure and Properties of Matter**
Editor: T. Matsubara

29 **Electron Correlation and Magnetism in Narrow-Band Systems** Editor: T. Moriya

30 **Statistical Physics I** Equilibrium Statistical Mechanics 2nd Edition
By M. Toda, R. Kubo, N. Saito

31 **Statistical Physics II** Nonequilibrium Statistical Mechanics 2nd Edition
By R. Kubo, M. Toda, N. Hashitsume

32 **Quantum Theory of Magnetism**
2nd Edition By R. M. White

33 **Mixed Crystals** By A. I. Kitaigorodsky

34 **Phonons: Theory and Experiments I**
Lattice Dynamics and Models of Interatomic Forces. By P. Brüesch

35 **Point Defects in Semiconductors II**
Experimental Aspects
By J. Bourgoin and M. Lannoo

36 **Modern Crystallography III**
Crystal Growth 2nd Edition
By A. A. Chernov

37 **Modern Crystallography IV**
Physical Properties of Crystals
Editor: L. A. Shuvalov

38 **Physics of Intercalation Compounds**
Editors: L. Pietronero and E. Tosatti

39 **Anderson Localization**
Editors: Y. Nagaoka and H. Fukuyama

40 **Semiconductor Physics** An Introduction
5th Edition By K. Seeger

41 **The LMTO Method**
Muffin-Tin Orbitals and Electronic Structure
By H. L. Skriver

42 **Crystal Optics with Spatial Dispersion, and Excitons** 2nd Edition
By V. M. Agranovich and V. L. Ginzburg

43 **Structural Analysis of Point Defects in Solids**
An Introduction to Multiple Magnetic Resonance Spectroscopy
By J.-M. Spaeth, J. R. Niklas, and R. H. Bartram

44 **Elastic Media with Microstructure II**
Three-Dimensional Models By I. A. Kunin

45 **Electronic Properties of Doped Semiconductors**
By B. I. Shklovskii and A. L. Efros

46 **Topological Disorder in Condensed Matter**
Editors: F. Yonezawa and T. Ninomiya

47 **Statics and Dynamics of Nonlinear Systems**
Editors: G. Benedek, H. Bilz, and R. Zeyher

48 **Magnetic Phase Transitions**
Editors: M. Ausloos and R. J. Elliott

49 **Organic Molecular Aggregates** Electronic Excitation and Interaction Processes
Editors: P. Reineker, H. Haken, and H. C. Wolf

MIX
Papier aus verantwortungsvollen Quellen
Paper from responsible sources
FSC® C105338

If you have any concerns about our products,
you can contact us on
ProductSafety@springernature.com

In case Publisher is established outside the EU,
the EU authorized representative is:
Springer Nature Customer Service Center GmbH
Europaplatz 3, 69115 Heidelberg, Germany

Printed by Libri Plureos GmbH
in Hamburg, Germany